MATHEMATICAL PHYSICS AND MATRIX REPRESENTATIONS

The Multiple Applications of Stochastic, Circulant and Bell Matrices

MATHEMATICAL PHYSICS AND MATRIX REPRESENTATIONS

The Multiple Applications of Stochastic, Circulant and Bell Matrices

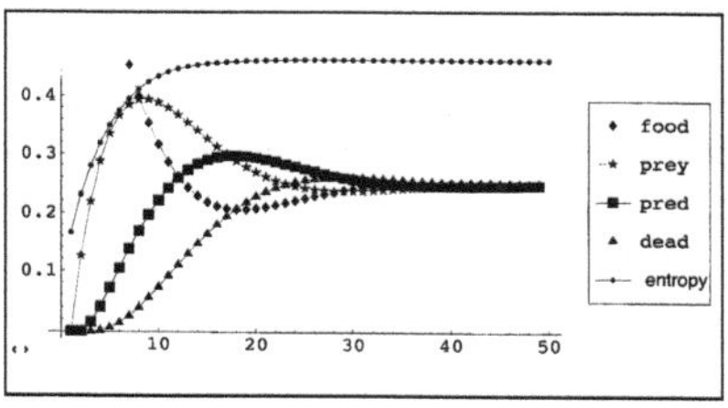

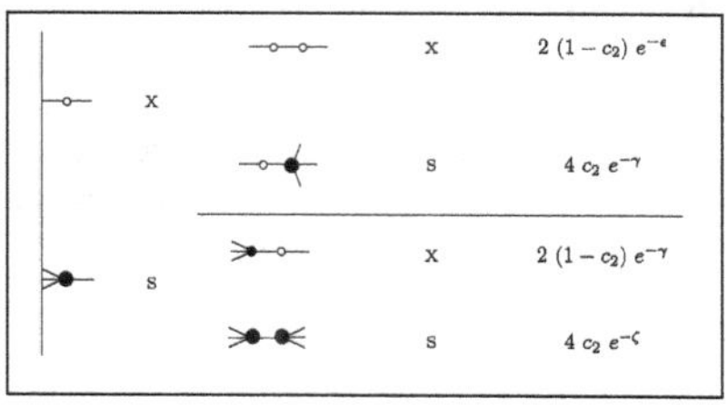

Ruben Aldrovandi

IFT - Institute of Theoretical Physics

State University of São Paulo (UNESP), Brazil

World Scientific

NEW JERSEY • LONDON • SINGAPORE • BEIJING • SHANGHAI • HONG KONG • TAIPEI • CHENNAI • TOKYO

Published by

World Scientific Publishing Co. Pte. Ltd.
5 Toh Tuck Link, Singapore 596224
USA office: 27 Warren Street, Suite 401-402, Hackensack, NJ 07601
UK office: 57 Shelton Street, Covent Garden, London WC2H 9HE

British Library Cataloguing-in-Publication Data
A catalogue record for this book is available from the British Library.

MATHEMATICAL PHYSICS AND MATRIX REPRESENTATIONS
The Multiple Applications of Stochastic, Circulant and Bell Matrices

ISBN 978-981-98-1388-9 (hardcover)
ISBN 978-981-98-1389-6 (ebook for institutions)
ISBN 978-981-98-1390-2 (ebook for individuals)

For any available supplementary material, please visit
https://www.worldscientific.com/worldscibooks/10.1142/14335#t=suppl

Dedication Page

to Ana Lúcia

Foreword

Preamble: The present book is as unique and original as was its Author, my late friend Ruben Aldrovandi. To write a Foreword for posthumous edition of his last book is a privilege bestowed on me by the Editors and his widow Ana Lucia Barbosa, who helped for completion of this final version. This book is an enlarged and improved version of Ruben Aldrovandi's book "Special Matrices of Mathematical Physics" edited in 2001 by the World Scientific Company. Ruben worked on new chapters and sections until his untimely death in May 2022. The final result became so enlarged and different from the first edition, that it fully deserves a new title.It suffices to say that the former book was about 300 pages long, whereas this one has 400 pages, and contains many new illustrative graphs and diagrams.

I decided to divide my Foreword in two parts, about the book itself first, highlighting its originality and scientific and pedagogical quality, and about the Author, his biography and scientific itinerary, containing also some of my personal recollections.

About the book: One of the most important threads in mathematics consists in finding ways enabling us to reduce problems involving continuum to the discrete ones, infinities to finite set of parameters, and complex non-linear problems to some number of linear ones. The history of mathematics, especially the part applied to physics, is full of such examples. To name a few: the Fourier analysis, which provides a discrete basis permitting to represent any continuous periodic function as superposition of elementary sinusoidal modes; the Legendre polynomials which stem from a multipole expansion of electric or gravitational potential; Bessel functions which were introduced in order to inverse the Kepler equation relating time with the azimuth angle, and so forth. And who could have supposed in the XIX-th century that the states of physical systems, no matter how complicated they might appear, could be represented by vectors in some linear space over complex numbers, with denumerable basis? — while nowadays we know it for a fact, using Hilbert spaces in Quantum Mechanics and Quantum Field Theory.

Mathematical objects known as *Matrices* appear naturally as transformations of linear spaces, real or complex ones. The most important case being the *square matrices*, which appear as linear transformations of spaces of the same dimension, or simply automorphisms of such spaces. The square matrices of a given dimension form a non-commutative ring, but they also can represent associative algebras. A very useful notion of determinant provides a criterion of inversibility of transformation: matrices with non-vanishing determinants are invertible. The eigenvalues are solutions of the N-th order characteristic equation, and the eigenvectors can be ascribed to the eigenvalues.

The importance of matrices can be measured by evoking Ado's theorem, according to which any finite-dimensional Lie algebra can be represented by matrices belonging to some associative algebra, with commutator representing the skew-symmetric Lie algebra product. Most often such associative structure, called *the enveloping algebra*, must be of higher dimension than the original Lie algebra. Any associative algebra of finite dimension can be represented by matrices, and the representation theory became a separate branch of mathematics.

All this was known since the 19th century, when matrices were used mostly in geometry and algebra, and later on, in mechanics. But the most striving entrance of matrices into physics was due to the first versions of Quantum Mechanics, elaborated by W. Heisenberg, M. Born and P. Jordan in the 1920-ties. Since then, matrices were used in many other branches of physics, including condensed matter physics and biological physics.

The three types of mathematical objects explored by Ruben Aldrovandi in this book are the Stochastic, Circular and Bell and Carleman matrices, which have vast applications not only in many branches of physics, but also in other sciences, such as chemistry or biology. Let us briefly comment of each of those.

Stochastic matrices were introduced by Russian mathematician Kolmogorov in his developments in the Theory of Probability. Their entries are positive or null, with each column normalized to 1. This property ensures that such a matrix acting on a column representing a probability distribution of N independent issues produces another normalized column, representing a modified probability distribution. Finite probability distributions arranged in columns obviously do not form a linear space, but linear operators can act on them nevertheless, leading to new distributions. Stochastic matrices do not form a group either, although the $N \times N$ unit matrix satisfies the definition; however, their determinant is not always different from zero (the simplest example in two dimensions being the matrix with all its entries equal to $\frac{1}{2}$). One of the most important properties is that among their eigenvalues there is at least one equal to 1, whereas the other ones, be they real or complex, are of absolute value lesser than 1. This means that after many consecutive iterations, a stochastic matrix acting on any initial probability distribution will produce asymptotically the unit eigenvector, describing the final steady configuration of the system under consideration.

Stochastic matrices are therefore an ideal tool to represent discrete or continuous changes in probability distributions, the Markov chains and their generalizations, and many other dynamical processes in statistical physics, especially in the non-equilibrium processes, chemistry, biology and population dynamics, and the book contains multiple original and novel applications ranging from the glass transition models to chemical equilibrium and genetics.

The *circulant matrices* are representations of cyclic groups Z_N and can be defined as matrix-valued N-th roots of the corresponding $N \times N$ identity matrix. They provide a discrete version of the Fourier transform, which acts on the discrete space of matrices. This circumstance led to generalized discrete geometries, known also as non-commutative geometries, and in discrete quantum mechanics. Another application of circulants is their use in generalized quantum statistics, e.g., generalized Pauli's exclusion principle based on the Z_3 symmetry instead of the usual Z_2 group action.

Of particular interest are circular matrices which are at the same time stochastic, called also *stochastic circulants.* They are used in the description of random walks, discrete diffusion equations and certain forms of Yang–Baxter equation. Two substantial Appendices to this part are devoted to the modelling of chemical reactions and to genetics, including original models of occurrence of blood groups in various populations.

Bell polynomials do appear as coefficients of the Taylor series of a function which itself is a composition of two functions of real variable, say $f(x)$ and $g(x)$, whose Taylor expansions are given by coefficients f_i and g_i. Consider now the function $F(x) = f(g(x))$, which is the composition of f with g. Its Taylor expansion can be the n written as follows:

$$F(x) = \sum_{n=1}^{\infty} \frac{x^n}{n!} F_n$$

with coefficients F_n depending on the Taylor expansion coefficients of both functions f and g in the following manner:

$$F_n[f \mid g] = \sum_{k=1}^{n} B_{nk}(g_1, g_2, \dots g_{n-k+1}) f_k$$

The coefficients B_{nk} are called *Bell polynomials*, denoted also by $B\{g_j\}$. Bell polynomials of a given function f give its representation in terms of a matrix $\mathbf{B} \mid f \mid$, called a *Bell matrix*, its entries being Bell polynomials. This enables one to represent the operation of composition of two functions by a product of corresponding Bell matrices — another example of reducing continuous problems to discrete ones.

Applications of Bell matrices are as various as they are numerous: in applied mathematics, they enable to discretize various operations, e.g., finding the inverse of a function and the Leibniz rule formulas. They are also applied in models of turbulence, statistical physics and dynamics of gaseous state. A generalization of Bell matrices called *Carleman matrices* is also introduced at the end of this section.

Ruben Aldrovandi's book on matrices is a unique blend of mathematical rigour, pedagogical skill and a plethora of novel and often unexpected applications in Physics, Chemistry, Biology and Genetics. It should be read and studied accordingly, by the graduates, doctoral students and the high-level researchers alike.

About the Author: One of the leading figures of the Theoretical and Mathematical Physics in Brazil, Ruben Aldrovandi was born on January 13, 1942, in Piracicaba, a town in the Brazilian State São Paulo, situated at 165 kilometres from the capital. Of Methodist parents, Ruben could read at the age of four, and got an excellent primary and secondary education at the Colégio Piracicabano. Besides the languages — English, Spanish and French — he would follow classes of mathematics, physics and chemistry, history and geography, notwithstanding the study of Holy Scriptures. A wide and solid basis for the forthcoming University studies. Ruben was taking piano lessons — he was a music lover since his earliest years — and was literally devouring books.

As a teenager, Ruben decided to become a scientist, although his father, Sr. José, insisted on his son pursuing rather a more practical career. After some hesitation and a year spent in the Faculty of Agronomy, young Ruben entered the University of São Paulo, joining the Faculty of Physics. He completed his studies in 1964, and soon after got his first teaching position at the Institute of Theoretical Physics. He preferred however to enlarge his knowledge of theoretical physics abroad, and went first to start doctoral studies in Trieste in 1967, and even started to work on a Ph.D. under the guidance of Christian Fronsdal, but after attending a seminar given by Roland Omnès he decided to move to France and to work under his direction, changing the subject quite radically. The French system at that time comprised two doctoral titles, the "small one" called "Doctorat du troisième cycle", and the final one, much more important, corresponding to the German "Habilitationsschrift", called "Doctorat d'Ètat", which was Ruben's thesis. Entitled "La coalescence dans le modèle d'univers symétrique", it was defended in 1973. He returned to Brazil in 1974, recovering his position at the Theoretical Physics Institute. The Institute occupied then a charming old building at 145, Rua Pamplona, where Ruben spent most of his scientific and pedagogical career.

Ruben's interests in Theoretical Physics covered many subjects, mostly in General Relativity and Field Theory, always with strong inkling towards mathematical models. Besides teaching, he devoted much of his time to the Institute, of which he became the Director in 1984, and remained in this position until 1990. It was a decade full of trouble in Brazil, political and economical, and Ruben had to fight in order to preserve the IFT, threatened with suppression. He succeeded at the end of the day, signing a convention with the University Estadual Paulista "Julio de Mesquita Filho", which ensured financial participation since then. During the same period Ruben married and had two children, a son and a daughter. Later on, his son Leonardo became a professional musician (piano), while the daughter Cibele devoted herself to archaeology.

Ruben was very demanding, towards himself in the first place, and towards his students. He would not publish a paper unless it was polished and faultless. His lectures, entitled "Elementary Introduction to Classical Fields" is a pedagogical masterpiece of clarity. With his best student and collaborator José Geraldo Pereira he initiated an original approach to gravitation called "the teleparallel gravity", which resulted in many articles and a book written in common. Among his doctoral students many became professional researchers and teachers. He was universally respected and admired as one of the most original minds of Brazilian theoretical physics, and this fact found its confirmation in a beautiful book "Tribute to Ruben Aldrovandi" edited recently by his colleagues and friends ([1]).

My first acquaintance with Ruben Aldrovandi goes back to late seventies, when he was on preparing his doctoral thesis in France.

During Ruben's stay in Paris we met several times, usually at theoretical and mathematical physics seminars organized at Collège de France in École Normale, or in Orsay, at the Paris-XI University. We discussed sometimes after a seminar, but rather casually. Already then I was impressed by his perfect mastership of French language and a vast knowledge of French literature and history. He had an extraordinarily wide range of interests; world literature, art and music, philosophy, psychoanalysis, history and religion. His tastes in literature and art were rather classical, but he cherished Brazilian music, especially Bossa Nova whose advent coincided with Ruben's adolescence and youth.

When Ruben returned to São Paulo, our contacts ceased for about a decade. But when in 1993 I visited Brazil for the first time participating at a workshop on theoretical physics at the CNPq in Rio de Janeiro, Ruben invited me to visit his Institute at 145 rua Pamplona in São Paulo, and give a talk at the seminar. I spent a week in Ruben's company, delighted by his kind hospitality, discovering common taste not only in science, but also in philosophy, music and literature. And we became real friends since then.

Our friendship really flourished during the two last years of XX-th century, 1998–2000, when Ruben and his wife Ana Lucia Barbosa spent two years in Paris, in the Laboratoire de Gravitation et Cosmologie Relativistes which I was directing from 1991 till 2002. Ana Lucia was on a post-doctoral visit, while Ruben was on sabbatical leave of absence from IFT. Very soon they became an integral part of our research group, appreciated by all its members. Those were the times when nobody objected smoking a pipe in the office, especially the excellent aromatic tobacco brands he was using.

During Ana Lucia and Ruben's stay in my Lab, another friend of mine, Dmitry V. Gal'tsov from the Moscow State University, spent three months there, too, as an invited professor. Ana Lucia's PhD thesis concerned the Yang-Mills fields, so it was quite natural that we started to work together and produced a common paper devoted to a non-abelian generalization of the Born-Infeld non-linear electrodynamics [2].

While continuing his research on teleparallel gravity, which he led since many years with his friend and collaborator José G. Pereira [3], [4], Ruben was writing a new book on various matrices used in mathematical physics [5]. By one of those happy coincidences that occur sometimes in our scientific development, I was working at that time on models of agglomeration and glass transition, with my friends Rafael Barrio of Mexico and Jean-Paul Duruisseau from my Paris-VI University, and our former PhD students Matthieu Micoulaut (Paris) and Gerardo G. Naumis (Mexico) who spent a year in my Lab as a post-doctoral visitor.

Ruben got interested in our stochastic matrix approach to the glass forming problem, and incorporated it in his book on matrices [5]. Moreover, we wrote a paper together which he presented in September 1999 at the theoretical physics workshop in Karpacz, Poland [6]. He produced a beautiful figure showing the agglomeration of two kinds of atoms — Se and As to be precise — on the rim of a cluster already formed, and used it as illustration for the cover of his book.

Having observed how Ruben worked on his book inspired me enough to consider writing my own book on "Model's of Agglomeration and Glass Transition" [7] One of the chapters of this book contains the discussion on stochastic matrices borrowed from Ruben's book.

After Ruben and Ana Lucia departure we continued to keep in touch. Ruben visited Paris again in 2002 when he participated in the 24th International Colloquium on Group Theoretical Methods in Physics. It was a wonderful occasion to meet again, in the company of many common friends from around the world. At the conference, Ruben presented a paper coauthored with Ana Lucia, on the algebraic skeleton of space-time [8]. Since then I visited São Paulo and met Ruben and Ana Lucia twice, in 2004 and in 2007. I was invited again to give a talk at the same charming building more than half-century old, and met Ruben's colleagues. We spent some time together walking through the streets of São Paulo, and in the evening shared meals in good restaurants. I was also invited to Ruben's home and tried his piano. Ruben played piano much better than me, his favorites being Bach and Mozart.

The last paper we authored in common was published in 2010, in the Proceedings of the Biomat Conference in Rio de Janeiro [9]. We exchanged mails and kept contact during Ruben's illness, on which he would not comment being very stoical and keeping rather silent on his ordeal. His passing away left me full of sorrow and sadness, and I miss him badly. Writing this text had a soothing effect on me, because it revived the memory of happy times spent together.

Richard Kerner,

Emeritus Professor, Sorbonne-University, Paris, France.

Preface

This book is a new edition which is extended from the previous title *Special Matrices of Mathematical Physics: Stochastic, Circulant and Bell Matrices*. Besides corrections, some rearrangements and additions have been performed with respect to the previous one. Some topics have been so enlarged that it became convenient to dedicate chapters to them. This is the case of Random Walks, Chemical Reactions and Genetics. The last subject may seem specially out of place in "Mathematical Physics". Our excuse is that mendellism and blood types are here presented by using just the same methods of the other chapters.

The Web revolution, still at its beginnings at the time of the first edition, has since become inescapable. It has been found necessary to add references to a great many online resources.

The same people — Colleagues and Masters — to which gratitude has been expressed in the Preface to the First edition on the next page are to be thanked again. An additional case is that of Ana Lúcia Barbosa, to whom the new figures are due.

Perhaps a comment on nomenclature should have place here: we adopt the names Bell matrices and polynomials, Carleman matrices and polynomials for certain objects. There is no generally accepted nomenclature in the literature, many authors just exchanging these two names. We here just stick to a particular choice.

Preface to the First edition

Notwithstanding their previous appearances in elasticity and fluid problems, matrices have made their definitive, triumphant entrance in Physics with the advent of Quantum Mechanics. Only then have they shown themselves as unavoidable, essential tools to the understanding of the basic ways of Nature. It is not only ironic that the man responsible for that inauguration had no previous knowledge of their existence — it is also an eloquent pointer to what a sane relationship between Mathematics and Physics would be. Physics will remain Physics as long as it remains an experimental science. The physicist must look for the Mathematics he needs *after* doing some Physics, or *while* trying to do so. The parts of Mathematics of use in Physics have, for the physicist and the mathematician, something of the fiddle and the violin: two distinct instruments in one same object. Only a Poincaré can play both as a Geraldine O'Grady does. The author is no Poincaré, and has not the pretension of teaching their jobs to other people, but he had the opportunity, as a young student, of hearing Heisenberg himself expressing just that opinion.

This point of view fixes the limits of the present text. Three kinds of matrices will be described which have shown to be, and promise to remain, of physical interest. The presentation is a physicist's introduction addressed to physicists, though some mathematicians may be interested in knowing which questions can be of physical relevance. It intends to be as simple as possible, centered on examples, with the aim of making concepts and properties intuitive enough to make them applicable in Physics. The exposition will have, in consequence, an alternating character. A mathematical notion must become intuitive to acquire physical applicability, but that intuition comes necessarily from some previous application. The basic properties of a matrix family are first introduced, some applications made, then more formal aspects presented, and so on.

The three types of matrices have not much in common, except for the fact that they appear to open new ways to further developments. In each case old ideas appear under new, simpler visages. Kolmogorov's stochastic matrices are introduced as instruments to describe in a surprisingly simple way dynamical systems of a wide variety of behaviors. Circulants appear as the bedrock of Fourier analysis and as a most natural gateway into Quantum Mechanics and Noncommutative Geometry. Bell polynomials, after turning up in Differential Geometry and in the theory of gases, promise to be instrumental in the study of chaotic mappings.

Stochastic matrices govern, through repeated iteration, the time evolution of probability distributions of which they are independent. The example chosen has been glass transition, a fascinating phenomenon for which there is no complete theory and for whose description they have made important contributions. More is said on glasses than on other subjects, because not much about the recent advances is found in textbooks. Some physical feeling once got, their spectral properties are presented and shown to epitomize all the wealth of possible evolution courses. The matrices finally provide toy models illustrating involved questions like ergodicity, multi-phase systems, dissipation and non-equilibrium.

Circulants, once applied to Quantum Mechanics, lead to phase spaces whose points are matrices. Spaces formed by such noncommutative points have noncommutative geometries. With a differential calculus on spaces of matrices as a preliminary, quantum groups turn up as algebras of matrices whose entries are themselves matrices. A matrix symplectic structure comes out in a natural way and sheds new light on the underlying texture of phase space.

Bell polynomials, and the matrices of which they are the entries, lead to closed expressions for many formulas usually presented with ellipses in the end, a true nightmare for algebraic computation. Those expressions include relationships between determinants and traces, coefficients and roots of polynomials, virial coefficients and configuration integrals, besides Lie algebra invariants of various types, from Casimir operators to characteristic classes and characters. Furthermore, a Bell polynomial is related to a mapping. A matrix formed with the Bell polynomials of a mapping is a linear representation of that mapping. Function composition is thereby reduced to matrix multiplication. In particular, the map iterations describing discrete-time dynamical systems become powers of matrices and can be interpolated into continuous versions of the time evolution. The theory of real gases provides the best physical illustrations of the polynomials and these, in retribution, reveal the presence of a matrix structure in the backstage of the theory.

The text is divided into four Parts: one for the general background and one for each kind of matrix. Up to references to the background introduction, the Parts are tentatively self–contained, which accounts for a certain amount of repetition. A formulary presents the main expressions of the text, besides formulas used in finding and/or checking them. The only way we break with standard editing is by the presence of Comments, which contain information additional to the main text and may be omitted in a first reading. Matrices, square or not, are indicated by boldface symbols as $\mathbf{M}$, $\mathbf{v}$, etc., unless that make the notation too charged. In particular, matrix functions of matrices are typed like $\mathbf{F}(\mathbf{M})$. Entries, however, are noted in common type. Only a few items have deserved some special notation. Thus, $\mathbb{B}_{nk}[f]$ stands for Bell polynomials in the variables $f_1, f_2, \ldots, f_{n-k+1}$.

The author owes a lot to some people and institutions.

On the institutional side, he is deeply thankful to the Fundação de Amparo à Pesquisa do Estado de São Paulo (FAPESP), Brazil, whose financial support has been essential. To R. Kerner for hospitality at the Laboratoire de Gravitation et Cosmologie Relativistes of the Université Pierre et Marie Curie and CNRS, France. And to J. Lukierski, whose invitation to the Institute for Theoretical Physics, University of Wroclaw, Poland, spurred him to put in order his notes on stochastic matrices.

On the editorial side, he is very much indebted to S. Sundaresan for assistance with the files. To the Editor, E.H. Chionh, for patient support and a sound checking of the manuscript which has greatly improved it.

On the professional and personal sides, warm thanks are due to D. Galetti, R. Kerner, M. Dubois-Violette and J. Madore for many helpful discussions on quantization and non–commutative geometry. To R. Kerner for showing him how stochastic matrices can be of high interest in Physics. To K. Fujii, Keith M. Briggs and M.A. Ozório de Almeida for consequential suggestions. To J.G. Pereira for so much help and understanding. To the masters from which he has so ineptly borrowed: Gantmacher and Comtet, Riordan and the senior Faddeevs, and Kirillov, and Pathria, and so many others. And, finally, to the creator mathematicians Kolmogorov, Weyl and Bell, despite their standing above all gratitude.

Contents

Chapter 1

Some fundamental notions

We recall here some basic concepts on matrices, with the main purpose of fixing notation and terminology. Only ideas and results necessary to the discussion of the particular kinds of matrices we shall be concerned with in this book will be considered. General Matrix Theory is covered in Gantmacher's classic [10].

1.1 Definitions

(1) Given an $N \times N$ matrix $\mathbf{M}$, its *characteristic matrix* is $\lambda\mathbf{I} - \mathbf{M}$, a function of $\mathbf{M}$ itself, of the $N \times N$ identity matrix $\mathbf{I}$ and of the complex variable λ.

(2) The *characteristic polynomial* of $\mathbf{M}$ is the determinant of its characteristic matrix,

$$\Delta_{\mathbf{M}}(\lambda) = \det(\lambda\mathbf{I} - \mathbf{M}) = \prod_{k=1}^{N} (\lambda - \lambda_k) \,. \tag{1.1}$$

(3) The *characteristic equation* of $\mathbf{M}$ (or, in a rather loose terminology, its *secular equation*) is

$$\Delta_{\mathbf{M}}(\lambda) = 0 \,. \tag{1.2}$$

(4) The *spectrum* of an $N \times N$ matrix $\mathbf{M}$ is defined as the set $\mathrm{Sp}[\mathbf{M}] = \{\lambda_1, \lambda_2, \lambda_3, \ldots, \lambda_N\}$ of all complex values of λ for which the characteristic matrix $(\lambda\mathbf{I} - \mathbf{M})$ is not invertible. It is, consequently, the set of the roots of the characteristic polynomial. These roots λ_k, already introduced in the last equality of (1.1), are the *eigenvalues* of $\mathbf{M}$. For each k there exists an *eigenvector* $\mathbf{v}_k$, a non-vanishing $N \times 1$ matrix

(that is, a column) satisfying $\mathbf{M}\,\mathbf{v}_k = \lambda_k\,\mathbf{v}_k$. If the eigenvalues are all distinct, $\mathbf{M}$ is said to be *non-degenerate*. $\mathbf{M}$ is *degenerate* if at least one root is multiple.

(5) The matrix

$$\mathbf{R}_{\mathbf{M}}(\lambda) = \Delta_{\mathbf{M}}(\lambda)\,\mathbf{I}\,(\lambda\mathbf{I} - \mathbf{M})^{-1}$$

will be called the *resolvent* of $\mathbf{M}$.[1]

(6) A polynomial $p(x)$, conventionally normalized so that the highest–order coefficient is 1, is an *annihilating polynomial* of $\mathbf{M}$ if $p(\mathbf{M}) = 0$.

(7) The Cayley–Hamilton theorem [11] states that, for any matrix $\mathbf{M}$, the characteristic polynomial is an annihilating polynomial:

$$\Delta_{\mathbf{M}}(\mathbf{M}) = (\mathbf{M} - \lambda_1)(\mathbf{M} - \lambda_2)(\mathbf{M} - \lambda_3)\dots(\mathbf{M} - \lambda_{N-1})(\mathbf{M} - \lambda_N) = 0\,. \tag{1.3}$$

In simple words, a matrix satisfies its own characteristic equation. Each multiple eigenvalue contributes to the product above with as many factors as its multiplicity.

(8) Writing the characteristic equation in the form $\Delta_{\mathbf{M}}(\lambda) = \sum_{j=0}^{N} c_j\lambda^j = 0$ implies $c_N = 1$, which is a standard normalization; and that $c_0 = (-)^N \prod_{k=1}^{N}\lambda_k = (-)^N \det \mathbf{M}$. The matrix is invertible *iff* (if and only if) $\det \mathbf{M} = (-)^N c_0 \neq 0$. In that case, the Cayley–Hamilton theorem gives the inverse as

$$\mathbf{M}^{-1} = -\,\frac{1}{c_0}\,\sum_{j=1}^{N} c_j \mathbf{M}^{j-1}\,. \tag{1.4}$$

(9) It may happen that a number $< N$ of factors produce an annihilating polynomial. The lowest-order annihilating polynomial of $\mathbf{M}$ is the *minimal polynomial* of $\mathbf{M}$. The minimal polynomial divides the characteristic polynomial, and will be indicated by $\mu(\lambda)$.

(10) Given a complex $N \times N$ matrix $\mathbf{M}$, its *adjoint* (or Hermitian conjugate) $\mathbf{M}^\dagger$ is the matrix whose entries are ${M^\dagger}_{ij} = {M^*}_{ji}$. For real matrices, the adjoint coincides with the transpose: $\mathbf{M}^\dagger = \mathbf{M}^T$. To ease typewriting, we shall frequently indicate a column-vector by its transpose row-vector, for example $\mathbf{p} = (p_1, p_2, \dots, p_N)^T$.

---

[1] Notice that there are many different definitions of a resolvent in the literature: with or without the $\Delta_{\mathbf{M}}(\lambda)$ factor, with the denominator $\mathbf{I} - \lambda\mathbf{M}$, with the unit matrix subtracted, etc. We shall follow the most current use in each branch of activity on the subject. Here is given that predominant in Matrix Theory. We shall find another in Fredholm Theory (Section 18.6).

(11) $\mathbf{M}$ is *normal* when it commutes with $\mathbf{M}^\dagger$: $[\mathbf{M}, \mathbf{M}^\dagger] = 0$. In that case,

(i) $\mathbf{M}$ and $\mathbf{M}^\dagger$ share all the eigenvectors;

(ii) the eigenvectors constitute a complete orthonormal ($\mathbf{v}_k \cdot \mathbf{v}_{k'} = \delta_{kk'}$) vector basis for the N-dimensional space.

Important subcases are the unitary matrices, for which $M^\dagger = M^{-1}$, and the Hermitian matrices, which satisfy $M = M^\dagger$. Hermitian matrices have real eigenvalues and, as particular cases, so do the symmetric matrices. More about normal matrices is said below, in Section 1.4.

(12) An account of non-negative matrices is given in Section 3.1.

(13) An algebra is a vector space V on which an internal binary operation $*$ is defined:

$$* : V \otimes V \to V \quad ; \quad *(a, b) = a * b = c \in V \, .$$

It is a Lie algebra if the operation is antisymmetric, $a * b =$ - $b * a$, and the Jacobi identity

$$(a * b) * c + (c * a) * b + (b * c) * a = 0$$

holds. Matrices provide standard examples: they form an algebra with their usual product, and a Lie algebra with the commutator.

(14) The following terminology will be frequently used. We call an *alphabet* any ordered set of quantities $x_1, x_2, x_3, \ldots, x_N$ (its *letters*) and indicate it by the typical letter in bold, as for example in $\mathbf{x} = \{x_k\} = \{x_1, x_2, x_3, \ldots, x_N\}$. A monomial is a *word*. This nomenclature is current in the theories of finite groups and symmetric functions (Section 16.2). In principle, the alphabet can be infinite and the letter positioning (as with real words) can be noncommutative. The spectrum $\mathrm{Sp}[\mathbf{M}] = \boldsymbol{\lambda}$ is an example of commutative alphabet, whose letters are the eigenvalues of $\mathbf{M}$.

1.2 Components of a matrix

Every matrix $\mathbf{M}$ with spectrum $\boldsymbol{\lambda} = \{\lambda_1, \lambda_2, \lambda_3, \ldots, \lambda_N\}$ can be decomposed in terms of certain "components" $\mathbf{Z_k}$, which are eigenmatrices of $\mathbf{M}$ ($\mathbf{M}\,\mathbf{Z_k} = \lambda_k \mathbf{Z_k}$). This *spectral decomposition* $\mathbf{M} = \sum_k \lambda_k \mathbf{Z_k}$ is intimately related to the theory of functions of matrices. The treatment is distinct for nondegenerate and degenerate matrices.

Let us begin with a simple example. Consider the characteristic equation of a 3×3 matrix $\mathbf{M}$:

$$\Delta_{\mathbf{M}}(\lambda) = (\lambda - \lambda_1)(\lambda - \lambda_2)(\lambda - \lambda_3) = 0 \, .$$

The Cayley–Hamilton theorem takes, in this case, the form

$$\Delta_{\mathbf{M}}(\mathbf{M}) = (\mathbf{M} - \lambda_1 \mathbf{I})(\mathbf{M} - \lambda_2 \mathbf{I})(\mathbf{M} - \lambda_3 \mathbf{I}) = 0 \,. \tag{1.5}$$

There would be repeated factors if some eigenvalue were multiple, but we shall suppose that not to be the case. Let us isolate one of the factors, say that including the eigenvalue λ_1, corresponding to some eigenvector $\mathbf{v}_1$, and look at the properties of the remaining expression,

$$\hat{\mathbf{Z}}_1 = (\mathbf{M} - \lambda_2 \mathbf{I})(\mathbf{M} - \lambda_3 \mathbf{I}) \,. \tag{1.6}$$

To begin with,

$$\hat{\mathbf{Z}}_1 \mathbf{v}_1 = (\lambda_1 - \lambda_2)(\lambda_1 - \lambda_3)\; \mathbf{v}_1 \,.$$

As any other contribution to the trace vanishes, the trace of $\hat{\mathbf{Z}}_1$ is also $(\lambda_1 - \lambda_2)(\lambda_1 - \lambda_3)$. Furthermore, Eq.(1.5) says that $\hat{\mathbf{Z}}_1$ is an eigenmatrix of $\mathbf{M}$ with eigenvalue λ_1: $\mathbf{M}\,\hat{\mathbf{Z}}_1 = \lambda_1 \hat{\mathbf{Z}}_1$. Equation (1.6) will consequently lead to

$$\hat{\mathbf{Z}}_1 \hat{\mathbf{Z}}_1 = (\lambda_1 - \lambda_2)(\lambda_1 - \lambda_3)\; \hat{\mathbf{Z}}_1 = (\mathrm{tr}\; \hat{\mathbf{Z}}_1)\; \hat{\mathbf{Z}}_1 \,.$$

This means that $\hat{\mathbf{Z}}_1$ is a (as yet unnormalized) projector. When the eigenvalues are distinct, we can divide $\hat{\mathbf{Z}}_1$ by $(\lambda_1 - \lambda_2)(\lambda_1 - \lambda_3)$ and define a new projector

$$\mathbf{Z}_1 = \frac{(\mathbf{M} - \lambda_2 \mathbf{I})(\mathbf{M} - \lambda_3 \mathbf{I})}{(\lambda_1 - \lambda_2)(\lambda_1 - \lambda_3)} \,. \tag{1.7}$$

This is a true eigenprojector. It is an idempotent, $\mathbf{Z}_1 \mathbf{Z}_1 = \mathbf{Z}_1$, has trace $= 1$, is an eigenmatrix $\mathbf{M}\mathbf{Z}_1 = \lambda_1 \mathbf{Z}_1$ and is such that $\mathbf{Z}_1 \mathbf{v}_1 = \mathbf{v}_1$. If $\mathbf{v}_2$ and $\mathbf{v}_3$ are the eigenvectors of $\mathbf{M}$ with eigenvalues λ_2 and λ_3, then $\mathbf{Z}_1 \mathbf{v}_2 = 0$, $\mathbf{Z}_1 \mathbf{v}_3 = 0$.

The above considerations can be repeated for each factor in (1.5), leading to eigenmatrices $\hat{\mathbf{Z}}_2$, $\hat{\mathbf{Z}}_3$ and, as long as the eigenvalues are different, to the corresponding projectors $\mathbf{Z}_2$ and $\mathbf{Z}_3$. The product of any two distinct $\mathbf{Z}_k$'s is proportional to $\Delta_{\mathbf{M}}(\mathbf{M})$ and, consequently, vanishes. In short, they are orthogonal to each other:

$$\mathbf{Z}_k \; \mathbf{Z}_{k'} = \delta_{kk'} \; \mathbf{Z}_k \quad \text{and} \quad \mathrm{tr}\,(\mathbf{Z}_k \; \mathbf{Z}_{k'}) = \delta_{kk'} \,.$$

Direct computation shows that $\mathbf{I} = \sum_k \mathbf{Z}_k$, which multiplied by $\mathbf{M}$ gives

$$\mathbf{M} = \textstyle\sum_k \lambda_k \; \mathbf{Z}_k \,.$$

An orthogonal decomposition in matrix space is thereby achieved.

There is an extra bonus. From the properties above it follows that

$$\mathbf{M}^m = \textstyle\sum_k \lambda_k^m \; \mathbf{Z}_k \,. \tag{1.8}$$

Consequently, given any power series function $F(z) = \sum_m c_m z^m$, a matrix function $\mathbf{F}(\mathbf{M})$ can be formally defined by $\mathbf{F}(\mathbf{M}) = \sum_m c_m \mathbf{M}^m = \sum_k \sum_m c_m \lambda_k^m \, \mathbf{Z}_k$, or

$$\mathbf{F}(\mathbf{M}) = \textstyle\sum_k F(\lambda_k) \, \mathbf{Z}_k \, . \tag{1.9}$$

We say "formally" because there are, of course, necessary conditions to make of the above formula a meaningful, convergent expression. Notice on the other hand that, besides the Cayley–Hamilton theorem, which is valid for every matrix, the only assumption made has been that the eigenvalues are distinct. No use has been made of any further special property. What has been done above holds consequently for any nondegenerate 3×3 matrix.

The procedure can be directly generalized to an $N \times N$ matrix. We shall formalize it in what follows, but through a rather indirect, alternative approach. The point is that the components $\mathbf{Z}_k$ lead to a fair formulation of the question of matrix functions — as suggested by (1.9) — which in return provides a nice way to calculate them. For this reason, it is preferable to start from matrix functions and come back to the components. That is what we now proceed to do. We start with the nondegenerate case, as it prepares the way to the study of the degenerate case as well.

1.3 Matrix functions

The functions of a matrix $\mathbf{A}$ are defined by functions on its spectrum. This requires a decomposition of $\mathbf{A}$ in terms of components, which are eigenmatrices, one along each eigenvalue. For nondegenerate matrices, these components are orthonormal projectors. Such eigenprojectors provide, under a simple condition, an immediate adaptation to matrices of any function defined by a Taylor series.

Suppose a function $F(\lambda)$ is given which can be expanded as a power series

$$F(\lambda) = \sum_{k=0}^{\infty} c_k (\lambda - \lambda_0)^k$$

inside a convergence circle $|\lambda - \lambda_0| < r$. Then the function $\mathbf{F}(\mathbf{M})$, whose argument is some given $N \times N$ matrix $\mathbf{M}$, is defined by

$$\mathbf{F}(\mathbf{M}) = \sum_{k=0}^{\infty} c_k (\mathbf{M} - \lambda_0)^k$$

and has a sense (here the announced condition) when the eigenvalues of $\mathbf{M}$ lie within the convergence circle. Given the eigenvalues $\lambda_1, \lambda_2, \ldots, \lambda_N$, there exists a set of matrices $\{\mathbf{Z}_j[\mathbf{M}]\}$ constituting a matrix basis in terms of which $\mathbf{M}$ can be written as

$$\mathbf{M} = \sum_{j=1}^{N} \lambda_j \, \mathbf{Z}_j[\mathbf{M}] \,. \tag{1.10}$$

The $\mathbf{Z}_j[\mathbf{M}]$'s are the *components* of $\mathbf{M}$, and the above expression is the "decomposition of $\mathbf{M}$ in terms of its components", or its *spectral decomposition.* The function $\mathbf{F}(\mathbf{M})$, defined as above, can be equivalently written as the $N \times N$ matrix

$$\mathbf{F}(\mathbf{M}) = \sum_{j=1}^{N} F(\lambda_j) \, \mathbf{Z}_j[\mathbf{M}] \,. \tag{1.11}$$

A few examples:

(1) the exponential

$$e^{\mathbf{M}} = \sum_{j=1}^{N} e^{\lambda_j} \, \mathbf{Z}_j[\mathbf{M}] \tag{1.12}$$

(2) the arbitrary power

$$\mathbf{M}^{\alpha} = \sum_{j=1}^{N} \lambda_j{}^{\alpha} \, \mathbf{Z}_j[\mathbf{M}] \tag{1.13}$$

(3) the r-th root

$$\mathbf{M}^{1/r} = \sum_{j=1}^{N} \lambda_j{}^{1/r} \, \mathbf{Z}_j[\mathbf{M}] \tag{1.14}$$

(4) the resolvent

$$\mathbf{R}_{\mathbf{M}}(\lambda) = \Delta_{\mathbf{M}}(\lambda) \sum_{j=1}^{N} \frac{\mathbf{Z}_j[\mathbf{M}]}{\lambda - \lambda_j}$$

(5) the inverse, when it exists:

$$\mathbf{M}^{-1} = \sum_{j=1}^{N} \lambda_j{}^{-1} \, \mathbf{Z}_j[\mathbf{M}] \,. \tag{1.15}$$

We are here, of course, forcing the hand a bit. The given definition of the function $\mathbf{F}(\mathbf{M})$ of a matrix assumes that $F(z)$ be given by a Taylor series. This is not the case, for instance, for the function $1/z$ which backs the last example. Experiments with small matrices show, however, that the formula works, provided the inverse exists. This suggests that the theory can be extended, and indeed it can [10]. We shall use (1.11) as the *definition* of $\mathbf{F}(\mathbf{M})$, to be used whenever well-defined.

1.3.1 *Nondegenerate matrices*

As seen in the introductory example, things are very simple when all the eigenvalues of $\mathbf{M}$ are distinct. We shall first describe that case, and later adapt the results to the degenerate case.

The way to get at the $\mathbf{Z}_k$'s is that shown in the 3×3 example given above. Let us list the main properties of the eigenmatrices $\mathbf{Z}_k[\mathbf{M}]$ of a nondegenerate matrix $\mathbf{M}$:

(1) the $\mathbf{Z}_k$'s are linearly independent and nonvanishing;

(2) the basis $\{\mathbf{Z}_k[\mathbf{M}]\}$ depends on $\mathbf{M}$, but is the same for every function $\mathbf{F}(\mathbf{M})$; thus, any $\mathbf{F}(\mathbf{M})$ will have the same components as $\mathbf{M}$;

(3) the $\mathbf{Z}_k$'s commute with each other, and with $\mathbf{M}$;

(4) the $\mathbf{Z}_k$'s are projectors, that is, idempotents, objects satisfying $\mathbf{Z}_k^2 = \mathbf{Z}_k$; this equation, in the form

$$\textstyle\sum_n [\mathbf{Z}_{(k)}]_{rn} \; [\mathbf{Z}_{(k)}]_{ns} = [\mathbf{Z}_{(k)}]_{rs} \; ,$$

shows that $\mathbf{Z}_{(k)}$ has for columns its own eigenvectors with eigenvalue 1; indeed, define the vectors $\mathbf{v}_{(ks)}$ with components $(\mathbf{v}_{(ks)})_j = [\mathbf{Z}_{(k)}]_{js}$; the equation is then

$$\textstyle\sum_n [\mathbf{Z}_{(k)}]_{rn} (\mathbf{v}_{(ks)})_n = (\mathbf{v}_{(ks)})_r \; ; \tag{1.16}$$

(5) the $\mathbf{Z}_k$'s can be normalized to tr $(\mathbf{Z}_k) = 1$ for each k;

(6) once thus normalized the eigenmatrices, besides satisfying $\mathbf{Z}_i \; \mathbf{Z}_j = \delta_{ij} \mathbf{Z}_i$, are orthonormal by the trace: tr $(\mathbf{Z}_i \mathbf{Z}_j) = \delta_{ij}$;

(7) $\mathrm{tr}\;[\mathbf{F}(\mathbf{M})] = \sum_{j=1}^{N} F(\lambda_j)$; $\mathrm{tr}\;[\mathbf{M}^k \mathbf{Z}_j] = \lambda_j^k$; $\mathrm{tr}\;[\mathbf{F}(\mathbf{M})\;\mathbf{Z}_j] = F(\lambda_j)$;

(8) since they are eigenmatrices of $\mathbf{M}$,

$$\textstyle\sum_b M_{ab} (Z_k)_{bc} = \lambda_k (Z_k)_{ac} \; , \tag{1.17}$$

it is clear that each column of $\mathbf{Z}_k$ is an eigenvector of $\mathbf{M}$ corresponding to the eigenvalue λ_k;

(9) if $\mathbf{M}$ is a normal matrix and $\mathbf{U}^{-1}\mathbf{M}\ \mathbf{U} = \mathbf{M}_{diagonal}$, then the entries of $\mathbf{Z}_k$ are given by $(Z_k)_{rs} = U_{rk}U^{-1}{}_{ks}$ (no summation, of course);

(10) in the above case, the eigenvector $\mathbf{v}_{(k)}$ corresponding to $\mathbf{Z}_{(k)}$ is unique, and has components $v_{(k)r} = U_{rk}$; in consequence, $(Z_k)_{rs} = v_{(k)r}v^*_{(k)s}$, with the eigenvectors normalized according to the completeness relation $\sum_{r=1}^{N} v^*_{(k')r}v_{(k)r} = \delta_{kk'}$.

The last two items show the projectors frequently used in Quantum Mechanics. Their simplicity comes from the supposed normal character of $\mathbf{M}$ in that case. As already announced, we shall say a little more on normal matrices in Section 1.4.

A set of N powers of $\mathbf{M}$ is enough to fix the projector basis. Using for $\mathbf{F}(\mathbf{M})$ in (1.11) the power functions $\mathbf{M}^0 = \mathbf{I}, \mathbf{M}^1, \mathbf{M}^2, \ldots, \mathbf{M}^{N-1}$, we have

$$\mathbf{I} = \sum_{j=1}^{N} \mathbf{Z}_j \ ; \ \mathbf{M} = \sum_{j=1}^{N} \lambda_j \mathbf{Z}_j \ ; \ \mathbf{M}^2 = \sum_{j=1}^{N} \lambda_j{}^2 \mathbf{Z}_j \ ; \ \ldots;$$

$$\mathbf{M}^k = \sum_{j=1}^{N} \lambda_j{}^k \mathbf{Z}_j \ ; \ \ldots \ ; \ \mathbf{M}^{N-1} = \sum_{j=1}^{N} \lambda_j{}^{N-1} \mathbf{Z}_j \ .$$

For $k \geq N$, the powers $\mathbf{M}^k$ are no more independent. This comes again from the Cayley–Hamilton theorem: the equation $\Delta_{\mathbf{B}}(\mathbf{M}) = 0$ provides $\mathbf{M}^N$ in terms of lower powers of $\mathbf{M}$, so that all the higher powers of $\mathbf{M}$ can be computed from those between $\mathbf{M}^0$ and $\mathbf{M}^{N-1}$. Inversion of the above expressions for the powers of $\mathbf{M}$ in terms of the $\mathbf{Z}_j$'s leads to a closed form for each $\mathbf{Z}_j$,

$$\mathbf{Z}_j[\mathbf{M}] = \tfrac{\mathbf{M}-\lambda_1\mathbf{I}}{\lambda_j-\lambda_1}\ \tfrac{\mathbf{M}-\lambda_2\mathbf{I}}{\lambda_j-\lambda_2} \cdots \tfrac{\mathbf{M}-\lambda_{j-1}\mathbf{I}}{\lambda_j-\lambda_{j-1}}\ \tfrac{\mathbf{M}-\lambda_{j+1}\mathbf{I}}{\lambda_j-\lambda_{j+1}} \cdots \tfrac{\mathbf{M}-\lambda_{N-1}\mathbf{I}}{\lambda_j-\lambda_{N-1}}\ \tfrac{\mathbf{M}-\lambda_N\mathbf{I}}{\lambda_j-\lambda_N}\ . \tag{1.18}$$

The function $\mathbf{F}(\mathbf{M})$ is consequently given by

$$\mathbf{F}(\mathbf{M}) = \sum_{j=1}^{N} \left\{ \prod_{k\neq j} \frac{\mathbf{M} - \lambda_k \mathbf{I}}{\lambda_j - \lambda_k} \right\} F(\lambda_j)\ . \tag{1.19}$$

This means that a matrix function $\mathbf{F}(\mathbf{M})$ is always written in terms of the Lagrange interpolating polynomial

$$F(\lambda) = \sum_{j=1}^{N} \left\{ \prod_{k \neq j} \frac{\lambda - \lambda_k}{\lambda_j - \lambda_k} \right\} F(\lambda_j) \tag{1.20}$$

of the function $F(\lambda)$, in which we replace the variable λ by $\mathbf{M}$. An alternative expression uses the characteristic polynomial and its derivatives at each eigenvalue:

$$F(\lambda) = \sum_{j=1}^{N} \frac{\Delta_{\mathbf{M}}(\lambda)}{(\lambda - \lambda_j)\, \Delta'_{\mathbf{M}}(\lambda_j)}\, F(\lambda_j)\ . \tag{1.21}$$

Expression (1.19) can be put into an elegant form, stating the vanishing of a formal determinant:

$$\begin{vmatrix} \mathbf{F(M)} & F(\lambda_1) & F(\lambda_2) & F(\lambda_3) & \cdots\cdots\cdots & F(\lambda_N) \\ \mathbf{I} & 1 & 1 & 1 & \cdots\cdots\cdots & 1 \\ \mathbf{M} & \lambda_1 & \lambda_2 & \lambda_3 & \cdots\cdots\cdots & \lambda_N \\ \mathbf{M}^2 & \lambda_1^2 & \lambda_2^2 & \lambda_3^2 & \cdots\cdots\cdots & \lambda_N^2 \\ \mathbf{M}^3 & \lambda_1^3 & \lambda_2^3 & \lambda_3^3 & \cdots\cdots\cdots & \lambda_N^3 \\ \cdots & \cdots & \cdots & \cdots & \cdots\cdots\cdots & \cdots \\ \mathbf{M}^{N-1} & \lambda_1^{N-1} & \lambda_2^{N-1} & \lambda_3^{N-1} & \cdots\cdots\cdots & \lambda_N^{N-1} \end{vmatrix} = 0\ . \tag{1.22}$$

This is again an expression, with $\mathbf{M}$ substituting λ, of the Lagrange interpolating polynomial for $F(\lambda)$. The projectors turn up as soon as the determinant is Laplace-expanded along the first row.

We have seen that, in order to obtain $\mathbf{F(M)}$, it is necessary to find the eigenvalues of $\mathbf{M}$ and the detailed form of its first $(N-1)$ powers. What we have done is to start from a basis $\{\mathbf{M}^k\}$ for the functions of $\mathbf{M}$, and go to the projector basis. Actually, any set of independent polynomials in $\mathbf{M}$ could be used as the starting basis. In particular, once we know them, the projectors themselves can be taken as the basis polynomials and the above determinant becomes

$$\begin{vmatrix} \mathbf{F(M)} & F(\lambda_1) & F(\lambda_2) & F(\lambda_3) & \cdots\cdots\cdots & F(\lambda_N) \\ \mathbf{Z}_1[\mathbf{M}] & 1 & 0 & 0 & \cdots\cdots\cdots & 0 \\ \mathbf{Z}_2[\mathbf{M}] & 0 & 1 & 0 & \cdots\cdots\cdots & 0 \\ \mathbf{Z}_3[\mathbf{M}] & 0 & 0 & 1 & \cdots\cdots\cdots & 0 \\ \cdots & \cdots & \cdots & \cdots & \cdots\cdots\cdots & \cdots \\ \mathbf{Z}_N[\mathbf{M}] & 0 & 0 & 0 & \cdots\cdots\cdots & 1 \end{vmatrix} = 0\ , \tag{1.23}$$

which is clearly the same as (1.11).

In the above formal determinants, only the first column contain matrices. In the Laplace expansion, these matrices will never multiply each other. Consequently, the same determinant expression is valid for each matrix element.

For N not too large, the $\mathbf{Z}_j[\mathbf{M}]$'s can be directly computed in this way. We shall later give closed expressions for them. Such expressions will involve the elementary symmetric functions of the eigenvalues of $\mathbf{M}$, which will be examined in Section 16.2 [see Eq.(16.46), for example]. The projectors themselves will be presented in Section 17.1.

Commentary 10.1. ► That functions of matrices are completely determined by their spectra is justified on much more general grounds. Matrix algebras are very particular kinds of von Neumann algebras and it is a very strong result of the still more general theory of Banach algebras [12] that functions on such spaces, as long as they can be defined, are fixed by the spectra. ◄

Commentary 10.2. ► **Rotations** Take the three generators of rotations in the Euclidean 3-dimensional space, in the forms

$$\mathbf{J}_1 = \begin{pmatrix} 0 & 0 & 0 \\ 0 & 0 & -1 \\ 0 & 1 & 0 \end{pmatrix} \; ; \; \mathbf{J}_2 = \begin{pmatrix} 0 & 0 & 1 \\ 0 & 0 & 0 \\ -1 & 0 & 0 \end{pmatrix} \; ; \; \mathbf{J}_3 = \begin{pmatrix} 0 & -1 & 0 \\ 1 & 0 & 0 \\ 0 & 0 & 0 \end{pmatrix} .$$

$\mathbf{J}_1$ generates rotations around the axis Ox, $\mathbf{J}_2$ generates those around Oy and $\mathbf{J}_3$ those around the third axis Oz. A generic member of the rotation group Lie algebra will be

$$\mathbf{W} = \alpha_1 \mathbf{J}_1 + \alpha_2 \mathbf{J}_2 + \alpha_3 \mathbf{J}_3 = \begin{pmatrix} 0 & -\alpha_3 & \alpha_2 \\ \alpha_3 & 0 & -\alpha_1 \\ -\alpha_2 & \alpha_1 & 0 \end{pmatrix} ,$$

where the variables α_k are angular parameters. In terms of $\alpha = \sqrt{\alpha_1^2 + \alpha_2^2 + \alpha_3^2}$, the eigenvalues of $\mathbf{W}$ are $\lambda_1 = 0$, $\lambda_2 = -i\,\alpha$ and $\lambda_3 = +i\,\alpha$. It is also convenient to introduce the unit vector $\mathbf{u}$ of components $u_k = \alpha_k/\alpha$. The projectors are easily obtained,

$$\mathbf{Z}_1 = \mathbf{I} + \frac{\mathbf{W}^2}{\alpha^2} \; ; \; \mathbf{Z}_2 = -\frac{\mathbf{W}^2}{2\alpha^2} + i\,\frac{\mathbf{W}}{2\alpha} \; ; \; \mathbf{Z}_3 = -\frac{\mathbf{W}^2}{2\alpha^2} - i\,\frac{\mathbf{W}}{2\alpha} .$$

For example, $\mathbf{Z}_1 = \begin{pmatrix} {u_1}^2 & u_1 u_2 & u_1 u_3 \\ u_1 u_2 & {u_2}^2 & u_2 u_3 \\ u_1 u_3 & u_2 u_3 & {u_3}^2 \end{pmatrix}$. The group element $\mathbf{R}$ corresponding to a rotation of angles $(\alpha_1, \alpha_2, \alpha_3)$ will be the exponential of $\mathbf{W}$:

$$\begin{aligned} \mathbf{R}(\alpha_1, \alpha_2, \alpha_3) &= e^{\lambda_1}\, \mathbf{Z}_1 + e^{\lambda_2}\, \mathbf{Z}_2 + e^{\lambda_3}\, \mathbf{Z}_3 = \mathbf{Z}_1 + e^{-i\alpha}\, \mathbf{Z}_2 + e^{i\alpha}\, \mathbf{Z}_3 \\ &= \mathbf{I} + \frac{\mathbf{W}^2}{\alpha^2}\,(1 - \cos\alpha) + \frac{\mathbf{W}}{\alpha}\,\sin\alpha = \end{aligned}$$

$$\begin{pmatrix} \cos\alpha + (1-\cos\alpha)u_1u_1 & (1-\cos\alpha)u_1u_2 - \sin\alpha\, u_3 & (1-\cos\alpha)u_1u_3 + \sin\alpha\, u_2 \\ (1-\cos\alpha)u_1u_2 + \sin\alpha\, u_3 & \cos\alpha + (1-\cos\alpha)u_2u_2 & (1-\cos\alpha)u_2u_3 - \sin\alpha\, u_1 \\ (1-\cos\alpha)u_1u_3 - \sin\alpha\, u_2 & (1-\cos\alpha)u_2u_3 + \sin\alpha\, u_1 & \cos\alpha + (1-\cos\alpha)u_3u_3 \end{pmatrix} . \blacktriangleleft$$

1.3.2 *Degenerate matrices*

As a rule, physicists dislike degenerate matrices and/or operators. In Quantum Mechanics, for example, they proceed to increase the number of operators up to the point at which all degeneracy is removed. In that case, they stop only at a complete set of intercommuting operators, whose eigenvectors provide the complete quantum description of the states. This means that matrix indices are multiple, and the entries indicated by expressions like $\langle I'I_3'Y';j'm'|M|II_3Y;jm\rangle$.

In some cases, however, a partial knowledge is enough to the ends in view, and degenerate matrices are used. We shall therefore briefly discuss the case in which there are multiple roots of the characteristic equation. Let us start from the nondegenerate case, and go back to the 3×3 example (1.5). The Lagrange interpolating polynomial (1.20) would be

$$F(\lambda)=F(\lambda_1)\tfrac{(\lambda-\lambda_2)(\lambda-\lambda_3)}{(\lambda_1-\lambda_2)(\lambda_1-\lambda_3)}+F(\lambda_2)\tfrac{(\lambda-\lambda_1)(\lambda-\lambda_3)}{(\lambda_2-\lambda_1)(\lambda_2-\lambda_3)}+F(\lambda_3)\tfrac{(\lambda-\lambda_1)(\lambda-\lambda_2)}{(\lambda_3-\lambda_1)(\lambda_3-\lambda_2)}\,.$$

To consider the case in which $\lambda_1 = \lambda_2$, we actually examine the limit $\lambda_1 \to \lambda_2$. Take $\lambda_2 = \lambda_1+\epsilon$ into the expression above and expand everything, keeping only terms up to the first order in ϵ: $F(\lambda_2) = F(\lambda_1)+\epsilon F'(\lambda_1)$, etc. It turns out that the divergence of the first term is just compensated by a term coming from the second, and the result is

$$F(\lambda)=F(\lambda_1)\tfrac{(\lambda-\lambda_3)(2\lambda_1-\lambda_3-\lambda)}{(\lambda_1-\lambda_3)^2}+F'(\lambda_1)\tfrac{(\lambda-\lambda_1)(\lambda-\lambda_3)}{(\lambda_1-\lambda_3)}+F(\lambda_3)\tfrac{(\lambda-\lambda_1)^2}{(\lambda_3-\lambda_1)^2}\,.$$

The component matrices of $\mathbf{M}$, and the new form of the matrix functions, are promptly recognized: replacing λ by $\mathbf{M}$, we have

$$\mathbf{F}(\mathbf{M})=F(\lambda_1)\frac{(\mathbf{M}-\lambda_3\mathbf{I})(2\lambda_1\mathbf{I}-\lambda_3\mathbf{I}-\mathbf{M})}{(\lambda_1-\lambda_3)^2}$$
$$+F'(\lambda_1)\tfrac{(\mathbf{M}-\lambda_1\mathbf{I})(\mathbf{M}-\lambda_3\mathbf{I})}{(\lambda_1-\lambda_3)}+F(\lambda_3)\tfrac{(\mathbf{M}-\lambda_1\mathbf{I})^2}{(\lambda_3-\lambda_1)^2}\,,$$

and, with a hopefully obvious notation,

$$\mathbf{Z}_{11}=\tfrac{(\mathbf{M}-\lambda_3\mathbf{I})(2\lambda_1\mathbf{I}-\lambda_3\mathbf{I}-\mathbf{M})}{(\lambda_1-\lambda_3)^2};\ \mathbf{Z}_{12}=\tfrac{(\mathbf{M}-\lambda_1\mathbf{I})(\mathbf{M}-\lambda_3\mathbf{I})}{(\lambda_1-\lambda_3)};\ \mathbf{Z}_3=\left(\tfrac{\mathbf{M}-\lambda_1\mathbf{I}}{\lambda_3-\lambda_1}\right)^2.$$

The same expressions come from

$$\begin{vmatrix}\mathbf{F}(\mathbf{M}) & F(\lambda_1) & F'(\lambda_1) & F(\lambda_3)\\ \mathbf{I} & 1 & 0 & 1\\ \mathbf{M} & \lambda_1 & 1 & \lambda_3\\ \mathbf{M}^2 & \lambda_1^2 & 2\lambda_1 & \lambda_3^2\end{vmatrix}=0\,. \tag{1.24}$$

Comparison with the nondegenerate case (1.22), which here reads

$$\begin{vmatrix} \mathbf{F(M)} & F(\lambda_1) & F(\lambda_2) & F(\lambda_3) \\ \mathbf{I} & 1 & 1 & 1 \\ \mathbf{M} & \lambda_1 & \lambda_2 & \lambda_3 \\ \mathbf{M}^2 & \lambda_1^2 & \lambda_2^2 & \lambda_3^2 \end{vmatrix} = 0 \,, \tag{1.25}$$

gives a rule-of-thumb: the case $\lambda_2 = \lambda_1$ comes from the nondegenerate case by taking, in the λ_2-column of the determinant, the derivatives with respect to λ_2 and then putting $\lambda_2 = \lambda_1$. This is a general rule: the one-time degenerate case $F(\lambda; \lambda_1, \lambda_1, \lambda_3)$ comes from the nondegenerate case as $\lim_{\epsilon \to 0} F(\lambda; \lambda_1, \lambda_1 + \epsilon, \lambda_3)$.

As pointed out previously, it is possible to use more general polynomials $p_k(\mathbf{M})$ instead of the simple monomials $\mathbf{M}^k$. To ensure that the determinant which multiplies $\mathbf{F(M)}$ remains nonvanishing, these polynomials should be of the same order as above: order of $p_k(\mathbf{M}) = k$. Thus:

$$\begin{vmatrix} \mathbf{F(M)} & F(\lambda_1) & F'(\lambda_1) & F(\lambda_3) \\ p_0(\mathbf{M}) & p_0(\lambda_1) & p_0'(\lambda_1) & p_0(\lambda_3) \\ p_1(\mathbf{M}) & p_1(\lambda_1) & p_1'(\lambda_1) & p_1(\lambda_3) \\ p_2(\mathbf{M}) & p_2(\lambda_1) & p_2'(\lambda_1) & p_2(\lambda_3) \end{vmatrix} = 0 \,. \tag{1.26}$$

Suppose now that also $\lambda_3 = \lambda_1$. The same kind of calculation shows that we must take, in the λ_3-column of the determinant, the second derivatives with respect to λ_3 and after that put $\lambda_3 = \lambda_1$.

All this generalizes to higher degeneracies (with corresponding higher derivatives) and to higher-order matrices. Suppose that there are m distinct eigenvalues, and denote by m_j the multiplicity of the j-th eigenvalue λ_j. For each such eigenvalue λ_j there will be components $\mathbf{Z}_{j1}, \mathbf{Z}_{j2}, \ldots \mathbf{Z}_{jm_j}$. The general form of $\mathbf{F(M)}$ will be

$$\begin{aligned} \mathbf{F(M)} &= \sum_{j=1}^{m} \left[F(\lambda_j)\mathbf{Z}_{j1} + F'(\lambda_j)\mathbf{Z}_{j2} + F''(\lambda_j)\mathbf{Z}_{j3} + \ldots + F^{(m_j-1)}(\lambda_j)\mathbf{Z}_{jm_j} \right] \\ &= \sum_{j=1}^{m} \sum_{r=0}^{m_j-1} F^{(r)}(\lambda_j)\mathbf{Z}_{j,r+1} \,. \end{aligned} \tag{1.27}$$

The components are identified by applying this formula to the polynomials,

$$p_k(\mathbf{M}) = \sum_{j=1}^{m} \left[p_k(\lambda_j)\mathbf{Z}_{j1} + p_k'(\lambda_j)\mathbf{Z}_{j2} + p_k''(\lambda_j)\mathbf{Z}_{j3} + \ldots + p_k^{(m_j-1)}(\lambda_j)\mathbf{Z}_{jm_j} \right] . \tag{1.28}$$

The whole thing is summarized in

$$\begin{vmatrix} \mathbf{F}(\mathbf{M}) & F(\lambda_1) & \dots & F^{(m_1-1)}(\lambda_1) & \dots & F(\lambda_m) & \dots & F^{(m_m-1)}(\lambda_m) \\ p_0(\mathbf{M}) & p_0(\lambda_1) & \dots & p_0^{(m_1-1)}(\lambda_1) & \dots & p_0(\lambda_m) & \dots & p_0^{(m_m-1)}(\lambda_m) \\ p_1(\mathbf{M}) & p_1(\lambda_1) & \dots & p_1^{(m_1-1)}(\lambda_1) & \dots & p_1(\lambda_m) & \dots & p_1^{(m_m-1)}(\lambda_m) \\ p_2(\mathbf{M}) & p_2(\lambda_1) & \dots & p_2^{(m_1-1)}(\lambda_1) & \dots & p_2(\lambda_m) & \dots & p_2^{(m_m-1)}(\lambda_m) \\ \dots & \dots & \dots & \dots & \dots & \dots & \dots & \dots \\ p_m(\mathbf{M}) & p_m(\lambda_1) & \dots & p_m^{(m_1-1)}(\lambda_1) & \dots & p_m(\lambda_m) & \dots & p_m^{(m_m-1)}(\lambda_m) \end{vmatrix} = 0 \,. \quad (1.29)$$

For the simple monomials, the above expression reduces to

$$\begin{vmatrix} \mathbf{F}(\mathbf{M}) & F(\lambda_1) & \dots & \dots F(\lambda_m) & \dots & F^{(m_m-1)}(\lambda_m) \\ \mathbf{M}^0 & \lambda_1^0 & 0 & \dots & \lambda_m^0 & \dots & \dots \\ \mathbf{M}^1 & \lambda_1^1 & 1 & \dots & \lambda_m^1 & \dots & \dots \\ \mathbf{M}^2 & \lambda_1^2 & 2\lambda_1 & \dots & \lambda_m^2 & 2\lambda_2 & \dots \\ \dots & \dots & \dots & \dots & \dots & \dots & \dots \\ \mathbf{M}^m & \lambda_1^m & m\lambda_1^{m-1} & \dots & \lambda_m^m & m\lambda_m^{m-1} & \dots \end{vmatrix} = 0 \,. \quad (1.30)$$

The components $\mathbf{Z}_{jk}$ share with the components of the nondegenerate case the first three properties listed on page 7: they are linearly independent, nonvanishing and commute with each other and with $\mathbf{M}$. The remaining properties are, however, absent in the generic degenerate case. In particular, the trace properties — and their consequences — hold no more. It is immediate to verify that failure in the above introductory example: tr $\mathbf{Z}_3$ $= 1$, but tr $\mathbf{Z}_{12} = 0$.

Commentary 13.1. ▶ We have seen in Comment 10.2 the calculation of a group element, corresponding to a finite transformation. That case, concerning the group of rotations, was nondegenerate. An example whose Lie algebra generic matrix is degenerate turns up in the Galilei group [13]. ◀

1.4 Normal matrices

As they are specially simple and of fundamental importance to Quantum Mechanics, normal matrices deserve particular attention. They are equivalent to diagonal matrices, which means that they can always be diagonalized by similarity transformations so as to exhibit their eigenvalues. When these eigenvalues are real, the matrices represent quantum mechanical observables. An $N \times N$ matrix $\mathbf{M}$, we repeat, is normal if $\mathbf{M}^\dagger\mathbf{M} = \mathbf{M}\mathbf{M}^\dagger$. In that case, there always exists a unitary matrix $\mathbf{U}$ with the property

$\mathbf{MU} = \mathbf{UM}_{diag}$, which in consequence makes $\mathbf{M}$ diagonal according to

$$\mathbf{M}_{diag} = \mathbf{U}^{-1}\,\mathbf{M}\,\mathbf{U}\,.$$

Conversely, any matrix for which such a $\mathbf{U}$ exists is normal. The matrices $\mathbf{Z}_k[\mathbf{M}]$ with entries

$$[\mathbf{Z}_k[\mathbf{M}]]_{rs} = U_{rk}{U^{-1}}_{ks} \tag{1.31}$$

(one for each fixed k) are eigenprojectors of $\mathbf{M}$. Actually,

$$\mathbf{M} = \sum_k \lambda_k \mathbf{Z}_k[\mathbf{M}]$$

and

$$\sum_{r,s} {U^{-1}}_{jr} M_{rs} U_{sk} = \lambda_k \delta_{jk}\ ; \tag{1.32}$$

$$\sum_{r,s} M_{rs} U_{sk} {U^{-1}}_{kr} = \sum_{r,s} M_{rs}\ Z_{(k)}[\mathbf{M}]_{sr} = \lambda_k\ . \tag{1.33}$$

The eigenvectors $\mathbf{v}_{(k)}$ of $\mathbf{M}$ have components

$$v_{(k)r} = U_{rk}$$

and are eigenvectors with unit eigenvalue of the corresponding projector: $\mathbf{Z}_k\ \mathbf{v}_{(k')} = \delta_{kk'}\ \mathbf{v}_{(k)}$. One of the most important properties of normal matrices comes out at this point: the eigenvectors $\mathbf{v}_{(k)}$ are necessarily linearly independent and, in consequence, constitute a base in terms of which any N-vector can be expanded.

The vectors of components $v^*_{(k)r} = {U^{-1}}_{kr}$ are dual to them, by which we mean that $\sum_r v^*_{(k')r} v_{(k)r} = \delta_{kk'}$. They are left-eigenvectors of $\mathbf{M}$, by which we mean that $\sum_r v^*_{(k)r} M_{rs} = v^*_{(k)s}$. Such left-vectors are really covectors, belonging to the space dual to the space of the column-vectors. Usual matrix notation can lead to some confusion in this case.

The best way out is Dirac's "bra–ket" notation. Indicate the component $v_{(k)r}$ by $\langle r|k\rangle\rangle$. Interpret it as follows: the vector $\mathbf{v}_{(k)}$ is the "ket" $|k\rangle\rangle$. It belongs to a vector space, on which we choose a basis $\{|i\rangle\}$. Thus, $\mathbf{v}_{(k)} = |k\rangle\rangle = \sum_i \langle i|k\rangle\rangle |i\rangle$ and $U_{rk} = \langle r|k\rangle\rangle$.

Every vector space V has another vector space V^* which is its dual: it is the set of linear mappings taking V into the field it is built upon. Each linear mapping is a covector, and sends a vector into a number. To every basis like $\{|i\rangle\}$ corresponds a basis in the dual space, whose members are indicated by the "bras" $\langle j|$, and are such that $\langle j|i\rangle = \delta_{ij}$. This says that each $\langle j|$ sends $|j\rangle$ into the number 1, and $|i \neq j\rangle$ into the number 0. A member of the dual space is written $\langle\langle k| = \sum_j \langle\langle k|j\rangle\langle j|$. This is just the above left-vector, whose components are got by "bracketing" it with a basis vector $|i\rangle$: $v^*_{(k)i} = U^{-1}_{ki} = \langle\langle k|i\rangle$. The above projectors are, in this notation, just $\mathbf{Z}_k = |k\rangle\rangle\langle\langle k|$.

We shall see later (Chapter 9) that circulant matrices are normal, and also that they play a role of "basic bricks" in Quantum Mechanics. The importance of normal operators in quantum physics is not strange to this fact. The decomposition into components, however, holds also when $\mathbf{M}$ is not normal (as a Bell matrix, see Chapter 15), though in that case, as said, the components may lose some of their most useful properties. Properties involving determinants, traces and minors will be examined in detail in Chapter 16.

PART 1
STOCHASTIC MATRICES

Chapter 2

Evolving systems

Stochastic matrices can be introduced in many ways, according to different fields of interest. We shall present them, and the Markov chains they are related to, as a means of describing time evolution of physical systems whose future history is completely determined by the present state, without any influence of the past.

Time evolution of physical systems is prevalently described through one of two standard approaches [14]: by continuous solutions of differential equations or by discrete mappings.

- The approach by differential equations takes its paradigm from Classical Mechanics — states correspond to points on phase space, whose continuous displacement in time is given by the solution of dynamical differential equations [15].
- The mapping procedure models evolution by the successive iterations of a well-chosen map ([16], [17]) so that the state is known at each step, as if the time parameter of the system were only defined at discrete values.

It is possible to go from the first kind of description to the second through the snapshots leading to a Poincaré map [18], and from the second to the first by the notion of continuous iteration [19] (see Section 17.2). Abrupt behavior, as that found in many phase transitions, is usually supposed to be connected with singular limits.

These two formalisms do not cover all the studied cases of dynamical evolution. Our subject here is precisely a third approach, based on Markov chains, which has been successful in the description of general processes involving *growth by agglomeration*. We shall use the best existing model for the formation of glasses — a beautiful example of that kind of process — as a fellow traveller in the presentation which follows.

Suppose a system is described by some set of variables. A state will be given by the values of those variables at some time. The system is said to follow a Markov process when its evolution from a given state depends on that state only, and not on its previous history. Stochastic Matrices are deeply related to Markov Chains, which are discrete Markov processes.

Markovian systems provide a nice sample of dynamical systems. They exhibit very simple examples of several characteristic phenomena found in general evolutive processes, however intricate. The stochastic approach has, in consequence, both a great qualitative appeal and a considerable classifying power. Besides its aptitude to solve important problems, it provides illuminating toy versions of many great questions, such as

- transience
- dissipation
- recurrence
- the *éternel retour*
- insensibility (or not) to initial conditions
- time reversal
- equilibrium, stable and unstable
- ergodicity
- non-equilibrium
- phase transitions and multi-phase systems.

The main characteristics of the approach are:

- the physical system is supposed to be in some of the states $a, b, c, \ldots$ of a discrete state-space S;
- knowledge is limited to the probabilities of the system being in each state; the most complete characterization of the real state of the system allowed by the formalism is given by a distribution $\mathbf{p}$ of probabilities, which says that the system has probability p_a of being in state a, probability p_b of being in state b, etc.;
- such distributions describing collective states will be column vectors $\mathbf{p} = (p_1, p_2, \ldots, p_N)^T$, whose components are non-negative and add up to 1. If we so wish, we can call such vectors collective, or macroscopic states, to distinguish them from the microscopic states labelled by $a, b, c, \ldots$;
- evolution of the system is presided by a *transition matrix* $\mathbf{M}$ taking a

state-vector into another, in a step-by-step process; the matrix entries are obtained by physical considerations;
- to ensure the preservation of good probabilistic properties, the matrix must be a stochastic matrix, whose columns are statistical distributions;
- given an initial state $\mathbf{p}^{(0)}$, the triple $(S, \mathbf{M}, \mathbf{p}^{(0)})$ characterizes a Markov chain.

We shall see that much can be stated in terms of the transition matrix $\mathbf{M}$, its eigenvalues and eigenvectors. There are extreme cases in which the system is insensible to the initial state and any initial $\mathbf{p}^{(0)}$ is led, in the long run, to the same situation. In those cases, everything is determined by $\mathbf{M}$ alone.

It should be clear once and for all that the passage to the case in which the space of states S is a continuum can, with the additional topological and measure-theoretical assumptions it requires, modify the results and bring forth new features. The scenario is not without recalling what happens with the thermodynamic limit. This is of special importance in applications of stochastic matrices in the field-theoretical approach to critical phenomena [20]. On the other hand, though our presentation will assume a finite number N of states, all that follows can be extended without much ado to a discrete countable set of states.

Vitreous transition will, as announced, be used as a physical guide, a source of ideas and examples. After the basic notions on stochastic matrices have been introduced, the Kerner model for glass transition [21] will be presented. We shall be concerned only with the simplest case exhibiting the main characteristics of a typical system, that of Selenium glasses. The assumption of a discrete set of states S is quite realistic in the case of glasses, though of course only a very simplified version of the model will find place here. We shall then proceed to the more formal aspects, here and there taking examples from the vitrification case.

It so happens that, at least for the time being, no examples of the more "pathological" types of Markov chains have been found in the approach to glasses. For that reason some purely formal, unphysical examples will be added to illustrate them.

Chapter 3

Markov chains

A Markov chain is a succession of probabilistic states in a step-by-step progression conducted by a state-independent stochastic matrix.

The entries of stochastic matrices are probabilities, real numbers in the closed interval $[0, 1]$. Consequently, stochastic matrices are non-negative matrices in the generic case. In many cases the matrices turning up are strictly positive. Positive and non-negative matrices govern Markov chains of different brands.

3.1 Non-negative matrices

Let us start by introducing a more precise terminology.

(1) Given two $N \times L$ real matrices $\mathbf{A} = (A_{ij})$ and $\mathbf{B} = (B_{ij})$, we write $\mathbf{A} \geq \mathbf{B}$ if $A_{ij} \geq B_{ij} \;\; \forall\, i = 1, 2, \ldots, N, \; \forall\, j = 1, 2, \ldots, L$. If $A_{ij} > B_{ij}$ instead, we write $\mathbf{A} > \mathbf{B}$. A matrix $\mathbf{M}$ whose entries are all strictly positive is itself said to be *positive*. If we indicate by $\mathbf{0}$ the matrix with all entries equal to 0, this is written $\mathbf{M} > \mathbf{0}$.

(2) Similarly, a general $N \times L$ matrix $\mathbf{M}$ of real elements is said to be *non-negative* (indicated $\mathbf{M} \geq \mathbf{0}$) when all its entries are non-negative: $M_{ij} \geq 0 \;\; \forall\, i = 1, 2, \ldots, N, \; \forall\, j = 1, 2, \ldots, L$. For example, a column-vector $\mathbf{v} = (v_1, v_2, \ldots, v_N)^T$ will be positive (written simply $\mathbf{v} > \mathbf{0}$) if all $v_j > 0$.

(3) An example: introduce the matrix $\mathbf{H}$ whose entries are all equal to 1: $H_{ij} = 1 \;\; \forall\, i, = 1, 2, \ldots, N$ and $\forall\, j = 1, 2, \ldots, L$. A particular case is the row $\mathbf{u} = (1, 1, 1, \ldots, 1)$ (or the column $\mathbf{u}^T$). If $\mathbf{M}$ is a matrix whose entries are all real and ≤ 1, we write $\mathbf{M} \leq \mathbf{H}$.

(4) An $N \times 1$ matrix $\mathbf{v}$, which is simply a column vector, is a *probability distribution* if $\mathbf{v} \geq \mathbf{0}$ and $\sum_{a=1}^{N} v_a = 1$.

(5) An $N \times N$ non-negative matrix $\mathbf{P}$ is a *permutation* if each row and column has a single entry $= 1$ and all the other $= 0$. Each matrix of this kind will represent an element of the symmetric group S_N. This group is an always-present symmetry of the formalism of stochastic matrices, as its action causes only a reshuffling of the entries which preserve their sum in a column.

(6) An $N \times N$ matrix $\mathbf{M}$ is *reducible* by permutations if there exists a permutation $\mathbf{P}$ which puts it in the form

$$\widetilde{\mathbf{M}} = \mathbf{P}^{-1}\mathbf{M}\mathbf{P} = \begin{pmatrix} \mathbf{B} & \mathbf{C} \\ \mathbf{0} & \mathbf{D} \end{pmatrix},$$

where $\mathbf{B}$ and $\mathbf{D}$ are lower-order square matrices.

(7) $\mathbf{M}$ is *completely reducible* if some $\mathbf{P}$ as above exists such that all the entries in $\mathbf{C}$ are 0 (that is, $\mathbf{C} = \mathbf{0}$). In that case $\mathbf{M}$ can be seen as composed of two quite independent matrices $\mathbf{B}$ and $\mathbf{D}$, which can be dealt with separately.

(8) Positive matrices are irreducible. The effect of a permutation is only to move the entries of the matrix it acts upon from one position to another. No zero entries are created and, in consequence, a positive matrix cannot be reduced. Thus, $\mathbf{M} > 0$ is a particular case of irreducible $\mathbf{M} \geq 0$.

To illustrate the difference between positive and non-negative matrices, let us advance two results: (i) a positive matrix will always govern a Markov chain which tends to a *single* equilibrium state, which is not necessarily the case if the matrix has some vanishing entry; (ii) the ergodic theorem only applies when the equilibrium probability distribution is strictly positive.

Commentary 24.1. ▶ A permutation is always a product of elementary transpositions. There are $N-1$ such transpositions $\mathbf{s}_r$ for $N \times N$ matrices. These matrices have, for $r = 1, 2, 3, \ldots, N-1$, the entries

$$(\mathbf{s}_r)_{ij} = \delta_{ij}\left(1 - \delta_{ir} - \delta_{i,r+1}\right) + \delta_{ir}\delta_{j,r+1} + \delta_{i,r+1}\delta_{j,r} \ .$$

Applied to a column vector, $\mathbf{s}_r$ simply exchanges rows r and $r+1$, leaving the remaining rows untouched. In consequence, $\mathbf{s}_r^2 = \mathbf{I}$. A little more about transpositions will be said in Comment 312.2. Permutations are further discussed in Section 18.1.1.1. ◀

Commentary 24.2. ▶ There are two other, equivalent definitions of a reducible matrix, more or less convenient according to the particular application in view:

- A square matrix $\mathbf{M} = (M_{ij})$ is reducible if the set of values of indices i, j in M_{ij} can be separated into two disjoint sets and $M_{ij} = 0$ if the first and second indices belong to different sets.
- An $N \times N$ matrix $\mathbf{M} = (M_{ij})$ is reducible if, seen as a linear operator on a space E of N-component column vectors, it leaves invariant some d-dimensional subspace ($d < N$) of E. This will lead to yet another definition of reducibility in the specific case of stochastic matrices (see Comment 27.1 below). An elementary transposition is completely reducible when $N > 2$. This is not the case for a generic permutation. ◀

3.2 General properties

An $N \times N$ matrix $\mathbf{M}$ is a *stochastic matrix* if its entries are real numbers satisfying the two following conditions:

(1) in the notation introduced in item 3 of the previous section,

$$\mathbf{0} \leq \mathbf{M} \leq \mathbf{H}, \quad \text{or}$$

$$0 \leq M_{ab} \leq 1 \ , \forall \ a, b = 1, 2, \dots N \ ; \tag{3.1}$$

(2) each column of $\mathbf{M}$ is a probability distribution, that is,

$$\sum_{a=1}^{N} M_{ab} = 1, \quad \forall \ b = 1, 2, \dots N \ . \tag{3.2}$$

A first consequence of these conditions is that all the successive powers $\mathbf{M}^m$ are also stochastic matrices:

$$\sum_{a=1}^{N} {M^m}_{ab} = \sum_{a=1}^{N} {M^{m-1}}_{ab} = \dots = \sum_{a=1}^{N} {M^2}_{ab} = \sum_{a=1}^{N} \sum_{c=1}^{N} M_{ac} M_{cb} = 1 \ . \tag{3.3}$$

A second consequence is that, when acting on a probability distribution, a stochastic matrix produces another probability distribution. In the evolutive processes we shall be interested in, evolution is given by the successive state distributions obtained by successive multiplications by $\mathbf{M}$. The entry M_{ab} is the probability of transition from state b to state a.[1] A permutation

[1] There is on this point a systematic difference between the standard mathematical and physical conventions. Mathematicians interpret the entry M_{ij} in a transition matrix as the probability to have state j if the system is initially in state i. Each row is a distribution and evolution is given by matrix right-multiplication. Most frequently, a physicist (which we pretend to be) does just the opposite, reading M_{ij} right to left in time, with i representing the final state. Evolution is then given by left-multiplication.

$\mathbf{P}$ is an extreme case, a matrix of certitudes. An initial state distribution is indicated by $\mathbf{p}^{(0)} = (p^{(0)}{}_1, p^{(0)}{}_2, \ldots, p^{(0)}{}_N)^T$.

If $p^{(n)}{}_a$ is the probability of the system being in state a at the n-th step of evolution (or at "time" n), the evolution equation is

$$p^{(n+1)}{}_a = \sum_{b=1}^{N} M_{ab}\, p^{(n)}{}_b \,. \tag{3.4}$$

We have in that case a *Markov chain* (that is, a discrete Markovian process) with transition matrix $\mathbf{M}$. The power n is the "time" parameter and the last of equations (3.3) is (a discrete, finite version of) the Chapman–Kolmogorov equation. Condition (3.2) means that probability is conserved, as it implies

$$\sum_{a=1}^{N} p^{(n+1)}{}_a = \sum_{b=1}^{N} p^{(n)}{}_b = \ldots = \sum_{c=1}^{N} p^{(0)}{}_c \,. \tag{3.5}$$

A Markov process [22] is, therefore, characterized by

(i) a transition matrix $\mathbf{M}$,
(ii) an initial distribution $\mathbf{p}^{(0)}$, and
(iii) the *Markov condition* embodied in (3.4).

Qualitatively, the Markov condition states that $\mathbf{p}^{(n+1)}$ depends directly on its immediate predecessor $\mathbf{p}^{(n)}$, but not on any other previous distribution $\mathbf{p}^{(r<n)}$. This "lack of memory" is the basic property of such processes. Given the state-space $S = \{a, b, c, \ldots\}$, the chain is alternatively indicated by $(S, \mathbf{M}, \mathbf{p}^{(0)})$, $(\mathbf{M}, \mathbf{p}^{(0)})$, or simply by the set $\{\mathbf{p}^{(n)}\}$ of successive probability distributions.[2]

To get the sum of all the elements of a matrix, it is enough to sandwich it between a row- and a column-vector, both with all components equal to 1. If we only multiply the matrix from the left by a row-vector with unit components, each component of the resulting row-vector is just the sum over a column. Condition (3.2) is, consequently, equivalent to the statement that the vector $\mathbf{u}$ with components $u_a = 1$, $\forall\, a$, is a *left*-eigenstate of $\mathbf{M}$ with eigenvalue 1:

$$\sum_{a=1}^{N} u_a M_{ab} = \sum_{a=1}^{N} M_{ab} = 1 = u_b \,. \tag{3.6}$$

This means, of course, that also the "flat", or homogeneous distribution $p_a = u_a/N = 1/N$ is a left-eigenstate. A probability distribution $\mathbf{p}$ is an *equilibrium* distribution if it is a *right*-eigenvector of $\mathbf{M}$ with eigenvalue 1:

[2] Some simple properties, stemming from the Cayley–Hamilton theorem, can be seen in Comment 274.1.

$$p_a = \sum_{b=1}^{N} M_{ab}\, p_b\ . \tag{3.7}$$

Therefore, an equilibrium distribution is a stationary solution of the Markov condition (3.4), a fixed point in the evolution process.

In most applications what is looked for is an equilibrium solution in the *asymptotic* evolution of a given initial distribution $\mathbf{p}^{(0)}$. The asymptotic distribution is defined as

$$p^{(\infty)}{}_a = \lim_{n\to\infty} p^{(n)}{}_a = \lim_{n\to\infty} \sum_{b=1}^{N} M^{(n)}{}_{ab}\, p^{(0)}{}_b = \sum_{b=1}^{N} M^{(\infty)}{}_{ab}\, p^{(0)}{}_b\ . \tag{3.8}$$

The equilibrium is attained asymptotically if

$$p^{(\infty)}{}_a = \sum_{b=1}^{N} M_{ab}\, p^{(\infty)}{}_b\ . \tag{3.9}$$

The *average* of a function $f : S \to \mathbb{R}$ on a distribution $\mathbf{p}$ is defined by

$$\bar{f} = \sum_a f_a p_a\ . \tag{3.10}$$

Once equilibrium is attained at step r, the average of *any* function is preserved: $\bar{f}^{(n)} = \sum_a f_a p_a^{(n)}$ will be the same for any step-label $n > r$. Operational criteria for equilibrium will be seen later (note 10.3.1, page 111).

From $\sum_{a=1}^{N}(p_a - \frac{1}{N})^2 \geq 0$ it follows that $\sum_{a=1}^{N} p_a^2 \geq \frac{1}{N}$.

Commentary 27.1. ▶ Notice that a permutation preserves the stochastic character. Complete reducibility means disjointness of the space of states, but mere reducibility does not (see a counterexample in Comment 47.2). For stochastic matrices, the second definition in Comment 24.2 leads to a more specific criterion for reducibility (or, more precisely, for irreducibility): $\mathbf{M} = (M_{ij})$ is irreducible if, given *any* two states a and b, there exists a "path" connecting them, that is, a positive product

$$M_{i_0 i_1} M_{i_1 i_2} \ldots M_{i_{n-1} i_n} > 0$$

with $i_0 = a$ and $i_n = b$. We say in that case that the space of states is (path-)connected. ◀

3.2.1 *Flow diagrams*

A diagram indicating which states are led into the other by a matrix $\mathbf{M}$ provides an immediate qualitative vision of the type of evolution it describes. Such a diagram is obtained by (i) labelling the states in some order, (ii) drawing arrows indicating which state may come from which state, simply by looking at the matrix entries — draw an arrow from j to i

whenever $M_{ij} \neq 0$. Take as an example a 3-state case, with the probability distributions labelled (x, y, z) and the stochastic matrix

$$\mathbf{M}(A,B) = \begin{pmatrix} A & \frac{B}{2} & B \\ 1-A & \frac{1-B}{2} & 1-B \\ 0 & \frac{1}{2} & 0 \end{pmatrix}, \tag{3.11}$$

depending on two parameters A and B. The corresponding diagram is

$$\begin{array}{ccc} x & \longleftrightarrow & y \\ \nwarrow & & \nearrow\swarrow \\ & z & \end{array} .$$

With the help of such a "connectivity diagram", or "flow diagram", it is possible to monitor the effects of the successive actions of $\mathbf{M}$.

Commentary 28.1. ► We can guess, for instance, that the system above tends to spend less time in states z and x, and to stay a bit more in y. This is confirmed by numerical analysis, for different values of the parameters A and B. The matrices $\mathbf{M}(1/3,1/2)$, $\mathbf{M}(1/2,1/3)$, $\mathbf{M}(1/5,1/2)$ and $\mathbf{M}(1/5,1/2)$ are shown below, with the respective asymptotic distributions:

$$\begin{pmatrix} \frac{1}{3} & \frac{1}{4} & \frac{1}{2} \\ \frac{2}{3} & \frac{1}{4} & \frac{1}{2} \\ 0 & \frac{1}{2} & 0 \end{pmatrix} \Rightarrow \mathbf{p}^{(\infty)} = \begin{pmatrix} \frac{3}{9} \\ \frac{4}{9} \\ \frac{2}{9} \end{pmatrix} ; \begin{pmatrix} \frac{1}{2} & \frac{1}{6} & \frac{1}{3} \\ \frac{1}{2} & \frac{1}{3} & \frac{2}{3} \\ 0 & \frac{1}{2} & 0 \end{pmatrix} \Rightarrow \mathbf{p}^{(\infty)} = \begin{pmatrix} \frac{4}{13} \\ \frac{6}{13} \\ \frac{3}{13} \end{pmatrix} ;$$

$$\begin{pmatrix} \frac{1}{5} & \frac{1}{4} & \frac{1}{2} \\ \frac{4}{5} & \frac{1}{4} & \frac{1}{2} \\ 0 & \frac{1}{2} & 0 \end{pmatrix} \Rightarrow \mathbf{p}^{(\infty)} = \begin{pmatrix} \frac{5}{17} \\ \frac{8}{17} \\ \frac{4}{17} \end{pmatrix} ; \begin{pmatrix} \frac{1}{9} & \frac{1}{6} & \frac{1}{3} \\ \frac{8}{9} & \frac{1}{3} & \frac{2}{3} \\ 0 & \frac{1}{2} & 0 \end{pmatrix} \Rightarrow \mathbf{p}^{(\infty)} = \begin{pmatrix} \frac{3}{15} \\ \frac{8}{15} \\ \frac{4}{15} \end{pmatrix} .$$

The probability distributions show precisely the trends qualitatively indicated by the diagram. ◄

The case above describes a very simple, well-behaved irreducible system. Flow diagrams are still more useful in intricate situations. We shall frequently, in what follows, show the diagram alongside the matrix.

Chapter 4

Glass transition

One of the more recent successes of the stochastic approach has come from its application, by Richard Kerner, to the problem of glass transition. Glasses are complicated, amorphous systems, for which there is, to the moment, no complete theory.

Glasses — misleading materials from many points of view — are better conceived as supercooled liquids. They are not real solids neither by origin nor by internal structure [23]. The absence of any lattice symmetry precludes the use of the methods of solid state physics, most of which suppose some kind of crystal structure, leading to a dual lattice in momentum space. It is enough to recall their baffling variety in appearance, from prosaic man-made bottle and window glasses to volcano-produced basalt and obsidian. What they have in common is precisely the supercooled status. Furthermore, one same substance (as silica) can appear in Nature both as a crystal and as a glass.

A glass is produced from a basic original constituent, the role played by silica in common glass. In general, this initial constituent can form a crystal by itself, and also a glass by itself. This pure glass can be strong (as with silica) or fragile (as with Selenium). Its fusion temperature can be too high (1470^oK for silica) or too low (316^oK for Se) for unsophisticated human manipulations. These characteristics are changed by impurities. Glasses produced under control have precise quantities of these impurities, empirically found to give them some desired qualities: to make them harder or softer, for example. Above all, impurities are added in order to change the fusion temperature. Silica with some 20% of sodium oxide has the fusion temperature reduced to 770^oK, Selenium poisoned with Arsenic can have a much higher fusion point than that given above.

Thus, there is a host material, or *glass former*, plus impurities, called *modifiers.* Modifiers sometimes behave as if their single atoms were the actors. It is the case when Arsenic and/or Germanium are added to Selenium. Other times, they behave as if molecules, or some ideal pieces of them, played the role of elementary constitutive units. This is the case when sodium oxide (Na_2O) is added to silica (SiO_2). In any case, the properties of the resulting glass depend strongly on the modifier relative concentration.

In general, this concentration is indicated by c, or x in the form

$$(\text{glass-former})_{1-x}(\text{modifier})_x$$

when only one modifier is present (for example, $Se_{(1-x)}Ge_x$). When two modifiers are at work, the concentrations are indicated by x and y, the composite material itself being represented by the symbol

$$(\text{glass-former})_{1-x-y}(\text{first modifier})_x(\text{second modifier})_y \ .$$

A glass is obtained from a melt, carefully doped with the necessary modifiers in the controlled case, through a well controlled cooling procedure. At some point in the middle of the process a transition takes place. The temperature at which this happens is the "vitrification point" or "glass temperature", usually denoted T_g. Characterization of this transition point in terms of the microscopic, atomic or molecular, properties of the material is not an easy task.

One of the main microscopic characteristics of a glass is the average coordination number, or connectivity, usually denoted $\langle r \rangle$. The main aspects of glass transitions are:

- a strong dependence on the modifier concentration; there are glasses for which the transition temperature increases with the modifier concentration, and glasses for which that temperature decreases;
- a rather slow evolution to stiffness, which renders rather uncertain what to take as the real temperature of phase transition;
- an abrupt change in Raman scattering at modifier concentrations around 20%, related to an "optimal" value $\langle r \rangle \approx 2.4$ of the mean coordination number;
- rapid increase in viscosity;
- the measured glass temperature depends on the rate of cooling;
- glasses exhibit quite different characters according to the nature of the glass-forming material; there are roughly two classes:

  the bonds involved are all covalent; to this type belong the chalcogenide glasses like $Se_{1-x}Ge_x$; these are the simplest cases;

there are ionic bonds, which lead to a more complicated medium-range structure; to this category belongs our venerable companion, old glass:

$$(SiO_2)_{1-x-y}(NaO_2)_x(CaO)_y \ ,$$

with $x \approx$ 12–16%, $y \approx$ 7–17%.

In the absence of an organizing theory, the subject is a realm of empirical or semi-empirical "laws", mostly valid for rather small amounts of the modifiers. There are, for instance, laws describing the qualitative behavior of viscosity as a function of the modifier concentration and of the temperature (Arhenius law and variants). The aim of microscopic theories is to explain those laws as consequences of the properties of the atomic or molecular constituents. The stochastic approach has been specially successful in explaining the so-called modified Gibbs–Di Marzio law, which describes the variation of the transition temperature T_g with the increase of the modifier concentration, for not too large values of the latter.

In simple (mainly, chalcogenide) cases, a glass is conceived as a network of string-like chains (formed, for example, of valence-2 Selenium atoms), which constitute the glass former, to which cross-linking elements (say, Germanium atoms of valence 4) are added. The main role of the modifier is just that: to increase the cross-linking. Gibbs and Di Marzio [24], studying polymers, successfully related the increase of T_g to this effect of modifiers. This was latter adapted by Di Marzio to glasses. In terms of a certain "cross-linking density" X, the glass temperature is expressed as

$$T_g(x) = \frac{T_g(0)}{1-\kappa X} \ ,$$

where $T_g(0)$ is the glass-temperature of the pure glass-former and κ is a constant, possibly independent of the system. Phillips concluded from his constraint theory [25] that a transition (later called by Thorpe [26] "rigidity transition") should take place at $< r > = 12/5$. This is the "optimal" value mentioned above, and the prediction has been experimentally confirmed by Feng, Bresser and Boolchand [27]. Sreeram, Swiler and Varshneya [28] defined the cross-linking density in terms of $< r >$. They conceived X as the difference between the mean coordination numbers of the modified and the unmodified networks. As Se has valence = 2, $X = < r > -2$ for Se glasses. This has led to the widely accepted "modified Gibbs–Di Marzio law"

$$T_g(x) = \frac{T_g(0)}{1-\beta(< r > -2)} \ . \tag{4.1}$$

The parameter β is now system-dependent. This law is microscopically explained by Kerner's approach [29], which is based on stochastic processes. We start presenting that model now. The final explanation of the modified Gibbs–Di Marzio law, however, will only be given in a later Section (7.3.1).

Chapter 5

The Kerner model

A beautiful illustration of the stochastic method comes from its application to Selenium glasses, modified by the presence of Arsenic or Germanium. The resulting model successfully describes simple glasses with small concentrations of modifiers.

5.1 A simple example: Se–As glass

In practice, a glass is obtained through the slow cooling of a hot liquid. Pure Selenium, with a vitrification temperature around $316°K$, produces a poor quality glass, which easily breaks down under strain. Resistance to strain is, however, very easily improved by the addition of modifiers like Arsenic and/or Germanium. We shall start with the Selenium plus Arsenic case, $Se_{(1-c)}As_c$.

The basic assumptions of the model are the following:

- starting from the hot liquid in equilibrium, destabilization is sparked by the appearance of nucleation centers;
- once such seed agglomerates are present in the medium, growth takes place by the accretion of free atoms to the available bonds at their surfaces; the sum of the surfaces of all the growing agglomerates will be called the *rim*;
- the state of the system at a certain time is given by the probability distribution $\mathbf{p}$ of the bonds at the rim;
- each step in the evolution process is described by the action, on the state vector, of a stochastic matrix $\mathbf{M}$;
- the progressive growth of the agglomerates is represented by the successive powers of $\mathbf{M}$ acting on an initial distribution $\mathbf{p}^{(0)}$;
- the final configuration, an intricate lace-work of atoms and bonds, is

provided by the equilibrium asymptotic matrix

$$\mathbf{M}^{\infty} = \lim_{m\to\infty} \mathbf{M}^m \ ;$$

- the basic units are Selenium (valence 2) and Arsenic (valence 3) atoms, symbolically represented by (—○—) and (—●<);
- the relative concentration of free As atoms in the medium will be c and that of Se, $(1-c)$;
- at the rim, there are three possible kinds of offered bonds, which will be indicated by $x =$ ○—, $y =$ ●< and $z =$ ●—;
- to each free bond one of the two basic units attaches itself, reproducing one of the initial bonds;
- a new macroscopic state is arrived at when a new shell is formed by the complete saturation of all the bonds of the previous state;
- the entries of the stochastic matrix are probabilities depending on
  the number of bonds of the attaching particle
  the available valences of the rim particle
  the concentration of the attaching particle
  the effective binding energies between each pair of atoms, through effective Boltzmann factors; these will be of three kinds:

$$\circ\!\!-\!\!\circ \iff e^{E_{SeSe}/kT} = e^{-\epsilon},\ \circ\!\!-\!\!\bullet \iff e^{E_{SeAs}/kT} = e^{-\eta}$$
$$\bullet\!\!-\!\!\bullet \iff e^{E_{AsAs}/kT} = e^{-\alpha}\ .$$

Figure 5.1 below shows:

- in the left column, the 3 possible microscopic states at the rim, with their labels x, y, and z;
- in the second column, the pairs that can be formed; their left parts are considered to be below the newly-formed rim; their right parts represent the state after one step, relabelled as the initial states;
- in the last column, the unnormalized probabilities, proportional to the concentration, the number of valences and the Boltzmann effective factor.

We build then a matrix with the transition elements $x \to x$, $x \to y$, $x \to z$, $y \to x$, etc., taking into account the multiple appearance of states in the second column:

$$\begin{pmatrix} 2(1-c)e^{-\epsilon} & 8(1-c)^2e^{-2\eta} + 12c(1-c)e^{-\eta-\alpha} & 2(1-c)e^{-\eta} \\ 3ce^{-\eta} & 12c(1-c)e^{-\eta-\alpha} + 18c^2e^{-2\alpha} & 3ce^{-\alpha} \\ 0 & 0 & 0 \end{pmatrix}.$$

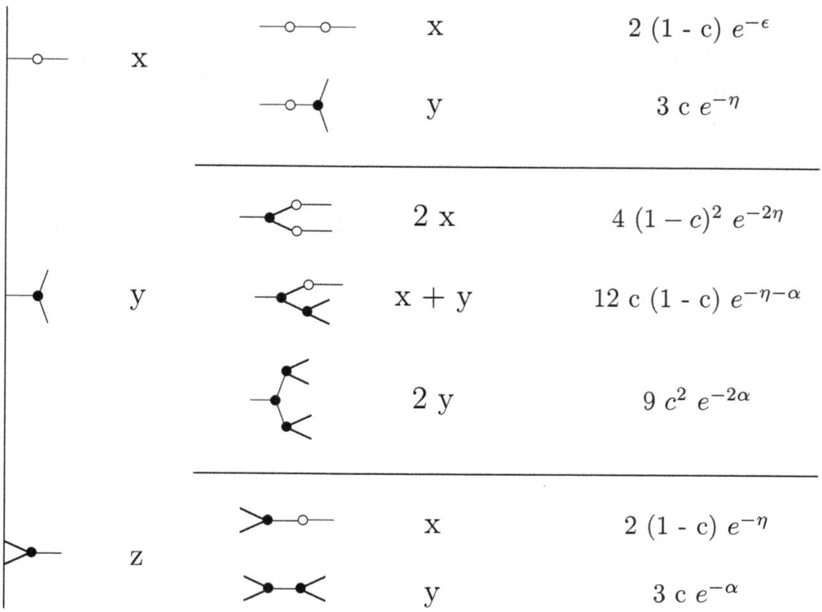

Fig. 5.1 *States, steps and unnormalized matrix entries for the Se + As case.*

We have now to normalize each column to 1. Any common factor will disappear. In a first step, it is convenient to multiply each exponential by $e^{+\epsilon}$, as follows:

$$\begin{pmatrix} 2(1-c) & 8(1-c)^2 e^{2\epsilon-2\eta} + 12c(1-c)e^{\epsilon-\eta+\epsilon-\alpha} & 2(1-c)e^{\epsilon-\eta} \\ 3ce^{\epsilon-\eta} & 12c(1-c)e^{\epsilon-\eta+\epsilon-\alpha} + 18c^2 e^{2\epsilon-2\alpha} & 3ce^{\epsilon-\alpha} \\ 0 & 0 & 0 \end{pmatrix}.$$

Introducing the variables $\xi = e^{\epsilon-\eta}$ and $\mu = e^{\epsilon-\alpha}$, the matrix acquires the form

$$\begin{pmatrix} 2(1-c) & 2(1-c)\ \xi\ [2(1-c)\xi + 3c\mu] & 2(1-c)\xi \\ 3c\xi & 3c\mu\ [2(1-c)\xi + 3c\mu] & 3c\mu \\ 0 & 0 & 0 \end{pmatrix}.$$

We now actually normalize, by dividing each entry by the sum of the entries in the same column:

$$\begin{pmatrix} \frac{2(1-c)}{2(1-c)+3c\xi} & \frac{2(1-c)\ \xi}{2(1-c)\xi+3c\mu} & \frac{2(1-c)\xi}{2(1-c)\xi+3c\mu} \\ \frac{3c\xi}{2(1-c)+3c\xi} & \frac{3c\mu}{2(1-c)\xi+3c\mu} & \frac{3c\mu}{2(1-c)\xi+3c\mu} \\ 0 & 0 & 0 \end{pmatrix}. \tag{5.1}$$

Notice that, once normalized, the entries of the transition matrix depend actually only on the differences between the effective binding energies. Notice also that there is no final state z in Figure 5.1. That state vanishes from the scene after the first step. From the second step onwards, everything goes on as if it did not exist, and the problem reduces to a (x, y) 2-state case. State z is an extreme example of transient state (see Section 7.1, page 62). The final matrix, which will really govern the evolution of the system, is

$$\mathbf{M} = \begin{pmatrix} 1-A & B \\ A & 1-B \end{pmatrix}, \tag{5.2}$$

where we have introduced the convenient variables

$$A = \frac{3c\xi}{2(1-c)+3c\xi} \;\; ; \;\; B = \frac{2(1-c)\xi}{2(1-c)\xi + 3c\mu} \, .$$

The eigenvalues of $\mathbf{M}$ are found to be 1 and $1 - A - B$. The asymptotic form is[1]

$$\mathbf{M}^{\infty} = \frac{1}{A+B}\begin{pmatrix} B & B \\ A & A \end{pmatrix}, \tag{5.3}$$

and the stationary eigenvector is

$$\mathbf{v}^{(\infty)} = \frac{1}{A+B}\begin{pmatrix} B \\ A \end{pmatrix}. \tag{5.4}$$

There are here two remarkable facts:

(1) $(\mathbf{M}^{\infty})^2 = \mathbf{M}^{\infty}$; this means that the asymptotic matrix is a projector.

(2) $\mathbf{v}^{(\infty)}$ appears as the columns of $\mathbf{M}^{\infty}$.

There are deep reasons for both facts, which will be explained in Section 6.3.

Figure 5.1 makes it easy to understand why Selenium alone makes such a poor glass. With their valence = 2, the Se atoms can only link to each other to produce string-like filaments. Addition of the valence-3 Arsenic modifier creates the cross-linkings which increase the material resistance to stress and strain (later we shall examine the addition of valence-4 Germanium, which is still more efficient). As we have said, the density of cross-linking X is a very important parameter and appears in the Gibbs–Di Marzio law.

[1] We shall later develop a systematic method to find out asymptotic forms of stochastic matrices and their eigenvectors. Ready-to-use results for 2×2 and 3×3 matrices are given in Comment 52.1.

The model finally imposes its most important condition, the so-called[2] "no-fluctuation condition":

> at the glass transition point, the asymptotic state is fixed by the medium concentration, which means that the above equilibrium eigenvector must equal the medium distribution vector $(1-c, c)$.

By equating $\mathbf{v}^{(\infty)}$ to this medium distribution, we find an equation relating the modifier concentration c to A and B. The solutions are $c = 0$, $c = 1$ and the nontrivial one

$$c = \frac{A}{A+B} = \frac{4-6\xi}{4+9\mu-12\xi}\ . \tag{5.5}$$

This gives the transition temperature in terms of the concentration, and leads to the alluded successes of the model. The most prominent is the Gibbs–Di Marzio law. We evaluate the derivative $\frac{\partial T}{\partial c} = \left(\frac{\partial c}{\partial T}\right)^{-1}$ for a given value of c. In particular, as $c \to 0$ and we can neglect the *As*–*As* bond creation (equivalent to putting $\mu = 0$ in (5.5)), we get

$$\left[\frac{\partial T}{\partial c}\right]_{c=0} = \frac{T_{g_0}}{\ln(3/2)}\ ,$$

(where T_{g_0} is the glass transition temperature of pure *Se*). This is the present-case expression of the general formula given by the model (see Section 7.3.1 below),

$$\left[\frac{\partial T}{\partial c}\right]_{c=0} = \frac{T_{g_0}}{\ln(m'/m)}$$

(where m and m' are the valences of the glass former and of the modifier), which is in very good agreement with experimental data.

5.1.1 *A simplified version*

We might wonder whether it would not be possible to consider, instead of the 3 states, only two. It seems that, for all that matters, state $y =$ •< is just twice state $z =$ •—. In that case, instead of what appears in Figure 5.1, we would have what is shown in Figure 5.2.

Notice that the states in the central column take into account the type and number of atoms remaining outside the rim, but not their valences. On

[2] *Nec fluctuat, mergitur* is a private joke with the city of Paris, where the model was conceived and whose heraldic motto says just the opposite.

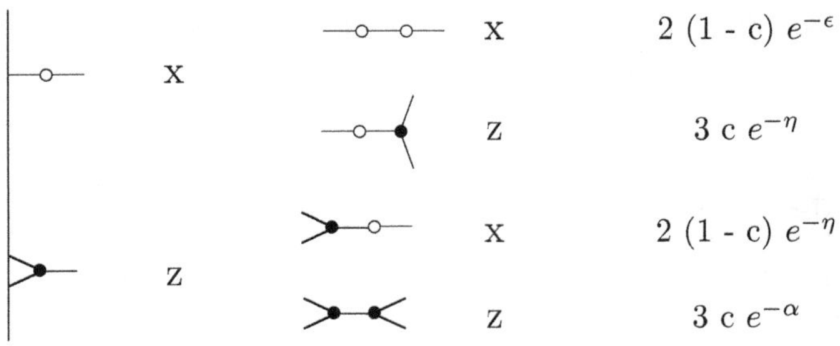

Fig. 5.2 *States, steps and matrix entries in the simplified version.*

the averages we may be willing to take (concentration, eventually coordination number, etc.), we count each atom once and put in the values of the variables for each atom. The provisional, unnormalized matrix turns out to be

$$\begin{pmatrix} 2(1-c)e^{-\epsilon} & 2(1-c)e^{-\eta} \\ 3ce^{-\eta} & 3ce^{-\alpha} \end{pmatrix}. \tag{5.6}$$

Once normalized, we fall back into matrix (5.2), with the same parameters A and B. The same consequences, in particular those of the no-fluctuation condition, follow naturally.

The Selenium–Arsenic case reduces, as seen, to a two-dimensional problem. After a short discussion of this aspect towards the end of the next section, we shall see later (Section 7.2.1) that irreducible 2×2 matrices are actually always in step-by-step equilibrium with the asymptotic state. A detailed calculation of the average coordination number of the $Se + As$ glass, both taking into account the *nec fluctuat* condition and not, will be shown at the end of Section 7.3.

5.1.2 *The Selenium–Germanium glass*

We have shown above how to calculate the matrix in the simplified version. The crux is to take only the kind of atom to fix the state, neglecting any valence counting. As a further exercise in this line, we repeat the computation for a $Se + Ge$ glass, whose pattern is summarized in Figure 5.3 below. Let us first agree about notation: the Boltzmann factors will be indicated by

 $\Longleftrightarrow e^{E_{SeGe}/kT} = e^{-\gamma}$;

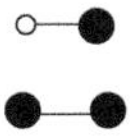 $\Longleftrightarrow e^{E_{GeGe}/kT} = e^{-\zeta}$.

We shall furthermore use c_2 for the Ge concentration, and write $\rho = e^{\epsilon - \gamma}$, $\omega = e^{\epsilon - \zeta}$.

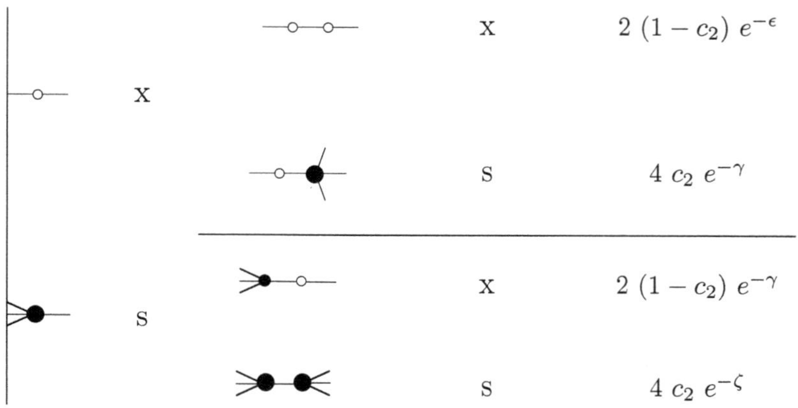

Fig. 5.3 *States, steps and matrix entries for the Se + Ge case.*

The evolution matrix is now

$$\mathbf{M} = \begin{pmatrix} 1 - A' & B' \\ A' & 1 - B' \end{pmatrix} ,$$

where

$$A' = \frac{2c_2\rho}{1 - c_2 + 2c_2\rho} \quad ; \quad B' = \frac{(1 - c_2)\rho}{(1 - c_2)\rho + 2c_2\omega} .$$

The eigenvector corresponding to the unit eigenvalue is

$$\mathbf{v}^{(\infty)} = \frac{1}{A' + B'} \begin{pmatrix} B' \\ A' \end{pmatrix}$$

$$= \frac{1}{(1 - c_2)^2\rho + 4c_2^2\omega\rho + 4c_2(1 - c_2)\rho^2} \begin{pmatrix} (1 - c_2)^2\rho + 2c_2(1 - c_2)\rho^2 \\ 4c_2^2\omega\rho + 2c_2(1 - c_2)\rho^2 \end{pmatrix} .$$

Imposition of the no-fluctuation condition $\mathbf{v}^{(\infty)} = \begin{pmatrix} 1 - c_2 \\ c_2 \end{pmatrix}$ leads now to

$$\frac{A'}{B'} = \frac{c_2}{1 - c_2} , \tag{5.7}$$

an equation for c_2 with 3 solutions: $c_2 = 0, \quad c_2 = 1$, and

$$c_2 = \frac{1 - 2\rho}{1 - 4\rho + 4\omega} \, . \tag{5.8}$$

The last is, of course, the solution of interest.

The *nec fluctuat* condition actually has an extra significance. Take the vector $\mathbf{v}^{(\infty)}$ as above. Then, (5.7) is just the condition

$$M_{12}\, v_2^{(\infty)} = M_{21}\, v_1^{(\infty)} \, .$$

The same is true of condition (5.5). It will be seen later (Section 7.2) that, given a general $N \times N$ matrix and an arbitrary distribution N-column $\mathbf{v}$, the equation

$$M_{ij}\, v_j = M_{ji}\, v_i$$

expresses the condition of detailed balancing, which implies automatic step-by-step equilibrium. Consequently, the no-fluctuation condition is, at least in the two cases presented, just the detailed balancing condition. We shall also see later (Section 7.2.1) that a 2×2 stochastic matrix is always in detailed balancing with its equilibrium eigenvector. What is new in the no-fluctuation condition is the identification of this eigenvector with the medium distribution vector. It imposes a step-by-step conservation of the medium concentrations.

Commentary 40.1. ► A few words on the limitations of the model will be in good place here. We have seen that the effective Boltzmann factors involve the (effective) binding energies between the different kinds of constituents. These energies appear as free parameters, but they are not so many in the examples above. Their values are phenomenologically predicted by the model and can, in principle, be measured. When the modifier concentration increases, new ways of binding turn up. Closed rings become important, and they appear in a wide variety: they can be formed with 3, 4 and more constituents. The binding energies of each pair of constituents change according to the type of ring they belong to. The number of independent non-measured parameters become too large and it is possible to fit anything. The model loses its predictive power. It will be recovered if, in future, some method to measure directly the binding energies is found. ◄

Finally, a comment on more formal aspects. We have seen how the stochastic matrices can be mounted from physical considerations. All the cases shown above are, however, of the best-behaved kinds. We shall (mainly in Section 7.1) supplement them with some formal ill-behaved examples to fully exhibit the possibilities of the formalism.

Chapter 6

Formal developments

There is no complete, clear-cut classification of the stochastic matrices, not even a well-established terminology. We shall say below that they can be reducible or irreducible, regular or not regular, that they are eventually primitive or imprimitive, but these properties do not always exclude each other. Roughly speaking, there are two main tentative classifications. One limited to the irreducible cases, another more general. It so happens that the limitation to irreducible matrices would exclude many cases of physical interest, specially those involving transients. In this chapter a general description of the standard notions will be presented. We shall, in particular, see how the properties of a matrix appear encapsulated in its spectrum.

6.1 Spectral aspects

The spectrum of $\mathbf{M}$ is of the utmost interest. A matrix has the same eigenvalues of its transpose, though not the same eigenvectors. In consequence, Eq.(3.6) — which is, we recall, a version of the conservation of probability — implies that the number 1 will be a particular right-eigenvalue of any stochastic matrix $\mathbf{M}$.

Commentary 41.1. ▶ The same result comes trivially from the Cayley–Hamilton theorem. If the characteristic equation is written as

$$\Delta_{\mathbf{M}}(\lambda) = \prod_{k=1}^{N} (\lambda - \lambda_k) = \sum_{j=0}^{N} c_j \lambda^j = 0 ,$$

then $\Delta_{\mathbf{M}}(\mathbf{M}) = 0$ holds for each matrix element:

$$\sum_{j=0}^{N} c_j M^j{}_{ab} = 0 .$$

Taking $\sum_a$ and using (3.3), we find $\sum_{j=0}^{N} c_j = 0$, which precisely says that the number 1 is a root of the characteristic equation of $\mathbf{M}$. ◀

Thus, a stochastic matrix always has — as a consequence of probability conservation — at least one eigenvector corresponding to an eigenvalue equal to 1. If that eigenvector is a distribution (see Eq.(6.4) below), it will be an equilibrium distribution.

The unit eigenvalue is maximal, in the sense that its absolute value is the largest possible for any eigenvalue of $\mathbf{M}$. In order to see why, consider the eigenvalue equation

$$\sum_{b=1}^{N} M_{ab}\, v_b = \lambda\, v_a \ . \tag{6.1}$$

A first consequence comes from summing over the subscript a: $\sum_{b=1}^{N} v_b = \lambda\ \sum_{a=1}^{N} v_a$. An alternative presents itself: either $\lambda = 1$ or $\sum_{a=1}^{N} v_a = 0$. This means that eigenvectors of the transition matrix corresponding to $\lambda \neq 1$ cannot be probability distributions. A second consequence comes from taking absolute values in (6.1). We then have

$$\sum_{b=1}^{N} M_{ab}\, |v_b| \geq |\lambda|\ |v_a| \ . \tag{6.2}$$

An eigenvector is nonvanishing by definition, so that at least one of the $|v_a|$'s is a positive number. In consequence, summation over a now leads forcibly to

$$|\lambda| \leq 1 \ . \tag{6.3}$$

We have seen that a stochastic matrix has indeed at least one eigenvalue $= 1$, which is therefore the maximal eigenvalue. Notice that the equality in (6.3) holds only if

$$\sum_{b=1}^{N} M_{ab}\, |v_b| = \ |v_a| \ , \ \ \forall a \ , \tag{6.4}$$

that is, when the distribution of components $|v_a|$ is an equilibrium distribution.

The spectrum of positive and non-negative matrices like $\mathbf{M}$ is the subject of two theorems going after the names of Perron and Frobenius [30], which add important details to the results above. We shall quote them in versions adapted to stochastic matrices.

Theorem (Perron, 1907)

A stochastic matrix $\mathbf{M} > 0$ has always a real unit eigenvalue, which is a *simple* root of the characteristic equation of $\mathbf{M}$ and is *larger* than the absolute value of any other eigenvalue; to this maximal eigenvalue corresponds a *positive* eigenvector $\mathbf{v}$ (this last statement involves a convention: the eigenvector is actually defined up to a factor, but all the components have the same sign and can be chosen to give $\mathbf{v} > \mathbf{0}$).

Thus, for a positive stochastic matrix, the unit eigenvalue is single, has no companion with the same absolute value, and corresponds to a positive eigenvector.

Theorem (Frobenius, 1912)

(1) An *irreducible* stochastic matrix $\mathbf{M} \geq 0$ always has a real unit eigenvalue, which is a *simple root* of the characteristic equation of $\mathbf{M}$ and is *larger than or equal to* the absolute value of any other eigenvalue; to this maximal eigenvalue corresponds a positive eigenvector.

(2) If $\mathbf{M}$ has h eigenvalues $\lambda_0 = 1, \lambda_1, \lambda_2, \ldots, \lambda_{h-1}$ with absolute values 1, then these numbers are all distinct and are roots of the unity, $\lambda_k = e^{i2k\pi/h}$.

(3) The spectrum $\{\lambda_0 = 1, \lambda_1, \lambda_2, \ldots, \lambda_{N-1}\}$ of $\mathbf{M}$, as a set of points on a disk of center 0 and radius 1 on the complex plane, is mapped into itself by a rotation of angle $2\pi/h$. This means that the Galois group of the characteristic polynomial includes the cyclic group Z_h.

Notice the differences between the two cases. For an irreducible $\mathbf{M}$ which is also > 0, there is only one eigenvalue of unit modulus, the number 1 itself. For a generic irreducible $\mathbf{M} \geq 0$, many may exist with unit *absolute* value, though "1" itself is single. An irreducible $\mathbf{M} \geq 0$ differs from $\mathbf{M} > 0$ because it can have other unit-modulus eigenvalues. Notice that the converse is not true: having a single eigenvalue equal to 1, and even a unique eigenvalue with unit modulus, does not mean that the matrix is irreducible (see Comment 46.1). However, in any case, finding two or more eigenvalues strictly equal to 1 is a signal that $\mathbf{M}$ is reducible.

If $\mathbf{M} \geq 0$ is irreducible and has h eigenvalues with maximal modulus 1, $\lambda_0 = 1 = |\lambda_1| = |\lambda_2| = \ldots |\lambda_{h-1}|$, $\mathbf{M}$ is said to be *primitive* if $h = 1$ and *imprimitive* if $h > 1$. The number h of unit-modulus eigenvalues for an irreducible stochastic matrix is called the *imprimitivity index*. Figure 6.1

shows a spectrum of a primitive matrix, as well as an imprimitive spectrum with $h = 5$.

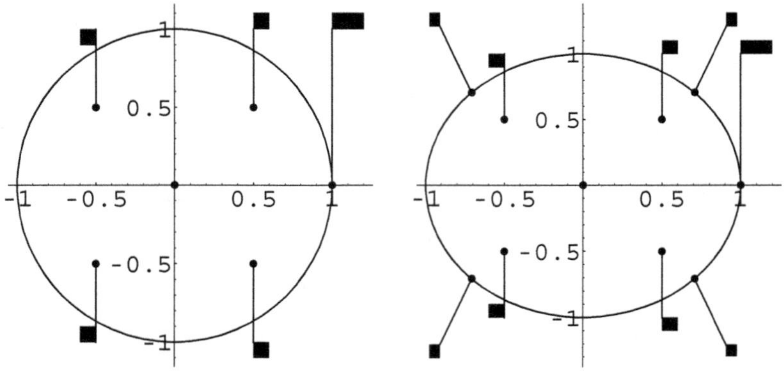

Fig. 6.1 *Examples of spectra of irreducible matrices: one primitive (left), the other imprimitive (right). The flags mark the eigenvalues.*

There are several theorems giving important information on imprimitivity. Let us here only list some results on this topic:

(1) A power of a primitive matrix is always primitive;
(2) Any $\mathbf{M} \geq 0$ is primitive *iff* there exists a natural number $r \geq 1$ such that $\mathbf{M}^r > 0$. This means that $\mathbf{M}^\infty > 0$, and also provides an interpretation: a primitive matrix is some high-order root of some positive matrix;
(3) If $\mathbf{M}$ is imprimitive with index h, then $\mathbf{M}^h$ decomposes into h primitive matrices with the same maximal eigenvalue;
(4) There is a means of obtaining h from the characteristic polynomial. Write the polynomial in the form $\Delta_{\mathbf{M}}(\lambda) = \lambda^N + a_1\lambda^{n_1} + a_2\lambda^{n_2} \ldots + a_s\lambda^{n_s}$, with only the coefficients $a_j \neq 0$ present and ordered according to $N > n_1 > n_2 > \ldots > n_s$. Take all the successive differences $N - n_1$, $n_1 - n_2$, $n_2 - n_3$, $\ldots$, $n_{s-1} - n_s$. Then, $h =$ the greatest common divisor of these differences.

Commentary 44.1. ▶ If $h > 1$, $\mathbf{M}$ can be put (by some permutation $\mathbf{P}$) into the "cyclic" form

$$\widetilde{\mathbf{M}} = \mathbf{P}^{-1}\mathbf{M}\mathbf{P} = \begin{pmatrix} \mathbf{0} & \mathbf{A}_{12} & \mathbf{0} & \ldots & \mathbf{0} \\ \mathbf{0} & \mathbf{0} & \mathbf{A}_{23} & \ldots & \mathbf{0} \\ \ldots & \ldots & \ldots & \ldots & \mathbf{0} \\ \mathbf{0} & \mathbf{0} & \mathbf{0} & \ldots & \mathbf{A}_{h-1,h} \\ \mathbf{A}_{h1} & \mathbf{0} & \mathbf{0} & \ldots & \mathbf{0} \end{pmatrix}, \tag{6.5}$$

where the $\mathbf{A}_{ij}$'s are square matrices. As a consequence, the power $\mathbf{M}^h$ is similar by a permutation to a block-diagonal matrix of the form

$$\widetilde{\mathbf{M}}^h = \begin{pmatrix} \mathbf{A}_{12}\mathbf{A}_{23}\dots\mathbf{A}_{h1} & 0 & 0\ \dots & 0 \\ 0 & \mathbf{A}_{23}\mathbf{A}_{34}\dots\mathbf{A}_{h1}\mathbf{A}_{12} & 0\ \dots & 0 \\ \dots & \dots & \dots\dots & 0 \\ 0 & 0 & 0\ \dots & 0 \\ 0 & 0 & 0\ \dots & \mathbf{A}_{h1}\mathbf{A}_{12}\dots\mathbf{A}_{h-1,h} \end{pmatrix} .$$

Written as above, as an $h \times h$ matrix, $\widetilde{\mathbf{M}}$ is the first example we meet of a hypermatrix, a matrix of matrices. The entries $\mathbf{A}_{ij}$, each one a matrix, are noncommutative. Such objects belong to Hopf algebras. We shall see more of them later (for example, in Section 14.2). Here, $\widetilde{\mathbf{M}}$ is only a shorthand for a larger $N \times N$ matrix. ◀

Commentary 45.1. ▶ The Perron and the Frobenius theorems hold for positive and non-negative matrices in general, with a single difference: instead of the number 1, the maximal eigenvalue is a general positive number λ_m.

(1) A positive matrix has a maximal eigenvalue $\lambda_m > 0$ which is single and corresponds to a positive eigenvector.

(2) An irreducible non-negative matrix $\hat{\mathbf{M}} \geq 0$ always has a real eigenvalue, which is a simple root of the characteristic equation and is larger than *or equal to* the absolute value of any other eigenvalue; to this maximal eigenvalue corresponds a positive eigenvector.

We can obtain a stochastic matrix from a non-negative matrix. In effect, consider a non-negative matrix $\hat{\mathbf{M}}$, its maximal eigenvalue λ_m and the corresponding positive eigenvector $\hat{\mathbf{v}}$:

$$\sum_b \hat{M}_{ab}\ \hat{v}_b = \lambda_m\ \hat{v}_a \ .$$

Then, the matrix $\mathbf{M}$ with the entries

$$M_{ab} = \frac{1}{\lambda_m}\ \hat{v}_b^{-1}\ \hat{M}_{ba}\ \hat{v}_a$$

is stochastic. Notice that $\mathbf{M}^n{}_{ab} = \frac{1}{\lambda_m^n}\ \hat{v}_b^{-1}\ \hat{\mathbf{M}}^n{}_{ba}\ \hat{v}_a$ and the traces are related by tr $\hat{\mathbf{M}}^n = \lambda_m^n$ tr $\mathbf{M}^n$.

For an application of this result to the Ising model, see Section 6.3.1 below. ◀

6.2 Reducibility and regularity

Consider now a reducible matrix $\mathbf{M}$. Then, what can be said is that

> a stochastic matrix $\mathbf{M} \geq \mathbf{0}$ always has a maximal real eigenvalue 1: the absolute values of all the other eigenvalues are ≤ 1. To the maximal eigenvalue corresponds a *non-negative* eigenvector.

Notice that these statements, if compared with those made on the irreducible case, show two differences:

(1) the number 1 is not said to be a single root;
(2) the corresponding eigenvector is not said to be positive, but only non-negative.

We have said in page 24 that a completely reducible matrix $\mathbf{M}$ can be decomposed into a number of independent matrices $\mathbf{M}_i$. If $\mathbf{M}$ is stochastic, each $\mathbf{M}_i$ will be stochastic and will have an eigenvalue equal to 1. The system described is, in that case, single only in appearance. It can be decomposed into disjoint subsystems ignoring each other, each described by one of the $\mathbf{M}_i$'s.

There are more results of interest on a general stochastic matrix $\mathbf{M}$:

(1) for a real variable $\lambda \geq 1$, the resolvent $\mathbf{R}_{\mathbf{M}}(\lambda)$ is $\geq \mathbf{0}$ and $\frac{\partial \mathbf{R}_{\mathbf{M}}(\lambda)}{\partial \lambda} \geq \mathbf{0}$.
(2) in the above condition in terms of the resolvent, the diagonal elements are $(R_{\mathbf{M}})_{ii} \geq 0$ $(i = 1, 2, \ldots, N)$. $\mathbf{M} \geq \mathbf{0}$ is reducible *iff* there is at least one vanishing diagonal entry, *i.e.*, one index "i" for which the equality $(R_{\mathbf{M}})_{ii} = 0$ is realized.
(3) if the maximal eigenvalue 1 for $\mathbf{M} \geq \mathbf{0}$ is simple, and positive right-eigenvectors correspond to 1 both for $\mathbf{M}$ and $\mathbf{M}^T$, then $\mathbf{M}$ is irreducible.

When $\mathbf{M}^T$ is also stochastic, $\mathbf{M}$ is said to be *doubly-stochastic*. In that case, of course, its rows also add up to 1. Notice that this is by no means the same as $\mathbf{M}^T = \mathbf{M}$. We have seen around Eq.(3.6) that probability conservation is equivalent to the statement that the homogeneous distribution $p_a = u_a/N = 1/N$, with $\mathbf{u} = (1, 1, 1, ..., 1)$, is a left-eigenvector. In the doubly-stochastic case, it is also a right-eigenvector. Thus, the matrix

$$\begin{pmatrix} 1-e-f & h & e+f-h \\ e & 1-h-j & h+j-e \\ f & j & 1-f-j \end{pmatrix},$$

which is doubly-stochastic though not symmetric, has the flat distribution $\mathbf{p} = (1/3, 1/3, 1/3)$ for right-eigenvector. More examples will be seen in the discussion of simple random walks (see 10.2.2).

Commentary 46.1. ▶ An irreducible stochastic matrix has, by the Frobenius theorem, a single eigenvalue $= 1$. The converse statement is not true: the fact that a stochastic matrix has an isolated, single unit eigenvalue does not mean that the matrix is irreducible. Take for example the reducible matrix

$$\mathbf{M} = \begin{pmatrix} 1 & 1/2 & 1/4 \\ 0 & 1/4 & 0 \\ 0 & 1/4 & 3/4 \end{pmatrix} \qquad \begin{array}{ccc} 1 & \longleftarrow & 2 \\ & \nwarrow \quad \swarrow & \\ & 3 & \end{array},$$

whose eigenvalues are $1/4, 3/4, 1$, corresponding to eigenvectors $(1, -2, 1)$, $(-1, 0, 1)$ and $(1, 0, 0)$. Notice that the eigenvectors are not positive. The transpose of $\mathbf{M}$ has the same eigenvalues, but corresponding to the eigenvectors $(0, 1, 0)$, $(0, 1/2, 1)$ and $(1, 1, 1)$. As said above, $\mathbf{M}$ would be irreducible if 1 were simple and the corresponding eigenvector were positive and also a unit eigenvector of $\mathbf{M}^T$ (see more below, Comment 50.1). In the present case, the system will tend to abandon states "2" and "3", and concentrate in "1". The main interest of reducible but not-completely-reducible chains is precisely the presence of evanescent states, called transients (see page 62). ◀

There are two notions of fundamental importance:

(1) If the number 1 is the only unit-modulus eigenvalue of $\mathbf{M}$ (which can be multiple), then $\mathbf{M}$ (and the Markov chain) is said to be *regular*.[1] Such a matrix can have several eigenvalues "1", but none of type $e^{i\phi}$. If "1" is multiple indeed, $\mathbf{M}$ is completely reducible.

(2) If furthermore the number 1 is a single root of the characteristic polynomial, then $\mathbf{M}$ (and the Markov chain) is said to be *totally regular*; in that case (a single eigenvalue = 1, and no eigenvalue of type $e^{i\phi}$), the permuted form of Comment 47.1 below reduces to the matrix $\mathbf{C}_1$; examples of regular and totally regular spectra are shown in Figure 6.2, page 50.

The Markov chain $(\mathbf{M}, \mathbf{p}^{(0)})$ is said to be:

(1) irreducible if $\mathbf{M}$ is irreducible;
(2) reducible if $\mathbf{M}$ is reducible;
(3) cyclic if $\mathbf{M}$ is imprimitive;
(4) non-cyclic (particular case of regular) if $\mathbf{M}$ is primitive;
(5) regular if $\mathbf{M}$ is regular;
(6) totally regular if $\mathbf{M}$ is totally regular.

Commentary 47.1. ▶ A regular matrix can be permuted into the form

$$\begin{pmatrix} \mathbf{C}_1 & 0 & 0 & \dots & \dots\dots\dots & 0 \\ 0 & \mathbf{C}_2 & 0 & \dots & \dots\dots\dots & 0 \\ \dots & \dots & \dots & \dots & \dots\dots\dots & 0 \\ 0 & 0 & \dots & \mathbf{C}_m & \dots\dots\dots & 0 \\ \mathbf{C}_{m+1,1} & \mathbf{C}_{m+2,2} & \mathbf{C}_{m+1,m} & \mathbf{C}_{m+1} & \dots\dots\dots & 0 \\ \dots & \dots & \dots & \dots & \dots\dots\dots & 0 \\ \mathbf{C}_{n1} & \mathbf{C}_{n2} & \dots & \mathbf{C}_{nm} & \dots\dots\dots & \mathbf{C}_n \end{pmatrix},$$

where all the $\mathbf{C}$'s are block matrices, and the blocks in the diagonal are primitive. ◀

Commentary 47.2. ▶ We have said in Comment 27.1 that reducibility does not mean

[1] Not to be confused with the other meaning (of invertibility) frequently given to that word.

disjointness. An example is the matrix

$$\mathbf{M} = \begin{pmatrix} \frac{1}{2} & 0 & \frac{1}{2} & 0 & 0 & 0 \\ \frac{1}{2} & 0 & 0 & 0 & 0 & 0 \\ 0 & 1 & 0 & 0 & 0 & 0 \\ 0 & 0 & \frac{1}{2} & \frac{1}{2} & 0 & 0 \\ 0 & 0 & 0 & \frac{1}{2} & 0 & 1 \\ 0 & 0 & 0 & 0 & 1 & 0 \end{pmatrix} . \tag{6.6}$$

It has eigenvalues $\lambda_1 = 1$, $\lambda_2 = -1$, $\lambda_3 = 1/2$, $\lambda_4 \approx 0.847$ and $\lambda_{5,6} \approx -0.174 \pm 0.51i$. The presence of a single eigenvalue $= 1$ indicates that there are no disjoint sub-chains. The presence of two unit-modulus eigenvalues signals some cyclic behavior. This is shown in the flow diagram

$$\begin{array}{ccccc} 1 & & & & 4 \\ & \nwarrow & & \nearrow & \\ \downarrow & & 3 & & \downarrow \\ & \nearrow & & \searrow & \\ 2 & & & & 5 \longleftrightarrow 6 \end{array}$$

We see immediately that, in the long run, the system will tend to settle itself in the subspace $(5, 6)$. There is some probability to get there starting from any state, but no way to get out, once there. ◀

Commentary 48.1. ▶ An identity matrix is stochastic, and also the product of two stochastic matrices is stochastic. It is immediate to see that, *if* the inverse $\mathbf{M}^{-1}$ exists, condition (3.2) holds for it. We might ask whether a group can be formed with $N \times N$ stochastic matrices. Actually, spectral considerations greatly limit an eventual group. Condition (6.3) should be respected by each $\mathbf{M}$ *and* its inverse, so that the eigenvalues must have unit absolute values for all members (imprimitivity index $h = N$). From the Frobenius theorem, those eigenvalues will be roots of unity. Examples are the members of basis of circulant matrices, formed by powers $\mathbf{U}^n$ of a matrix

$$\mathbf{U} = \begin{pmatrix} 0\,0\,0 \ldots\ldots 1 \\ 1\,0\,0 \ldots\ldots 0 \\ 0\,1\,0 \ldots\ldots 0 \\ \ldots\ldots\ldots\ldots \\ 0\,0\,0 \ldots\ldots 1\,0 \end{pmatrix} , \tag{6.7}$$

whose entries are $(U^n)_{rs} = \delta_{r,s+n}$. The groups involved are then the cyclic groups Z_k, whose rings are constituted by the circulant matrices [see the chapters on circulants, mainly below Eq.(9.20)]. In that case they are, of course, included in the group of permutations. ◀

A very special method leads to the equilibrium distribution of an irreducible stochastic matrix [22]. Consider those two entities referred to in page 23, whose entries are all equal to 1: (i) the $N \times N$ matrix $\mathbf{H}$ and (ii) the column N-vector $\mathbf{u}$. It is easily verified that $\mathbf{H}^2 = N\,\mathbf{H}$. The equilibrium eigenvector $\mathbf{p}^\infty$ will satisfy $(\mathbf{I} + \mathbf{H} - \mathbf{M})\, \mathbf{p}^\infty = \mathbf{u}$. We have, of course, that $\mathbf{H}\,\mathbf{p} = \mathbf{u}$ for any distribution $\mathbf{p}$ and $\mathbf{HM} = \mathbf{H}$ for any $\mathbf{M}$ stochastic. Clearly $\mathbf{H}$ has no inverse. If there are two equilibrium eigenstates $\mathbf{p}^\infty$ and

$\mathbf{p}'^{\infty}$, their difference will be an eigenvector of $\mathbf{K} = \mathbf{I} + \mathbf{H} - \mathbf{M}$ with eigenvalue 0 and $\mathbf{K}^{-1}$ does not exist. When $\mathbf{M}$ is irreducible, this inverse does exist and $\mathbf{p}^{\infty}$ can be obtained as $\mathbf{p}^{\infty} = (\mathbf{I} + \mathbf{H} - \mathbf{M})^{-1}\mathbf{u}$.

Commentary 49.1. ▶ $\mathbf{I} + \mathbf{H}$ is a circulant with one eigenvalue $\lambda_N = N + 1$ and all the others $\lambda_{k \neq N} = 1$ [see Eq.(9.5)]. A stochastic matrix can be, with some pedantry, defined as a solution of $\mathbf{HM} = \mathbf{H}$ such that $\mathbf{0} \leq \mathbf{M} \leq \mathbf{H}$. ◀

6.3 Projectors and asymptotics

Applied to a stochastic matrix, the decomposition into components [see Eq.(1.13)]

$$\mathbf{M}^m = \sum_{j=1}^{N} \lambda_j{}^m \, \mathbf{Z}_j[\mathbf{M}] \tag{6.8}$$

shows that, as m increases, all components related to eigenvalues with absolute values < 1 will be progressively damped. Roots strictly equal to 1 are the *main roots*. They remain fixed in the evolution process, as they are also roots of the characteristic polynomials of all the powers of $\mathbf{M}$. Eigenvalues with absolute values less than 1 are *evanescent roots*, which tend to zero in the process of evolution.

If $\mathbf{M}$ is totally regular, the powers of $\mathbf{M}$ will tend to the single projector corresponding to $\lambda = 1$,

$$\mathbf{M}^{\infty} = \mathbf{Z}_{(\lambda=1)}[\mathbf{M}] \; . \tag{6.9}$$

We see in (1.17) that each column of $\mathbf{Z}_{(k)}[\mathbf{M}]$ is an eigenvector corresponding to eigenvalue λ_k. As the unit eigenvalue is single in the totally regular case, $\mathbf{Z}_{(\lambda=1)}[\mathbf{M}]$ will be a very special stochastic matrix: all its columns are equal to each other and to $\mathbf{p}^{(\infty)}$. In that case, there is a damping towards equilibrium. The system will tend to the sole eigenstate $\mathbf{p}^{(\infty)}$ of $\mathbf{Z}_{(\lambda=1)}[\mathbf{M}]$, the final equilibrium state. Figure 6.2 shows examples of a regular and of a totally regular spectrum.

Both the matrix $\mathbf{M}^{\infty}$ and the eigenstate $\mathbf{p}^{(\infty)}$ are well-defined only if $\mathbf{M}$ is regular. If $\mathbf{M}$ has r eigenvalues $= 1$, both $\mathbf{M}^{\infty}$ and $\mathbf{p}^{(\infty)}$ decompose into r independent pieces. If the main root is single, the picture described in the previous paragraph holds. If $\mathbf{M}$ is imprimitive of index h, the system will tend to an "attractor" formed by h probability distributions. The successive powers $\mathbf{M}^n$, for n large, will also be different, and will take the members of the "attractor" one into the other [an example will be given in page 61, Eq.(7.2)].

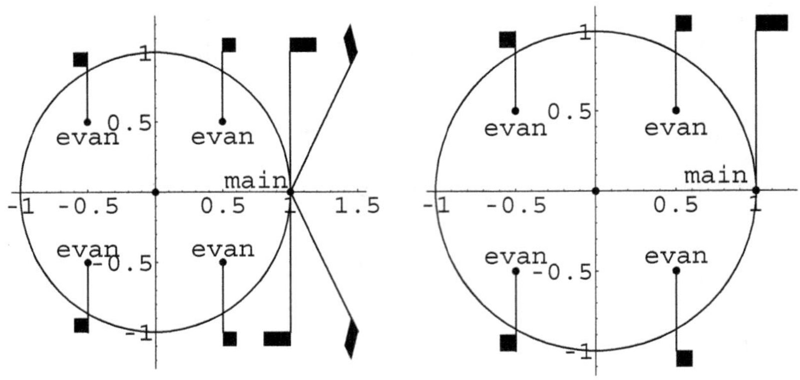

Fig. 6.2 *Spectra of a regular (left) and a totally regular matrix. The flags indicate the main and the evanescent roots.*

Consider again Eq.(1.18), for a nondegenerate case. If we take the matrix elements and sum over the row index, we find

$$\sum_a Z_{(k)ab} =$$

$$\frac{(1-\lambda_1)(1-\lambda_2)...(1-\lambda_{k-1})(1-\lambda_{k+1})...(1-\lambda_{N-1})(1-\lambda_N)}{(\lambda_k-\lambda_1)(\lambda_k-\lambda_2)...(\lambda_k-\lambda_{k-1})(\lambda_k-\lambda_{k+1})...(\lambda_k-\lambda_{N-1})(\lambda_k-\lambda_N)}.$$

We see once again how special is the eigenvalue 1. If $\lambda_k = 1$, then $\sum_a Z_{(k)ab} = 1$. We know that at least one such eigenvalue exists for $\mathbf{M}$ stochastic. For any eigenvalue $\lambda_k \neq 1$, "1" will be the value of one of the λ_j's in the numerator above and, consequently, $\sum_a Z_{(k)ab} = 0$. Thus, of all the projectors, only that corresponding to a unit eigenvalue can be a stochastic matrix. This generalizes to eigenmatrices what has been said previously on eigenvectors [see the discussion below Eq.(6.1)].

Commentary 50.1. ▶ From what has been said of the matrix of Comment 46.1, we can immediately say that its limiting projector will have all columns equal to the equilibrium distribution $(1, 0, 0)$:

$$\mathbf{Z}_{(1)}[\mathbf{M}] = \mathbf{M}^\infty = \begin{pmatrix} 1 & 1 & 1 \\ 0 & 0 & 0 \\ 0 & 0 & 0 \end{pmatrix}.$$

Any $N \times N$ stochastic matrix of the form $M_{ij} = \frac{c_i}{\sum_{k=1}^N c_k}$, for example

$$\mathbf{C} = \frac{1}{c_1 + c_2 + c_3} \begin{pmatrix} c_1 & c_1 & c_1 \\ c_2 & c_2 & c_2 \\ c_3 & c_3 & c_3 \end{pmatrix},$$

is a projector: $\mathbf{C}^2 = \mathbf{C}$. This is always the case when all columns are the same probability distribution. ◀

A negative aspect of practical nature is visible from (1.18): in order to obtain the asymptotic transition matrix, all the eigenvalues of $\mathbf{M}$ must be known. For higher-order matrices, the characteristic polynomial may be impossible to solve analytically. It is then necessary to resort to numerical methods. Furthermore we have seen that, when multiple roots appear, the procedure to obtain the components is far from simple.

Let us go back to the beginning. "State" columns are distributions, acted upon by the basic stochastic matrix $\mathbf{M}$ of the problem. $\mathbf{M}$ takes each distribution into its successor. The system proceeds from some distribution $\mathbf{p}^{(0)}$, taken as the initial configuration, through its "orbit", the succession of distributions obtained by repeated left-multiplications by $\mathbf{M}$.

Now, each column in $\mathbf{M}$ is a possible distribution vector, and the effect of multiplying $\mathbf{M}$ by itself consists in replacing each column by its successor in the evolution process. Thus, the powers of $\mathbf{M}$ can be seen as samples of possible developments — one per column — of the system starting from those different "virtual" initial states which are represented by the very columns of $\mathbf{M}$. It may happen that, for some power $\mathbf{M}^m$, the j-th column turns out to be an equilibrium distribution. That column will remain the same from that step on. Suppose further that, after n more steps, the k-th column turns out to be also an eigenstate with eigenvalue 1. The k-th column will change no more. And so on.

It may also happen that no such eigenstates ever turn up as columns in any power of $\mathbf{M}$. In the regular case, however, $\mathbf{M}^\infty$ will consist of such columns, each one representing an equilibrium state. All the columns will be the same in the totally regular case, in which there exists a single eigenvalue equal to 1 and no other unit-modulus eigenvalues. Otherwise, in the general regular case, $\mathbf{M}^\infty$ will be an exhibit of possible equilibrium states.

In the totally regular case, any initial state will tend to the same equilibrium state. That is: the final state is independent of the initial state. A theorem says that the converse is also true:

> $\mathbf{M}$ is totally regular *iff* $\mathbf{p}^{(\infty)}$ does not depend on the initial state.

This establishes the only case in which the system becomes asymptotically independent of any initial state. Another theorem, of great interest in checking calculations, says that

> If $\mathbf{M}$ is regular, then $\mathbf{M}^{(\infty)} > 0$ *iff* $\mathbf{M}$ is non-cyclic.

Finally, if $\mathbf{M}^h$ is regular and $\mathbf{M}^{h-j}$ is not, for $j = 1, 2 \ldots, h-1$, then h is the *period* of the chain and $\mathbf{M}^k$ tends to a periodic repetition when $k \to \infty$.

Thus, in order to become presently independent of the initial state, the Markov chain must have a governing matrix with only one eigenvalue λ of modulus $|\lambda| = 1$, and that will be a single main root. This case is just the opposite of the hypersensitivity to initial conditions found in chaotic systems.

Let us try to summarize the spectral characteristics, related terminology and properties. The eigenvalues of a stochastic matrix $\mathbf{M}$ are all situated on a closed unit disk around the origin of the complex plane. The eigenvalues on the circle interior ($|\lambda| < 1$) are evanescent: they are progressively damped down as the evolution proceeds by increasing powers of $\mathbf{M}$. The long-run behavior is determined by those eigenvalues which stand on the border, those of unit modulus. These can be either the number 1 — a main root — or some phase factors like $e^{i\phi}$. If there are no phase factors in the spectrum, $\mathbf{M}$ is regular. If $\mathbf{M}$ is regular with a single main root, then $\mathbf{M}$ is totally regular. Positive matrices are always totally regular.

One of the eigenvalues is necessarily a main root, which can be multiple or not. If it is multiple, $\mathbf{M}$ is reducible into independent sub-matrices governing independent sub-chains. In this completely reducible situation, the study is better done by analyzing each subcase separately. We remain consequently with the case in which 1 is a single eigenvalue. In that case, $\mathbf{M}$ can be either (simply-)reducible or irreducible. Positive matrices are always irreducible.

In the irreducible case, $\mathbf{M}$ can be imprimitive (if there are phase factors $e^{i\phi}$ as eigenvalues) or primitive (if there are no phase factors as eigenvalues, a particular case of totally regular). The powers of primitive matrices tend to positive matrices. Imprimitivity implies cyclic behavior.

Commentary 52.1. ▶ It is always good to do some calculations by hand with lower-order matrices to get a feeling of how things work. We give here a few ready-to-use results on 2×2 and 3×3 stochastic matrices.

<u>2×2 matrices</u>

$\mathbf{M}$ has the generic form

$$\mathbf{M} = \begin{pmatrix} 1-a & b \\ a & 1-b \end{pmatrix}.$$

The eigenvalues are, with the corresponding eigenvectors and projectors:

$$\lambda_1 = 1 \Rightarrow \mathbf{v}_1 = \tfrac{1}{a+b}\begin{pmatrix} b \\ a \end{pmatrix} \; ; \; \mathbf{M}^\infty = \mathbf{Z}_1 = \tfrac{1}{a+b}\begin{pmatrix} b & b \\ a & a \end{pmatrix} \; ;$$

$$\lambda_2 = 1 - a - b \Rightarrow \mathbf{v}_2 = \begin{pmatrix} -1 \\ 1 \end{pmatrix} \; ; \; \mathbf{Z}_2 = \tfrac{1}{a+b}\begin{pmatrix} a & -b \\ -a & b \end{pmatrix}.$$

λ_2 can only have absolute value 1 if $\mathbf{M}$ is the identity matrix. An example of matrix function is (for $a+b<1$)

$$\mathbf{Q} = \ln \mathbf{M} = \tfrac{\ln(1-a-b)}{a+b} \begin{pmatrix} a & -b \\ -a & b \end{pmatrix} = \ln(1-a-b)\ \mathbf{Z}_2 \ .$$

3 × 3 matrices

The general 3×3 stochastic matrix has the form

$$\mathbf{M} = \begin{pmatrix} 1-a-b & c & e \\ a & 1-c-d & f \\ b & d & 1-e-f \end{pmatrix} .$$

The main root $\lambda_1 = 1$ corresponds to the eigendistribution

$$\mathbf{v}_{\lambda_1} = \frac{1}{a\ (d+e+f)+b\ (c+d+f)+c\ (e+f)+d\,e} \begin{pmatrix} d\,e+c\ (e+f) \\ b\,f+a\ (e+f) \\ a\,d+b\ (c+d) \end{pmatrix} .$$

The projector $\mathbf{Z}_{\lambda_1}$ (and, generically, the asymptotic matrix) will have 3 columns equal to $\mathbf{v}_{\lambda_1}$. In terms of the trace T (≤ 3) and the determinant D (≤ 1) of $\mathbf{M}$, the other eigenvalues are

$$\lambda_2 = \frac{T-1-i\sqrt{4D-(T-1)^2}}{2} \ ; \ \ \lambda_3 = \frac{T-1+i\sqrt{4D-(T-1)^2}}{2} \ .$$

◄

6.3.1 *The 1-dimensional Ising model*

We have been interpreting the power of the ruling stochastic matrix as a discrete time. This is its real physical meaning in most usual cases, and is anyhow of great intuitive help. Let us examine an application to the Ising model, in which it has a different interpretation. In $\mathbf{M}^N$, the increase in N will represent the progress toward the thermodynamic limit.

The model consists ([31], [32]) of a single line of spins $\sigma_i = \pm\frac{1}{2}$, disposed on N sites. Periodic boundary conditions are taken, so that $\sigma_{N+1} = \sigma_1$. Interaction takes place only among nearest neighbors, and the interaction energy is equal to $-J\,\sigma_i\sigma_j$ for each pair. We can furthermore include an external magnetic field H. Each state of the system is fixed by a configuration $\sigma = (\sigma_1, \sigma_2, \ldots, \sigma_N)$ of spins along the closed line. The total energy for the state σ will be

$$E[\sigma] = -J \textstyle\sum_{i=1}^{N} \sigma_i\sigma_{i+1} - H \sum_{i=1}^{N} \sigma_i \ , \tag{6.10}$$

and the partition function,

$$Q_N(J, H) = \textstyle\sum_\sigma e^{-\ \beta E[\sigma]} \ . \tag{6.11}$$

It is convenient to introduce the notation $K = \beta J/2$, $h = \beta H/2$ and define the transfer matrix

$$\mathbf{W} = \begin{pmatrix} e^{K+h} & e^{-K} \\ e^{-K} & e^{K-h} \end{pmatrix} . \qquad (6.12)$$

Call λ_1 and λ_2 the eigenvalues of $\mathbf{W}$. Then, $Q_N = tr\ \mathbf{W}^N = \lambda_1^N + \lambda_2^N$. The eigenvalues are

$$\lambda_1 = e^K \cosh(h) + \sqrt{e^{2K}\cosh(h)^2 - 2\sinh(2K)} \quad \text{and}$$

$$\lambda_2 = e^K \cosh(h) - \sqrt{e^{2K}\cosh(h)^2 - 2\sinh(2K)} \ , \qquad (6.13)$$

leading to the partition function

$$Q_N(K,h) = \left(e^K \cosh(h) + \sqrt{e^{2K}\cosh(h)^2 - 2\sinh(2K)}\right)^N$$
$$+ \left(e^K \cosh(h) - \sqrt{e^{2K}\cosh(h)^2 - 2\sinh(2K)}\right)^N$$

$$= \sum_{r=0}^{N} \binom{N}{r} \left[1 + (-)^r\right] e^{(N-r)K} \cosh^{N-r}(h) \left(e^{2K}\cosh(h)^2 - 2\sinh(2K)\right)^{r/2} .$$

Only even r's contribute in the last summation. Thermodynamic quantities can be obtained from the Helmholtz free energy $A_N(H,T) = -\ kT \ln Q_N$. For example, the internal energy $U = -\ kT^2 \frac{\partial}{\partial T}\ (\frac{A_N}{kT})$; the heat capacity $C_N(H,T) = \frac{\partial U_N}{\partial T}$; and the magnetization $M_N(H,T) = < \sum_{i=1}^N \sigma_i > = -\ \frac{\partial}{\partial H} \frac{A_N}{kT}$. When $M_N(0,T) \neq 0$, there is spontaneous magnetization. This is actually not the case for the 1-dimensional model, but it does happen in 2 dimensions.

The transfer matrix $\mathbf{W}$ is real positive, so that the largest eigenvalue $\lambda_{max} = \lambda_1$ will correspond to a positive eigenvector $\mathbf{v}$:

$$\textstyle\sum_{b=1}^2 W_{ab} v_b = \lambda_{max} v_a \ .$$

$\mathbf{W}$ is also symmetric, so that $\mathbf{W}$ and $\mathbf{W}^\dagger = \mathbf{W}^T$ have the same eigenvectors. Take now the matrix $\mathbf{M}$ with the entries

$$M_{ab} = v_b^{-1} \frac{W_{ab}}{\lambda_{max}} v_a \qquad (6.14)$$

(see Comment 45.1). Notice that this can be written due to the positivity of the components v_a. If we introduce the diagonal matrix $\mathbf{D}$ with elements $D_{ab} = v_a \delta_{ab}$, this is simply

$$\mathbf{M} = \mathbf{D}^{-1} \ \frac{\mathbf{W}}{\lambda_{max}} \ \mathbf{D} \ .$$

In consequence, $\mathbf{M}^N = \mathbf{D}^{-1} \ \frac{\mathbf{W}^N}{\lambda_{max}^N} \ \mathbf{D}$ and

$$Q_N = \text{tr} \ \ \mathbf{W}^N = \lambda_{max}^N \ \text{tr} \ \mathbf{M}^N \ .$$

Now, $\mathbf{M}$ is a stochastic matrix, consequently with a maximal eigenvalue $= 1$. The corresponding equilibrium eigenvector $\mathbf{p}$ has components p_a proportional to v_a^2. As $\mathbf{p}$ must be a distribution, we normalize each component as $p_a = v_a^2 / \sum_c v_c^2$, and this will have the detailed form

$$\mathbf{p} = \begin{pmatrix} \frac{1}{1+[\lambda_1 e^K - e^{2K+h}]^2} \\ \frac{[\lambda_1 e^K - e^{2K+h}]^2}{1+[\lambda_1 e^K - e^{2K+h}]^2} \end{pmatrix}.$$

The stochastic matrix corresponding to $\mathbf{W}$ is

$$\mathbf{M} = \frac{1}{\lambda_1} \begin{pmatrix} e^{K+h} & \frac{e^{-K}}{[\lambda_1 e^K - e^{2K+h}]} \\ e^{-K}[\lambda_1 e^K - e^{2K+h}] & e^{K-h} \end{pmatrix}. \tag{6.15}$$

Is there a meaning for the "evolution" presided by $\mathbf{M}$? Or to averages taken over $\mathbf{p}$? We shall see later (Section 7.3) that when the conditions for ergodicity hold, averages over time and over the equilibrium distribution are the same. Those conditions (totally regular chain plus strictly positive equilibrium distribution) are satisfied here, so that averages over $\mathbf{p}$ coincide with averages over the involved "time". Here, the "time" parameter is related to the approach to the thermodynamic limit. As N increases, the system approaches more and more that limit. Thus, what we have is "evolution towards the thermodynamic limit". By the way, we read from $\mathbf{p}$ that

$$\mathbf{M}^\infty = \begin{pmatrix} \frac{1}{1+[\lambda_1 e^K - e^{2K+h}]^2} & \frac{1}{1+[\lambda_1 e^K - e^{2K+h}]^2} \\ \frac{[\lambda_1 e^K - e^{2K+h}]^2}{1+[\lambda_1 e^K - e^{2K+h}]^2} & \frac{[\lambda_1 e^K - e^{2K+h}]^2}{1+[\lambda_1 e^K - e^{2K+h}]^2} \end{pmatrix}.$$

Of course, tr $\mathbf{M}^\infty = 1$. The average over $\mathbf{p}$ is an average along the scaling towards the thermodynamic limit. We shall see later that the system is actually in step by step equilibrium, and that that average is constant as N increases. This is no surprise, as $\mathbf{p}$ is N-independent (see Comment 66.1; as a curiosity, see also Comment 82.1).

6.4 Continuum time

When we say that a Markov chain is a discrete Markov process, we mean that the set of states is discrete. In a general Markov process, the states can form a continuum, a case we are not considering. A Markov chain can, however, represent a continuous evolution in time. In this case, the system goes from collective state to collective state (from one probability distribution to another) in a continuous way.

6.4.1 *Passage to the continuum*

We shall find it convenient to isolate the non-changing piece of the evolution operator **M**, the identity operator **I**. Take accordingly $M_{ab} = \delta_{ab} + \epsilon\, Q_{ab}$, with $0 < \epsilon \ll 1$. It follows from (3.2) that necessarily $\sum_{a=1}^{N} Q_{ab} = 0$. Notice that this is the same as $Q_{bb} = -\sum_{a\neq b}^{N} Q_{ab}$. From (3.1), we find $Q_{aa} \leq 0$ and $Q_{a,b\neq a} \geq 0$ for each pair of states $(a, b \neq a)$.

We can borrow from the language of scattering theory, and call Q_{bb} the "elastic" contribution, in opposition to the "inelastic" terms Q_{ab} taking from a state b into a different state a. We have then a timid version of the "optical theorem": there is no inelastic process without an elastic $b \to b$ contribution. Indeed, $Q_{bb} = 0$ implies $\sum_{a\neq b}^{N} Q_{ab} = 0$. As each $Q_{a,b\neq a} \geq 0$, it follows that each $Q_{a,b\neq a}$ must vanish independently.

The evolution equation becomes a finite-difference equation,

$$\frac{1}{\epsilon}[p^{(n+1)}{}_a - p^{(n)}{}_a] = \textstyle\sum_{b=1}^{N} Q_{ab}\, p^{(n)}{}_b\ , \tag{6.16}$$

which leads, in the limit $\epsilon \to 0$, to

$$\frac{d}{dt}\, p_a(t) = \textstyle\sum_{b=1}^{N} Q_{ab}\, p_b(t)\ . \tag{6.17}$$

An immediate and expected consequence is a continuous version of probability conservation $\sum_a \frac{dp_a(t)}{dt} = 0$. The expression above can be rewritten as $\frac{dp_a}{dt} = Q_{aa}p_a + \sum_{b\neq a} Q_{ab}p_b = -\sum_{b\neq a}^{N} Q_{ba}p_a + \sum_{b\neq a} Q_{ab}p_b$, or

$$\frac{dp_a}{dt} = \sum_{b\neq a} [Q_{ab}\, p_b - Q_{ba}\, p_a]\,. \tag{6.18}$$

An equation of this type is a *master equation*, involving gains and losses. The first term in the right-hand side represents the gains of p_a coming from transitions from all the other states; the second represents the losses due to transitions from a into the other states.

Commentary 56.1. ▶ Master equations are of special interest in the kinetics of chemical reactions, in which case the **Q** matrices have for entries the corresponding reaction rates. For an example, see Section 11.1.1. ◀

6.4.2 *Hamiltonian language*

The rather sloppy approach given above can be justified by using exponential matrices [22]. We shall instead rephrase it in a more "physical" language. Take the evolution Eq.(3.4) in the form

$$p_a^{(n)} = \sum_{b=1}^{N} M^n{}_{ab}\, p_b^{(0)}\ . \tag{6.19}$$

It gives the state at a time n, given the initial distribution at time 0. This can be put into a "Hamiltonian" form if we define the Hamiltonian matrix by $i\,\mathbf{H} = \ln \mathbf{M}$. From (1.11), it is immediate that

$$i\,\mathbf{H} = \sum_{c=1}^{N} \ln \lambda_c \; \mathbf{Z}_{(c)}[\mathbf{M}] \; , \tag{6.20}$$

where the λ_c's are the eigenvalues of $\mathbf{M}$. It follows also from the definition of $\mathbf{H}$ that

$$p_a^{(n)} = \sum_{b=1}^{N} \left(e^{in\mathbf{H}}\right)_{ab} \, p_b^{(0)} \; . \tag{6.21}$$

Passage to continuum time is made by defining

$$p_a^{(t)} = \sum_{b=1}^{N} \left(e^{it\mathbf{H}}\right)_{ab} \, p_b^{(0)} \tag{6.22}$$

which, together with (6.17), leads to $\mathbf{Q} = \mathrm{i}\,\mathbf{H}$. This Hamiltonian formalism coincides with the previous "Q-matrix" approach at the first order. Of course, the factor i, introduced by analogy with the usual evolution operators in Physics, hinders the sign considerations made in that case.

Commentary 57.1. ▶ Write $\sum_a M_{ab} = 1$ in the form $\sum_a [M - I]_{ab} = 0$. Then, $\sum_a [M-I]^2_{ab} = \sum_a \sum_c [M-I]_{ac}[M-I]_{cb} = \sum_c \left(\sum_a [M-I]_{ac}\right)[M-I]_{cb} = 0$. The same is true for any power of $(\mathbf{M} - \mathbf{I})$, $\sum_a [M-I]^k_{ab} = 0$. As a matrix function is defined by its formal series, $i\mathbf{H} = \ln \mathbf{M} = \sum_k \frac{(-)^{k-1}}{k}[\mathbf{M} - \mathbf{I}]^k$, from which it follows that $\sum_a H_{ab} = 0$, independently from the relation between $\mathbf{H}$ and $\mathbf{Q}$. ◀

Now, an eigenvalue $= 1$ for $\mathbf{M}$ means an eigenvalue $= 0$ for $\mathbf{H}$. The corresponding eigenvector is a "zero-energy" state, a "vacuum", and does not suffer any evolution. "Equilibrium state" is in this way translated into "vacuum". The existence of at least one vacuum is ensured by the Perron–Frobenius theorem for an irreducible $\mathbf{M}$. If $\mathbf{M}$ is reducible and has several main roots, the set of the corresponding eigenvectors will constitute a multiple vacuum. Only those components will remain which are along the vacuum projectors when $t \to \infty$. The others are damped.

The physical view becomes clearer now. The vacuum will be the asymptotic limit of the system, the equilibrium state. If multiple, the system will tend to constitute several unconnected domains. In other words, once equilibrium is attained, each vacuum will determine a *phase* of the system. In principle, this situation would correspond to a multi-phase system. Notice, however, that such phases can be intermingled microscopically so as to elude macroscopic observation.

In the irreducible primitive case, there is a single vacuum to which the system tends. In the irreducible imprimitive case, $\mathbf{H}$ will have $(h - 1)$ extra eigenvalues of type $\frac{2\pi}{h}k$. The corresponding eigenvectors will never be probability distributions, and there will be no well-defined equilibrium distribution. As the non-unit-modulus directions will be damped down, for high enough m we shall have $\mathbf{M}^m = \sum_{k=1}^{h} e^{i\frac{2\pi}{h}km}\, \mathbf{Z}_k$. The system will maintain fixed the real vacuum, but will take the other unit-modulus vectors one into the other at each step, by transformations of the cyclic group Z_h.

Commentary 58.0. ▶ Equation (6.17), in matrix notation

$$\frac{d}{dt}\, \mathbf{p}(t) = \mathbf{Q}\, \mathbf{p}(t)\ , \tag{6.23}$$

has the obvious solution

$$\mathbf{p}(t) = e^{t\mathbf{Q}}\, \mathbf{p}(0)\ .$$

By the comments given below Eq.(6.22) concerning the control of signs, it is simpler to use directly the matrix $\mathbf{Q}$ (and the above equation), taking it as a non-hermitian "Hamiltonian". In that case, $\mathbf{M} = e^{\mathbf{Q}}$ and the solution takes on the expected form $\mathbf{p}(t) = \mathbf{M}^t\, \mathbf{p}(0)$. It is interesting to attribute a rather elastic meaning to the expression "$\mathbf{Q}$-matrix". It can be made more precise in each context. In Eqs.(6.16), (6.17), it was used as the difference between the evolution matrix $\mathbf{M}$ and the identity $\mathbf{I}$. ◀

Commentary 58.1. ▶ We have above considered the continuum limit in the "time" parameter, actually the power of the ruling stochastic matrix. Continuum time-derivatives turn up naturally. The space of states remains discrete, and that is why we keep on using the name Markov "chain". Analogous evolutive systems with continuum state spaces are Markov "processes". We could consider discrete derivatives on the space of states. These can be introduced through their finite-difference versions. As examples, the first and second derivatives of a distribution $\mathbf{p} = (p_1, p_2, \ldots, p_N)^T$ will have components

$$p'_a = p_{a+1} - p_a;\ p''_a = p_{a+1} - 2\, p_a + p_{a-1}\ .$$

Notice a peculiarity of the finite-difference formalism: the second derivative is not obtained by simply applying twice the first derivative. Applying twice the first derivative would give $p_{a+2} - 2p_{a+1} + p_a$, which is centered at point p_{a+1}. To have it at point p_a, it is necessary to shift this result back one step. The same holds for the third derivative,

$$p'''_a = p_{a+1} - 3\, p_a + 3p_{a-1} - p_{a-2}\ .$$

There are obvious difficulties at the end-points (for example, p_{N+1} appears in p'_N, p_0 and p_{-1} in p''_1). It is necessary to impose some extra condition for the case $N < \infty$, the most natural being periodicity: $p_0 = p_N$, $p_{N+1} = p_1$. Rather surprisingly, the matrices defining such derivatives turn out to be circulant matrices. We shall come back to the subject later (Section 10.3.1). ◀

Chapter 7

Equilibrium, dissipation and ergodicity

The wealth of possibilities in the evolution of a Markov process is very great. A microscopic state can remain as an actor indefinitely, or revisit the scene from time to time, or still vanish from it. Some systems stick to equilibrium in a compulsive, step-by-step way, others refuse it from the start. Equilibrium itself can be stable or not. In the long run, a chain can attain one equilibrium distribution, or many, or none. It can also circulate along several distributions indefinitely. And, among those chains which do arrive at an equilibrium state, a few do recover some memory of things past.

7.1 Recurrence, transience and periodicity

Many distinct Markov chains can be defined on a given space of microstates, differing by the stochastic matrix and/or the initial collective state. A Markov chain K is to be seen as one sole entity, comprising the space of states, the ruling stochastic matrix and the initial distribution. The triple $K = (S, \mathbf{M}, \mathbf{p}^0)$ should always be kept in mind, even when some shorter notation is used.

Given two states a and b, it may happen that evolution takes one into the other, or not. It may also happen that some power of $\mathbf{M}$ does take a into b, but no power of $\mathbf{M}$ exists taking b into a. All this depends on the properties of $\mathbf{M}$. When we say that the space S of states can have a non-trivial composition, we suppose a given, fixed chain. The definition of reducibility as path-connectedness, given in Comment 27.1, for example, gives not a property of S, but of the Markov chain, including $\mathbf{M}$. And S is said to be "disconnected" when some state is unattainable from others by successive applications of $\mathbf{M}$. Let us examine a few formal, physically unrealistic but illustrative examples on a "microscopic" 4-state space $S = \{a, b, c, d\}$ whose macroscopic state is given by some distribution $\mathbf{p} = (p_a, p_b, p_c, p_d)^T$.

(1) The matrix

$$\mathbf{U}_4 = \begin{pmatrix} 0\,0\,0\,1 \\ 1\,0\,0\,0 \\ 0\,1\,0\,0 \\ 0\,0\,1\,0 \end{pmatrix}$$

will take every state into the other in at most 3 steps. Thus, all the states are recurrent in the chain $(S, \mathbf{U}_4, \mathbf{p})$. Such properties are clearly summarized in the chain flow diagram: the scheme

$$\begin{array}{ccc} a & \longrightarrow & b \\ \uparrow & & \downarrow \\ d & \longleftarrow & c \end{array}$$

describes all the connectivity relationships on S, as well as the cyclic kind of evolution induced by the "circulant" $\mathbf{U}_4$ (see Comment 84.1 below). The characteristic polynomial is $x^4 - 1$, with roots $1, -1, i$ and $-i$ corresponding to eigenvectors $\frac{1}{4}(1,1,1,1)^T$, $(1,-1,1,-1)^T$, $(1,-i,-1,i)^T$ and $(1,i,-1,-i)^T$ (normalized at convenience). The degree of imprimitivity is clearly 4, and there will be no equilibrium state. Notice the expected properties: only the eigenvector corresponding to eigenvalue 1 can be, if normalized as above, a statistical distribution. The others have entries adding up to zero, and eventually complex components.

(2) A matrix like

$$\mathbf{U}_{22} = \begin{pmatrix} 0\,1\,0\,0 \\ 1\,0\,0\,0 \\ 0\,0\,0\,1 \\ 0\,0\,1\,0 \end{pmatrix}$$

(characteristic polynomial $= x^4 - 2x^2 + 1$, roots $1, 1, -1, -1$) will preserve the subspaces (a, b) and (c, d), and keep them separated indefinitely. The chain $(S, \mathbf{U}_{22}, \mathbf{p})$ is completely reducible into two independent chains and each state is recurrent. This is seen in the diagram

$$\begin{array}{c} a \longleftrightarrow b \\ d \longleftrightarrow c \end{array}$$

(3) Some chains exhibit states to which the system comes back from time to time, and states which it progressively tends to avoid. A (still reducible,

but no more completely reducible) "perturbation" of $\mathbf{U}_{22}$,

$$\mathbf{U}'_{22} = \begin{pmatrix} 0 & 1 & 1/2 & 0 \\ 1 & 0 & 0 & 0 \\ 0 & 0 & 0 & 1 \\ 0 & 0 & 1/2 & 0 \end{pmatrix}, \qquad \begin{array}{ccc} a & \longleftrightarrow & b \\ \uparrow & & \\ c & \longleftrightarrow & d \end{array} \tag{7.1}$$

given with the corresponding diagram, shows a rather involved behavior: a and b keep on constituting a closed subset, but c and d are taken into all other states. The eigenvalues are $(1, -1, -1/\sqrt{2}, 1/\sqrt{2})$, with eigenvectors $\mathbf{p}_{(\lambda=1)} = \mathbf{p}^{(\infty)} = (1/2, 1/2, 0, 0)^T$, $(-1, 1, 0, 0)^T$, $(-1, \sqrt{2}, -\sqrt{2}, 1)^T$ and $(-1, -\sqrt{2}, \sqrt{2}, 1)^T$ respectively. The diagram shows a trend to concentration in states a and b. This is confirmed by the asymptotic behavior: high powers of $\mathbf{U}'_{22}$ alternate between two matrices,

$$\begin{pmatrix} 1 & 0 & 0 & 1 \\ 0 & 1 & 1 & 0 \\ 0 & 0 & 0 & 0 \\ 0 & 0 & 0 & 0 \end{pmatrix} \quad \begin{array}{cc} a & b \\ \uparrow & \uparrow \\ d & c \end{array} \quad \text{and} \quad \begin{pmatrix} 0 & 1 & 1 & 0 \\ 1 & 0 & 0 & 1 \\ 0 & 0 & 0 & 0 \\ 0 & 0 & 0 & 0 \end{pmatrix} \quad \begin{array}{ccc} a & \longleftrightarrow & b \\ \uparrow & & \uparrow \\ c & & d \end{array},$$

changing from one to the other at each step. The system will ultimately desert the subset $\{c, d\}$ and establish itself in a two-state "attractor" $\{a, b\}$, announced by the two unit-modulus eigenvalues (imprimitivity index $h = 2$). The subset (c, d) is said to be "dissipated" in the evolution process. For a high-enough power m, only the two unit-modulus states remain, with

$$\mathbf{U}'^{m}_{22} = \mathbf{Z}_{(\lambda=1)} + (-)^m \mathbf{Z}_{(\lambda=-1)} \,. \tag{7.2}$$

The state $\mathbf{p}_{(\lambda=1)}$ is a fixed point under the action of $\mathbf{U}'_{22}$, the other unit eigenvector (not a possible state distribution!) changes sign at each step. All this is characteristic of imprimitivity: the state $\mathbf{p}_{(\lambda=1)}$ remains as it is under the action of an imprimitive matrix, while the other unit-modulus eigenvectors change into each other at every step by the action of the cyclic group Z_h.

This means that we are speaking of the generic case when we say that there is no equilibrium. If the initial state is $\mathbf{p}^0 = \mathbf{p}^\infty = (1/2, 1/2, 0, 0)^T$, the system will remain in that state forever. That equilibrium is, however, unstable. If we take the initial state $\mathbf{p}^0 = (1/2 + \epsilon, 1/2 - \epsilon, 0, 0)^T$, for ϵ as small as we may wish, the system will asymptotically oscillate between $(1/2 + \epsilon, 1/2 - \epsilon, 0, 0)^T$ and $(1/2 - \epsilon, 1/2 + \epsilon, 0, 0)^T$ indefinitely. There is a qualitative change in the behavior for any "perturbation" $\epsilon > 0$. This is a toy example of hypersensibility to the initial conditions.

Let us borrow from the mathematicians a few more notions, which help putting the comments above into a more precise setting:

- We say that b *leads* to a, and note $b \to a$, when $M^n{}_{ab} > 0$ for some n.
- We say that b *communicates* with a, and note $b \leftrightarrow a$, if $b \to a$ and $a \to b$.
- The set D of all the a's for which there exists some b such that $a \to b$ but $b \not\to a$ is the *dissipative* part of S. Chains $(S, \mathbf{U}_4, \mathbf{p})$ and $(S, \mathbf{U}_{22}, \mathbf{p})$ have no dissipative parts. In chain $(S, \mathbf{U}'_{22}, \mathbf{p})$, the subset $D = (c, d)$ is the dissipative part, and $S - D = (a, b)$ is the non-dissipative part.
- A state a is *recurrent* if $M^n{}_{aa} = 1$ for infinitely many values of n, a condition which can be alternatively written $\sum_{n=0}^{\infty} M^n{}_{aa} = \infty$. We see that the system keeps coming back to a recurrent state. In the naïve cases above, a and b are recurrent in the three chains.
- A state a is *transient* if $M^n{}_{aa} = 0$ for infinitely many values of n; alternatively, $\sum_{n=0}^{\infty} M^n{}_{aa} < \infty$; it is an ephemeral state, one which the system eventually leaves forever. This is the case of c and d in chain $(S, \mathbf{U}'_{22}, \mathbf{p})$. The flow diagram of the matrix in Comment 46.1 shows two transient states, "2" and "3". As noticed in Comment 50.1, that system will tend inevitably to state "1". We have seen an extreme case of transient case in matrix (5.1): there, state z drops out at the first step.
- A chain is *recurrent* if every point of S is *recurrent.* This is the case in which the *éternel retour* takes place. Chain $(S, \mathbf{U}_4, \mathbf{p})$ is clearly recurrent. Recurrence is one of the two conditions for the ergodic theorem (to be seen in Section 7.3).
- A state a is *aperiodic* if the diagonal entries $M^n{}_{aa}$ are > 0 for all powers n sufficiently large.
- The greatest common divisor d of the set $\{n > 1 \text{ such that } M^n{}_{aa} > 0\}$ is the *period* of a; the above aperiodicity happens when $d = 1$. In the chain $(S, \mathbf{U}'_{22}, v)$, a and b have period 2.

Communication is an equivalence relation on the non-dissipative part $S - D$. The corresponding classes are called *ergodic* classes. A class C such that the two propositions $b \in C$ and $b \to a$ imply $a \in C$ is *closed.* There is no scape from a closed class. Parts (a, b) and (c, d) in $(S, \mathbf{U}_{22}, \mathbf{p})$ are closed, as is (a, b) in $(S, \mathbf{U}'_{22}, \mathbf{p})$. If the whole space of states constitutes a single class, the Markov chain is frequently said to be *irreducible*, or *ergodic.* This terminology is current, but rather misleading. That a chain is ergodic does not mean that the ergodic theorem applies. The theorem (see Section 7.3) requires the existence of an equilibrium state, which must furthermore be single. The chain $(S, \mathbf{U}_4, \mathbf{p})$ is ergodic, but does not tend to an equilibrium state.

Transience, recurrence and periodicity are class properties:

- the states of an ergodic class C are all either recurrent or transient
- every recurrent class is closed
- every finite closed class is recurrent
- all states in an ergodic class have the same period.

7.2 Detailed balancing and reversibility

The equilibrium condition

$$\sum_b M_{ab}\, p_b - p_a = 0$$

can evidently be rewritten as

$$\sum_b [M_{ab}\, p_b - M_{ba}\, p_a] = 0\ ,$$

as the M_{ba}'s in the second summand yield simply a factor 1. All that is needed for equilibrium is that the above summation vanish. A very particular way of enforcing it is the requirement

$$M_{ab}\, p_b = M_{ba}\, p_a\ . \tag{7.3}$$

This is a strong imposition: the summation vanishes term by term. It is called the *detailed balancing condition*, and has one advantage: it provides a stochastic process which has a certain distribution $\mathbf{p} = (p_a)$, given *a priori*, as equilibrium distribution. In effect, it is immediate that $M^n_{ab}\, p_b = M^n_{ba}\, p_a\ \forall\ n > 1$, and consequently

$$p^{(n+1)}{}_a = \sum_{b=1}^{N} M_{ab}\, p^{(n)}{}_b = \sum_{b=1}^{N} M_{ba}\, p^{(n)}{}_a = p^{(n)}{}_a\ . \tag{7.4}$$

When this condition holds, we say that $\mathbf{M}$ and $\mathbf{p}$ "are in detailed balance", or that the chain $(\mathbf{M}, \mathbf{p})$ is in detailed balance. In that case the average $\bar{f}^{(n)} = \sum_a f_a p_a^{(n)}$ of any quantity f is, of course, stationary.

A matrix $\mathbf{M}$ in detailed balance with a positive distribution is closely related to a symmetric matrix, and its spectrum is purely real. In order to see it, first define $\hat{p}^{(n)}{}_a$ by

$$p^{(n)}{}_a = \sqrt{p_a}\ \hat{p}^{(n)}{}_a\ .$$

Then (7.4) takes the form

$$\hat{p}^{(n+1)}{}_a = \sum_{b=1}^{N} \bar{M}_{ab}\, \hat{p}^{(n)}{}_b\ , \tag{7.5}$$

with

$$\bar{M}_{ab} = p_a^{-1/2}\ M_{ab}\ p_b^{1/2}\ . \tag{7.6}$$

Using (7.3),

$$\bar{M}_{ab} = p_a^{-1/2}\ M_{ab}\ p_b\ p_b^{-1/2} = p_a^{1/2}\ M_{ba}\ p_b^{-1/2} = \bar{M}_{ba}\ .$$

$\bar{\mathbf{M}}$ is a real symmetric matrix, and consequently all its eigenvalues are real. But the eigenvalue equation

$$\sum_{b=1}^{N} \bar{M}_{ab}\ \hat{v}_b = \lambda\ \hat{v}_a$$

is the same as

$$\sum_{b=1}^{N} M_{ab}\ p_b^{1/2} \hat{v}_b = \lambda\ p_a^{1/2}\ \hat{v}_a\ ,$$

which shows that $\mathbf{M}$ has the same real eigenvalues, with eigenvectors of components $v_a = p_a^{1/2} \hat{v}_a$. Of course, a stochastic matrix which is in detailed balance with the flat distribution $\mathbf{u} = \frac{1}{N}(1,1,1,\ldots,1)^T$ is symmetric, $\mathbf{M}^T = \mathbf{M}$. And, of course, also the converse holds: every symmetric stochastic matrix $\mathbf{M}$ is in detailed balancing with the flat distribution. If $\mathbf{M}$ is totally regular, its Markov chain will tend fatally to $\mathbf{u}$. This will be the case of odd-dimensional standard random walks (Section 10.2.3).

Commentary 64.1. ▶ Notice that detailed balancing imposes $\frac{N(N-1)}{2}$ conditions on the entries of $\mathbf{M}$. With the N stochasticity conditions (3.2), there are $\frac{N(N+1)}{2}$ conditions. From Comment 52.1, it is easy to check that a 2×2 stochastic matrix is always in detailed balancing with its asymptotic equilibrium state. ◀

Let us change the order of the assumptions made. Given $(\mathbf{M}, \mathbf{p})$ *in equilibrium*, with $\mathbf{p} > 0$, suppose there is another matrix $\hat{\mathbf{M}}$ such that

$$\hat{M}_{ab}\ p_b = M_{ba}\ p_a \qquad \forall a, b\ . \tag{7.7}$$

Then,

- $\hat{\mathbf{M}}$ is also a stochastic matrix:

$$\sum_a \hat{M}_{ab} = \tfrac{1}{p_b} \sum_a M_{ba} p_a = 1\ .$$

- $(\hat{\mathbf{M}}, \mathbf{p})$ is also in equilibrium:

$$\sum_b \hat{M}_{ab}\, p_b = \sum_b M_{ba} p_a = p_a \ .$$

- if $\mathbf{M}$ is irreducible, then $\hat{\mathbf{M}}$ is also irreducible. In effect, the irreducibility of $\mathbf{M}$ says (see Comment 27.1) that, given any two states a and b, there exists a chain of states $i_1, i_2, \ldots i_{n-1}$ with

$$M_{bi_1} M_{i_1 i_2} \ldots M_{i_{n-2} i_{n-1}} M_{i_{n-1} a} p_a > 0 \ .$$

Using (7.7) successively, we can move p_a leftwards, changing $\mathbf{M}$ into $\hat{\mathbf{M}}$ at each step:

$$\begin{aligned} & M_{bi_1}\ M_{i_1 i_2} \ldots M_{i_{n-2} i_{n-1}}\ p_{i_{n-1}}\ \hat{M}_{a i_{n-1}} \\ &= M_{bi_1}\ M_{i_1 i_2} \ldots p_{i_{n-2}}\ \hat{M}_{i_{n-1} i_{n-2}}\ \hat{M}_{a i_{n-1}} \\ &= \hat{M}_{a i_{n-1}}\ \hat{M}_{i_{n-1} i_{n-2}} \ldots \hat{M}_{i_2 i_1}\ \hat{M}_{i_1 b}\ p_b > 0 \ . \end{aligned}$$

Notice that the order of states in the path is just reversed. The chains $(\hat{\mathbf{M}}, \mathbf{p})$ and $(\mathbf{M}, \mathbf{p})$ are the *time-reversals* of each other. In the generic case, $\hat{\mathbf{M}} \neq \mathbf{M}$. $(\mathbf{M}, \mathbf{p})$ is said to be *reversible* when $\hat{\mathbf{M}} = \mathbf{M}$. Thus, if the matrix $\mathbf{M}$ is irreducible, then

$$(\mathbf{M},\ \mathbf{p}) \text{ is in detailed balance} \longleftrightarrow (\mathbf{M},\ \mathbf{p}) \text{ is reversible.}$$

Notice that reversibility presupposes equilibrium.

Commentary 65.1. ► If we define the Boltzmann function $H^{(n)} = \sum_a p_a^{(n)} \ln p_a^{(n)}$, then detailed balancing implies, as it would be expected, $H^{(n+1)} = H^{(n)}$. As it is, the above expression for $H^{(n)}$ seems to be defined only for a positive distribution. It can be extended to non-negative distributions by noticing that $\lim_{p \to 0} p \ln p = 0$. The vanishing components simply do not contribute. A particularly simple illustration of the Boltzmann theorem's working turns up when we suppose the distribution to be initially concentrated at a point (say, at point 1: $p_a^{(n)} = \delta_{a1}$). In that case, $H^{(n)} = 0$. As $H^{(n+1)} \leq 0$ (being a sum of non-positive terms), $H^{(n+1)} \leq H^{(n)}$. ◄

Let us remark that detailed balancing of $(\mathbf{M},\ \mathbf{p})$ by itself does not ensure equilibrium, unless $\mathbf{p}$ is taken as the initial state. If the initial state is $\mathbf{p}$, the system remains in $\mathbf{p}$. If, however, the system starts from another state, it may happen that final equilibrium be absent. Matrix $\mathbf{U}'_{22}$ of Eq.(7.1) is a good example: it is in detailed balancing with its $\lambda = 1$ eigenvector, which is its fixed-point. Thus, the particular chain $(\mathbf{U}'_{22},\ \mathbf{p}_{\lambda=1} = (1/2, 1/2, 0, 0)^T)$ will be in permanent equilibrium. This is not, as has been observed, a stable case. Any other chain $(\mathbf{U}'_{22},\ (1/2+\epsilon, 1/2-\epsilon, 0, 0)^T)$ with an initial state as close as possible to the equilibrium state, will fail to attain any equilibrium.

Commentary 66.1. ▶ If the matrix $\hat{\mathbf{M}}$ of Comment 45.1 is symmetric, then the pair $(\mathbf{M}, \mathbf{p})$, with $p_a = \frac{\hat{v}_a^2}{\sum_c \hat{v}_c^2}$, is in detailed balance. In the Ising model of Section 6.3.1, the transfer matrix $\mathbf{W}$ is symmetric. Its transformed $\mathbf{M}$ of Eq.(6.14) is consequently in detailed balancing with its equilibrium distribution $\mathbf{p}$, and the averages over $\mathbf{p}$ are constant. ◀

Commentary 66.2. ▶ **A 1-dimensional "crystal"** Let us once again look for some formally interesting behaviors. We shall forget the physical considerations leading to the stochastic matrix $\mathbf{M}$, but even so use the suggestive terminology of Section 5.1. Thus, let us call "pure Selenium" the distribution $\mathbf{p}^{(0)} = \begin{pmatrix} 1 \\ 0 \\ 0 \end{pmatrix}$, which we shall take as initial state. Take an arbitrary 3×3 stochastic matrix

$$\mathbf{M} = \begin{pmatrix} 1-a-b & c & e \\ a & 1-c-d & f \\ b & d & 1-e-f \end{pmatrix}.$$

Then, $\mathbf{p}^{(1)} = \mathbf{M}\ \mathbf{p}^{(0)} = \begin{pmatrix} 1-a-b \\ a \\ b \end{pmatrix}$. Impose now that $\mathbf{p}^{(1)}$ be some pure Arsenic state, say $\mathbf{p}^{(1)} = \begin{pmatrix} 0 \\ a \\ 1-a \end{pmatrix}$. Applying $\mathbf{M}$ to $\mathbf{p}^{(1)}$, we find some complicated state. We impose then that this state be again $\mathbf{p}^{(0)}$, so as to get back a pure Selenium state. This can be done by a convenient choice of the entries in $\mathbf{M}$, which now becomes

$$\mathbf{M} = \begin{pmatrix} 0 & c & (1-ac)/(1-a) \\ a & 1-c-d & a(1-c-d)/(a-1) \\ 1-a & d & ad/(a-1) \end{pmatrix}.$$

We verify then that an oscillatory behavior is established: $\mathbf{p}^{(1)} = \mathbf{M}\,\mathbf{p}^{(0)}$, $\mathbf{p}^{(0)} = \mathbf{M}\,\mathbf{p}^{(1)}$. And we have our "crystal": a Selenium shell is succeeded by an Arsenic shell, which is covered by a new Selenium shell, and so on. The parameters a, c, d are still free to be chosen. For instance, taking $a = 0$, meaning that $\mathbf{p}^{(1)}$ is a pure Arsenic state of type $\mathbf{p}^{(1)} = \begin{pmatrix} 0 \\ 0 \\ 1 \end{pmatrix}$, leads to $\mathbf{M} = \begin{pmatrix} 0 & c & 1 \\ 0 & 1-c-d & 0 \\ 1 & d & 0 \end{pmatrix}$. We find then the eigenvalues $-1, 1, 1-c-d$. The eigenvector corresponding to eigenvalue -1 is $\begin{pmatrix} -1 \\ 0 \\ 1 \end{pmatrix}$. That corresponding to eigenvalue 1 is $\begin{pmatrix} 1/2 \\ 0 \\ 1/2 \end{pmatrix}$, half Selenium, half Arsenic. We cannot speak of asymptotic state, as the eigenstate corresponding to eigenvalue $= -1$ is not damped.

It turns out to be impossible to find a non-negative stochastic matrix if $a = 2/3$. We learn in the process something of practical interest: it is very difficult to satisfy condition (3.1). A possible, though rather constrained, situation comes out if we choose $a = 1/3$. The system then oscillates between $\begin{pmatrix} 1 \\ 0 \\ 0 \end{pmatrix}$ and $\begin{pmatrix} 0 \\ 1/3 \\ 2/3 \end{pmatrix}$, but the matrix found,

$$\mathbf{M} = \begin{pmatrix} 0 & c & 3(1-c/3)/2 \\ 1/3 & 1-c-d & (c+d-1)/2 \\ 2/3 & d & -d/2 \end{pmatrix},$$

is not non-negative for arbitrary c, d. The only possible choice is $d = 0$ and $c = 1$, for which

$$\mathbf{M} = \begin{pmatrix} 0 & 1 & 1 \\ 1/3 & 0 & 0 \\ 2/3 & 0 & 0 \end{pmatrix}.$$

The eigenvalues are now $-1, 0, 1$, with respective eigenvectors $\begin{pmatrix} -3/2 \\ 1/2 \\ 1 \end{pmatrix}$, $\begin{pmatrix} 0 \\ -1 \\ 1 \end{pmatrix}$ and $\begin{pmatrix} 1/2 \\ 1/6 \\ 1/3 \end{pmatrix}$. It can be checked that neither here nor in the previous case does the oscillation mean that $\mathbf{M}^2 = \mathbf{I}$. In the present case, actually, $\mathbf{M}^2 = \begin{pmatrix} 1 & 0 & 0 \\ 0 & 1/3 & 1/3 \\ 0 & 2/3 & 2/3 \end{pmatrix}$. The components are

$$\mathbf{Z}_{\lambda=-1} = \begin{pmatrix} 1/2 & -1/2 & -1/2 \\ -1/6 & 1/6 & 1/6 \\ -1/3 & 1/3 & 1/3 \end{pmatrix}, \quad \mathbf{Z}_{\lambda=0} = \begin{pmatrix} 0 & 0 & 0 \\ 0 & 2/3 & -1/3 \\ 0 & -2/3 & 1/3 \end{pmatrix}$$

$$\mathbf{Z}_{\lambda=1} = \begin{pmatrix} 1/2 & 1/2 & 1/2 \\ 1/6 & 1/6 & 1/6 \\ 1/3 & 1/3 & 1/3 \end{pmatrix},$$

and verify all the good projector properties. Notice that $\mathbf{Z}_{\lambda=1}$ has for columns the distribution corresponding to $\lambda = 1$. This is not enough to get equilibrium, because there is another eigenvector with unit-modulus eigenvalue. The characteristic polynomial is $P(\lambda) = (\lambda^2 - 1)\lambda$, giving the expected imprimitivity index $h = 2$. By the way, this suggests how to proceed to get a matrix with a given index: write the desired polynomial, substitute $\mathbf{M}$ for λ and use the Cayley–Hamilton theorem. Solve then the resulting system of equations for the matrix elements. ◄

7.2.1 *General binary glasses*

Let us return to the glasses of Chapter 5. We may consider a general binary case, with states 1 and 2, with concentrations c_1 and c_2, valences m_1 and m_2, Boltzmann exponents $\epsilon = \epsilon_{11}$, $\eta = \eta_{12}$ and $\alpha = \alpha_{22}$. The evolution matrix will be

$$\mathbf{M} = \begin{pmatrix} \frac{m_1(1-c_2)e^{-\epsilon}}{m_1(1-c_2)e^{-\epsilon}+m_2c_2e^{-\eta}} & \frac{m_1(1-c_2)e^{-\eta}}{m_1(1-c_2)e^{-\eta}+m_2c_2e^{-\alpha}} \\ \frac{m_2c_2e^{-\eta}}{m_1(1-c_2)e^{-\epsilon}+m_2c_2e^{-\eta}} & \frac{m_2c_2e^{-\alpha}}{m_1(1-c_2)e^{-\eta}+m_2c_2e^{-\alpha}} \end{pmatrix}$$

$$= \begin{pmatrix} \frac{m_1(1-c_2)}{m_1(1-c_2)+m_2c_2\xi} & \frac{m_1(1-c_2)\xi}{m_1(1-c_2)\xi+m_2c_2\mu} \\ \frac{m_2c_2\xi}{m_1(1-c_2)+m_2c_2\xi} & \frac{m_2c_2\mu}{m_1(1-c_2)\xi+m_2c_2\mu} \end{pmatrix}, \tag{7.8}$$

with $\xi = e^{\epsilon-\eta}$ and $\mu = e^{\epsilon-\alpha}$. The asymptotic eigenvector $\mathbf{v}^{(\infty)}$ is

$$\begin{pmatrix} p_1 \\ p_2 \end{pmatrix}^{(\infty)} = \begin{pmatrix} \frac{M_{12}}{M_{12}+M_{21}} \\ \frac{M_{21}}{M_{12}+M_{21}} \end{pmatrix} = \begin{pmatrix} \frac{m_1^2(1-c_2)^2\xi+m_1m_2c_2(1-c_2)\xi^2}{m_1^2(1-c_2)^2\xi+m_2^2c_2^2\xi\mu+2m_1m_2c_2(1-c_2)\xi^2} \\ \frac{m_2^2c_2^2\xi\mu+m_1m_2c_2(1-c_2)\xi^2}{m_1^2(1-c_2)^2\xi+m_2^2c_2^2\xi\mu+2m_1m_2c_2(1-c_2)\xi^2} \end{pmatrix}. \tag{7.9}$$

We might have implied, from our previous comments, that the no-fluctuation condition is related to detailed balancing. In the glass cases examined up to present time, the no-fluctuation condition has indeed been found to be equivalent to detailed balancing.

Actually, for 2×2 matrices detailed balancing of $\mathbf{M}$ and $\mathbf{v}^{(\infty)}$ holds automatically. It reduces to only one condition (see Comment 64.1). From $M_{11}v_1^{(\infty)} + M_{12}v_2^{(\infty)} = v_1^{(\infty)}$ and $M_{11} + M_{21} = 1$, we have $M_{12}v_2^{(\infty)} = (1 - M_{11})v_1^{(\infty)} = M_{21}v_1^{(\infty)}$. In the generic case, every 2×2 stochastic matrix is in detailed balance with its equilibrium eigendistribution.

The no-fluctuation condition leads to the equation (we shall write simply c for c_2)

$$c\,(1-c)\,\left[m_1^2 - m_1 m_2 \xi - c\left(m_1^2 - 2m_1 m_2 \xi + m_2^2 \mu\right)\right] = 0\,.$$

The solutions are $c = 0$, $c = 1$ and the non-trivial

$$c = \frac{m_1^2 - m_1 m_2 \xi}{m_1^2 + m_2^2 \mu - 2m_1 m_2 \xi}\,. \tag{7.10}$$

This leads to the results previously found for the *Se–As* case ($m_1 = 2$, $m_2 = 3$),

$$c = \frac{4 - 6\xi}{4 + 9\mu - 12\xi}\,,$$

and for the *Se–Ge* case ($m_1 = 2$, $m_2 = 4$),

$$c = \frac{1 - 2\xi}{1 + 4\mu - 4\xi}\,.$$

As a bonus, we would have the case of a hypothetical *As–Ge* glass ($m_1 = 3$, $m_2 = 4$):

$$c = \frac{9 - 12\xi}{9 + 16\mu - 24\xi}\,.$$

In the limit of small modifier concentration, $c \to 0$, $T \to T_{g_0}$, we have $\xi = e^{(E_{11} - E_{12})/kT_{g0}} = m_1/m_2$ and

$$E_{11} - E_{12} = kT_{g_0} \ln\left(\frac{m_1}{m_2}\right). \tag{7.11}$$

A few general remarks:

- If $E_{11} - E_{12} > 0$, then the bond 1–2 is energetically favored with respect to bond 1–1. Modifiers will only be effective if this condition is satisfied.
- The value of $(M^n)_{22}$ is the probability to have, beginning with "2", the same state again after n steps.
- On the other hand, $(M_{22})^n$ is the probability to have an n-step path entirely composed of "2".

- The notion of a rim is necessary to the use of a stochastic matrix. If we start with free molecules, we do not recover the same set of states at each step. The first step, for example, leads to doublets. If we try to identify pieces of these doublets with the initial states, the *nec fluctuat* condition becomes impossible (for example, in the *Ge–As* case).
- The valences at the rim are actually irrelevant. Each cell at the rim contributes to a particular column in the matrix and its valence factor is cancelled out in the normalization.

7.3 Ergodicity

Consider again a space S of states and suppose there is a function

$$f : S \longrightarrow \mathbb{R}\ , \quad f : a \longmapsto f_a\ ,$$

which attributes some value to each state. We can think of the coordination number, or valence, or of spin, magnetic moment or polarizability, or whatever else property that can be defined for each unit. The average of f in a distribution $\mathbf{p}$ is given by Eq.(3.10), which we shall write as

$$\bar{f}(\mathbf{p}) = \bar{f}_{\mathbf{p}} = \sum_a f_a\ p_a\ .$$

Given $p_a^{(k)} = \sum_b M^k{}_{ab}\ p_b^{(0)}$, the time average of f up to step n will be

$$\frac{1}{n}\sum_{k=0}^{n-1} \bar{f}_{\mathbf{p}^{(k)}} = \frac{1}{n}\sum_{k=0}^{n-1}\sum_{a,b} f_a\ M^k{}_{ab} p_b^{(0)}\ .$$

Commentary 69.1. ▶ Notice that, in the case of the glass evolution, this is an average over the successive shell averages, consequently over the agglomerates. ◀

The ergodic theorem says that the long-term time average equals the average on the equilibrium distribution:

$$\lim_{n\to\infty}\frac{1}{n}\sum_{k=0}^{n-1} \bar{f}_{\mathbf{p}^{(k)}} = \bar{f}_{\mathbf{p}^{(\infty)}} = \sum_a f_a\ p_a^{(\infty)}\ . \tag{7.12}$$

This makes sense only if the system does tend to a single equilibrium state. Thus, an applicability condition for the theorem is that the chain be totally regular. If irreducible, it must be primitive.

Nevertheless, not every totally regular chain admits the theorem. In order to state under which conditions the theorem does hold, a supplementary notion is necessary.

Given a state a, we can define its *expected return time* as

$$t_a = \inf\{n \mid M^n{}_{aa} > 0\} .$$

If a is recurrent and t_a is finite, then the state a is said to be *positive-recurrent*. If every state is positive-recurrent, we say that the *chain* is positive-recurrent.

The ergodic theorem holds for positive-recurrent totally regular chains.

If the chain is irreducible and has an equilibrium distribution $\mathbf{p}^{(\infty)}$, then a theorem [22] states that

$$t_a = \frac{1}{p_a^{(\infty)}} .$$

We see thus the importance of the property $p_a^{(\infty)} > 0$ turning up in the Perron–Frobenius theorems: it means that t_a is finite and, consequently, a can be positive-recurrent. But there is more: another theorem [22] says that, if the chain is irreducible, the existence of one positive-recurrent state in the equilibrium distribution implies that every state is positive-recurrent, so that the chain itself is positive-recurrent.

Markov chains are characterized by their forgetfulness. Evolution from a certain macroscopic state onwards depends on that state and not on the way through which it has been attained. It is independent of its past history. This lack of memory is mitigated in the conditions of the ergodic theorem. The asymptotic distribution keeps a memory of the average behaviors along the whole history of the system.

A simple example can be found in the case *Se–As* of Section 5.1. Take for f the coordination number: $f = 2$ for the *Se* atom and $f = 3$ for *As*. With an initial state $\mathbf{v}^{(0)} = (\alpha, 1 - \alpha)$, the starting average coordination number will be $2\alpha + 3(1 - \alpha)$. To calculate it at the k-th step, it will be necessary to have $\mathbf{M}^k$. Given the eigenvalues $\lambda_1 = 1$ and $\lambda_2 = A - B$,

$$\mathbf{M}^k = \lambda_1^k\, \mathbf{Z}_{(1)} + \lambda_2^k\, \mathbf{Z}_{(2)} = \mathbf{Z}_{(1)} + (A - B)^k \mathbf{Z}_{(2)} .$$

From $\mathbf{Z}_{(1)} = \frac{1}{1+B-A} \begin{pmatrix} B & B \\ 1-A & 1-A \end{pmatrix}$, we obtain

$$\mathbf{Z}_{(2)} = \mathbf{I} - \mathbf{Z}_{(1)} = \frac{1}{1+B-A} \begin{pmatrix} 1-A & -B \\ A-1 & B \end{pmatrix} .$$

Consequently,

$$\mathbf{M}^k = \frac{1}{1+B-A} \left[\begin{pmatrix} B & B \\ 1-A & 1-A \end{pmatrix} + (A-B)^k \begin{pmatrix} 1-A & -B \\ A-1 & B \end{pmatrix} \right] .$$

The average on the k-th distribution in the chain is

$$\bar{f}_{\mathbf{v}^{(k)}} = \bar{f}(\mathbf{Z}_{(1)}\mathbf{v}^{(0)} + (A-B)^k \mathbf{Z}_{(2)}\mathbf{v}^{(0)}) \ .$$

The first term on the right-hand side is the average on the asymptotic distribution, $\bar{f}(\mathbf{Z}_{(1)}\mathbf{v}^{(0)}) = \bar{f}(\mathbf{v}^{(\infty)})$. Thus,

$$\bar{f}_{\mathbf{v}^{(k)}} = \bar{f}(\mathbf{v}^{(\infty)}) + (A-B)^k \bar{f}(\mathbf{Z}_{(2)}\mathbf{v}^{(0)}) \ . \tag{7.13}$$

The average in discrete time, up to the n-th step, will be

$$\frac{1}{n}\sum_{k=0}^{n-1} \bar{f}_{\mathbf{v}^{(k)}} = \frac{1}{n}\sum_{k=0}^{n-1} \bar{f}(\mathbf{Z}_{(1)}\mathbf{v}^{(0)} + (A-B)^k \mathbf{Z}_{(2)}\mathbf{v}^{(0)}) \ .$$

The contribution coming from the asymptotic distribution is the same for any $\mathbf{v}^{(k)}$. Consequently,

$$\frac{1}{n}\sum_{k=0}^{n-1} \bar{f}_{\mathbf{v}^{(k)}} = \bar{f}(\mathbf{v}^{(\infty)}) + \frac{1}{n}\sum_{k=0}^{n-1} (A-B)^k \bar{f}(\mathbf{Z}_{(2)}\mathbf{v}^{(0)}) \ .$$

When $n \to \infty$, as $|A-B| < 1$

$$\frac{1}{n}\sum_{k=0}^{n-1} (A-B)^k = \frac{1}{n}\,\frac{1-(A-B)^n}{1-A+B} \to 0 \ .$$

Consequently,

$$\frac{1}{n}\sum_{k=0}^{n-1} \bar{f}_{\mathbf{v}^{(k)}} = \bar{f}(\mathbf{Z}_{(1)}\mathbf{v}^{(0)}) = \bar{f}(\mathbf{v}^{(\infty)}) \ .$$

Thus, the average value of the coordination number will be

$$\bar{f} = \frac{2B + 3(1-A)}{1+B-A} \ .$$

It is always interesting to look at the extreme cases: for $c = 1$, $A = B = 0$, $\bar{f} = 3$, $t_1 = \infty$, $t_2 = 1$; for $c = 0$, $A = B = 1$, $\bar{f} = 2$, $t_1 = 1$, $t_2 = \infty$.

With the no-fluctuation condition, the value of $\frac{B}{1+B-A}$ becomes $1-c$ and

$$\bar{f} = 2(1-c) + 3c \ . \tag{7.14}$$

This gives again the expected results in the extreme cases $c = 1$ and $c = 0$. We can profit to see some more detail. Given $\mathbf{v}^{(\infty)}$, the expected return times are $t_1 = (1+B-A)/B$ and $t_2 = (1+B-A)/(1-A)$. With the no-fluctuation condition, $t_1 = 1/(1-c)$, $t_2 = 1/c$. At the high-temperature limit, $A = B = 2(1-c)/(2+c)$ and

$$t_1 = 1/B = (2+c)/2(1-c) \to 1, \infty \text{ when } c \to 0, 1 \ ;$$

$$t_2 = 1/(1-A) = (2+c)/3c \to \infty, 1 \text{ when } c \to 0, 1 \ .$$

These asymptotic or average values are the equilibrium values. Equation (7.13) gives the average for the sole k-th shell. It shows an extra, fluctuation term

$$- (A-B)^k \left[\alpha - \frac{B}{1+B-A}\right] ,$$

which depends on k and on the initial distribution.

7.3.1 *The Gibbs–Di Marzio Law*

Let us consider again the more general case of Section 7.2.1. The average coordination number has the general expression

$$<r> = m_1 + (m_2 - m_1)\, c \,. \tag{7.15}$$

We find (7.14) for the $(m_1, m_2) = (2,3)$ case, and $<r> = 4 - c$ when $(m_1, m_2) = (4,3)$. For the *Se–Ge* case, the result is $<r> = 2 + 2c$. If the rigidity threshold is $<r> = 12/5$, we might expect something special to happen at concentrations slightly above $c \approx 0.2$. And something happens indeed. As Feng, Bresser and Boolchand have found [27], there is a discontinuity in the Raman spectrum near that point. This point is interpreted as a stiffness threshold, beyond which rigidity is firmly established.

Let us now consider the behavior of $\left(\frac{\partial T_g}{\partial c_2}\right)$ at very low modifier concentrations. We take derivatives in Eq.(7.10), as we have done for Eq.(5.5). When $c_2 \to 0$,

$$\left(\frac{\partial T_g}{\partial c_2}\right)_{c_2 \to 0} = \frac{T_{g0}}{\ln(\frac{m_2}{m_1})} \,. \tag{7.16}$$

This formula explains a remarkable experimental fact. The glass temperature increases with the modifier concentration when the latter has coordination number larger than that of the glass-former: $m_2 > m_1$. This is the case of the *Se* glasses. And the glass temperature decreases with the modifier concentration in the opposite case, as happens with common glass. But we can go further. As long as a first Taylor approximation is acceptable, the transition temperature will depend on the modifier concentration according to

$$T_g(c) \simeq T_g(0) + \frac{T_{g0}}{\ln(\frac{m_2}{m_1})}\, c = T_{g0}\left[1 + \frac{c}{\ln(\frac{m_2}{m_1})}\right] \simeq \frac{T_{g0}}{1 - \frac{c}{\ln(\frac{m_2}{m_1})}} \,.$$

Use of (7.15) leads then to

$$T_g(c) \simeq \frac{T_{g0}}{1 - \frac{<r> - m_1}{(m_2 - m_1)\ln(\frac{m_2}{m_1})}} \,, \tag{7.17}$$

which is the Gibbs–Di Marzio law (4.1) with the parameter β fixed as

$$\beta = \frac{1}{(m_2 - m_1)\ln(m_2 - m_1)} . \tag{7.18}$$

This would give $\beta = 0.40$ for *Se–As* and $\beta = 0.72$ for *Se–Ge*. There are experimental measurements for the last value, standing in the range 0.65–0.74.

We see that the Kerner model is very successful in explaining the Gibbs–Di Marzio law. First, it gives the value and origin of the factor $< r > - 2$, with the number "2" predicted to hold only for *Se*-like glass formers. Second, it shows why the temperature increases or decreases with the addition of modifier, according to the different former × modifier relative valences. Finally, it explains the meaning and fixes the value of the parameter β.

Commentary 73.1. ▶ **Random walks** A general random walk [30] is a Markov chain associated to a stochastic matrix **M** which is also a circulant matrix. Circulant matrices are our next topic. It will be more convenient to postpone the subject (see Section 10.2.1). ◀

PART 2

CIRCULANT MATRICES

Chapter 8

Prelude

Making a Fourier transformation is equivalent to diagonalizing a circulant matrix, and an inverse Fourier transformation takes a diagonal matrix into a circulant matrix. In this way, circulant matrices — and their diagonalized counterparts — underpin elementary harmonic analysis. Though the formalism may attain its greatest interest in the continuum case, we shall content ourselves with the discrete finite case, in which the central roles played by certain special matrices are easier to ascertain. There is a direct relationship between those matrices and the Weyl–Heisenberg groups. Consequently, in our policy of always using some Physics steersman, Quantum Mechanics on phase space will be the main character on the scene. In the discrete case all the most important properties, with the exception of those involving topological and measure-theoretical questions, turn up in a very simple, algebraic way. Everything can be deduced from the properties of a single matrix, the shift operator.

Fourier transformations take a function space into another, but they cannot be introduced on arbitrary function spaces. They require some kind of homogeneity on the underlying space. Even in the simplest cases, they can only act on functions defined on Lie groups and their homogeneous spaces. Furthermore, the target spaces of Fourier transformations are not of that same kind in the general case, spaces of functions on Lie groups or homogeneous spaces. They are spaces of functions defined on larger structures — quantum groups, or better, Hopf algebras. And inverse Fourier transformations take functions on these Hopf algebras into functions on other Hopf algebras, which include the original groups. Two spaces whose functions are related by a Fourier transformation and its inverse are said to be "Fourier-dual" to each other. Ultimately, this duality is a relationship between Hopf algebras.

Things are simpler when the original group is abelian. The Fourier-dual of an abelian group is another abelian group. Thus, the circle and the additive group of integer numbers are dual to each other. The real line $\mathbb{R}$ is self-dual, and so are all the cyclic groups Z_N, which will be our main subject (discrete groups can be made into Lie groups by using the discrete topology).

The simplest non-trivial examples of quantum groups are provided by matrices of matrices, that is, matrices whose entries are other matrices. From the geometrical point of view, the entries of an $N \times N$ real matrix $\mathbf{M}$ are the coordinates of $\mathbf{M}$ in the N^2-dimensional space $\mathbf{M}$ belongs to. Noncommutative entries mean noncommutative coordinates, a signal of noncommutative geometry. In consequence, spaces formed by hypermatrices have noncommutative geometries.

The relation of matrices to Fourier transformations is obscured by our inclination to look at them as sets of rows and columns. This propensity is clear in the usual algorithm to obtain matrix products. It is actually only a convenient convention, and comes from the preferred use of a special ("canonical") basis in matrix space. Other bases, in particular that formed by Weyl's operators, give the prominent role to the diagonals. Matrix product acquires a new aspect and many a Fourier-related property come to the fore.

We shall, accordingly, look at matrices "sideways" in this Part. This oblique view will bring us nearer to Quantum Mechanics, whose noncommutativity reduces to a simple phase factor. We shall see how a matrix space can be made into a differentiable manifold, and how a natural symplectic (phase space) setup turns up. This quantum phase space tends, in the due limit, to the classical phase space, and is at the origin of its Hamiltonian structure. Finally, we shall see that the Yang–Baxter braid equation reflects the preservation of associativity in a space of hypermatrices and has non-trivial solutions in quantum phase space.

Chapter 9

Definition and main properties

An $N \times N$ matrix is a circulant if it has the "cyclic" form

$$\mathbf{A} = \begin{pmatrix} a_N & a_{N-1} & a_{N-2} & \dots & a_1 \\ a_1 & a_N & a_{N-1} & \dots & a_2 \\ a_2 & a_1 & a_N & \dots & a_3 \\ \cdot & \cdot & \cdot & \dots & \cdot \\ \cdot & \cdot & \cdot & \dots & \cdot \\ a_{N-1} & a_{N-2} & a_{N-3} & \dots & a_N \end{pmatrix}, \tag{9.1}$$

where the a_k's are complex numbers. All the entries in a given diagonal, principal or secondary, are the same. The chosen label ordering is only a suitable convention, but it is extremely convenient to impose a cyclic notation on the numbers a_k: we shall actually make the identifications

$$a_N = a_0 \quad \text{and} \quad a_{N+r} = a_r = a_{r-N} \; .$$

With this notation, the matrix entries can be written as

$$A_{ij} = a_{i-j} \tag{9.2}$$

and exhibit the "circulating" property: for any integer r,

$$A_{i+r,j+r} = A_{ij} \; . \tag{9.3}$$

Each matrix element is invariant under a common shift in both indices. A Fourier transformation always supposes some homogeneity, an invariance under some kind of translation. The above property is just that, an invariance under cyclic translations.

We shall indicate the set of eigenvalues of $\mathbf{A}$ by $\{\lambda_k\}$. What is more frequently called "circulant" in the literature[1] is the determinant

[1]See [34], which sends to an exercise in [35]. A brief mention is made in [30], which call them "cyclic determinants". The inversion of circulant matrices has been object of some interest in the fifties. As presented here, the subject is well-known in the Signal Analysis community, see [36]. The author has notice of the more recent book by Davis [37], but has been unable to consult it.

$$\begin{vmatrix} a_N & a_{N-1} & a_{N-2} & \cdots & a_1 \\ a_1 & a_N & a_{N-1} & \cdots & a_2 \\ a_2 & a_1 & a_N & \cdots & a_3 \\ \cdot & \cdot & \cdot & \cdots & \cdot \\ \cdot & \cdot & \cdot & \cdots & \cdot \\ a_{N-1} & a_{N-2} & a_{N-3} & \cdots & a_N \end{vmatrix} = \prod_{k=1}^{N} \lambda_k \ . \tag{9.4}$$

We shall use the word for the matrix and call it a circulant matrix, or simply a circulant.

The identity matrix is a circulant, with $a_N = 1$ and $a_{k \neq N} = 0$. If $\mathbf{A}$ is a circulant, then the transpose $\mathbf{A}^T$ and the hermitian conjugate $\mathbf{A}^\dagger$ are also circulants. And circulant matrices are always normal, as

$$(\mathbf{A}\mathbf{A}^\dagger)_{ij} = \sum_{k=1}^{N} a_k^* a_{k+i-j} = \sum_{n=1}^{N} a_{n-i}^* a_{n-j} = (\mathbf{A}^\dagger\mathbf{A})_{ij} \ .$$

It is easily verified that the addition and the product of two circulants are circulants, and that the product-algebra of circulant $N \times N$ matrices is abelian. Consequently, powers and sums of powers of a circulant matrix are circulants. The inverse of an $N \times N$ matrix $\mathbf{A}$, when it exists, can be written in the form of a polynomial of order $(N-1)$ in $\mathbf{A}$ by the Cayley–Hamilton theorem [see Eq.(1.4)]. Thus, the inverse of a circulant, when existent, is also a circulant. It follows that the invertible circulant matrices (those with all $\lambda_k \neq 0$) constitute a group under multiplication.

The foremost properties of circulants come from the fact that their eigenvalues are given by a Z_N–Fourier transform of the entries a_j:

$$\lambda_k = \sum_{j=1}^{N} a_j \ e^{i(2\pi/N)kj} \ . \tag{9.5}$$

This is seen by applying (9.1) on the (conveniently normalized) eigenvectors, which are the column-vectors of components $u_{(k)s} = N^{-1/2} \ e^{-i(2\pi/N)ks}$. Expression (9.5) for the eigenvalues represents indeed a Fourier transform $\lambda_k = [F(a)]_k$ on the cyclic group Z_N. The cyclic convention has actually defined each $a_j = a(j)$ as a function on Z_N (or as a function on the space of inequivalent unitary irreducible representations of Z_N, which is the same). The inverse $[F^{-1}(\lambda)]_r$ will be given by

$$\frac{1}{N}\sum_{k=1}^{N} \lambda_k e^{-i(2\pi/N)kr} = \frac{1}{N}\sum_{s=1}^{N} a_s \sum_{k=1}^{N} e^{-i(2\pi/N)kr} e^{i(2\pi/N)ks}$$

$$= \sum_{s=1}^{N} a_s \ \frac{1}{N}\sum_{k=1}^{N} e^{i(2\pi/N)k(s-r)} = a_r \ ,$$

where use has been made of the Kronecker delta representation

$$\delta^{rs} = \tfrac{1}{N}\sum_{k=1}^{N} e^{i(2\pi/N)k(r-s)} \tag{9.6}$$

(for which a demonstration is given below, Comment 269.4). We have, consequently, just a Fourier transform and its inverse:

$$\lambda_k = \sum_{j=1}^{N} e^{i(2\pi/N)kj} a_j \ ; \tag{9.7}$$

$$a_j = \tfrac{1}{N}\sum_{k=1}^{N} e^{-i(2\pi/N)kj} \lambda_k \ . \tag{9.8}$$

The Parseval identity $\frac{1}{N}\sum_{k=1}^{N}|\lambda_k|^2 = \sum_{j=1}^{N}|a_j|^2$ follows without difficulty. Z_N-Fourier transforms have recently received some attention in the area of quantum computation [38], but they are of permanent interest in other fields, such as coding theory, modular representations and cyclic Galois extensions. Notice the particular values

$$\lambda_0 = \lambda_N = \sum_{j=1}^{N} a_j \tag{9.9}$$

and

$$a_0 = a_N = \tfrac{1}{N}\sum_{k=1}^{N} \lambda_k = \tfrac{1}{N}\mathrm{tr}\mathbf{A} \ . \tag{9.10}$$

Let us recall that a usual Fourier transform takes a function (here, a: $j \mapsto a(j) = a_j$) on a group into another function (here λ: $k \mapsto \lambda_k$) on another group. This property — the Pontryagin duality — holds for abelian groups, which is our case. The circulant matrix is fixed by the alphabet $\mathbf{a}$ $= (a_1, a_2, a_3, \cdots, a_N)$. The eigenvalues constitute another alphabet, $\boldsymbol{\lambda} = (\lambda_1, \lambda_2, \lambda_3, \cdots, \lambda_N)$. The Fourier transformations take these alphabets one into the other, reflecting the fact that the cyclic groups Z_N are Pontryagin-self-dual (more about that will be said in Section 13.4).

We shall sometimes use the notation $\tilde{a}_k$ for the Fourier transform. For a fixed N, it is convenient to call $\omega = e^{i(2\pi/N)}$ the N-th primitive root of 1, so that

$$\lambda_k = \tilde{a}_k = \sum_{j=1}^{N} a_j\, \omega^{kj} \ ; \tag{9.11}$$

$$a_j = \frac{1}{N} \sum_{k=1}^{N} \lambda_k \, \omega^{-kj} \ . \tag{9.12}$$

The eigenvectors

$$u_{(k)s} = \frac{\omega^{-ks}}{\sqrt{N}} \tag{9.13}$$

are orthogonal by the scalar product given by $\langle u_{(k)}, u_{(j)} \rangle = \sum_{r=1}^{N} u^*_{(k)r} u_{(j)r} = \delta_{kj}$. It is remarkable that they are independent of the entries a_j and are, consequently, the same for all circulants. $\mathbf{A}$ is diagonalized to $\mathbf{A}_{diag} = \mathbf{M}^{-1}\mathbf{AM}$ by the unitary matrix $\mathbf{M}$ with the entries

$$M_{rs} = \frac{\omega^{-rs}}{\sqrt{N}} \ . \tag{9.14}$$

This leads (see Section 1.4) to self-adjoint eigenprojectors $\mathbf{P}_{(k)}$ such that

$$\mathbf{A} = \sum_{k=1}^{N} \lambda_k \, \mathbf{P}_{(k)} \tag{9.15}$$

and with entries which are also independent of $\mathbf{A}$:

$$\left[P_{(k)}\right]_{rs} = u_{(k)r} u^*_{(k)s} = M_{rk} \, M^{-1}_{ks} = \tfrac{1}{N} \, \omega^{-k(r-s)} \ . \tag{9.16}$$

Such projectors have the properties

$$\mathbf{P}^{\dagger}_{(k)} = \mathbf{P}_{(k)}, \ tr \ \mathbf{P}_{(k)} = 1, \ \mathbf{P}_{(k)}\mathbf{P}_{(j)} = \delta_{kj}\mathbf{P}_{(k)}, \sum_k \mathbf{P}_{(k)} = \mathbf{I} \ . \tag{9.17}$$

Clearly, their entries satisfy

$$(P_{(k)})_{r+j,s+j} = (P_{(k)})_{rs} \ ,$$

so that each $\mathbf{P}_{(k)}$ is also a circulant. A circulant matrix can be expanded in a projector base whose members are themselves circulants. The projector $\mathbf{P}_{(k)}$ will have, in our cyclic notation, entries $p_j = \frac{1}{N} \, \omega^{-kj}$, and eigenvalues $\lambda_j[\mathbf{P}_{(k)}] = \delta_{jk}$.

Commentary 82.1. ▶ An exercise: the transfer matrix of 1-dimensional Ising model of Section 6.3.1 reduces to a circulant in the absence of an external magnetic field:

$$\mathbf{W} = \begin{pmatrix} e^K & e^{-K} \\ e^{-K} & e^K \end{pmatrix}.$$

The eigenvalues are then easily calculated:

$$\lambda_k = a_1 e^{i\pi k} + a_2 e^{i2\pi k} = a_2 + (-)^k a_1 \ .$$

As $a_2 = e^K$ and $a_1 = e^{-K}$,

$$\lambda_1 = a_2 - a_1 = e^K - e^{-K} = 2\sinh K \ ; \lambda_2 = a_2 + a_1 = e^K + e^{-K} = 2\cosh K$$

(different conventions lead an exchange $\lambda_1 \Leftrightarrow \lambda_2$ with respect to the notation of Section 6.3.1). ◀

Commentary 82.2. ▶ Every normal matrix is similar to a diagonal matrix and, consequently, to a circulant. If $\mathbf{B}$ is normal and diagonalized by some unitary matrix $\mathbf{R}$, so that $\mathbf{B}_{diag} = \mathbf{R}^{-1}\mathbf{B}\mathbf{R}$, then $\mathbf{B}$ is similar to the circulant $\mathbf{A} = (\mathbf{M}\mathbf{R}^{-1})\mathbf{B}(\mathbf{M}\mathbf{R}^{-1})^{-1}$. The similarity transformation matrix has the entries $(\mathbf{M}\mathbf{R}^{-1})_{rs} = N^{-1/2}\sum_j \omega^{-jr} R^*_{sj}$. ◀

Let us now look for the expression of the convolution. Always with the cyclic notation in mind, we define the first convolution as

$$(a^{*2})_r = (a * a)_r = \sum_{s=1}^{N} a_s\, a_{r-s} \;.$$

Convolution is, of course, "global" on the alphabet $\mathbf{a}$: it involves the whole of it. The third order is given by

$$(a^{*3})_r = (a * a * a)_r = \sum_{k=1}^{N} (a^{*2})_{r-k}\, a_k \;.$$

The higher orders are easily obtained. With the notation $a^{*m} = (a{*}a{*}\ldots{*}a)$ for the convolution taken $(m-1)$ times [and $(a^{*1})_r = a_r$], we find

$$\lambda_k^m = \sum_{j=1}^{N} \omega^{kj} a_j^{*m} \;; \tag{9.18}$$

$$a_j^{*m} = \tfrac{1}{N} \sum_{k=1}^{N} \omega^{-kj} \lambda_k^m \;. \tag{9.19}$$

Notice the particular cases $\lambda_N^m = \sum_{j=1}^{N} a_j^{*m}$; $N(a^{*m})_N = \sum_{k=1}^{N} \lambda_k^m = tr\mathbf{A}^m$. Idempotents correspond to convolution-invariants: $(a^{*2})_r = a_r$ would imply $\mathbf{A}^2 = \mathbf{A}$. We can easily check that the powers of $\mathbf{A}$ have entries $(\mathbf{A}^m)_{rs} = (a^{*m})_{r-s}$, a manifestation of the fact that the power $\mathbf{A}^m$ is a circulant. With the converse convolution

$$(\lambda^{*2})_r = \tfrac{1}{N} \sum_{k=1}^{N} \lambda_k \lambda_{r-k} = \sum_{j=1}^{N} a_j^2\, \omega^{rj} \;,$$

we find

$$(\lambda^{*n})_r = \sum_{j=1}^{N} a_j^n\, \omega^{rj}$$

and

$$a_j^m = \tfrac{1}{N} \sum_{k=1}^{N} (\lambda^{*m})_k\, \omega^{-kj} \;.$$

We are here specially interested in powers of one same object, but all definitions and consequences above can be generalized without much ado. For example, the Fourier transform of a product is a convolution of the Fourier transforms, $\sum_{j=1}^{N} a_j b_j {\omega_N}^{sj} = \sum_{k=1}^{N} \tilde{a}_k \tilde{b}_{s-k} = (\tilde{a} * \tilde{b})_s$.

An outstanding basis for the circulant matrices can be obtained as follows. Consider the circulant corresponding to the alphabet $a_j = \delta_{1j}$. It will be a unitary matrix $\mathbf{U}$ with entries $U_{rs} = \delta_{r,s+1}$:

$$\mathbf{U} = \begin{pmatrix} 0 & 0 & 0 & \cdots & \cdots & \cdots & \cdots & \cdots & 1 \\ 1 & 0 & 0 & \cdots & \cdots & \cdots & \cdots & \cdots & 0 \\ 0 & 1 & 0 & \cdots & \cdots & \cdots & \cdots & \cdots & 0 \\ 0 & 0 & 1 & 0 & \cdots & \cdots & \cdots & \cdots & 0 \\ \cdots & \cdots & \cdots & \cdots & \cdots & \cdots & \cdots & \cdots & \cdots \\ 0 & 0 & 0 & \cdots & \cdots & \cdots & \cdots & 1 & 0 \end{pmatrix}. \tag{9.20}$$

$\mathbf{U}$ is a displacement operator, or a shift operator: on any matrix $\mathbf{A}$, $(\mathbf{UA})_{ij} = A_{i-1,j}$. It is a good thing to have a visual idea of its effect: left-multiplication by $\mathbf{U}$ lowers every row of a matrix $\mathbf{A}$ one step down, but cyclically: 1st row becomes 2nd, 2nd becomes 3rd, etc., and the last row becomes the first. If $\mathbf{A}$ is a circulant, this corresponds to making, in each entry, the substitution $a_k \to a_{k-1}$. The same cyclic lowering happens when $\mathbf{U}$ multiplies a column-vector. Right-multiplication by $\mathbf{U}$ displaces every column of a matrix $\mathbf{A}$ to the left, also cyclically: 2nd column becomes first, first becomes last.

The set of powers $\mathbf{U}$, $\mathbf{U}^2$, $\mathbf{U}^3$, ..., $\mathbf{U}^{N-1}$ and $\mathbf{U}^N = \mathbf{I}$ constitutes a circulant basis. In effect, the general expression (9.1) is simply a polynomial in $\mathbf{U}$:

$$\mathbf{A} = a_1\mathbf{U}^1 + a_2\mathbf{U}^2 + a_3\mathbf{U}^3 + \ldots + a_N\mathbf{U}^N = \sum_{n=1}^{N} a_n \mathbf{U}^n . \tag{9.21}$$

The powers of $\mathbf{U}$ provide an $N \times N$ matrix representation of the group Z_N. The circulant matrices are the members of that group's ring. The numbers a_k are the components of $\mathbf{A}$ along the $\mathbf{U}^k$'s, and will be called "the letters of $\mathbf{A}$", or "the circulant components of $\mathbf{A}$".

Commentary 84.1. ► We see here more clearly why the stochastic matrix U_4 of Section 7.1 leads to cyclic behavior. When applied to any initial distribution, its powers simply realize a representation of the cyclic group Z_4. ◄

Note. The eigenvalues of the circulant

$$\mathbf{A} = \sum_{n=1}^{N} n\, U^n = \begin{pmatrix} N & N-1 & N-2 & \ldots & 1 \\ 1 & N & N-1 & \ldots & 2 \\ 2 & 1 & N & \ldots & 3 \\ \cdot & \cdot & \cdot & \cdots & \cdot \\ N-1 & N-2 & N-3 & \ldots & N \end{pmatrix}$$

will be the Fourier transforms $\lambda_k = \sum_{n=1}^{N} n e^{-i\frac{2\pi}{N}kn}$. If $k = N$, $\lambda_N =$ sum of the first N integers $= \frac{N(N+1)}{2}$. The finite series $\sum_{n=1}^{N} z^n = z\frac{1-z^N}{1-z}$ provided $z \neq 1$. When $k \neq N$, $z = e^{-i\frac{2\pi}{N}k} \neq 1$ though $z^N = 1$. The sum can be calculated as $\sum_{n=1}^{N} nz^n = z\frac{\partial}{\partial z}\sum_{n=1}^{N} z^n$, which is $= \frac{z}{z-1}$ for $z^N = 1$. The result is $\lambda_k = N\frac{e^{-i\frac{\pi}{N}k}}{2\sin\frac{\pi}{N}k}$ for $k \neq N$. In an analogous way, the eigenvalues for $\mathbf{A} = \sum_{n=1}^{N} n^2 U^n$ can be obtained by applying twice $z\frac{\partial}{\partial z}$ to $\sum_{n=1}^{N} z^n$, except for $\lambda_N =$ sum of the first N squared integers $= \frac{N(N+1)(2N+1)}{6}$. See subsection 10.4.1.

$\mathbf{U}$ is unitary, $\mathbf{U}^\dagger = \mathbf{U}^T = \mathbf{U}^{-1}$. Laplace expansion along the first row gives trivially det $\mathbf{U} = (-)^{N-1}$. We have evidently $(\mathbf{U}^n)_{rs} = \delta_{r,s+n}$. That a circulant matrix $\mathbf{A}$ be diagonal in the basis in which $\mathbf{U}$ is diagonal is a matter of course. $\mathbf{U}$ has eigenvalues ω^k, and $(\mathbf{U}^j_{diag})_{rs} = \delta_{r,s}\ \omega^{jr}$. The eigenvectors (9.13) are actually the eigenvectors of $\mathbf{U}$, which explains why they are the same for all circulants.

As the projectors are themselves circulants, they should be of the form (9.21). They are, to be sure, given by

$$\mathbf{P}_{(k)} = \tfrac{1}{N}\sum_{n=1}^{N} \omega^{-kn}\mathbf{U}^n\ , \tag{9.22}$$

an expression whose inverse is

$$\mathbf{U}^n = \sum_{k=1}^{N} \omega^{kn}\mathbf{P}_{(k)}\ . \tag{9.23}$$

The entries are circulant-typical, $\mathbf{P}_{(k)ab} = \frac{\omega^{k(b-a)}}{N}$. The two last equations are related to (9.11) and (9.12), and could be seen as operator Fourier transforms. The a_j's are the components of a given matrix $\mathbf{A}$ in basis $\{\mathbf{U}^j\}$ and the eigenvalues are its components in basis $\{\mathbf{P}_{(k)}\}$. By the Fourier transform $a_j = N^{-1}\sum_{k=1}^{N}\lambda_k\ \omega^{-kj} = N^{-1}tr(\mathbf{A}\mathbf{U}^{N-j})$, the function a on Z_N is expanded in the basis of orthogonal functions $\{N^{-1}\omega^{-kj}\}$. But, as $A_{r+j,r} = a_j = \sum_{k=1}^{N}\lambda_k\ P_{(k)r+j,r}$, with $[P_{(k)}]_{r+j,r} = N^{-1}\omega^{-kj}$, the Fourier basis is actually given by the projector matrix elements: $\{N^{-1}\omega^{-kj}\} = \{[P_{(k)}]_{r+j,r}\}$. The Fourier inverse is then $\lambda_k = \sum_{j=1}^{N} a_j\omega^{kj} = tr(\mathbf{A}\mathbf{P}_{(k)})$. Let us retain two of these expressions: the Fourier transforms are actually related to the two bases $\{\mathbf{U}^j\}$ and $\{\mathbf{P}_{(k)}\}$ according to

$$a_j = N^{-1} tr(\mathbf{A}\ \mathbf{U}^{N-j})\ ; \tag{9.24}$$

$$\lambda_k = tr(\mathbf{A}\ \mathbf{P}_{(k)})\ . \tag{9.25}$$

Given two circulants **A** and **B**, the components of their product in the **U**-basis are convolutions,

$$\mathbf{AB} = \sum_{n=1}^{N} (a * b)_n \mathbf{U}^n\ , \tag{9.26}$$

with

$$(a * b)_n = \sum_{m=1}^{N} a_m b_{n-m}\ . \tag{9.27}$$

In particular, as mentioned previously, $\mathbf{A}^m = \sum_{n=1}^{N} (a^{*m})_n \mathbf{U}^n$.

If we define a function **f** of a circulant matrix **A** by Eq.(1.11), which here reads $\mathbf{f}(\mathbf{A}) = \sum_k \mathrm{f}\,(\lambda_k)\mathbf{P}_{(k)}$, we find

$$\mathbf{f}(\mathbf{A}) = \tfrac{1}{N} \sum_n \left[\sum_k f(\lambda_k)\ \omega^{-kn} \right] \mathbf{U}^n\ . \tag{9.28}$$

We arrive thus at another remarkable result: any function of a circulant is also a circulant, with alphabet given by $f_n = \frac{1}{N} \sum_k f(\lambda_k)\ \omega^{-kn}$. And, clearly, functions of **A** are functions of **U** alone, so that the projectors of **U** are the same for them all. A special case is $\mathbf{f}(\mathbf{U}) = \frac{1}{N} \sum_n [\sum_k f(\omega^k)\ \omega^{-kn}]$ $\mathbf{U}^n$. Other formulas of interest are

$$e^{i\mathbf{A}} = \tfrac{1}{N} \sum_{n=1}^{N} \left[\sum_{k=1}^{N} e^{i\lambda_k}\ \omega^{-kn} \right] \mathbf{U}^n\ ; \tag{9.29}$$

$$(\ln[\mathbf{I} + \mathbf{A}])_{rs} = \sum_{k=1}^{\infty} \frac{(-)^{k-1}}{k}\ (a^{*k})_{r-s}\ ; \tag{9.30}$$

$$\ln \mathbf{U} = \tfrac{1}{N} \tfrac{2\pi i}{N} \sum_{n=1}^{N} \left[\sum_{k=1}^{N} k\ \omega^{-kn} \right] \mathbf{U}^n\ ; \tag{9.31}$$

$$(\ln \mathbf{U})_{rs} = \tfrac{1}{N} \tfrac{2\pi i}{N} \sum_{k=1}^{N} k\ \omega^{-k(r-s)}\ . \tag{9.32}$$

The inverse of a circulant **A**, if it exists, will have the alphabet of letters

$$a_j^{<-1>} = \tfrac{1}{N} \sum_{k=1}^{N} \lambda_k^{-1} \omega^{-jk} \tag{9.33}$$

as components in the base $\{\mathbf{U}^j\}$, and the entries

$$(A^{-1})_{rs} = \sum_{k=1}^{N} \lambda_k^{-1} \; P_{(k)rs} = \tfrac{1}{N} \sum_{k=1}^{N} \frac{\omega^{-k(r-s)}}{\sum_j^N a_j \; \omega^{kj}} = (A^{-1})_{r+j,s+j} \; . \qquad (9.34)$$

Note. Notice

$$\sum_{j=1}^{N} a_j^{<-1>} = \sum_{k=1}^{N} \lambda_k^{-1} \tfrac{1}{N} \sum_{j=1}^{N} \omega^{-jk} = \lambda_N^{-1} = \frac{1}{\sum_{j=1}^{N} a_j} \; .$$

It is immediate to verify that

$$\mathbf{U}_{diag} \; \mathbf{U} \; = \omega \; \mathbf{U} \; \mathbf{U}_{diag} \; . \qquad (9.35)$$

We shall denote $\mathbf{U}_{diag}$ by $\mathbf{V}$. Thus,

$$\mathbf{V} = \begin{pmatrix} \omega & 0 & 0 & 0 & \cdots & \cdots & \cdots & \cdots & 0 \\ 0 & \omega^2 & 0 & \cdots & \cdots & \cdots & \cdots & \cdots & 0 \\ 0 & 0 & \omega^3 & 0 & \cdots & \cdots & \cdots & \cdots & 0 \\ 0 & 0 & 0 & \omega^4 & 0 & \cdots & \cdots & \cdots & 0 \\ \cdots & \cdots & \cdots & \cdots & \cdots & \cdots & \cdots & \cdots & \cdots \\ 0 & 0 & 0 & 0 & \cdots & \cdots & \cdots & 0 & \omega^N \end{pmatrix} . \qquad (9.36)$$

The eigenvector $v_{(k)}$ of $\mathbf{V}$ corresponding to the eigenvalue ω^k has components $v_{(k)s} = \delta_{ks}$. As $V_{rs} = \delta_{rs} \; \omega^r$, it follows that $\mathbf{V}^n{}_{rs} = \delta_{rs} \; \omega^{rn}$ and $\mathbf{V}^N = I$. We shall rewrite (9.35) as

$$\mathbf{V} \; \mathbf{U} = \omega \; \mathbf{U} \; \mathbf{V} \; . \qquad (9.37)$$

It follows immediately that

$$\mathbf{V}^m \; \mathbf{U}^n = \omega^{nm} \; \mathbf{U}^n \; \mathbf{V}^m \; . \qquad (9.38)$$

Summing it all up, circulants are matrices underlying elementary Fourier analysis. Those which are invertible constitute a group. They have projectors which are also circulants, and every well-defined function of a circulant is another circulant. The powers of a circulant $\mathbf{A}$ have for entries convolutions of the entries of $\mathbf{A}$. There is a basic matrix $\mathbf{U}$, which is the cornerstone of all the theory and is a shift operator. Every circulant is a sum of weighted shifts.

We have seen that the matrices $\mathbf{A}$ and $\mathbf{A}_{diag}$ contain just the same amount of information, and that performing a Fourier transform is equivalent to diagonalizing a circulant matrix. A question coming naturally to the mind is whether such circulant matrices have some deeper role in more general Fourier analysis.[2] In particular, whether such matrices also stay in the background of Fourier analysis on the line. The answer, which is *yes*, is suggested by Weyl's approach to Quantum Mechanics on phase space [51], in which the above pair of matrices **U**, **V** plays a crucial role. Equation (9.38) is fundamental for (discrete) Quantum Mechanics, as it gives a representation of the discrete Heisenberg group. The passage into the continuum case, at least in the simple **q**, **p** case, is well-known. We shall leave these topics to Chapter 13, devoted to discrete Quantum Mechanics.

Commentary 88.0. ▶ **Combing** Compare the exponential and the cosh series,

$$e^x = \sum_{n=0}^{\infty} \frac{x^n}{n!} \qquad \text{and} \qquad \cosh(x) = \sum_{p=0}^{\infty} \frac{x^{2p}}{(2p)!} \ .$$

The second comes from the first by simply "picking up" in the summation only those terms whose indices n are even numbers, or multiples of 2. We shall say in that case that the exponential series has been "multisectionned", or "combed" by 2, or still "2-combed", and use the notation

$$\text{Comb}_2[\exp](x) = \cosh(x) \ .$$

A comb is a practical device restricting a summation to a subset. For example, the cosh series can be obtained as

$$\tfrac{1}{2}\sum_{k=1}^{2} e^{x\omega_2} \ ,$$

with $\omega_2 = \exp[i2\pi/2] = -1$. The general "N-combed exponential" will be

$$\text{Comb}_N[\exp](x) = \tfrac{1}{N}\sum_{k=1}^{N} e^{x\omega_N} = \sum_{p=0}^{\infty} \frac{x^{pN}}{(pN)!}$$

with $\omega_N = \exp[i2\pi/N]$.

Notice that, in the infinite summation, there will appear one Kronecker delta at each interval of size N. Let us be more precise. Suppose an $N \times N$ circulant **M** of eigenvalues $\lambda_k = \frac{1}{1-a\omega^k}$, with $\omega = \exp[i2\pi/N]$. The circulant component $a_N = \frac{1}{N}$ tr **M** will be

$$\tfrac{1}{N}\sum_{k=1}^{N} \frac{1}{1-a\omega^k} = \tfrac{1}{N}\sum_{k=1}^{N}\sum_{n=0}^{\infty} a^n \omega^{kn} \ .$$

A naïve exchange of the summations order would give the result $N^{-1}a^N$, which is wrong. Actually all those n's which are multiples of the integer N are to be retained in the infinite series, so that

$$\tfrac{1}{N}\sum_{k=1}^{N} \omega^{kn} = \delta_{kN} \Longrightarrow \sum_{p=0}^{\infty} \delta_{n,pN} \ .$$

In consequence,

$$\tfrac{1}{N}\sum_{k=1}^{N} \frac{1}{1-a\omega^k} = \sum_{n=0}^{\infty}\sum_{p=0}^{\infty} \delta_{n,pN} a^n = \sum_{p=0}^{\infty} a^{pN} = \frac{1}{1-a^N} \ ,$$

[2] We are here concerned with discrete abelian harmonic analysis. The general case, including discrete nonabelian groups, is beautifully treated in [39].

the correct result. It all happens as if the commutation rule for the summations were

$$\tfrac{1}{N}\sum_{k=1}^{N}\sum_{n=0}^{\infty} f_n \omega^{kn} = \sum_{n=0}^{\infty}\sum_{p=0}^{\infty} \delta_{n,pN} f_n = \sum_{p=0}^{\infty} f_{pN} \;.$$

Conversely, take the geometric series $f(x) = \frac{1}{1-x} = \sum_{n=0}^{\infty} x^n$ and suppose the summation is to be confined to those n's which are multiples of the integer N. Take then $\omega = \exp[i2\pi/N]$, replace $x \to x\omega^k$, and take the "average":

$$\tfrac{1}{N}\sum_{k=1}^{N} f(x\omega^k) = \tfrac{1}{N}\sum_{k=1}^{N}\sum_{n=0}^{\infty} x^n \omega^{nk} = \sum_{n=0}^{\infty}\sum_{p=0}^{\infty} \delta_{n,pN} x^n = \frac{1}{1-x^N} \;,$$

the "N-combed geometric series". We see, by the way, that combs can be useful in obtaining finite summations. For an example in the continuum case, see Comment 233.1. Applications and developments will turn up in Comments 112.1 and 233.2. ◀

9.1 Bases

The set of all complex $N \times N$ matrices constitutes a vector space, of which each matrix is a "vector". The N^2 matrices $\{\mathbf{U}^m\mathbf{V}^n\}$ are linearly independent and constitute a general vector basis for that space.

The common practice of describing an $N\times N$ matrix $\mathbf{B}$ whose entries (or "matrix elements") are B_{ab} by the very set of these elements corresponds to using a preferred matrix basis $\{\mathbf{\Delta}_{ab}\}$, in which every matrix has for components just its entries. This means that $\mathbf{B} = \sum_{a,b} B_{ab}\mathbf{\Delta}_{ab}$. The matrices $\mathbf{\Delta}_{ab}$ have entries $(\Delta_{ab})_{ij} = \delta_{ai}\delta_{bj}$ and constitute the so-called complete "canonical" basis for the algebra of $N \times N$ matrices, with the simplest of multiplication tables, $\mathbf{\Delta}_{ab}\mathbf{\Delta}_{cd} = \delta_{bc}\mathbf{\Delta}_{ad}$. The matrices $\mathbf{U}$ and $\mathbf{V}$, for example, are

$$\mathbf{U}^j = \sum_{i=1}^{N} \mathbf{\Delta}_{i,N+i-j}; \quad \mathbf{V}^k = \sum_{r=1}^{N} \omega^{rk}\, \mathbf{\Delta}_{rr} \;.$$

The transformations

$$\mathbf{\Delta}_{ab} = \tfrac{1}{N}\sum_{n=1}^{N} \omega^{-nb}\, \mathbf{U}^{a-b}\, \mathbf{V}^n \;,\quad \mathbf{U}^i\mathbf{V}^j = \sum_{s=1}^{N} \omega^{js}\, \mathbf{\Delta}_{s+i,s} \tag{9.39}$$

have nonvanishing Jacobians, and consequently any matrix can be written also in the basis $\{\mathbf{U}^m\mathbf{V}^n\}$:

$$\mathbf{B} = \tfrac{1}{N}\sum_{a,b=1}^{N} B_{ab} \sum_{n=1}^{N} \omega^{-nb}\, \mathbf{U}^{a-b}\, \mathbf{V}^n \;. \tag{9.40}$$

We shall adopt the lowercase notation b_{ij} for the components of $\mathbf{B}$ along the members of the unitary basis $\mathbf{U}^i\mathbf{V}^j$:

$$\mathbf{B} = \sum_{i,j=1}^{N} b_{ij} \mathbf{U}^i \mathbf{V}^j \ . \tag{9.41}$$

These components can be calculated by

$$N\, b_{ij} = tr(\mathbf{V}^{-j}\, \mathbf{U}^{-i}\, \mathbf{B}) \ .$$

To compare the two bases, let us write a general matrix in both of them:

$$\mathbf{A} = \sum_{a,b=1}^{N} A_{ab}\, \mathbf{\Delta}_{ab} = \sum_{i,j=1}^{N} a_{ij}\, \mathbf{U}^i \mathbf{V}^j = \sum_{i,j,k=1}^{N} a_{ij}\, \omega^{jk}\, \mathbf{\Delta}_{k+i,k} \ . \tag{9.42}$$

Some relations involving the coefficients are:

$$tr(\mathbf{U}^{-i}\mathbf{A}) = \sum_{s=1}^{N} A_{s+i,s} \ ; \quad A_{rs} = \sum_{j=1}^{N} a_{r-s,j}\, \omega^{js} \ ;$$

$$a_{ij} = \tfrac{1}{N} \sum_{s=1}^{N} \omega^{-js} A_{s+i,s} \ . \tag{9.43}$$

The base $\{\mathbf{\Delta}_{ab}\}$ is responsible for our viewing matrices as formed by rows and columns. Basis $\{\mathbf{U}^m\mathbf{V}^n\}$ teaches us to look at them as formed by diagonal lines. This is the "oblique" view announced in the introduction to this Part. Notice that the summation in the middle expression of (9.42) takes place along the i-th (cyclic) diagonal. Fix i: in that case, (9.8) tells us that, for each fixed i, $a_j^{(i)} = a_{ij}$ will be the entries of a circulant with eigenvalues $\lambda_k^{(i)} = A_{k+i,k}$. The circulants correspond to the diagonals, as indicated by

$$\begin{pmatrix} \searrow & \cdot & \cdot & \cdot \\ \cdot & \searrow & \cdot & \cdot \\ \cdot & \cdot & \searrow & \cdot \\ \cdot & \cdot & \cdot & \searrow \end{pmatrix}, \begin{pmatrix} \cdot & \cdot & \cdot & \searrow \\ \searrow & \cdot & \cdot & \cdot \\ \cdot & \searrow & \cdot & \cdot \\ \cdot & \cdot & \searrow & \cdot \end{pmatrix}, \begin{pmatrix} \cdot & \cdot & \searrow & \cdot \\ \cdot & \cdot & \cdot & \searrow \\ \searrow & \cdot & \cdot & \cdot \\ \cdot & \searrow & \cdot & \cdot \end{pmatrix}.$$

Or still

$$\mathbf{A} = \begin{pmatrix} \lambda_1^{(N)} & \lambda_2^{(N-1)} & \lambda_3^{(N-2)} & \dots & \lambda_{N-1}^{(2)} & \lambda_N^{(1)} \\ \lambda_1^{(1)} & \lambda_2^{(N)} & \lambda_3^{(N-1)} & \dots & \dots & \lambda_N^{(2)} \\ \lambda_1^{(2)} & \lambda_2^{(1)} & \lambda_3^{(N)} & \dots & \dots & \lambda_N^{(3)} \\ \lambda_1^{(3)} & \lambda_2^{(2)} & \lambda_3^{(1)} & \dots & \dots & \lambda_N^{(4)} \\ \dots & \dots & \dots & \dots & \dots & \dots \\ \lambda_1^{(N-1)} & \lambda_2^{(N-2)} & \lambda_3^{(N-3)} & \dots & \dots & \lambda_N^{(N)} \end{pmatrix}.$$

A matrix **A** "contains" N circulants. The fact of interest is that such circulants have eigenvalues related to the components of **A** in the $\mathbf{U}^i\mathbf{V}^j$ basis. Of course, this would be more interesting if some relationship could be found with the very eigenvalues of **A**.

Note. The "oblique" view has some extra advantages. As the operator **U** lowers all diagonals in a cyclic way, matrix **UA** will have for main diagonal the first upper diagonal of **A**, matrix $\mathbf{U}^2\mathbf{A}$ will have for main diagonal the second upper diagonal of **A**, and so on. Of course, $\mathbf{U}^N\mathbf{A} = \mathbf{A}$. The trace tr **A** is the sum of the entries in the principal diagonal of **A**, tr **UA** will give the sum of the entries in the first upper diagonal, etc. Of course, $\mathbf{U}^N\mathbf{A} = \mathbf{A}$. It is sometimes necessary to calculate the sum of all entries in a matrix. This will be given by tr $\sum_{j=1}^{N}\mathbf{U}^j\mathbf{A}$. Recall the unit-entries matrix **H** and column vector **u** used at the end of Section 6.2. Clearly $\mathbf{H} = \sum_{j=1}^{N}\mathbf{U}^j$ is a "flat" circulant with all components $a_j = 1$. The sum of all entries of **A** can be calculated by any of the expressions

$$\text{Sum }\mathbf{A} = \mathbf{u}^T\mathbf{A}\mathbf{u} = \text{tr}\left(\sum_{j=1}^{N}\mathbf{U}^j\mathbf{A}\right) = \text{tr }(\mathbf{HA})\ .$$

9.2 Double Fourier transform

The relations (9.43) between coefficients in bases $\{\mathbf{\Delta}_{ab}\}$ and $\{\mathbf{U}^m\mathbf{V}^n\}$ have the form of Fourier transforms in one of the entries indices. This fact suggests looking into such transforms in both indices.

The relationship between circulants and diagonal matrices is actually a particular case of a more general "duality". An $N \times N$ matrix **A** can be seen as a two-variable function on its indices or, if we use the cyclic convention, on the discrete torus $Z_N \otimes Z_N$, **A**: $(a, b) \mapsto A_{ab}$. Take for the time being an arbitrary **A** and, always with the cyclic notation, consider the double Fourier transform

$$\tilde{A}_{rs} = \tfrac{1}{N}\sum_{i,j=1}^{N} A_{ij}\ \omega^{ir}\ \omega^{-js}\ . \tag{9.44}$$

Take also an eigenvector $\mathbf{v}^{(k)} = (v_j^{(k)})$ of **A**, corresponding to the eigenvalue $\lambda_{(k)}$ and its transform $\tilde{\mathbf{v}}^{(k)}$ of components

$$\tilde{v}_s^{(k)} = \tfrac{1}{N}\sum_{i=1}^{N} v_i^{(k)}\ \omega^{is}\ . \tag{9.45}$$

The inverses of the above transformations are

$$A_{ij} = \tfrac{1}{N} \sum_{r,s=1}^{N} \tilde{A}_{rs}\, \omega^{-ir} \omega^{js} \quad \text{and} \quad v_i^{(k)} = \sum_{r=1}^{N} \omega^{-ir} \tilde{v}_r^{(k)} \,.$$

We find then that

$$\sum_{s=1}^{N} \tilde{A}_{rs}\, \tilde{v}_s^{(k)} = \lambda_{(k)}\, \tilde{v}_r^{(k)} \,. \tag{9.46}$$

The transformed $\tilde{\mathbf{A}}$ has, therefore, the same eigenvalues as the matrix $\mathbf{A}$, though with Fourier-transformed eigenvectors. The transformation $\mathbf{A} \Rightarrow \tilde{\mathbf{A}}$ commutes with hermitian conjugation, $\tilde{\mathbf{A}}^\dagger = \widetilde{\mathbf{A}^\dagger}$. We remark also that (9.44) is $(\mathbf{M}^{-1}\mathbf{A}\mathbf{M})_{rs}$, with $\mathbf{M}$ given by (9.14). The diagonal entries are $\tilde{A}_{rr} = \text{tr}\,(\mathbf{A}\mathbf{P}_{(r)})$, where $\mathbf{P}_{(r)}$ is the r-th circulant projector (9.16).

If now $\mathbf{A}$ is a circulant, $A_{i,j} = a_{i-j}$ and

$$\tilde{A}_{rs} = \delta_{rs} \sum_j a_j \omega^{js} = \delta_{rs}\, \lambda_{(s)} \,.$$

Conversely, if $\tilde{A}_{rs} = \tilde{a}_{r-s}$, then

$$A_{ij} = \delta_{ij} \sum_s \tilde{a}_s\, \omega^{-js} \,.$$

If one matrix of the pair $\mathbf{A}$, $\tilde{\mathbf{A}}$ is a circulant, the other is diagonal and vice versa.

Note. The same can be seen by noticing that $\tilde{\mathbf{U}}^n = \mathbf{V}^n$, so that $\tilde{A} = \sum_s a_s \tilde{\mathbf{U}}^s = \sum_s a_s \mathbf{V}^s$. The members of the canonical basis $\{\mathbf{\Delta}_{ab}\}$ for $N \times N$ matrices introduced in page 89 will have Fourier transforms with entries

$$(\tilde{\Delta}_{ab})_{rs} = \tfrac{1}{N}\, \omega^{ar} \omega^{-bs} \,.$$

Note. We can introduce the symmetric "Fourier matrix" $\mathbf{F}$ with entries $F_{ab} = e^{i\frac{2\pi}{N}ab}$. It will have for inverse $\mathbf{F}^{-1}$, with $F^{-1}_{ab} = \frac{1}{N} e^{-i\frac{2\pi}{N}ab}$. The Fourier transform of a general matrix $\mathbf{A}$ will then be given by $\tilde{\mathbf{A}} = \mathbf{F}\mathbf{A}\mathbf{F}^{-1}$. In particular, matrices (9.20) and (9.36) are related by

$$\mathbf{V}^i = \mathbf{F}\mathbf{U}^i\mathbf{F}^{-1}, \quad \mathbf{U}^{-j} = \mathbf{F}\mathbf{V}^j\mathbf{F}^{-1} \,.$$

In terms of the base $\{\mathbf{U}^i\mathbf{V}^j\}$, $\mathbf{A} = \sum_{i,j=1}^N a_{ij}\ \mathbf{U}^i\mathbf{V}^j$ and $\tilde{\mathbf{A}} = \sum_{i,j=1}^N a_{ij}\ \mathbf{V}^i\mathbf{U}^{-j}$.

These matrices are related to the diagonalizing matrix (9.14) for circulants: $\sqrt{N} F^{-1}_{ab} = M_{ab}$. They are also such that any circulant can be written in the form $\mathbf{A} = \mathbf{F}^{-1}\mathbf{A}_{diag}\mathbf{F}$, or $\mathbf{A}_{diag} = \tilde{\mathbf{A}}$. In consequence, $\mathbf{A}_{diag} = \mathbf{F}\mathbf{A}\mathbf{F}^{-1} = \mathbf{F}\left(\sum_n a_n \mathbf{U}^n\right)\mathbf{F}^{-1} = \sum_n a_n \mathbf{V}^n$.

The change in the normalizations (by factors $1/\sqrt{N}$) entails the loss of unitarity, but is convenient when the circulants are also stochastic matrices, a case which is the subject of the next chapter.

Note. In base $\{\mathbf{U}^i\mathbf{V}^j\}$,

$$\mathbf{F} = \tfrac{1}{N}\sum_{i,j=1}^{N}\sum_{a=1}^{N} e^{i\frac{2\pi}{N}a^2} e^{i\frac{2\pi}{N}a(i-j)}\mathbf{U}^i\mathbf{V}^j$$

has for components the Fourier transform of a Gaussian:

$$f_{ij} = \tfrac{1}{N}\sum_{a=1}^{N} e^{i\frac{2\pi}{N}a^2} e^{i\frac{2\pi}{N}a(i-j)} \ .$$

Notice that the trace of $\mathbf{F}$ is a finite Gaussian sum, tr $\mathbf{F} = \sum_{a=1}^{N} e^{i\frac{2\pi}{N}a^2}$; also tr $\mathbf{F}^3 = N^2$ tr $\mathbf{F}^{-1} = N\sum_{a=1}^{N} e^{-i\frac{2\pi}{N}a^2}$; while tr $\mathbf{F}^2 = N$ and tr $\mathbf{F}^4 = N^3$. Number "4" plays a special, "modular" role, as $\mathbf{F}^4 = N^2\ \mathbf{I}$ and, consequently, $\mathbf{F}^5 = N^2\ \mathbf{F}$, $\mathbf{F}^6 = N^2\ \mathbf{F}^2$, etc.

Note. The Fourier transform of a projector $\mathbf{Z}$ is also a projector: if $\tilde{\mathbf{Z}} = \mathbf{F}\mathbf{Z}\mathbf{F}^{-1}$, then $\tilde{\mathbf{Z}}^2 = \tilde{\mathbf{Z}}$ and tr $\tilde{\mathbf{Z}}$ = tr $\mathbf{Z}$. With the unitary Fourier transformers $\mathbf{M}$ of Eq.(9.14), also the eventual hermitian character is preserved: $\left[\mathbf{M}\mathbf{Z}\mathbf{M}^{-1}\right]^\dagger = \mathbf{M}\mathbf{Z}^\dagger\mathbf{M}^{-1}$.

Chapter 10

Random walks

Circulants, which are also stochastic matrices, represent discrete random walks of a very general type [30]. Before going into the subject of such stochastic circulants, we start with a brief section on the continuum case.

10.1 Continuum random walks

To recall the main ideas and introduce some terminology, let us summarize the standard treatment to a random walk[1] on a d-dimensional continuous space of microstates. A point on that space will be indicated by a d-column vector $\mathbf{x} = (x^1, x^2, \ldots, x^d)^T$. Time will be taken as discrete.

Let $p^{(n)}(\mathbf{x})$ be the probability density of finding the particle at $\mathbf{x}$ after n steps, supposing the walk to have started at $\mathbf{x} = 0$ at initial time $n = 0$: $p^{(0)}(\mathbf{x}) = \delta(\mathbf{x})$. If $K(\mathbf{x} - \mathbf{x}')$ is the probability of a jump from $\mathbf{x}'$ to $\mathbf{x}$ during a unit time interval, then

$$p^{(n+1)}(\mathbf{x}) = \int d^d x' \; K(\mathbf{x} - \mathbf{x}') \; p^{(n)}(\mathbf{x}') \; . \tag{10.1}$$

This means also

$$p^{(n+1)}(\mathbf{x}) = \int d^d x' \int d^d x'' \; K(\mathbf{x} - \mathbf{x}') \; K(\mathbf{x}' - \mathbf{x}'') \; p^{(n-1)}(\mathbf{x}'') =$$

$$= \int \underbrace{d^d x' d^d x'' \ldots d^d x^{(n)}}_{n \; times} \; \underbrace{K_{\mathbf{x}-\mathbf{x}'} K_{\mathbf{x}'-\mathbf{x}''} \ldots K_{\mathbf{x}^{(n-1)}-\mathbf{x}^{(n)}}}_{n \; times} p^{(0)}(\mathbf{x}^{(n)}) \; . \tag{10.2}$$

Let us remark that K is a transition matrix, and $K_{\mathbf{x}-\mathbf{x}'} = K_{(\mathbf{x}+\mathbf{y})-(\mathbf{x}'+\mathbf{y})}$. Were it a finite matrix, this property would be characteristic of a circulant. Furthermore, we can identify $K_{(\mathbf{x}+\mathbf{y}),\mathbf{y}} = K_{\mathbf{x}}$, and

[1] See [41], Chapter 12.

$K_{\mathbf{x}} = K(\mathbf{x})$ is a probability distribution. It corresponds to the components in a circulant basis, $\mathbf{K} = \sum_{\mathbf{x}} K_{\mathbf{x}} \mathbf{U}^{\mathbf{x}}$ or, in the discrete case, to $\mathbf{K} = \sum_j K_j \mathbf{U}^j$.

Equations like (10.1) and (10.2), involving convolutions, invite naturally a Fourier approach. In the case, Fourier transforms are usually called *characteristic functions* and take probability distributions into functions on the Fourier dual space, whose points will be indicated by $\mathbf{k} = (k_1, k_2, \dots, k_d)^T$. Thus, $\tilde{p}^{(n)}(\mathbf{k}) = \int d^d x \; e^{i\mathbf{k}\cdot\mathbf{x}} p^{(n)}(\mathbf{x})$ is the characteristic function of the distribution $p^{(n)}(\mathbf{x})$. The characteristic function

$$\tilde{K}(\mathbf{k}) = \int d^d x e^{i\mathbf{k}\cdot\mathbf{x}} K(\mathbf{x}) \tag{10.3}$$

of the transition probability is called *structure function.* It corresponds, in the finite discrete case, to the eigenvalue expression $\tilde{K}_k = \sum_j K_j \; e^{i\frac{2\pi k j}{N}}$.

The evolution equation (10.1) takes on the form

$$\tilde{p}^{(n+1)}(\mathbf{k}) = \tilde{K}(\mathbf{k}) \; \tilde{p}^{(n)}(\mathbf{k}) \; , \tag{10.4}$$

with the initial condition $\tilde{p}^{(0)}(\mathbf{k}) = 1$. We find *which* distribution $K(\mathbf{x})$ is: as $\tilde{p}^{(1)}(\mathbf{k}) = \tilde{K}(\mathbf{k})$, $K(\mathbf{x})$ is just $p^{(1)}(\mathbf{x})$ (recall that we are taking a very special initial distribution, concentrated at a point).

We find also $\tilde{p}^{(2)}(\mathbf{k}) = [\tilde{K}(\mathbf{k})]^2$, $\tilde{p}^{(3)}(\mathbf{k}) = [\tilde{K}(\mathbf{k})]^3$, and the general case

$$\tilde{p}^{(n)}(\mathbf{k}) = [\tilde{K}(\mathbf{k})]^n \; , \tag{10.5}$$

evidently the Fourier transform of (10.2). Its inverse is the solution,

$$p^{(n)}(\mathbf{x}) = (2\pi)^{-d} \int d^d k \; e^{-i\mathbf{k}\cdot\mathbf{x}} \; [\tilde{K}(\mathbf{k})]^n \; . \tag{10.6}$$

Notice that $\tilde{p}^{(n)}(\mathbf{k} = \mathbf{0}) = 1$ and $\tilde{K}(\mathbf{k} = \mathbf{0}) = 1$. Such solutions, for an initial distribution concentrated at a point, are called *fundamental solutions.*

All this leads to simple formal expressions for the r-point functions at instant n, or momenta. As examples, the average position is given by

$$< \mathbf{x} >^{(n)} = - \; i \; \left[\frac{\partial}{\partial \mathbf{k}} \; \tilde{p}^{(n)}(\mathbf{k}) \right]_{\mathbf{k}=0} \tag{10.7}$$

and the average of products of components,

$$< x_{a_1} x_{a_2} x_{a_3} ... x_{a_r} >^{(n)} = - \; i \; \left[\frac{\partial^r}{\partial k_{a_1} k_{a_2} k_{a_3} ... k_{a_r}} \; \tilde{p}^{(n)}(\mathbf{k}) \right]_{\mathbf{k}=0} \; . \tag{10.8}$$

In particular, the *mean square displacement* is

$$< \mathbf{x}^2 > = - \; \left[\frac{\partial}{\partial \mathbf{k}} \cdot \frac{\partial}{\partial \mathbf{k}} \; \tilde{p}^{(n)}(\mathbf{k}) \right]_{\mathbf{k}=0} \; . \tag{10.9}$$

We can also take averages in the distribution $K(\mathbf{x})$, of course. We shall indicate those averages by an index K: $< \mathbf{x} >_K$, $< \mathbf{x}^2 >_K$, etc.

10.1.1 *Symmetric and stable random walks*

A random walk is said to be *symmetric* when the probability distributions depend only on the modulus of $\mathbf{x}$: $p^{(n)}(\mathbf{x}) = p^{(n)}(|\mathbf{x}|) = p^{(n)}(x)$ and $\tilde{K}(\mathbf{k}) = \tilde{K}(k)$. In this symmetric case, the average displacement of (10.7) will be zero. The first nonvanishing momentum will be (10.9).

For reasons given below, a structure function is said to be *stable* when it has the property

$$\prod_{i=1}^{j} \tilde{K}(a_i\mathbf{k}) = \tilde{K}(a\mathbf{k}) \ , \tag{10.10}$$

where the a_k's are constants and a is some function of them. The solutions for these functional equations are the *Lévy distributions*

$$\tilde{K}(\mathbf{k}) = e^{-Ck^{\beta}} \ , \tag{10.11}$$

for which

$$a = \left[\sum_{i=1}^{j} a_i^{\beta}\right]^{1/\beta} \ . \tag{10.12}$$

There is a condition on β: to have K nonnegative, $0 < \beta \leq 2$. For $\beta = 1$, it corresponds to the *Cauchy-Lorentz distribution*

$$\tilde{p}^{(1)}(\mathbf{x}) = \frac{1}{\pi} \frac{1}{1 + |\mathbf{x}|^2} \ , \tag{10.13}$$

typically related to resonances in high-energy physics. The better known case is $\beta = 2$, the gaussian. For the general Lévy distribution (10.11), Eq.(10.5) takes the form

$$\tilde{p}^{(n)}(\mathbf{k}) = [\tilde{K}(\mathbf{k})]^n = [\tilde{K}(n^{1/\beta}k)] = e^{-Cnk^{\beta}} \ , \tag{10.14}$$

in which we see the reason for the name *stable*: the form of the characteristic function is maintained at each time-step. The middle equality, actually unnecessary to arrive at the result, makes use of condition (10.12).

For the gaussian ($\beta = 2$) the inverse Fourier transform can be obtained analitically:

$$p^{(n)}(\mathbf{x}) = \frac{1}{(4\pi nC)^{d/2}} e^{-\frac{|\mathbf{x}|^2}{4nC}} \ . \tag{10.15}$$

This leads to the mean square displacement

$$< x^2 > = 2dCn \ . \tag{10.16}$$

Recalling that n here plays the role of time, this establishes a bridge between random walks and diffusion.

The cases $\beta < 2$ exhibit intriguing features: the solution distributions have long tails, and in consequence the mean square displacement is infinite: $< x^2 > \propto \left[k^{\beta-2}\right]_{k \to 0}$.

10.1.2 *Non-symmetric random walks*

For symmetric distributions, $< \mathbf{x} > = 0$ and $< \mathbf{x} >_K = 0$. Let us examine, under rather strong simplifying assumptions, what happens when this is not the case. There will be a net displacement. It is interesting to introduce a time unit τ and put the time variable in the form $t = n\tau$. A measure of the net displacement can be given by the velocity $\mathbf{V} = \frac{<\mathbf{x}>_K}{\tau}$.

Take the expression of the characteristic function and expand the exponential:

$$\tilde{K}(\mathbf{k}) = \int d^d x e^{i\mathbf{k}\cdot\mathbf{x}} K(\mathbf{x}) \approx 1 + i\ \mathbf{k}\cdot < \mathbf{x} >_K - \tfrac{1}{2}\ k_r k_s < x^r x^s >_K + ...$$

Besides neglecting terms of order three and higher, suppose isotropy, that is, $< x^r x^s >_K = \frac{1}{d}\ \delta^{rs} < \sum_n x^n x^n >_K = \frac{1}{d}\ \delta^{rs} < r^2 >_K$. Then,

$$\tilde{K}(\mathbf{k}) \approx 1 + i\ \mathbf{k}\cdot < \mathbf{x} >_K - \frac{1}{2d}\ k^2 < r^2 >_K \ .$$

There is another current representation of a characteristic function, in terms of cumulants. The first cumulant is $<< \mathbf{x} >> = < \mathbf{x} >_K$, the second is $<< r^2 >> = < r^2 >_K - | < \mathbf{x} >_K |^2 = < |\mathbf{x} - < \mathbf{x} >_K |^2 >_K$, all of them defined in

$$\tilde{K}(\mathbf{k}) = e^{i\mathbf{k}\cdot <<\mathbf{x}>> - \frac{1}{2d}\ k^2 <<r^2>> + \ \cdots}.$$

Taking only terms up to second order, the probability density (10.6) will be

$$p^{(t)}(\mathbf{x}) = (2\pi)^{-d} \int d^d k\ e^{-i\mathbf{k}\cdot(\mathbf{x}-\mathbf{V}t)} e^{-k^2 Dt}\ , \qquad (10.17)$$

where we have introduced the diffusion coefficient $D = \frac{<<r^2>>}{2d\tau}$. This can be evaluated analitically, giving the moving gaussian packet

$$p^{(t)}(\mathbf{x}) = \frac{1}{(2\pi Dt)^{d/2}}\ e^{-\frac{|\mathbf{x}-\mathbf{V}t|^2}{4Dt}}\ . \qquad (10.18)$$

As time goes by, the packet spreads by diffusion while its peak moves with velocity $\mathbf{V}$. It is an evanescent soliton, if we wish. The motion of the peak is called *advection*. The packet represents the fundamental solution of the *diffusion-advection equation*

$$\frac{d}{dt} p^{(t)}(\mathbf{x}) = D\ \Delta\ p^{(t)}(\mathbf{x}) - \mathbf{V}\cdot\nabla\ p^{(t)}(\mathbf{x})\ . \qquad (10.19)$$

10.1.3 *Summing up*

We learn from the resumé above some points of general interest:

(1) the entries of K are probability transitions, so that K is a stochastic matrix;
(2) the entries of K depend only on the label differences, so that K is a circulant matrix;
(3) last conclusion is reinforced by the fact that the product of circulants correspond to convolutions of the corresponding circulant components.

Of course, K is an infinite matrix. Let us now address the question of finite random walks.

10.2 Stochastic circulants

It has been seen how the behavior of a Markov chain is codified in the spectrum of its stochastic matrix. That spectrum, however, is not analytically calculable for generic $N \times N$ matrices when $N \geq 5$. $N \times N$ circulants, on the other hand, are all functions of a single shift matrix $\mathbf{U}$, their component projectors are the same and their spectra are analytically known. Stochastic circulants are consequently of obvious formal interest. We shall see that they are also of great physical interest, as they underpin a whole family of processes, that related to random walks and diffusion. Notice that, if a circulant is a stochastic matrix, then it is automatically doubly-stochastic.

We shall be particularly concerned with a discrete version of the diffusion equation. Fourier analysis was born from the study of its twin, the heat equation, and we shall find that circulants are indeed instrumental in solving it. We prefer to talk about diffusion because the concerned variable, concentration, seems more intuitively akin to a probability than its analogous in the heat equation, temperature. Let us begin with some comments on Fourier transforms of probability distributions.

The presence of a stochastic circulant at work is signaled by the fact that, if an equilibrium distribution exists, then it is flat. In effect, given a circulant as in Eq.(9.21),

$$\mathbf{A} = \sum_{n=1}^{N} a_n \mathbf{U}^n ,$$

the probability conservation condition (3.2) is equivalent to

$$\lambda_N = \sum_{j=1}^{N} a_j = 1 \ . \tag{10.20}$$

Eigenvalue λ_N is a main root of $\mathbf{A}$. The N-th homogeneous distribution is $\mathbf{u} = \frac{1}{N}(1, 1, 1, \ldots, 1)^T$, invariant under the actions of the $\mathbf{U}^n$'s. Consequently, $\mathbf{A}\mathbf{u} = \sum_{n=1}^{N} a_n \mathbf{u} = \lambda_N \ \mathbf{u} = \mathbf{u}$. Clearly, any stochastic matrix $\mathbf{A}$ with equilibrium eigenvector $\mathbf{u}$ will be doubly-stochastic. As previously noticed (page 64), it will be symmetric if in detailed balance with $\mathbf{u}$.

The periodic boundary conditions $p_0 = p_N$, $p_{N+a} = p_a$ introduced in Note 58.1 turn the probability distributions into cyclic objects, with well-defined finite Fourier transforms. Given a distribution $\mathbf{p} = (p_1, p_2, \ldots, p_N)^T$, its Fourier transform will be

$$\tilde{p}_k = \sum_{a=1}^{N} e^{i\frac{2\pi}{N}ka} p_a, \ \text{ with inverse } \ p_a = \tfrac{1}{N} \sum_{k=1}^{N} e^{-i\frac{2\pi}{N}ak} \tilde{p}_k \ . \tag{10.21}$$

It should we be noticed, however, that the very first example of stochastic circulant has no equilibrium: it is the "nuclear" circulant matrix $\mathbf{U}$ itself.

Note. In terms of the matrix $\mathbf{F}$ of Note 9.2, $\tilde{\mathbf{p}} = \mathbf{F} \ \mathbf{p}$.

The $\tilde{p}_k$'s are generically complex numbers. They are not probabilities, of course, but the set $\tilde{\mathbf{p}} = (\tilde{p}_1, \tilde{p}_2, \ldots, \tilde{p}_N)^T$ has the same information content as the original distribution $\mathbf{p}$. For example, the average (3.10) of a function f can be calculated, if we have its transform $\tilde{f}_k = \frac{1}{N} \sum_a f_a e^{-i\frac{2\pi}{N}ak}$, as

$$\bar{f} = \sum_{a=1}^{N} f_a p_a = \sum_{k=1}^{N} \tilde{f}_k \tilde{p}_k \ .$$

It is immediate that

$$\sum_{k=1}^{N} \tilde{p}_k = \sum_{a=1}^{N} p_a \sum_{k=1}^{N} e^{i\frac{2\pi}{N}ak} = \sum_{a=1}^{N} N\delta_{Na} p_a = N p_N \ . \tag{10.22}$$

The condition $\sum_{n=1}^{N} p_a = 1$ translates, in terms of the Fourier transform components, as $\tilde{p}_N = 1$. The "uncertainty relation" between a probability distribution and its transformed is best illustrated by the example of a thoroughly concentrated distribution $p_a = \delta_{aN}$, whose transform is $\tilde{p}_k = 1 \ \forall \ k$. The flat distribution $p_a = \frac{1}{N} \ \forall \ a$, on the other hand, has $\tilde{p}_k = \delta_{kN}$.

Let us examine the transform of the t-th step in the evolutive process for a stochastic matrix which is also a circulant. Using the decomposition of the evolution matrix into projectors, and the expression (9.16) for the circulant case,

$$\tilde{p}_k^{(t)} = \textstyle\sum_{b=1}^N e^{i\frac{2\pi}{N}kb} p_b^{(t)} = \sum_{b=1}^N \sum_{c=1}^N e^{i\frac{2\pi}{N}kb} M^t{}_{bc} p_c^{(0)}$$

$$= \sum_{b=1}^N \sum_{c=1}^N \sum_{a=1}^N e^{i\frac{2\pi}{N}kb} \lambda_a^t P_{(a)bc} p_c^{(0)} = \sum_{c=1}^N \sum_{a=1}^N \lambda_a^t \tfrac{1}{N} \sum_{b=1}^N e^{i\frac{2\pi}{N}b(k-a)}\, e^{i\frac{2\pi}{N}ac} p_c^{(0)}$$

$$= \lambda_k^t\, \tilde{p}_k^{(0)} \quad \therefore \quad \tilde{p}_k^{(t)} = \lambda_k^t\, \tilde{p}_k^{(0)}\,. \tag{10.23}$$

The components of the Fourier-transformed distribution evolve independently of each other, and in a very simple way: at the t-th step, each one is the product of the initial component by the t-th power of the corresponding eigenvalue. This independence makes of them the "normal modes" of the evolution process.

Note. Again using the matrix $\mathbf{F}$ (Notes 9.2 and 10.2), this is

$$\tilde{\mathbf{p}}(t) = \mathbf{F}\ \mathbf{p}(t) = \mathbf{F}\ \mathbf{M}^t \mathbf{p}(0) = \mathbf{F}\ \mathbf{F}^{-1} \mathbf{M}^t_{diag} \mathbf{F} \mathbf{p}(0) = \mathbf{M}^t_{diag} \tilde{\mathbf{p}}(0)\,.$$

Note. Two particularly important averages come immediately from the expression of the Fourier transform:

(1) the average displacement

$$\bar{n}(t) = \sum_{n=1}^N n p_n(t) = \tfrac{N}{2\pi i}\left[\frac{\partial}{\partial k}\tilde{p}_k(t)\right]_{k=0} = \tfrac{N}{2\pi i}\left[\frac{\partial}{\partial k}\left(\lambda_k^t\, \tilde{p}_k^{(0)}\right)\right]_{k=0}\ ; \tag{10.24}$$

(2) and the mean square deviation

$$\bar{n^2}(t) = \sum_{n=1}^N n^2 p_n(t) = \left(\tfrac{N}{2\pi i}\right)^2 \left[\frac{\partial^2}{\partial k^2}\tilde{p}_k(t)\right]_{k=0}$$

$$= \left(\tfrac{N}{2\pi i}\right)^2 \left[\frac{\partial^2}{\partial k^2}\left(\lambda_k^t\, \tilde{p}_k^{(0)}\right)\right]_{k=0}\,. \tag{10.25}$$

For the case of an initial distribution concentrated at a point, $p_a^{(0)} = \delta_{aN}$, $\tilde{p}_k^{(0)} = 1\ \ \forall\ k$ and

$$\tilde{p}_k^{(t)} = \lambda_k^t = \sum_{j=1}^N a_j^{*t}\ e^{i\frac{2\pi}{N}kj}$$

[see Eq.(9.18)]. This means that the solution is $p_c^{(t)} = a_c^{*t}$, the entries in the last column of $\mathbf{M}^t$. In particular $p_c^{(1)} = a_c = M_{cN}$. For this *fundamental solution*, the evolution is entirely contained in the entries of the last column. Notice that the coincidence of the distributions $\{p_c\}$ and $\{a_c\}$ only occurs if the initial distribution is concentrated at a point. We shall find counterexamples in what follows [see, for instance, Eq.(10.60) and the comment below it].

The evolution equation will have the general form

$$p_a^{(t)} = \sum_{b=1}^{N} M_{ab}^t \; p_b(0) = \sum_{b=1}^{N} a_{a-b}^{*t} \; p_b(0) \; . \tag{10.26}$$

The average of a function f at time t will be

$$\bar{f}(t) = \textstyle\sum_{k=1}^{N} \tilde{f}_k \; \tilde{p}_k^{(t)} = \sum_{k=1}^{N} \lambda_k^t \; \tilde{f}_k \; \tilde{p}_k^{(0)} \; .$$

For the fundamental solution, $p_c^{(t)} = a_c^{*t}$ and

$$\bar{f}(t) = \textstyle\sum_{c=1}^{N} a_c^{*t} f_c \; .$$

Despite their difference, averages using distributions $\{p_c\}$ and $\{a_c\}$ coincide in this finite case, because of Eq.(10.26). In effect, as $p_c = \sum_{b=1}^{N} a_{c-b} p_b^{(0)}$, the average with p will be

$$\bar{f} = < f >_{(\text{with p})} = \sum_{c=1}^{N} f_c p_c = \sum_{b=1}^{N} p_b^{(0)} \sum_{c=1}^{N} f_c a_{c-b}$$

$$= \sum_{b=1}^{N} p_b^{(0)} < f >_{(\text{with a})} = < f >_{(\text{with a})} \sum_{b=1}^{N} p_b^{(0)} = < f >_{(\text{with a})} \; .$$

The time average of f up to step n will be

$$\frac{1}{n} \sum_{k=0}^{n-1} \bar{f}(k) = \sum_{a=1}^{N} \tilde{f}_a \; \tilde{p}_a^{(0)} \; \frac{1}{n} \sum_{k=0}^{n-1} \lambda_a^k = \begin{cases} \sum_{a=1}^{N} \tilde{f}_a \; \tilde{p}_a^{(0)} \frac{1-\lambda_a^n}{1-\lambda_a} & \text{if } \lambda_a \neq 1 \\ \sum_{a=1}^{N} \tilde{f}_a \; \tilde{p}_a^{(0)} & \text{if } \lambda_a = 1 \end{cases}$$

We have learned a few things of interest:

the Fourier component $\tilde{p}_a$ is the average of function $f(b) = e^{i \frac{2\pi}{N} ab}$ on $\mathbf{p}$.

the Fourier components are normal modes, evolving independently from each other;

they evolve by simple multiplication, each one by the time-power of one of the eigenvalues;

as one at least of the eigenvalues is equal to 1, one at least of the normal modes will remain constant.

10.2.1 *General random walks*

The Markov chain associated to an $N \times N$ stochastic matrix $\mathbf{R}$ is said to represent a general random walk when the steps depend only on the "label distance" between the states. In other words, the entries exhibit the further property

$$R_{ij} = a_{i-j} \tag{10.27}$$

for some set of N numbers $\{a_k\}$, besides satisfying conditions (3.1) and (3.2). This is just (9.2), so that $\mathbf{R}$ is a circulant matrix. There are two clear consequences: $\mathbf{R}$ will be doubly-stochastic (see page 46) and will have eigenvalues given by (9.5). Probability conservation (10.20) says that λ_N is a main root of $\mathbf{R}$: $\lambda_N = 1$.

The iterates $\mathbf{R}^m$, all of them stochastic circulants, will have eigenvalues λ_k^m, with $\lambda_N^m = 1$ fixed for all orders. Trivial cases are those in which $\mathbf{R} = \mathbf{U}^j$ for some j, which are in the intersection of the invertible stochastic and circulant matrices.

For crystals endowed with a simple symmetry, it is always possible to define a dual lattice in "momentum" space. Given the close relationship between circulants and Fourier transformation, we have here also a case in which a dual lattice can be defined. However, as the "walk" takes place in state space and not in configuration space, that lattice lies in a space dual to the space of states.

Commentary 103.1. ▶ An example is given by the entries and eigenvalues

$$a_j = \frac{2^N}{2^N - 1}\,\frac{1}{2^j} \quad ; \quad \lambda_k = \frac{\omega_N^k}{2 - \omega_N^k}$$

(where $\omega_N = \exp[i\ (2\pi/N)]$). For $N = 3$, the matrix and its asymptotic limit are

$$\begin{pmatrix} 1/7 & 4/7 & 2/7 \\ 2/7 & 1/7 & 4/7 \\ 4/7 & 2/7 & 1/7 \end{pmatrix} \Rightarrow \begin{pmatrix} 1/3 & 1/3 & 1/3 \\ 1/3 & 1/3 & 1/3 \\ 1/3 & 1/3 & 1/3 \end{pmatrix}.$$

Any initial distribution will be taken into $\frac{1}{3}(1,1,1)$. The eigenvalues are 1 and $\frac{-4 \mp i2\sqrt{3}}{14}$, the latter with absolute value $\frac{1}{\sqrt{7}}$. ◀

10.2.2 *Simple random walks*

A simple random walk has only transitions between nearest neighbors. We shall here force the things a bit, and add a cyclic condition. For a general N, $\mathbf{R} = (1-q)\mathbf{U} + q\mathbf{U}^{-1}$, so that $\lambda_k = (1-q)\omega^k + q\omega^{-k} = e^{\frac{i2\pi k}{N}} -$

$2iq\sin\left(\frac{2\pi k}{N}\right)$. Some general results come from a case study of the equation $|\lambda_k|^2 = (1-2q)^2 + 4q(1-q)\cos^2\left(\frac{2\pi k}{N}\right)$:

(1) clearly $\lambda_N = 1$ in all cases;
(2) cases $q = 0$ and $q = 1$ lead to $|\lambda_k|^2 = 1\ \forall k$;
(3) cases $q \neq 0, 1$, including $q = 1/2$, lead to $|\lambda_k|^2 = 1$ only when $N = k$ or $N = 2k$.

The last item points to a deep difference between even and odd values of N. When $N = 2p$, $\lambda_p = -1$, so that there is no asymptotic equilibrium. Even-dimensional matrices will exhibit an indefinite oscillation. Odd-dimensional are totally regular. Let us see two examples:

- $N = 4$:

$$\mathbf{R} = \begin{pmatrix} 0 & q & 0 & 1-q \\ 1-q & 0 & q & 0 \\ 0 & 1-q & 0 & q \\ q & 0 & 1-q & 0 \end{pmatrix}; \quad \begin{matrix} a & \longleftrightarrow & b \\ \updownarrow & & \updownarrow \\ d & \longleftrightarrow & c \end{matrix}$$

These are still circulants, for which it is very easy to calculate eigenvalues: they are $\lambda_k = \sum_{j=1}^{N} e^{ikj2\pi/N} a_j = (1-q)e^{ik\frac{\pi}{2}} + qe^{i3k\frac{\pi}{2}} = (1-q)i^k + q(-i)^k$, so that $\lambda_{1,2,3,4} = i(1-2q)$, -1, $-i(1-2q)$ and 1, with respective eigenvectors

$$\begin{pmatrix} -\frac{\sqrt{-(1-2q)^2}}{(1-2q)} \\ -1 \\ \frac{\sqrt{-(1-2q)^2}}{(1-2q)} \\ 1 \end{pmatrix}; \quad \begin{pmatrix} -1 \\ 1 \\ -1 \\ 1 \end{pmatrix}; \quad \begin{pmatrix} \frac{\sqrt{-(1-2q)^2}}{(1-2q)} \\ -1 \\ -\frac{\sqrt{-(1-2q)^2}}{(1-2q)} \\ 1 \end{pmatrix}; \quad \frac{1}{4}\begin{pmatrix} 1 \\ 1 \\ 1 \\ 1 \end{pmatrix}.$$

As the eigenvalues are all distinct, we can already guess that the projector related to the unit eigenvalue is

$$\mathbf{Z}_1 = \frac{1}{4}\begin{pmatrix} 1\,1\,1\,1 \\ 1\,1\,1\,1 \\ 1\,1\,1\,1 \\ 1\,1\,1\,1 \end{pmatrix}.$$

Due to the presence of more than one unit-modulus eigenvalue, there will be no well-defined asymptotic matrix.

- $N = 5$

$$\mathbf{R} = \begin{pmatrix} 0 & q & 0 & 0 & 1-q \\ 1-q & 0 & q & 0 & 0 \\ 0 & 1-q & 0 & q & 0 \\ 0 & 0 & 1-q & 0 & q \\ q & 0 & 0 & 1-q & 0 \end{pmatrix};$$

$$\begin{array}{ccccc} & a & \Longleftrightarrow & b & \\ \nearrow\!\swarrow & & & & \nwarrow\!\searrow \\ e & & & & c \\ \nwarrow\!\searrow & & & & \nearrow\!\swarrow \\ & & d & & \end{array}$$

The eigenvalues are $\lambda_k = \sum_{j=1}^{N} e^{ikj2\pi/5} a_j = (1-q)e^{ik\frac{2\pi}{5}} + qe^{i4k\frac{2\pi}{5}}$.

Both examples are "perturbations" of the shift operator: for $q = 0$, $\mathbf{R} = \mathbf{U}$, and for $q = 1$, $\mathbf{R} = \mathbf{U}^{-1}$. In these extreme cases the "walk" would not be random: it would be a steady march in one direction.

10.2.3 *Standard random walks*

The standard random walk takes $q = \frac{1}{2}$, with equal probabilities of jumping from one site to each of its two immediate neighbors. In that case, $\lambda_k = \cos\frac{2\pi k}{N}$ and $\mathbf{R} = \sum_j a_j U^j = \frac{1}{2}(\mathbf{U} + \mathbf{U}^{-1})$ has circulant components $a_j = \frac{1}{2}(\delta_{j1} + \delta_{j,N-1}) = \frac{1}{N}\sum_k e^{-i\frac{2\pi jk}{N}} \cos\frac{2\pi k}{N}$. This result can be used to obtain other finite Fourier transforms, such as

$$\frac{1}{N}\sum_{k=1}^{N} e^{-i\frac{2\pi jk}{N}} \sin\frac{2\pi k}{N} = \frac{i}{2}(\delta_{j,N-1} - \delta_{j1})$$

$$\frac{1}{N}\sum_{k=1}^{N} e^{-i\frac{2\pi jk}{N}} \sin^2\frac{\pi k}{N} = \frac{1}{2}\delta_{jN} - \frac{1}{4}(\delta_{j1} + \delta_{j,N-1}) .$$

The latter means, by the way, that the numbers $4\,\sin^2\frac{\pi k}{N}$ are, up to a sign, the eigenvalues of the matrix $\mathbf{\Delta} = \mathbf{U} + \mathbf{U}^{-1} - 2\,\mathbf{I} = 2(\mathbf{R} - \mathbf{I})$, which will be our model "Laplacian". This is a curious point: the Laplacian is (twice) a random walk minus the identity. Which means that, defined by $\mathbf{R} = \mathbf{I} + \mathbf{Q}$, the $\mathbf{Q}$-matrix related to $\mathbf{R}$ is just $\mathbf{Q} = \frac{1}{2}\,\mathbf{\Delta}$. Eqs.(6.16) and (6.17) take on the form of a diffusion equation:

$$\frac{1}{\epsilon}[p^{(n+1)}{}_a - p^{(n)}{}_a] = \frac{1}{2}\sum_{b=1}^{N} \mathbf{\Delta}_{ab}\; p^{(n)}{}_b\,, \tag{10.28}$$

$$\frac{d}{dt}\, p_a(t) = \frac{1}{2}\sum_{b=1}^{N} \mathbf{\Delta}_{ab}\; p_b(t)\;. \tag{10.29}$$

10.3 Discrete diffusion equation

Derivatives on the discrete space of states can be introduced by using finite differences. Of course, the notion of derivative assumes that space to be ordered in some sense. We shall suppose the natural order given by the labels in the probability distributions. Our main objective involves only the second derivative, but we start by examining derivatives in general. And that for a very good reason: once periodic conditions are imposed, derivatives turn out to be

(1) $\boldsymbol{Q}$ matrices, or better, logarithms of stochastic matrices;
(2) circulant matrices;

In consequence, also the associated stochastic matrices are circulants.

10.3.1 *Discrete derivatives*

We have considered in Section 6.4 the continuum limit in the "time" parameter of probability distributions, and introduced in Commentary 58.1 finite-difference derivatives on the space of states. With periodic conditions, the matrix $\mathbf{d}$ defining the first derivative $\mathbf{p}' = \mathbf{d}\ \mathbf{p}$, with $p'_a = p_{a+1} - p_a$, is given by

$$d_{ab} = \delta_{a+1,b} - \delta_{a,b} \quad \text{or} \quad \mathbf{d} = \mathbf{U}^{-1} - \mathbf{I}\ .$$

This is a circulant, $\begin{pmatrix} -1 & 1 \\ 1 & -1 \end{pmatrix}$ for $N = 2$, $\begin{pmatrix} -1 & 1 & 0 \\ 0 & -1 & 1 \\ 1 & 0 & -1 \end{pmatrix}$ for $N = 3$. The general form is

$$\mathbf{d} = \begin{pmatrix} -1 & 1 & 0 & 0 & 0 & \cdots\cdots & 0 \\ 0 & -1 & 1 & 0 & 0 & \cdots\cdots & 0 \\ 0 & 0 & -1 & 1 & 0 & \cdots\cdots & 0 \\ 0 & 0 & 0 & -1 & 1 & \cdots\cdots & 0 \\ \cdots & \cdots & \cdots & \cdots & \cdots & \cdots\cdots & \cdots \\ \cdots & \cdots & \cdots & \cdots & \cdots & \cdots\cdots & 1 \\ 1 & 0 & 0 & \cdots & \cdots & \cdots\ 0 & -1 \end{pmatrix}.$$

The eigenvalues are immediately found: $\chi_k = \omega^{-k} - 1$. The "exponential" distribution corresponding to this derivative, satisfying $\mathbf{d}\ \mathbf{p} = \mathbf{p}$, has entries $p_j = \frac{2^{j-1}}{2^N - 1}$. The "constant" distribution, satisfying $\mathbf{d}\ \mathbf{p} = \mathbf{0}$, is nearer to intuition: it is the "flat" distribution of entries $p_j = \frac{1}{N}$. Applied to a general matrix $\mathbf{A}$, $(\mathbf{dA})_{ab} = A_{a+1,b} - A_{a,b}$ and $[\mathbf{d}, \mathbf{A}]_{ab} = A_{a+1,b} - A_{a,b-1}$. Of course, $[\mathbf{d}, \mathbf{A}] \equiv 0$ if $\mathbf{A}$ is a circulant. Notice that, applied to matrices with the usual matrix product, $\mathbf{d}$ is not a true derivative: as $\mathbf{d}(\mathbf{AB}) = \mathbf{d}(\mathbf{A})\,\mathbf{B}$, the Leibnitz rule fails.

The hermitean conjugate $\mathbf{d}^\dagger = \mathbf{U} - \mathbf{I}$ has the expected behavior with respect to the trivial discrete "measure": $\sum_{a,b} f_a d_{ab}\, g_b = \sum_{a,b} g_a d^\dagger_{ab}\, f_b$.

The second derivative will be again a circulant,

$$\Delta_{ab} = \delta_{a+1,b} + \delta_{a-1,b} - 2\,\delta_{ab} \quad \text{or} \quad \mathbf{\Delta} = \mathbf{U}^{-1} + \mathbf{U} - 2\,\mathbf{I}\,,$$

with the general aspect

$$\mathbf{\Delta} = \begin{pmatrix} -2 & 1 & 0 & 0 & 0 & \cdots & \cdots & \cdots & 1 \\ 1 & -2 & 1 & 0 & 0 & \cdots & \cdots & \cdots & 0 \\ 0 & 1 & -2 & 1 & 0 & \cdots & \cdots & \cdots & 0 \\ 0 & 0 & 1 & -2 & 1 & 0 & \cdots & \cdots & 0 \\ \cdots & \cdots & \cdots & \cdots & \cdots & \cdots & \cdots & \cdots & \cdots \\ \cdots & \cdots & \cdots & \cdots & \cdots & \cdots & \cdots & -2 & 1 \\ 1 & 0 & 0 & \cdots & \cdots & \cdots & \cdots & 1 & -2 \end{pmatrix}.$$

the eigenvalues are $\chi_k = \omega^k + \omega^{-k} - 2 = -\,4\,\sin^2\frac{\pi k}{N}$. The above aspect fails for small N: for example, $\mathbf{\Delta} = \begin{pmatrix} -2 & 2 \\ 2 & -2 \end{pmatrix}$ for $N = 2$.

Notice that $\mathbf{\Delta} = \mathbf{d} + \mathbf{d}^\dagger = -\,\mathbf{d}\mathbf{d}^\dagger = \mathbf{\Delta}^\dagger$ and $\mathbf{\Delta} = \mathbf{U}\mathbf{d}^2$, with the operator $\mathbf{U}$ performing the shift announced in Commentary 58.1. We shall discuss this operator (already met at the end of Section 10.2.3) below, and the above notation will find its justification: $\mathbf{\Delta}$ will be assimilated to a (one-dimensional) Laplacian. Notice that this has nothing to do with the (also standard notation for the) canonical matrix basis $\{\mathbf{\Delta}_{ab}\}$ of Section 9.1.

Note. Path counting The matrix $\mathbf{A} = \mathbf{U} + \mathbf{U}^{-1} = \mathbf{\Delta} + 2\,\mathbf{I} = 2\,\mathbf{R}$ is an example of what is called, in graph theory, "adjacency operator". Let us try to understand its meaning through examples. Consider a discrete circle like that of Figure 10.1. Take $N = 7$ points, represented by the vector $(p_1, p_2, p_3, p_4, p_5, p_6, p_7)^T$. It is not important here whether it is a probability distribution or not: we shall suppose that the entries indicate simply the positions on the circle with respect to some arbitrarily chosen origin: p_3 represents the third point, etc. Suppose we start at point p_1 and perform a standard random walk. In the first step, we will attain either p_2 or $p_N = p_7$. In the second step there are more possibilities: we can attain p_3 or p_6, but also come back to point p_1, and that in two ways: going one step to the right and then coming back with one step to the left; or taking first one step to the left and then coming back with one step to the right. There is one 2-step path from p_1 to p_3; one 2-step path from p_1 to p_6 and

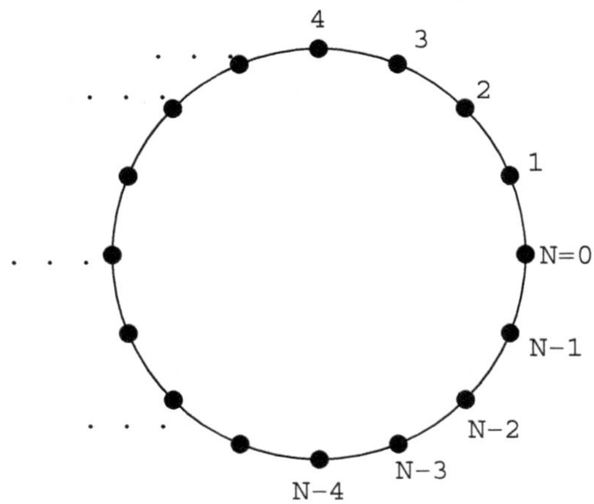

Fig. 10.1 *The discrete circle Z_N.*

two 2-step paths from p_1 to p_1 itself. The adjacency operator provides a means to calculate these numbers of paths. For $N = 7$, it is

$$\mathbf{A} = \mathbf{U} + \mathbf{U}^{-1} = \begin{pmatrix} 0&1&0&0&0&0&1 \\ 1&0&1&0&0&0&0 \\ 0&1&0&1&0&0&0 \\ 0&0&1&0&1&0&0 \\ 0&0&0&1&0&1&0 \\ 0&0&0&0&1&0&1 \\ 1&0&0&0&0&1&0 \end{pmatrix}.$$

Let us apply it to the initial vector $(p_1, p_2, p_3, p_4, p_5, p_6, p_7)^T$:

$$\begin{pmatrix} 0&1&0&0&0&0&1 \\ 1&0&1&0&0&0&0 \\ 0&1&0&1&0&0&0 \\ 0&0&1&0&1&0&0 \\ 0&0&0&1&0&1&0 \\ 0&0&0&0&1&0&1 \\ 1&0&0&0&0&1&0 \end{pmatrix} \begin{pmatrix} p_1 \\ p_2 \\ p_3 \\ p_4 \\ p_5 \\ p_6 \\ p_7 \end{pmatrix} = \begin{pmatrix} p_2+p_7 \\ p_1+p_3 \\ p_2+p_4 \\ p_3+p_5 \\ p_4+p_6 \\ p_5+p_7 \\ p_1+p_6 \end{pmatrix}.$$

Notice what happens: with one step, you can go from p_1 to either p_2 or p_7; from p_2 to either p_1 or p_3 and so on, for whatever chosen starting point. Looking at the components in the right-hand side, we see that each one just labels the possible 1-step paths from the initial left-hand entry. Applied twice, matrix $\mathbf{A}$ gives

$$\mathbf{A}^2 \begin{pmatrix} p_1 \\ p_2 \\ p_3 \\ p_4 \\ p_5 \\ p_6 \\ p_7 \end{pmatrix} = \begin{pmatrix} 2p_1+p_3+p_6 \\ 2p_2+p_4+p_7 \\ p_1+2p_3+p_5 \\ p_2+2p_4+p_6 \\ p_3+2p_5+p_7 \\ p_1+p_4+2p_6 \\ p_2+p_5+2p_7 \end{pmatrix}.$$

Now, each right-hand entry shows just the points that can be attained in 2 steps from the corresponding left-hand entry, but there is more: the coefficients are just the number of such paths (the total number of paths in each case can be found by putting $p_1 = 1$ after the calculation). There is, however, another way to find these numbers. Explicitly,

$$\mathbf{A}^2 = \begin{pmatrix} 2&0&1&0&0&1&0 \\ 0&2&0&1&0&0&1 \\ 1&0&2&0&1&0&0 \\ 0&1&0&2&0&1&0 \\ 0&0&1&0&2&0&1 \\ 1&0&0&1&0&2&0 \\ 0&1&0&0&1&0&2 \end{pmatrix}.$$

The numbers of 2-step paths appear as the entries of $\mathbf{A}^2$: ${A^2}_{13} = 1$ is the number of 2-step paths from p_1 to p_3; ${A^2}_{16} = 1$ is the number of 2-step paths from p_1 to p_6; the diagonal term ${A^2}_{11} = 2$ is the number of 2-step "closed" paths, those taking p_1 into p_1 itself; and so for all the other entries.

Going to higher orders, we verify that ${A^n}_{ij}$ is the number of n-step paths from point p_i to point p_j. Now, $\mathbf{A}$ is a circulant, and so is each $\mathbf{A}^n$. The number of possible 1-step paths starting from p_1 is the same as the number of 1-step paths starting from any other point to its corresponding neighbors; that number depends only on the "label distance". In other words, A_{ij} is the same as the letter a_{i-j} and can be calculated from the spectrum: $a_s = A_{i,i+s} = \frac{1}{N}\sum_{k=1}^{N} \lambda_k e^{i\frac{2\pi}{N}ks}$. For higher orders, it corresponds to the convolutions: ${A^n}_{ij} = a^{*n}_{i-j} = \frac{1}{N}\sum_{k=1}^{N} \lambda_k^n \; e^{i\frac{2\pi}{N}k(i-j)}$. In particular, the number of closed n-paths, from a point to itself, is a trace: ${A^n}_{ii} = a^{*n}_N = \frac{1}{N}\sum_{k=1}^{N} \lambda_k^n = \frac{1}{N}\,\mathrm{tr}\,\mathbf{A}^n$.

Take the simpler case $N = 4$. The first powers of the adjacency matrix are

$$\mathbf{A} = \begin{pmatrix} 0&1&0&1 \\ 1&0&1&0 \\ 0&1&0&1 \\ 1&0&1&0 \end{pmatrix}, \quad \mathbf{A}^2 = \begin{pmatrix} 2&0&2&0 \\ 0&2&0&2 \\ 2&0&2&0 \\ 0&2&0&2 \end{pmatrix}, \quad \mathbf{A}^3 = \begin{pmatrix} 0&4&0&4 \\ 4&0&4&0 \\ 0&4&0&4 \\ 4&0&4&0 \end{pmatrix}, \quad \mathbf{A}^4 = \begin{pmatrix} 8&0&8&0 \\ 0&8&0&8 \\ 8&0&8&0 \\ 0&8&0&8 \end{pmatrix},$$

$$\mathbf{A}^5 = \begin{pmatrix} 0&16&0&16 \\ 16&0&16&0 \\ 0&16&0&16 \\ 16&0&16&0 \end{pmatrix}, \quad \mathbf{A}^6 = \begin{pmatrix} 32&0&32&0 \\ 0&32&0&32 \\ 32&0&32&0 \\ 0&32&0&32 \end{pmatrix}.$$

The entries come easily from the $\mathbf{A}$ eigenvalues $(\lambda_1, \lambda_2, \lambda_3, \lambda_4) = (0, -2, 0, 2)$ by inverse Fourier transforms, as ${A^n}_{ij} = 2^{n-2}\left[1 + (-)^{n+i-j}\right]$. Notice the rapid growth in the number of possible paths. This interesting relation between paths and spectrum is very deep, but the actual calculation of entries and eigenvalues is quite trivial: evidently $\lambda_k = \omega_N^k + \omega_N^{-k} = 2\cos\frac{2\pi k}{N}$ and $a_j = \delta_{j1} + \delta_{j,N-1}$, with convolutions easily obtained: $a_j^{*2} = 2(\delta_{j2} + \delta_{j,N})$, $a_j^{*3} = 2(\delta_{j3} + \delta_{j1} + a_j)$, etc. Notice that $\lambda_N = 2$ and $\sum_{i,j=1}^{N} {A^n}_{ij} = 2^n N$.

Note. In graph theory, $\boldsymbol{\Delta}$ is just the "combinatorial Laplacian" when all graphs involved have each vertex with degree (coordination number, in Physics terminology) = 2. In our case, all graphs are Cayley graphs (and Ramanujan).[2]

Note. The Fourier transforms of $\mathbf{d}$ and $\mathbf{d}^\dagger$ are $\tilde{\mathbf{d}} = \mathbf{V}^{-1} - \mathbf{I}$, $\tilde{\mathbf{d}}^\dagger = \mathbf{V} - \mathbf{I}$; that of $\boldsymbol{\Delta}$ will be $\tilde{\boldsymbol{\Delta}} = \mathbf{V}^{-1} + \mathbf{V} - 2\,\mathbf{I}$, with the expected entries

$$(\tilde{\Delta})_{rs} = \delta_{rs}\chi_r = -\,4\,\delta_{rs}\sin^2\frac{\pi r}{N}\,.$$

The third derivative will be given by

$$T_{ab} = \delta_{a+1,b} - 3\,\delta_{a,b} + 3\,\delta_{a-1,b} - \delta_{a-2,b} \text{ or } \boldsymbol{T} = \mathbf{U}^{-1} - 3\,\mathbf{I} + 3\,\mathbf{U} - \mathbf{U}^2 = \mathbf{U}^2\mathbf{d}^3,$$

of type

$$\boldsymbol{T} = \begin{pmatrix} -3 & 1 & 0 & 0 & 0 & \cdots & \cdots & -1 & 3 \\ 3 & -3 & 1 & 0 & 0 & \cdots & \cdots & \cdots & -1 \\ -1 & 3 & -3 & 1 & 0 & \cdots & \cdots & \cdots & 0 \\ 0 & -1 & 3 & -3 & 1 & \cdots & \cdots & \cdots & 0 \\ \cdots & \cdots & \cdots & \cdots & \cdots & \cdots & \cdots & \cdots & \cdots \\ 0 & \cdots & \cdots & \cdots & \cdots & \cdots & 3 & -3 & 1 \\ 1 & 0 & 0 & \cdots & \cdots & \cdots & -1 & 3 & -3 \end{pmatrix}.$$

In all these derivatives, the sum of the entries in a column is always zero. These matrices satisfy the conditions for a $\boldsymbol{Q}$ matrix. They have one zero eigenvalue. Fittingly enough, the corresponding eigenvector for $\mathbf{d}$ is the constant, flat distribution, $u_{a+1} = u_a = \frac{1}{N}\ \forall\, a$.

Note. "Integration by parts" Recall that the average of a function on a distribution $\mathbf{p}$ has been defined in Eq.(3.10) as $\bar{f} = \sum_a f_a p_a$. Consider the derivatives

$$f'_a = f_{a+1} - f_a\,, \quad f''_a = f'_{a+1} - f'_a = f_{a+1} + f_{a-1} - 2f_a\,, \quad \text{etc.}$$

Then, it is immediate to verify that, with the cyclic notation,

$$\bar{f'} = \sum_a f'_a p_a = -\sum_a f_a p'_a\,, \quad \bar{f''} = \sum_a f''_a p_a = \sum_a f_a p''_a\,,$$

$$\bar{f'''} = \sum_a f'''_a p_a = -\sum_a f_a p'''_a\,, \quad \text{etc.}$$

[2] For these concepts, see Ref. [39], ps. 49 and 56.

Note. Time derivatives and equilibrium The average of a function on a distribution at the n-th step $\mathbf{p}^{(n)}$ will be $\bar{f} = \sum_a f_a p_a^{(n)}$. The time derivative of such an average will be given by a finite-difference equation like (6.16), which here takes on the form

$$\dot{\bar{f}}^{(n)} = \bar{f}^{(n+1)} - \bar{f}^{(n)} = \sum_a f_a \dot{p}_a{}^{(n)} \ .$$

In equilibrium, $\dot{p}_a{}^{(n)} = 0$, which means that $\dot{\bar{f}}^{(n)} = 0$ for all functions f.

For standard random walks, Eq.(10.29) says that "harmonic" distributions, satisfying

$$\textstyle\sum_{b=1}^{N} \mathbf{\Delta}_{ab}\ p_b(t) = 0 \ , \tag{10.30}$$

are in equilibrium. In such a harmonic distribution, each entry is the average of its two immediate neighbors: $p_a = \frac{1}{2}\left(p_{a+1} + p_{a-1}\right)$. This is satisfied, of course, by the flat distribution $p_a = 1/N\ \forall\ a$ (see Commentary 117.1).

The master equation (6.18) will give to equilibrium a very natural meaning: gains and losses just compensate each other.

10.3.2 *Discrete Poisson process*

The Poisson process[3] is defined by the equation

$$\frac{d}{dt}\ p_a(t) = v \textstyle\sum_{b=1}^{N} \mathbf{d}_{ab}\ p_b(t) \ , \quad \text{or} \quad \frac{d}{dt}\ \mathbf{p}(t) = v\ \mathbf{d}\ \mathbf{p}(t) \ , \tag{10.31}$$

where we have introduced a constant v for future convenience. The solution has the form

$$\mathbf{p}(t) = e^{vt\mathbf{d}}\ \mathbf{p}(0) = \mathbf{M}^t\ \mathbf{p}(0) \ , \tag{10.32}$$

with the ruling stochastic matrix $\mathbf{M} = e^{v\mathbf{d}}$, whose eigenvalues are

$$\lambda_k = e^{\chi_k} = e^{v(\omega^{-k}-1)} \ . \tag{10.33}$$

There is a single equilibrium value $\lambda_N = 1$.

In the circulant base,

$$\mathbf{M} = \tfrac{1}{N} \sum_{k=1}^{N} \sum_{n=1}^{N} e^{v(\omega^{-k}-1)} e^{-i\frac{2\pi}{N}kn} \mathbf{U}^n \ , \tag{10.34}$$

which means that its letters are

$$a_n = \tfrac{1}{N} \sum_{k=1}^{N} e^{v(\omega^{-k}-1)} e^{-i\frac{2\pi}{N}kn} \ . \tag{10.35}$$

[3] See [40], p. 73.

$\mathbf{M}^t$ can also be written in terms of the general circulant eigenprojectors $\mathbf{P}_{(k)}$ given by (9.16):

$$\mathbf{M}^t{}_{ab} = \sum_{k=1}^{N} \lambda_k^t \, \mathbf{P}_{(k)ab} = \tfrac{1}{N} \sum_{k=1}^{N} e^{(\omega^{-k}-1)vt} \, e^{-i\frac{2\pi}{N}k(a-b)} \, . \tag{10.36}$$

Suppose the distribution to be initially concentrated at the "origin" point $0 \equiv N : p_b(0) = \delta_{bN}$. Then, at time t, it will be

$$p_a(t) = \tfrac{1}{N} \sum_{k=1}^{N} e^{(\omega^{-k}-1)vt} \, e^{-i\frac{2\pi}{N}k(a-b)} p_b(0) = \tfrac{1}{N} \sum_{k=1}^{N} e^{(\omega^{-k}-1)vt} \, e^{-i\frac{2\pi}{N}ka} \, . \tag{10.37}$$

The average deviation from the initial concentration point is given by Eq.(10.24):

$$< n >_{(t)} = \sum_{n=1}^{N} n p_n(t) = \tfrac{N}{2\pi i} \left[\frac{\partial}{\partial k} \left(\lambda_k^t \right) \right]_{k=0} = - \, vt \, . \tag{10.38}$$

This shows an average motion along the microstate space, with velocity v. Distribution (10.37) is not symmetric with respect to the point of initial concentration, and the the motion has a preferred sense, determined by the sign of v. The mean square deviation is given by Eq.(10.25):

$$< n^2 >_{(t)} = \sum_{n=1}^{N} n^2 p_n(t) = \left(\tfrac{N}{2\pi i} \right)^2 \left[\frac{\partial^2}{\partial k^2} \left(\lambda_k^t \right) \right]_{k=0} = v^2 t^2 - vt \, ,$$

or

$$< n^2 >_{(t)} - < n >^2_{(t)} = vt \, . \tag{10.39}$$

Commentary 112.1. ▶ The circulant components for the Poisson process are, if calculated in detail, given by combed series (see Comment 88.0). For example:

$$\operatorname{tr} \mathbf{M} = N a_N = \sum_{k=1}^{N} e^{v(\omega^{-k}-1)} = e^{-\,v} \sum_{k=1}^{N} e^{v\omega^{-k}} = e^{-\,v} \sum_{k=1}^{N} \sum_{r=0}^{\infty} \frac{(v)^r}{r!} \omega^{-kr} \, .$$

By the combing procedure,

$$\operatorname{tr} \mathbf{M} = N e^{-\,v} \sum_{p=0}^{\infty} \sum_{r=0}^{\infty} \frac{(v)^r}{r!} \delta_{r,pN} = N e^{-\,v} \sum_{p=0}^{\infty} \frac{(v)^{pN}}{(pN)!}$$
$$= N e^{-\,v} \, \mathrm{Comb}_N[\exp](v) \, .$$

The general component will be

$$a_j = e^{-\,v} \sum_{p \geq [j/N]}^{\infty} \frac{(v)^{pN-j}}{(pN-j)!} \, .$$

◀

10.4 Discrete diffusion

We can now address the question we have been preparing for, the discrete diffusion equation

$$\frac{d}{dt}\, p_a(t) = D \sum_{b=1}^{N} \Delta_{ab}\, p_b(t)\ , \quad \text{or} \quad \frac{d}{dt}\, \mathbf{p}(t) = D\ \boldsymbol{\Delta}\ \mathbf{p}(t)\ , \tag{10.40}$$

which is Eq.(6.23) with $\mathbf{Q} = D\boldsymbol{\Delta}$, D being a diffusion coefficient. It is easily verified that it is identical to the master equation

$$\frac{d}{dt}\, p_a(t) = D \sum_{b\neq a}^{N} \left[\Delta_{ab}\, p_b(t) - \Delta_{ba}\, p_a(t)\right]. \tag{10.41}$$

The solution of (10.40) has the form

$$\mathbf{p}(t) = e^{tD\boldsymbol{\Delta}}\ \mathbf{p}(0) = \mathbf{M}^t\ \mathbf{p}(0)\ , \tag{10.42}$$

with the ruling stochastic matrix $\mathbf{M} = e^{D\boldsymbol{\Delta}}$, whose eigenvalues are

$$\lambda_k = e^{D\chi_k} = e^{-\ 4D \sin^2 \frac{\pi k}{N}}\ . \tag{10.43}$$

There will be a single main root $\lambda_N = 1$, and $|\lambda_k| < 1$ for all other values of k. Consequently, a single equilibrium distribution.

Once more, $\mathbf{M}^t$ is better written in terms of the general circulant eigenprojectors $\mathbf{P}_{(k)}$ given by (9.16):

$$\mathbf{M}^t{}_{ab} = \sum_{k=1}^{N} \lambda_k^t\ \mathbf{P}_{(k)ab} = \tfrac{1}{N} \sum_{k=1}^{N} e^{-\ 4Dt \sin^2 \frac{\pi k}{N}}\ e^{-i\frac{2\pi}{N}k(a-b)}\ . \tag{10.44}$$

This matrix plays here the role of the heat kernel.

By using (9.22) we find

$$\mathbf{M} = \sum_{k=1}^{N} \lambda_k\ \mathbf{P}_{(k)} = \tfrac{1}{N} \sum_{k=1}^{N} \sum_{n=1}^{N} e^{-\ 4D \sin^2 \frac{\pi k}{N}}\ e^{-i\frac{2\pi}{N}kn}\ \mathbf{U}^n\ , \tag{10.45}$$

which says that $\mathbf{M}$ has circulant components

$$a_n = \tilde{\lambda}_n = \tfrac{1}{N} \sum_{k=1}^{N} e^{-\ 4D \sin^2 \frac{\pi k}{N}}\, e^{-i\frac{2\pi}{N}kn}\ .$$

Or, if we prefer, $\mathbf{M}$ is circulant like (9.1) with entries $M_{ab} = a_{a-b}$. The expressions for the entries,

$$\mathbf{M}_{ab} = \sum_{k=1}^{N} \lambda_k\ \mathbf{P}_{(k)ab} = \tfrac{1}{N} \sum_{k=1}^{N} e^{-\ 4D \sin^2 \frac{\pi k}{N}}\ e^{-i\frac{2\pi}{N}k(a-b)}\ , \tag{10.46}$$

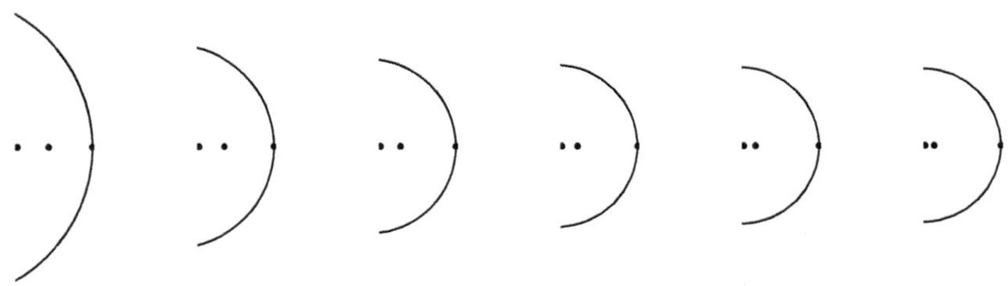

Fig. 10.2 *The first 6 steps in the evolution of the eigenvalues for $N = 4$, $D = 1/6$.*

show clearly that $\mathbf{M}$ is doubly-stochastic. As $\boldsymbol{\Delta}$ is symmetric, so is its exponential $\mathbf{M}$. The latter will be, consequently, in detailed balancing with the flat distribution. As an illustration: for $N = 4$ and $D = 1/6$, $\mathbf{M}$ has the eigenvalues $(0.716, 0.513, 0.716, 1)$. Figure 10.2 shows, within the unit circle, the six first steps in the evolution of these eigenvalues (value D = 1/6 has been chosen for graphical convenience). Two of them are coincident and eigenvalue $\lambda_4 = 1$ remains, of course, fixed.

The solution (10.42) takes on the form

$$p_a(t) = \tfrac{1}{N} \sum_{b=1}^{N} \sum_{k=1}^{N} e^{-\ 4Dt \sin^2 \frac{\pi k}{N}}\ e^{-i\frac{2\pi}{N}k(a-b)} p_b(0)\ . \tag{10.47}$$

10.4.1 *From a concentrated distribution*

We can learn something by considering point–concentrated initial distributions. If the distribution is initially concentrated at a point (say, at the point $0 \equiv N : p_b(0) = \delta_{bN}$), then after a time t it will be

$$p_a(t) = \tfrac{1}{N} \sum_{k=1}^{N} e^{-\ 4Dt \sin^2 \frac{\pi k}{N}}\ e^{-i\frac{2\pi}{N}ka}\ . \tag{10.48}$$

In this case the Fourier components are the eigenvalues themselves,

$$\tilde{p}_k(t) = e^{-\ 4Dt \sin^2 \frac{\pi k}{N}} = \lambda_k^t\ , \tag{10.49}$$

and matrix $\mathbf{M}^t$ of Eq.(10.44) can be written as

$$\mathbf{M}^t = \sum_{a=1}^{N} p_a(t)\ \mathbf{U}^a\ . \tag{10.50}$$

In other words, if we write $\mathbf{M} = \sum_{k=1}^{N} a_k \mathbf{U}^k$ for the governing stochastic circulant and $\mathbf{M}^t = \sum_{k=1}^{N} a_k(t)\ \mathbf{U}^k$ for its evolved at time t, then $a_k(t) =$

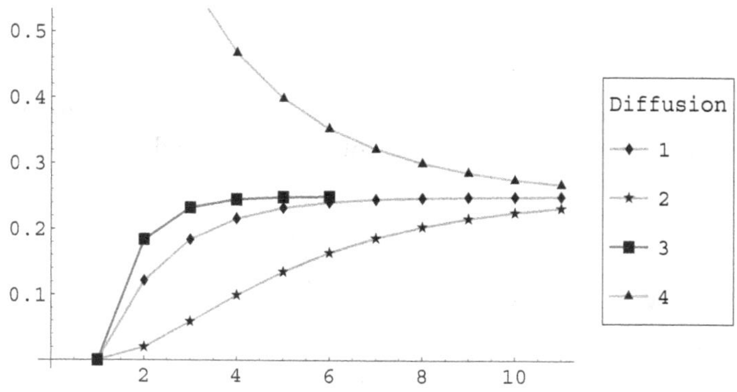

Fig. 10.3 *Case $N = 4$, $D = 1/6$. The first 6 steps in the evolution of the components of $\mathbf{p}(t)$ given by Eq.(10.48).*

$p_k(t)$. The letters of the basic stochastic matrix evolve like the probabilities of a distribution which is initially concentrated at a point.

We find the expected values $\tilde{p}_N(t) = \lambda_N^t = 1$. Using (10.22),

$$\sum_{a=1}^{N} \tilde{p}_a^t = \sum_{a=1}^{N} \lambda_a^t = N p_N^t \; . \tag{10.51}$$

The average of the deviation from the initial concentration point is easily obtained from Eq.(10.24):

$$< n > (t) = \sum_{n=1}^{N} n p_n(t) = \tfrac{N}{2\pi i} \left[\frac{\partial}{\partial k} \left(\lambda_k^t \right) \right]_{k=0} = \left[2iDt \sin \frac{2\pi k}{N} \right]_{k=0} = 0 \; . \tag{10.52}$$

This is consistent with the symmetric character of the distributions. The mean square deviation comes from Eq.(10.25):

$$< n^2 > (t) = \sum_{n=1}^{N} n^2 p_n(t) = \left(\tfrac{N}{2\pi i}\right)^2 \left[\frac{\partial^2}{\partial k^2} \left(\lambda_k^t \right) \right]_{k=0} = 2Dt \; . \tag{10.53}$$

Always for this initially concentrated (10.48) case, Figure 10.3 illustrates (for $N = 4$ and $D = 1/6$) the evolution of the components $p_a(t)$. We see the diffusion of the probability, initially concentrated in the fourth component, towards the common point of equilibrium. The velocity of this decline in concentration depends strongly on the value of D. More information comes from the profile of a given distribution at different times. Examples will be given below, in the left sides of Figures 10.6 and 10.7. The latter, in particular, shows the evolution of a gaussian distribution, a progressive flattenning towards homogeneity.

Note. It has been seen in a Note at page 91 how some Fourier transforms can be obtained by taking derivatives of the corresponding finite series. For $z^N = 1$ but $z \neq 1$,

$$\tfrac{1}{N}\sum_{n=1}^{N} n z^n = z\frac{\partial}{\partial z}\tfrac{1}{N}\sum_{n=1}^{N} z^n = \tfrac{z}{z-1} \tag{10.54}$$

$$\tfrac{1}{N}\sum_{n=1}^{N} n^2 z^n = z\frac{\partial}{\partial z}\left[z\frac{\partial}{\partial z}\tfrac{1}{N}\sum_{n=1}^{N} z^n\right] = \tfrac{z(Nz-N-2)}{(z-1)^2} \tag{10.55}$$

Notice that in the above final results we have used $z^N = 1$, so that the second does not come directly from the first. Considering separately the cases $z = e^{-i\frac{2\pi}{N}k} \neq 1$ for $k \neq N$ and $z = 1$ for $k = N$, we arrive at some expressions which will be useful in the following:

$$\tfrac{1}{N}\sum_{n=1}^{N} n e^{-i\frac{2\pi}{N}kn} = \begin{cases} i\,\frac{e^{-i\frac{\pi}{N}k}}{2\sin\frac{\pi}{N}k} & \text{for } k \neq N \\ \frac{(N+1)}{2} & \text{for } k = N\,. \end{cases} \tag{10.56}$$

$$\tfrac{1}{N}\sum_{n=1}^{N} n^2 e^{-i\frac{2\pi}{N}kn} = \begin{cases} \frac{1+iNe^{-i\frac{\pi}{N}k}\sin\frac{\pi}{N}k}{2\sin^2\frac{\pi}{N}k} & \text{for } k \neq N \\ \frac{(N+1)(2N+1)}{6} & \text{for } k = N\,. \end{cases} \tag{10.57}$$

10.4.2 *Normal modes*

Taking Fourier transforms diagonalize a circulant matrix. In consequence, each Fourier component will have an independent — diagonal — evolution, given simply by

$$\tilde{p}_a(t) = \sum_{b=1}^{N} e^{i\frac{2\pi}{N}ab} p_b(t) = \sum_{c=1}^{N}\sum_{k=1}^{N} e^{-\,4Dt\sin^2\frac{\pi k}{N}}\, e^{i\frac{2\pi}{N}kc} p_c(0)\tfrac{1}{N}\sum_{b=1}^{N} e^{i\frac{2\pi}{N}b(a-k)}$$

$$= e^{-\,4Dt\sin^2\frac{\pi a}{N}}\,\tilde{p}_a(0) = \lambda_a^t\,\tilde{p}_a(0)\,.$$

As in many mechanical problems, the Fourier components are here the normal modes. The dual lattice evolves in a simple multiplicative way. The N-th mode, which ultimately represents overall probability preservation, does not evolve at all: $\tilde{p}_N(t) = \tilde{p}_N(0) = 1$. It corresponds to the equilibrium eigenvector of $\mathbf{M}$, $\lambda_N = 1$. At large t, only that mode remains, so that

$$p_b(t) = \tfrac{1}{N}\sum_{k=1}^{N} e^{-i\frac{2\pi}{N}ab}\tilde{p}_a(t) \ \Rightarrow\ p_b(\infty) = \tfrac{1}{N}\ .$$

The homogeneous distribution will represent equilibrium.

We can alternatively use (9.22) to get

$$\mathbf{M}^t = \sum_{k=1}^{N} \lambda_k^t\, \mathbf{P}_{(k)} = \tfrac{1}{N} \sum_{k=1}^{N} \sum_{n=1}^{N} e^{-\ 4Dt\sin^2\frac{\pi k}{N}} e^{-i\frac{2\pi}{N}kn}\ \mathbf{U}^n$$

and

$$p_a(t) = \tfrac{1}{N} \sum_{k=1}^{N} \sum_{b=1}^{N} e^{-\ 4Dt\sin^2\frac{\pi k}{N}}\ e^{-i\frac{2\pi}{N}kb} p_{a-b}(0)$$

where use has been made of $(\mathbf{U}^n)_{ab} = \delta_{a,b+n}$. This is clearly the same as (10.47).

Solution (10.47) can, for $t = m$ integer, be written as a convolution:

$$p_r(m) = \tfrac{1}{N} \sum_{s=1}^{N} \sum_{k=1}^{N} \lambda_k^m\ e^{-i\frac{2\pi}{N}k(r-s)} p_s(0) = \sum_{s=1}^{N} a^{*m}_{(r-s)}\ p_s(0) \qquad (10.58)$$

because, by Eq.(9.19),

$$a^{*m}_{(r-s)} = \tfrac{1}{N}\sum_{k=1}^{N} \omega^{-k(r-s)}\ \lambda_k^m\ .$$

Or, again if we prefer, ${M^m}_{ab} = a^{*m}_{(a-b)}$, as it should be [see comments below Eq.(9.19)].

Commentary 117.1. ▶ As we are assimilating $\boldsymbol{\Delta}$ to a "Laplacian", we might ask about eventual "harmonic" distributions, those $\boldsymbol{p}$ for which $\boldsymbol{\Delta p} = 0$. It is very easy to see that, with the cyclic condition, the only solution is a constant distribution, just the above flat equilibrium distribution $p_a = \frac{1}{N}$, corresponding to the eigenvalue $\chi_N = 0$. This is reminiscent of the classical result on the complex plane. As $\boldsymbol{\Delta}$, and consequently $\boldsymbol{M}$, are symmetric matrices, $\boldsymbol{M}$ will be in detailed balance with that constant distribution. A last point worth noticing is that the Laplacian is essentially that part of a standard random walk which really changes something: $\frac{1}{2}\ \boldsymbol{\Delta} = \mathbf{R} - \mathbf{I}$. ◀

10.4.3 *Generalization*

A generalization including both the Poisson and the diffusion processes leads to some formal properties of interest. Take, instead of the previous $\mathbf{Q}$-matrix, $\mathbf{Q} = \alpha\mathbf{U} + \beta\mathbf{U}^{-1} - (\alpha+\beta)\mathbf{I}$. The corresponding master equation will be

$$\frac{d}{dt}\, p_a(t) = \sum_{b\neq a}^{N} \left[Q_{ab}\, p_b(t) - Q_{ba}\, p_a(t)\right] = \alpha p_{a-1}(t) + \beta p_{a+1}(t) - (\alpha+\beta) p_a(t). \tag{10.59}$$

The $\mathbf{Q}$ eigenvalues will be $\chi_k = \alpha\omega^k + \beta\omega^{-k} - (\alpha+\beta)$ with $\omega = e^{i\frac{2\pi}{N}}$. The solution will be $\mathbf{p}(t) = \mathbf{M}^t\mathbf{p}(0)$, with the stochastic matrix $\mathbf{M} = e^{\mathbf{Q}}$ of eigenvalues $\lambda_k = e^{\chi_k}$:

$$\mathbf{M}_{ab} = \sum_{k=1}^{N} \lambda_k\, \mathbf{P}_{(k)ab} = \tfrac{1}{N} \sum_{k=1}^{N} e^{\alpha\omega^k + \beta\omega^{-k} - (\alpha+\beta)}\, e^{-i\frac{2\pi}{N}k(a-b)}\ .$$

As a circulant,

$$\mathbf{M} = \tfrac{1}{N} \sum_{k=1}^{N} \sum_{n=1}^{N} e^{\alpha\omega^k + \beta\omega^{-k} - (\alpha+\beta)} e^{-i\frac{2\pi}{N}kn}\ \mathbf{U}^n$$

will have entries

$$a_n = \tfrac{1}{N} \sum_{k=1}^{N} e^{\alpha\omega^k + \beta\omega^{-k} - (\alpha+\beta)} e^{-i\frac{2\pi}{N}kn}\ .$$

In particular, $a_N = 1$. Notice $\frac{\partial}{\partial\alpha}\mathbf{M} = (\mathbf{U} - \mathbf{I})\mathbf{M} = \mathbf{d}^\dagger\mathbf{M}$, $\frac{\partial}{\partial\beta}\mathbf{M} = (\mathbf{U}^{-1} - \mathbf{I})\mathbf{M} = \mathbf{dM}$ and $\frac{\partial^2}{\partial\alpha\partial\beta}\mathbf{M} = -\ \mathbf{\Delta M}$. The equilibrium condition is met by the eigenstate related to λ_N, and all above reduces to the diffusion case when $\alpha = \beta = D$. The Poisson process corresponds to $\alpha = 0$, $\beta = v$. A "displaced" diffusion case comes out if $\alpha = \omega^{-K}$, $\beta = \omega^{K}$, with $\chi_k = \chi_K - 4\sin^2\frac{\pi}{N}(k-K)$.

In detail, the general solution is

$$p_a^{(t)} = \sum_{b=1}^{N} a^{*t}_{a-b}\, p_b(0) = \tfrac{1}{N} \sum_{b=1}^{N}\sum_{k=1}^{N} e^{t\left[\alpha\omega^k + \beta\omega^{-k} - (\alpha+\beta)\right]}\, e^{-i\frac{2\pi}{N}k(a-b)} p_b(0)\ . \tag{10.60}$$

Notice that only for an initial distribution concentrated at a point do the distributions $\{p_c\}$ and $\{a_c\}$ coincide.

The distribution is no more necessarily symmetric, and in principle we expect a nonvanishing first-momentum. In effect, from Eq.(10.24),

$$< n > (t) = \sum_{n=1}^{N} n p_n^{(t)} = \tfrac{N}{2\pi i} \left[\frac{\partial}{\partial k} \left(\lambda_k^t \right) \right]_{k=0} = (\alpha - \beta)\, t \,. \tag{10.61}$$

which vanishes when $\alpha = \beta$ and the distribution reduces to a symmetric one. The mean square deviation comes from Eq.(10.25):

$$< n^2 > (t) = \sum_{n=1}^{N} n^2 p_n^{(t)} = \left(\tfrac{N}{2\pi i}\right)^2 \left[\frac{\partial^2}{\partial k^2} \left(\lambda_k^t \right) \right]_{k=0} = [\alpha + \beta + (\alpha - \beta)^2]\, t \,. \tag{10.62}$$

A physically relevant example is given below, in the discussion of diffusion-advection chains.

10.5 Some phenomenology

We have up to this point stressed the formal aspects of both stochastic matrices and circulants. Let us now see how they can provide simple models for real systems. These are usually better described by models using differential equations, but the stochastic approach is suggestive enough to deserve attention. Even if in most cases they provide too rough versions of reality and should be classified as toy-models, they do convey, in some cases, real clues to complex behavior. Anyhow, the exercises which follow are intended to render more intuitive some of the most important features of these exciting objects.

10.5.1 *Diffusion-advection equation*

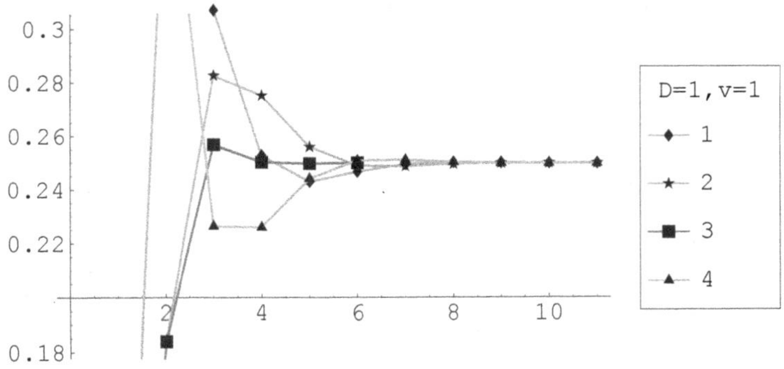

Fig. 10.4 *Case $N = 4$, $D = 1$, $v = 1$.*

The discrete version of the diffusion-advection equation (10.19) combines diffusion with a Poisson process (with opposite velocity). For an advection velocity v, the equation is

$$\frac{d}{dt}\,\mathbf{p}(t) = [D\,\mathbf{\Delta} - v\boldsymbol{d}]\,\mathbf{p}(t) = \left[D\mathbf{U} + (D-v)\mathbf{U}^{-1} + (v-2D)\mathbf{I}\right]\mathbf{p}(t)\ . \tag{10.63}$$

In components,

$$\frac{d}{dt}\,p_a(t) = (D-v)p_{a+1}(t) + Dp_{a-1}(t) + (v-2D)p_a(t)\ . \tag{10.64}$$

The eigenvalues of the $\boldsymbol{Q}$-matrix are $\chi_k = D\omega^k + (D-v)\omega^{-k} + (v-2D)$.

This is a particular case of the generalization of Section 10.4.3, with $\alpha = D$ and $\beta = D - v$. The solution will have once again the form

$$\mathbf{p}(t) = \mathbf{M}^t\ \mathbf{p}(0)\ , \tag{10.65}$$

but now with the ruling stochastic matrix of entries

$$\mathbf{M}_{ab} = \sum_{k=1}^{N} \lambda_k\, \mathbf{P}_{(k)ab} = \tfrac{1}{N}\sum_{k=1}^{N} e^{D\omega^k + (D-v)\omega^{-k} - (2D-v)}\ e^{-i\frac{2\pi}{N}k(a-b)}\ .$$

Note. As the eigenvalues are $\lambda_k = e^{-4D\sin^2\frac{\pi k}{N} - v(\omega^{-k}-1)}$,

$$\mathbf{M}^t_{ab} = \tfrac{1}{N}\sum_{k=1}^{N} e^{-4Dt\sin^2\frac{\pi k}{N} - vt(\omega^{-k}-1)}\omega^{k(b-a)}\ ;$$

$$p_a(t) = \tfrac{1}{N}\sum_{k=1}^{N} e^{-4Dt\sin^2\frac{\pi k}{N} - vt(\omega^{-k}-1)}\omega^{-ka}\tilde{p}_k(0)\ ;$$

$$\tilde{p}_n(t) = e^{-4Dt\sin^2\frac{\pi n}{N} - vt(\omega^{-n}-1)}\tilde{p}_n(0)\ .$$

This is an example of the "normal mode" behavior of the Fourier transformed distribution components, given by Eq.(10.23).

As a circulant,

$$\mathbf{M} = \tfrac{1}{N}\sum_{k=1}^{N}\sum_{n=1}^{N} e^{D\omega^k + (D-v)\omega^{-k} - (2D-v)} e^{-i\frac{2\pi}{N}kn}\ \mathbf{U}^n$$

will have entries

$$\begin{aligned} a_n &= \tfrac{1}{N}\sum_{k=1}^{N} e^{D\omega^k + (D-v)\omega^{-k} - (2D-v)} e^{-i\frac{2\pi}{N}kn} \\ &= \tfrac{1}{N}\sum_{k=1}^{N} e^{-4D\sin^2\frac{\pi k}{N} - v(\omega^{-k}-1)} e^{-i\frac{2\pi}{N}kn}\ , \end{aligned}$$

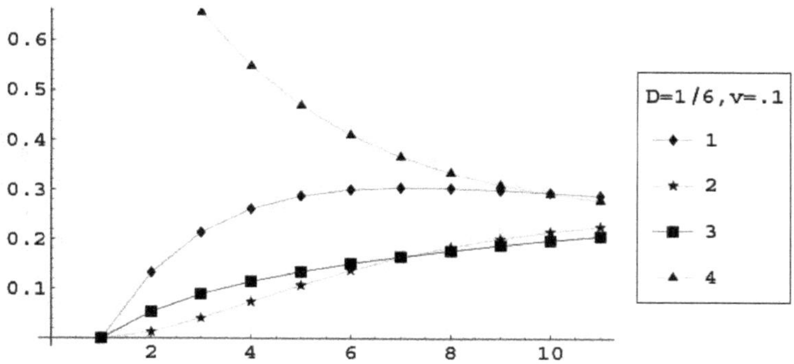

Fig. 10.5 *Case $N = 4$, $D = 1/6$, $v = 0.1$.*

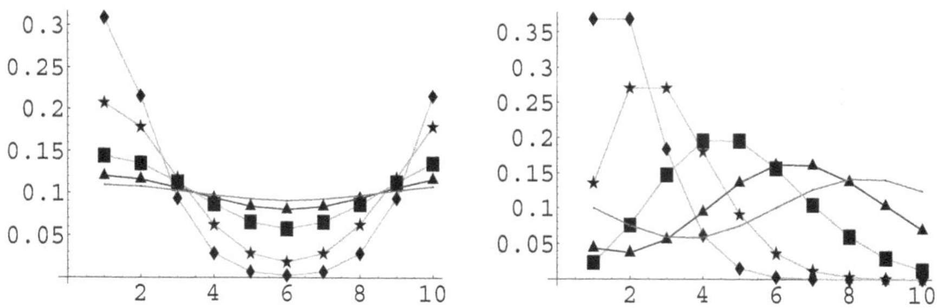

Fig. 10.6 *Profiles in function of time. At the right, $D = v = 1$ starting from concentrated at a point, $\mathbf{p}^{(0)} = (1, 0, 0, 0, 0, 0, 0, 0, 0, 0)$. At the left, no avection: $D = 1$ but $v = 0$.*

or

$$a_n = \frac{1}{N}\sum_{k=1}^{N} e^{2(v-2D)\sin^2\frac{\pi k}{N}} e^{i[v\sin\frac{2\pi k}{N} - \frac{2\pi}{N}kn]} .$$

In order to have for $\mathbf{M}$ a stochastic matrix, each a_n must satisfy $1 \geq a_n \geq 0$. The conditions on v and D are different for each value of N. For the case $N = 2$, it is necessary that $v \leq 2D$. Results of numerical computations for $N = 4$ are shown in Figures 10.4 and 10.5, for $D = v = 1$ and $D = 1/6$, $v = 0.1$. They should be compared with the pure diffusion of Figure 10.3. As expected, a large v/D ratio causes large departures, but the behavior tends to pure diffusion for smaller and smaller v's.

Figures 10.6 and 10.7 show (for $N = 10$) quite another aspect of the question: the time evolution of distributions. Figure 10.6 shows what hap-

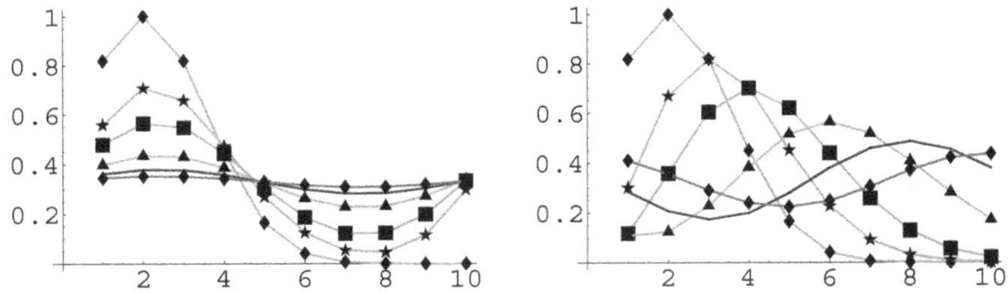

Fig. 10.7 *Evolution of distribution profiles, starting from a smoother, gaussian distribution.* $N = 10$, $D = 1$, $v = 1$.

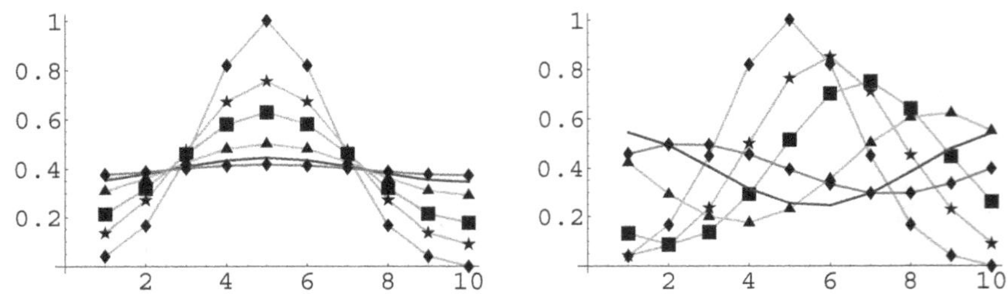

Fig. 10.8 *Same evolution as in the previous Figure, but starting from a centered gaussian distribution.*

pens to the picked distribution $\mathbf{p}^{(0)} = (1, 0, 0, 0, 0, 0, 0, 0, 0, 0)$. The left scheme shows its pure diffusion behavior, a simple flattenning. The curves in the right side show avection at work: the distribution does flatten by diffusion, but its maximum travels to the right. Figure 10.7 shows the analogous behavior for a smoother initial distribution, conveniently taken as the gaussian $f(x) = \exp(-\frac{(x-1)^2}{5})$. Figure 10.8 is the same case, but starting from a more centered distribution.

We expect a nonvanishing first-momentum. In effect, from Eqs.(10.24) or (10.61),

$$< n >_{(t)} = \sum_{n=1}^{N} n p_n(t) = \frac{N}{2\pi i} \left[\frac{\partial}{\partial k} \left(\lambda_k^t \right) \right]_{k=0} = vt \;, \tag{10.66}$$

which vanishes when $v = 0$ and the distribution reduces to a symmetric one. Pure diffusion does not contribute: this is just the effect of Eq.(10.38). The mean square deviation comes from Eqs.(10.25) or (10.62):

$$< n^2 >_{(t)} - < n >^2_{(t)} = \sum_{n=1}^{N} n^2 p_n(t) = \left(\tfrac{N}{2\pi i}\right)^2 \left[\frac{\partial^2}{\partial k^2}\left(\lambda_k^t\right)\right]_{k=0} = [2D - v]\, t\,. \tag{10.67}$$

In absence of diffusion, it reduces to the Poisson result (10.39).

10.5.2 *Prey-predator chains*

Let us examine some chains intended to model simple life-systems, in particular the so-called birth-death (or prey-predator) processes [42], [43], [44], [45].

In such processes there are 4 relevant protagonists: a predator (Y) feeds on a prey (X) and dies, thereby providing a death fund (D). The prey feeds on some other living character, represented by a source of food (A).

A first, very naïve stochastic model associates each protagonist to a microstate with a corresponding probability which is proportional to its quantity or concentration: p_1 for the food reservoir, p_2 for the prey, p_3 for the predator and p_4 for the dead remnant. Food is supposed to reproduce only by itself, according to the "chemical" reaction $A + A \to 2A$; besides self-reproduction, the prey reproduces by taking food, indicated by $X+A \to 2X$; the predator reproduces by eating the prey, $X + Y \to 2Y$; it also dies according to $Y \to D$; and the dead fund maintains itself, $D \to D$. Once normalized, the matrix describing those simple reactions is

$$\begin{pmatrix} 1/2 & 0 & 0 & 0 \\ 1/2 & 1/2 & 0 & 0 \\ 0 & 1/2 & 2/3 & 0 \\ 0 & 0 & 1/3 & 1 \end{pmatrix}. \tag{10.68}$$

Notice that this is a stochastic matrix, but not a circulant. The eigenvalues are $(1/2, 1/2, 2/3, 1)$, the last corresponding to the equilibrium distribution $(0, 0, 0, 1)$. This means that, in this model, pure death is the final state. As will be the case of all the Figures in this section, Figure 10.9 shows the evolution of each component of the successive states starting from a pure-food initial distribution $(1, 0, 0, 0)$. Probability of each living protagonist tends to zero at length. Notice, however, one point which is confirmed by observation, the predator maximum, which is retarded with respect to the prey maximum.

This model is exceedingly simplified and has too many obvious deffects. The whole set is supposed to constitute a closed system. The prey engenders

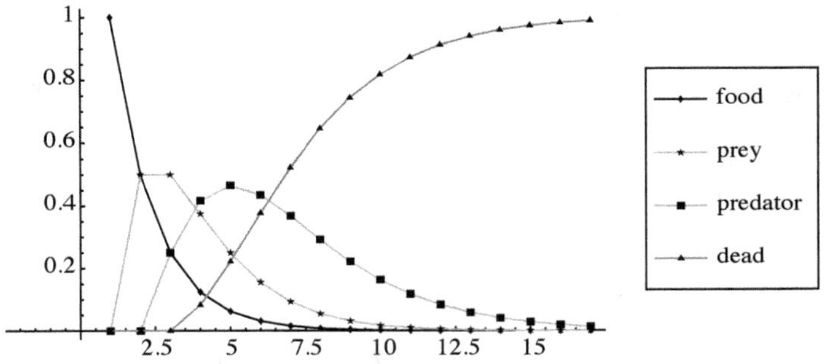

Fig. 10.9 *A simplest prey-predator model, in which food exhausts itself.*

predators from itself; food A engenders preys also, and exhausts itself. Also, there is no continuated oscillation as that found in the celebrated Lotka-Volterra model [45]. In at least one aspect it can be improved, by providing an extra food supply. Suppose in effect that part of the dead fund gives rise to food. The reactions become

$$A+A \to 2A\ ; \quad X+A \to 2X\ ; \quad X+Y \to 2Y\ ; \quad Y \to Y+D\ ; \quad D \to D+A\ .$$

The matrix is now

$$\begin{pmatrix} 1/2 & 0 & 0 & 1/2 \\ 1/2 & 1/2 & 0 & 0 \\ 0 & 1/2 & 2/3 & 0 \\ 0 & 0 & 1/3 & 1/2 \end{pmatrix}, \tag{10.69}$$

with a quite different equilibrium distribution $(2/9, 2/9, 3/9, 2/9)$. As illustrated in Figure 10.10, the simple fact that death creates food leads to amazingly new picture: the equilibrium has now a better "ecological", life-supporting, character. Though the predator keeps dominating, food and prey tend to a constant non-vanishing value, as does the dead fund.

We can proceed to some experiments. How to get more prey? It is enough to suppose a fair competition, the prey also feeding a bit from the predator. If, in the third column giving the contributions of the predator, we transfer half the diagonal probability to the prey row, we get

$$\begin{pmatrix} 1/2 & 0 & 0 & 1/2 \\ 1/2 & 1/2 & 1/3 & 0 \\ 0 & 1/2 & 1/3 & 0 \\ 0 & 0 & 1/3 & 1/2 \end{pmatrix}. \tag{10.70}$$

The evolution now, as shown in Figure 10.10, exhibits a prey dominance.

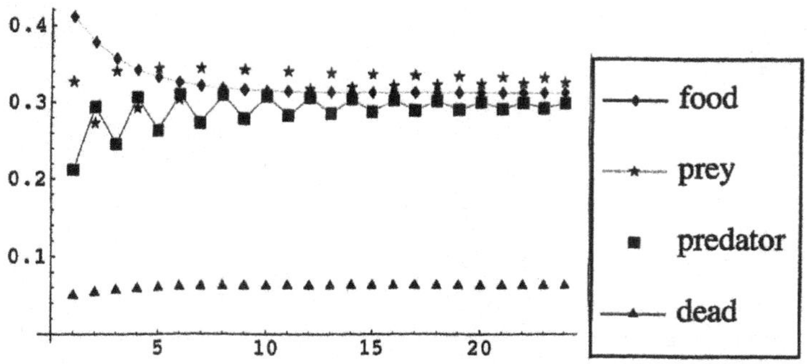

Fig. 10.10 *An experiment: some competition leads to prey dominance.*

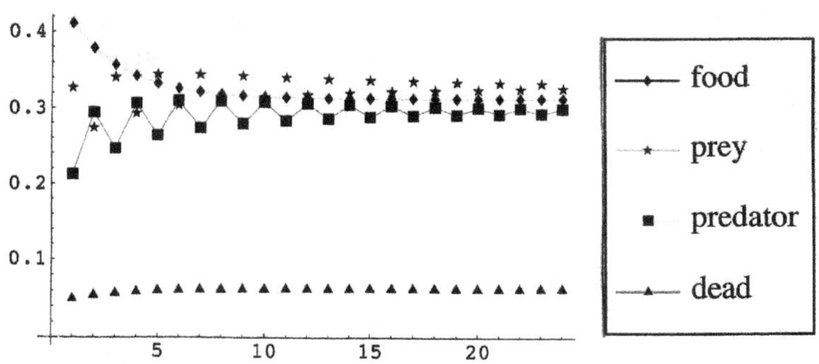

Fig. 10.11 *Oscillations, and all death becoming food.*

It is more difficult to get oscillations of the Lotka-Volterra type. We may try to match the $X \to Y$ and the $Y \to X$ contributions, and furthermore introduce a nearly cyclic relationship in the (X, Y) sector, represented by the central square in the matrix. For example,

$$\begin{pmatrix} 8/10 & 0 & 0 & 1 \\ 2/10 & 0 & 9/10 & 0 \\ 0 & 9/10 & 0 & 0 \\ 0 & 1/10 & 1/10 & 0 \end{pmatrix}$$

gives the behavior of Figure 10.11. We have imposed also that all death becomes food, a case to which we shall come back below. These more fairly competitive models stay already at a good distance from the prey-predator situation.

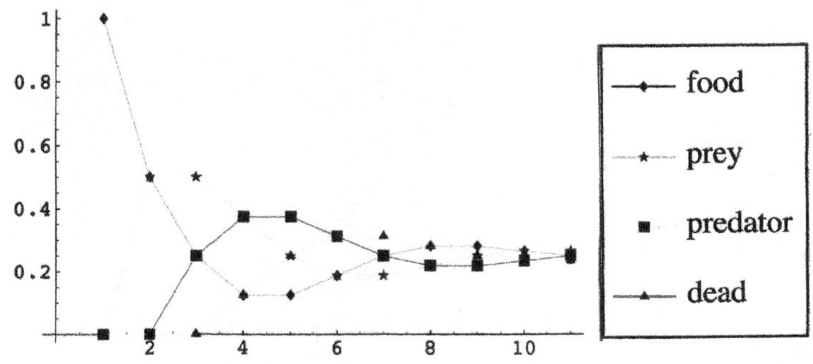

Fig. 10.12 *Forcing a circulant leads to a homogeneous equilibrium.*

The above matrices are not, of course, circulants. When equilibrium is possible, circulants are constrained by the forceful presence of final homogeneity (see comment below Eq.(10.20)). Let us take as an exercise a modification of matrix (10.70):

$$\begin{pmatrix} 1/2 & 0 & 0 & 1/2 \\ 1/2 & 1/2 & 0 & 0 \\ 0 & 1/2 & 1/2 & 0 \\ 0 & 0 & 1/2 & 1/2 \end{pmatrix}.$$

The predator now dies more quickly and regenerates itself less. This model leads, after some oscillations, to the flat distribution $\frac{1}{4}$ $(1,1,1,1)$: food, prey, predator and dead with the same probability (Figure 10.12).

There is no reason however, to have equilibrium in a system of the Lotka-Volterra type. Actually, a behavior of that kind — with coupled continuated oscillations — is given by the simplest of all random walks, the basic, pure circulant matrix $\mathbf{U}$. For the present four-state case,

$$\mathbf{U} = \begin{pmatrix} 0 & 0 & 0 & 1 \\ 1 & 0 & 0 & 0 \\ 0 & 1 & 0 & 0 \\ 0 & 0 & 1 & 0 \end{pmatrix}.$$

An example is given in Figure 10.13, for an initial distribution with 50% food, 40% prey and 10% predator. Notice that the matrix entries say that food comes exclusively from death. Prey, only from food. Predator, from prey. And death feeds only from predator. This is a good illustration of pure circulant behavior.

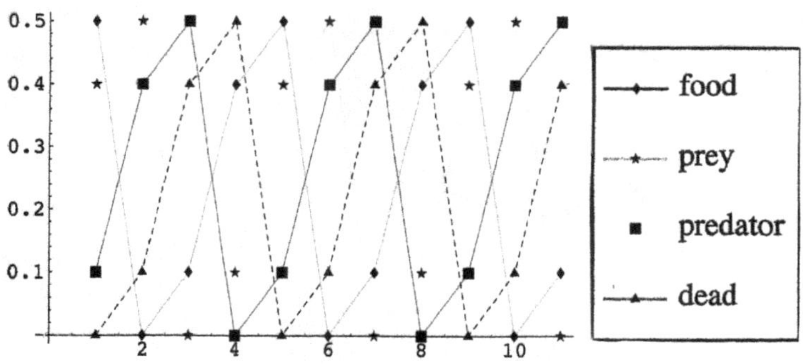

Fig. 10.13 *A Lotka-Volterra behavior coming from matrix* **U**.

Going back to our initial model: what happens when more food is supplied? We can see it by increasing the contribution $D \to A$, for example by taking

$$\begin{pmatrix} 1/2 & 0 & 0 & 3/4 \\ 1/2 & 1/2 & 0 & 0 \\ 0 & 1/2 & 2/3 & 0 \\ 0 & 0 & 1/3 & 1/4 \end{pmatrix},$$

which has equilibrium $(6/25, 6/25, 9/25, 4/25)$, much better for life.

We can let all death become food: matrix

$$\begin{pmatrix} 1/2 & 0 & 0 & 1 \\ 1/2 & 1/2 & 0 & 0 \\ 0 & 1/2 & 2/3 & 0 \\ 0 & 0 & 1/3 & 0 \end{pmatrix}$$

leads to an ecological equilibrium which is only slightly better, $\frac{1}{8}\,(2, 2, 3, 1)$.

We can now force things a bit so as to have again a circulant: matrix

$$\begin{pmatrix} 1/4 & 0 & 0 & 3/4 \\ 3/4 & 1/4 & 0 & 0 \\ 0 & 3/4 & 1/4 & 0 \\ 0 & 0 & 3/4 & 1/4 \end{pmatrix},$$

which once more exhibits the homogeneous equilibrium distribution, attained however after many oscillations, as shown in the Figure 10.14.

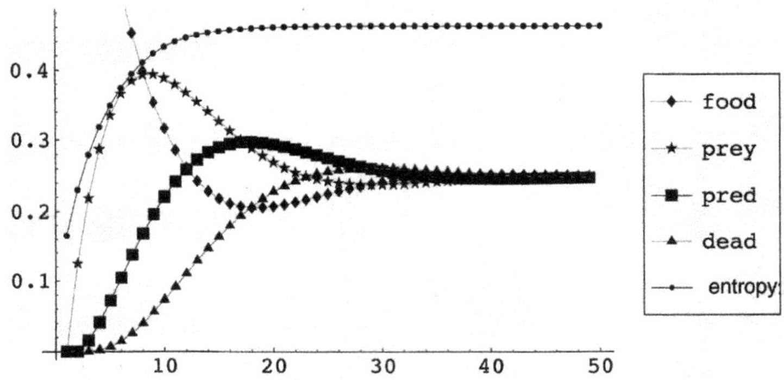

Fig. 10.14 *Less food, less oscillations. A flat equilibrium, and the growth of entropy. The latter has been downscaled by a factor* 1/3.

Practically the same equilibrium is found with less food feedback and oscillations if we take the circulant

$$\begin{pmatrix} 7/8 & 0 & 0 & 1/8 \\ 1/8 & 7/8 & 0 & 0 \\ 0 & 1/8 & 7/8 & 0 \\ 0 & 0 & 1/8 & 7/8 \end{pmatrix},$$

which presides on the behavior shown in the that Figure. Also the progression of an entropy appears there. That entropy is defined as the negative of the Boltzmann function introduced in Comment 65.1 as a function of the evolution time "n":

$$S^{(n)} = - \sum_{j=1}^{N} p_j^{(n)} \ln p_j^{(n)} \ . \qquad (10.71)$$

As everything above, it has been calculated for $N = 4$ and starting with $p^{(0)} = (1, 0, 0, 0)^T$.

Chapter 11

Applications I: Chemical reactions

The word "reaction" will be used here in a very broad sense, so as to include decays and processes involving either elementary particles, atoms or molecules.

11.1 General aspects

Reactions are usually taken to be processes

$$\nu_i A_i \longrightarrow \nu_k B_k \ ,$$

with "reactants" being written conventionally in the left-hand side and "products" at the right. Decays $A_i \longrightarrow \nu_k B_k$ are of interest to emphasize the non-equilibrium possibilities of the method, as long as the steps in the Markovian progress are interpreted as being taken along time. The probabilities used are those of finding certain kinds of particles in a large enough sample of the system and are thus ultimatelly related to their relative concentrations. This is not the most usual approach in Statistical Mechanics, which treats with probabilities of finding the system as a whole in certain (classical or quantum) mechanical states. Neither is the matrix, linear approach the best formulation for usual chemical reactions in gases and liquids, in which the reactant particles move to meet each other. The matrix approach can be expected to describe situations in which only one of the reactants has significant mobility.

We can think of some examples of such cases of unequal mobility:

- decays, in which there is a single reactant which can be taken at rest, so that only the products have mobility; and two cases in which the converse happens:

- when the process is preceded by nucleation: one of the two combining atoms or molecules is at rest on the rim of the nucleation center, while the other travels freely through the medium (case of the vitrification, seen in Chapter 4); once it interacts, the mobile element replaces a former element of the rim, becoming immobile: $A_{mobile}+A_{rim} \to A_{rim}$;
- when the molecules have distinct behaviors, as in the "reactions" involved in genetic pairing: one stays at the female gamete, waiting for one of the many many mobile molecules of the male gamete (this will be seen later, in Chapter 12).

To get a better insight into the conditions in which the treatment can be expected to be valid and, still more, the situations in which it should not hold at all, let us begin with a short review of the usual kinetic approach to chemical reactions in gases and liquids.

11.1.1 *Kinetic approach*

Consider two particles p and n, able to transmute into each other by some reaction. We shall take as a guide the well-defined case of protons and neutrons related by the "inverse β decay" reaction

$$n + e^{+} \longleftrightarrow p + \bar{\nu} \ , \tag{11.1}$$

of interest in cosmological nucleosynthesis. As the total baryon number is conserved, only the relative concentrations of neutrons and protons will be of interest. Let them be

$$X_n = \frac{n_n}{n} \ , \quad X_p = \frac{n_p}{n} \ ,$$

where n_p, n_n and $n = n_p + n_n$ are the proton, neutron and total baryon concentrations.

A reaction rate is the number of reactions per second. Thus, $R(n \to p)$ is the number of neutrons which produce protons per second. In the kinetic approach, the relative concentrations will be given as functions of time by the master equations (see Eq.(6.18) and comments around it)

$$\frac{d}{dt}X_p = R(n \to p)X_n - R(p \to n)X_p$$

$$\frac{d}{dt}X_n = R(p \to n)X_p - R(n \to p)X_n \ .$$

Variation in neutron abundance is the abundance of protons times the rate of proton to neutron transformation (this represents the gain) less the neutron abundance times the rate of its disappearance.

11.1.1.1 *Linear approximation*

As long as the reaction rates can be supposed to be constant, a purely stochastic approach can be used. How to correct it, at least partially, will be seen later.

We shall use the notation $R_{np} = R(p \to n)$, $R_{pn} = R(n \to p)$ and rewrite the above equation as

$$\tfrac{d}{dt}\mathbf{X} = \mathbf{R}\,\mathbf{X} = \tfrac{d}{dt}\begin{pmatrix} X_p \\ X_n \end{pmatrix} = \begin{pmatrix} -R_{np} & R_{pn} \\ R_{np} & -R_{pn} \end{pmatrix}\begin{pmatrix} X_p \\ X_n \end{pmatrix}. \tag{11.2}$$

Equilibrium corresponds to $\frac{d}{dt}\,\mathbf{X} = 0$. In detail, it means $R_{np}X_p = R_{pn}X_n$, which is just the detailed balance condition (7.3).

Matrix $\mathbf{R}$ respects the conditions for a $\mathbf{Q}$-matrix like that of Eq.(6.23). Its eigenvalues are $(0, -R_{pn} - R_{np})$, so that

$$\mathbf{M} = e^{\mathbf{R}} = \mathbf{I} + \frac{1 - e^{-(R_{np}+R_{pn})}}{R_{np} + R_{pn}}\,\mathbf{R}$$

will be a stochastic matrix with eigenvalues $(1, e^{-R_{pn}-R_{np}})$. The solution of (11.2) will be

$$\mathbf{X}(t) = \mathbf{M}^t\,\mathbf{X}(0)\;, \tag{11.3}$$

with

$$\mathbf{M}^t = \frac{1}{R_{np} + R_{pn}}\begin{pmatrix} e^{-(R_{np}+R_{pn})t}R_{np} + R_{pn} & R_{pn}(1-e^{-(R_{np}+R_{pn})t}) \\ R_{np}(1-e^{-(R_{np}+R_{pn})t}) & e^{-(R_{np}+R_{pn})t}R_{pn} + R_{np} \end{pmatrix}.$$

The first column gives the solution at time t if the initial state is pure proton, $\mathbf{X}(0) = \binom{1}{0}$. The equilibrium distribution will be $\mathbf{X}^{(\infty)} = \frac{1}{R_{pn}+R_{np}}\begin{pmatrix} R_{pn} \\ R_{np} \end{pmatrix}$, so that the concentration ratio tends to $\frac{n_p}{n_n} = \frac{R_{pn}}{R_{np}}$, as expected. The solution

$$X_p(t) = \left\{ e^{-(R_{np}+R_{pn})t}\left(X_p(0) - \frac{R_{pn}}{R_{np} + R_{pn}} \right) + \frac{R_{pn}}{R_{np} + R_{pn}} \right\} \tag{11.4}$$

shows clearly that equilibrium.

The error we are doing comes from the supposition that $\mathbf{M}$ is independent of the state on which it acts. This is clearly not the case for the reaction rates for reaction (11.1), which are given by velocity/(mean free path):

$$R(n \to p) = v_e\, n_n\, \sigma_{ne} = n\, v_e\, \sigma_{ne}\, X_n\;, \tag{11.5}$$

$$R(p \to n) = c\, n_p\, \sigma_{\bar{\nu}p} = n\, c\, \sigma_{\bar{\nu}p}\, X_p\;, \tag{11.6}$$

where v_e is the electron velocity, c the neutrino (supposed with zero mass) velocity, σ_{ne} and $\sigma_{\bar{\nu}p}$ the neutron-electron and neutrino-proton weak-interaction cross-sections provided by the standard Glashow–Weinberg–Salam electroweak theory.

Equation (11.2) shows its true, non-linear face when we use the values (11.5) and (11.6), $R_{np} = n\, c\, \sigma_{\bar{\nu}p}\, X_p$ and $R_{pn} = n\, v_e\, \sigma_{ne}\, X_n$:

$$\frac{d}{dt}\begin{pmatrix} X_p \\ X_n \end{pmatrix} = \begin{pmatrix} -\,n\,c\,\sigma_{\bar{\nu}p}\,X_p & n\,v_e\,\sigma_{ne}\,X_n \\ n\,c\,\sigma_{\bar{\nu}p}\,X_p & -\,n\,v_e\,\sigma_{ne}\,X_n \end{pmatrix}\begin{pmatrix} X_p \\ X_n \end{pmatrix} = \begin{pmatrix} -a & b \\ a & -b \end{pmatrix}\begin{pmatrix} X_p^2 \\ X_n^2 \end{pmatrix}, \tag{11.7}$$

where now we use a far better approximation, with the constants $a = n\, c\, \sigma_{\bar{\nu}p}$ and $b = n\, v_e\, \sigma_{ne}$.

11.1.1.2 *Full kinetics*

Consider now the differential equation

$$\frac{d}{dt}X_p = -aX_p^2 + bX_n^2 = -aX_p^2 + b(1-X_p)^2 = b + (b-a)\,X_p^2 - 2\,b\,X_p\,. \tag{11.8}$$

The second-order term is present only if $a \neq b$. In that case the solution is

$$X_p(t) = \frac{1}{a-b}\left\{-b + \sqrt{a\,b}\,\tanh\left[\sqrt{a\,b}\,t + \operatorname{arctanh}\left(\frac{b + (a-b)\;X_p(0)}{\sqrt{a\,b}}\right)\right]\right\}, \tag{11.9}$$

and, for the neutron,

$$X_n(t) = \frac{1}{b-a}\left\{-a + \sqrt{a\,b}\,\tanh\left[\sqrt{a\,b}\,t + \operatorname{arctanh}\left(\frac{a + (b-a)\;X_n(0)}{\sqrt{a\,b}}\right)\right]\right\}. \tag{11.10}$$

One goes from $X_p(t)$ to $X_n(t)$ by simply exchanging a and b.

Note. As an aside, let us rewrite these expressions as

$$\frac{b + (a-b)\,X_p(t)}{\sqrt{a\,b}} = \tanh\left[\sqrt{a\,b}\,t + \operatorname{arctanh}\left(\frac{b + (a-b)\;X_p(0)}{\sqrt{a\,b}}\right)\right],$$

$$\frac{a + (b-a)\,X_n(t)}{\sqrt{a\,b}} = \tanh\left[\sqrt{a\,b}\,t + \operatorname{arctanh}\left(\frac{a + (b-a)\;X_n(0)}{\sqrt{a\,b}}\right)\right],$$

and observe a curious fact: introducing the function

$$u_p(t) = \frac{b + (a-b)\,X_p(t)}{\sqrt{a\,b}}\,, \tag{11.11}$$

the solution can be rewritten as

$$u_p(t) = \tanh\left(\sqrt{a\,b}\,t + \operatorname{arctanh}[u_p(0)]\right). \tag{11.12}$$

This function is of iterative type for integer values of t, by which we mean that

$$u_p(n) = \tanh\left(\sqrt{a\,b} + \operatorname{arctanh}[u_p(n-1)]\right)$$
$$= \tanh\left(\sqrt{a\,b}\,n + \operatorname{arctanh}[u_p(0)]\right) = u_p^{<n>}(0) \ , \qquad (11.13)$$

where $u_p^{<n>}$ is the n-th iterate of function u_p. This means that also $X_p(t)$ is of iterative type. In effect, using the notation $f \circ g$ for the composition of functions f and g, define the function w by $w = u^{<-1>} \circ \tanh$,

$$w(z) = \frac{1}{a-b}\left(-b + \sqrt{a\,b}\,\tanh z\right) \qquad (11.14)$$

and its inverse

$$w^{<-1>}(x) = \operatorname{arctanh}\left[\frac{b + (a-b)\,x}{\sqrt{a\,b}}\right]. \qquad (11.15)$$

Introduce also f by

$$f(u) = \sqrt{a\,b} + u, \quad f^{<2>}(u) = 2\sqrt{a\,b} + u, \quad \text{etc.} \qquad (11.16)$$

Then,

$$X_p(2) = w \circ f \circ w^{<-1>} \circ X_p(1) = w \circ f^{<2>} \circ w^{<-1>} \circ X_p(0) \qquad (11.17)$$

$$X_p(m) = w \circ f^{<m>} \circ w^{<-1>} \circ X_p(0)\ . \qquad (11.18)$$

Putting the functions together in

$$v = w \circ f \circ w^{<-1>}\ , \qquad (11.19)$$

we arrive at the compact form

$$X_p(m) = v^{<m>}\left(X_p(0)\right). \qquad (11.20)$$

At each order of iteration the interval $[0,1]$, on which lie the values of X_p, is taken into itself. As $Y_p = 1 - X_p$, each point on $[0,1]$ fixes a state. The mapping involved, of type $g(x) = \tanh[c + \operatorname{arctanh}(x)]$ has a very simple behavior: its higher iterates tend more and more to a step-function, so that ultimately every point of $[0,1]$ will be taken into one same point (Figure 11.1, right). This characterizes a tendency to equilibrium and an independence of the initial state. We shall later (Section 17.2.2) meet the extreme logistic map (Figure 11.1, left), which is related to $g(x) = \cos[c \times \arccos(x)]$ and has quite different properties — in particular, each image-point has two inverses.

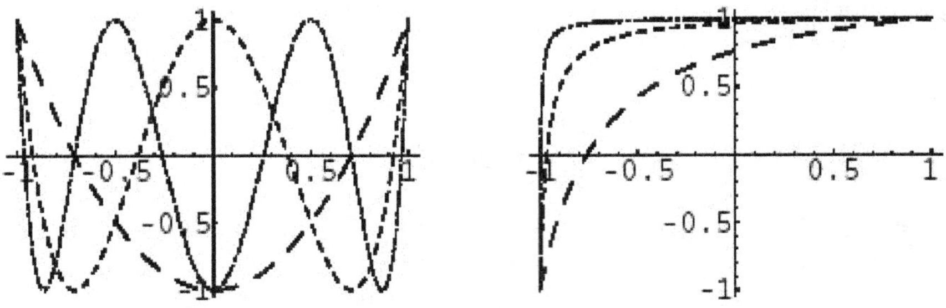

Fig. 11.1 *Maps and some of their first iterations, with increasing iteration orders indicated by finer dashings. Left: extreme logistic map* $g(x) = \cos[2 \times \arccos(x)]$. *Right: map* $g(x) = \tanh[c + \operatorname{arctanh}(x)]$.

Equation (11.9) shows that, for $t \to \infty$, equilibrium is attained with

$$X_p(t \to \infty) = \frac{\sqrt{a\,b} - b}{a - b} = \frac{\sqrt{b}}{\sqrt{a} + \sqrt{b}}\,, \tag{11.21}$$

one of the two possibilities suggested by Eq.(11.8). Now, as $a = nc\sigma_{\bar{\nu}p}$ and $b = nv_e\sigma_{ne}$,

$$X_p(t \to \infty) = \frac{\sqrt{v_e\sigma_{ne}}}{\sqrt{c\sigma_{\bar{\nu}p}} + \sqrt{v_e\sigma_{ne}}}\,.$$

Only to have a plot, let us put $a = 2$ and $b = 1/2$. Equilibrium is attained with $X_p(t \to \infty) = 1/3$ (Figure 11.2).

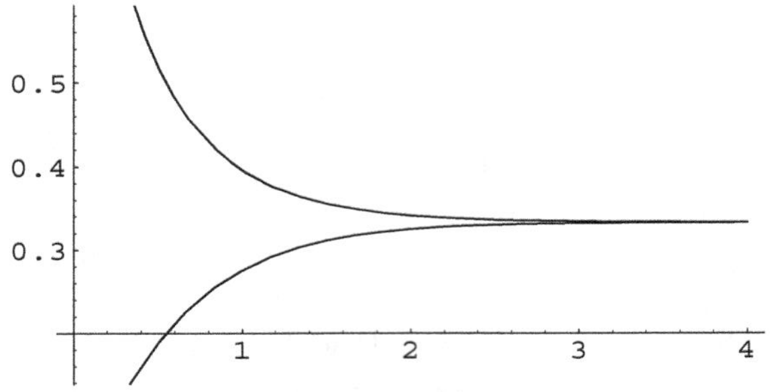

Fig. 11.2 *A toy example: relative "proton" concentration, with* $a = 2$ *and* $b = 1/2$. *Upper graph: starting with "protons" only; lower graph: starting only with "neutrons".*

Let us now formalize this 2×2 case a little more.

11.1.1.3 *The simplest of all cases*

Consider an evolution stochastic matrix $\mathbf{M} = \begin{pmatrix} 1-A & B \\ A & 1-B \end{pmatrix}$, acting on probability distributions denoted $\begin{pmatrix} X \\ Y=1-X \end{pmatrix}$. It has eigenvalues $\lambda_1 = 1$ [corresponding to equilibrium state $\begin{pmatrix} X^\infty \\ Y^\infty \end{pmatrix} = \frac{1}{A+B}\begin{pmatrix} B \\ A \end{pmatrix}$] and $\lambda_2 = 1 - A - B$, and its $\mathbf{Q}$-matrix is

$$\mathbf{Q} = \ln \mathbf{M} = \frac{\ln(1-A-B)}{A+B} \begin{pmatrix} A & -B \\ -A & B \end{pmatrix},$$

with eigenvalues $\chi_1 = 0$ and $\chi_2 = \ln(1 - A - B)$. It is convenient to introduce parameters a and b given by

$$a = \frac{B}{A+B}; \quad b = -\ln(1 - A - B) = -\ln \lambda_2$$
$$A = (1-a)(1-e^{-b}); \quad B = a(1-e^{-b}),$$

in terms of which

$$\mathbf{M} = \begin{pmatrix} e^{-b}+a(1-e^{-b}) & a(1-e^{-b}) \\ (1-a)(1-e^{-b}) & 1-a(1-e^{-b}) \end{pmatrix} \quad \text{and} \quad \mathbf{Q} = \begin{pmatrix} -b(1-a) & ab \\ b(1-a) & ab \end{pmatrix}.$$

Matrix $\mathbf{Q}$ corresponds to the linear differential equation

$$\frac{dX}{dt} = ab - bX = \ln(1 - A - B)X - \frac{B\ln(1-A-B)}{A+B}, \tag{11.22}$$

whose solution is

$$X_t = a + e^{-bt}(X_0 - a) = \frac{B+(1-A-B)^t[(A+B)X_0-B]}{A+B}. \tag{11.23}$$

Introducing the function $F(u) = \frac{B+u}{A+B} = \frac{a(1-e^{-b})+u}{(1-e^{-b})} = a + \frac{u}{(1-e^{-b})}$ and its inverse $F^{<-1>}(z) = (A+B)z - B = (1-e^{-b})(z-a)$, the solution takes up the form of an iterative mapping:

$$X_t = F\left[(1-A-B)^t \times F^{<-1>}[X_0]\right] \tag{11.24}$$
$$= F[\lambda_2^t \times F^{<-1>}[X_0]] = F[e^{-bt} \times F^{<-1>}[X_0]]. \tag{11.25}$$

For integer values of time,

$$X_{k+1} = F[e^{-b} \times F^{<-1>}[X_k]] = a + e^{-b}(X_k - a) = B + (1-A-B)X_k. \tag{11.26}$$

This is, of course, just one step in the evolution of component X, the effect of the action of matrix $\mathbf{M}$.

Notice from the equation that equilibrium ($\frac{dX}{dt} = 0$) is given by $X_\infty = a$, so that solution (11.23) is $X_t + X_\infty = e^{-bt}(X_0 + X_\infty)$. This is consistent with $\left(\begin{smallmatrix} X_\infty \\ Y_\infty \end{smallmatrix}\right) = \left(\begin{smallmatrix} a \\ 1-a \end{smallmatrix}\right)$. Recall (Comment 64.1) that any 2×2 stochastic matrix is always in detailed balance with its asymptotic equilibrium state. The parameter $\tau = 1/b$ has the character of a relaxation time or a mean lifetime. If $X_\infty = 0$, as happens in most usual decays, $\mathbf{M} = \left(\begin{smallmatrix} e^{-1/\tau} & 0 \\ 1-e^{-1/\tau} & 1 \end{smallmatrix}\right)$.

The case $A = B$ leads to a diffusion equation. Indeed, $\mathbf{\Delta} = 2\left(\begin{smallmatrix} -1 & 1 \\ 1 & -1 \end{smallmatrix}\right)$ in this case, and $\mathbf{M} = \mathbf{I} + \frac{B}{2}\,\mathbf{\Delta}$. The equations take up the form

$$\frac{d}{dt}\begin{pmatrix} X \\ Y \end{pmatrix} = \boldsymbol{Q}\begin{pmatrix} X \\ Y \end{pmatrix} = \frac{b}{4}\mathbf{\Delta}\begin{pmatrix} X \\ Y \end{pmatrix}.$$

This 2×2 case shows the main qualities and limitations of the linear approach. The matrix action corresponds to a linear map (11.26). We feel now, with all that in mind, prepared to apply stochastic matrices to more involved systems.

11.2 Binary decays

Suppose particles of type A decay into particles of two other types,

$$A \longrightarrow C + D\ .$$

This happens mainly in radioactive decays involving strong and electromagnetic interactions. The system state will be macroscopically characterized by the concentrations of the three especies. If N_A, N_C and N_D are the numbers of particles A, C and D, with total number $N = N_A + N_C + N_D$, the relative concentrations $c_A = \frac{N_A}{N}$, $c_C = \frac{N_C}{N}$ and $c_D = \frac{N_D}{N}$ are all non-negative and add up to 1. If we take one particle at random in the system, the probability of getting a molecule type A will be $p_A = c_A$ and so on. Thus, the statistical distribution $\left(\begin{smallmatrix} p_A \\ p_C \\ p_D \end{smallmatrix}\right)$ will represent the state. If the unstable particles A have just been produced or introduced, the system will be in state $\left(\begin{smallmatrix} 1 \\ 0 \\ 0 \end{smallmatrix}\right)$. This is a reasonable assumption for the initial state. The most general 3×3 stochastic matrix has been seen in Comment 52.1:

$$\mathbf{M} = \begin{pmatrix} 1-a-b & c & e \\ a & 1-c-d & f \\ b & d & 1-e-f \end{pmatrix}.$$

Let us examine the meaning of the entries: those in the first column represent the probabilities of A producing other particles. Using bracket notation, we can say that $a = \langle C|M|A\rangle$ and $b = \langle D|M|A\rangle$ are compensated by the probability $1 - a - b = \langle A|M|A\rangle$ of A staying itself. The entries in the second column are the probabilities of C producing A, remaining itself, and (a final state transition) producing D. Similarly for for the third column, in which $e = \langle A|M|D\rangle$ and $f = \langle C|M|D\rangle$.

Now, decay is a one-way trip: neither C nor D produce A, so that $c = \langle A|M|C\rangle = e = \langle A|M|D\rangle = 0$. For the time being we can leave open the question of whether or not there are final state transitions, represented by f and d.

There are still too many parameters. Some extra input is necessary, and can be provided by phenomenology. Phenomenology is a backward-thinking art. First it would recall that C and D are produced by A in equal amounts: no atom C can appear without an atom D turning up. This means $b = a$. Then it would tell that we should take profit of a fact of observation — we expect to have a final state $\begin{pmatrix} 0 \\ 1/2 \\ 1/2 \end{pmatrix}$: no more decaying particles, and equal amounts of C and D. By itself, this would give information only on the asymptotic matrix $\mathbf{M}^\infty$. To make life simpler we shall simply impose detailed balancing of $\mathbf{M}$ with that final state. A trivial calculation shows that the only ensuing requirement is $f = d$, with the possibility of forbidding CD transitions by imposing $d = 0$. Still leaving that possibility in the open, the arguments above lead to

$$\mathbf{M} = \begin{pmatrix} 1-2a & 0 & 0 \\ a & 1-d & d \\ a & d & 1-d \end{pmatrix}. \tag{11.27}$$

Let us recapitulate. This matrix shows equal probabilities a of transitions of A to C and D; no possibility of C or D regenerating A; and a probability d of $C \to D$ and $D \to C$. The eigenvalues are $(1, 1-2a, 1-2d)$, so that $d = 0$ would lead to a reducible case. As it stands, its powers have the form

$$\mathbf{M}^j = \begin{pmatrix} (1-2a)^j & 0 & 0 \\ \frac{1}{2}[1-(1-2a)^j] & \frac{1}{2}[1+(1-2d)^j] & \frac{1}{2}[1-(1-2d)^j] \\ \frac{1}{2}[1-(1-2a)^j] & \frac{1}{2}[1-(1-2d)^j] & \frac{1}{2}[1+(1-2d)^j] \end{pmatrix}.$$

Applied to the initial state $\mathbf{p}^{(0)} = \begin{pmatrix} 1 \\ 0 \\ 0 \end{pmatrix}$, any matrix gives its own first column. The state will be, at the j-th step, actually independent of whether

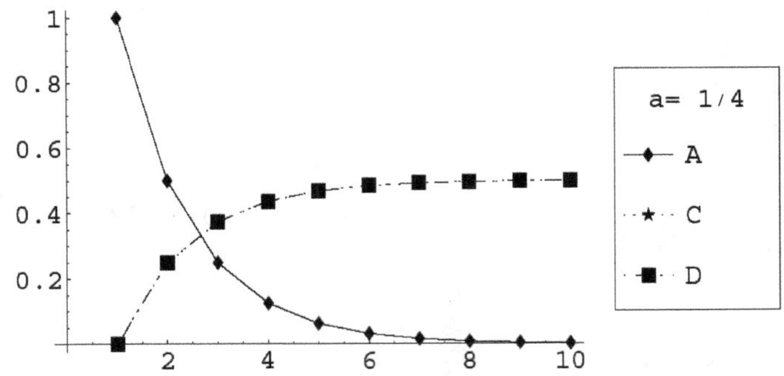

Fig. 11.3 *Decay $A \to C + D$ (parameter a with value 1/4).*

there are or not transitions CD:

$$\mathbf{p}^{(j)} = \mathbf{M}^j \mathbf{p}^{(0)} = \begin{pmatrix} (1-2a)^j \\ \frac{1}{2}[1-(1-2a)^j] \\ \frac{1}{2}[1-(1-2a)^j] \end{pmatrix}.$$

Notice the total irrelevance of parameter d. The value $a = 0$ would represent a stable A (infinite mean life): the initial state remains unmoved. Case $a = 1/2$ would give the final state already at the first step (mean life $<<$ one step time). Between these extremes, mean life progresses from 0 to ∞ when a decreases from 1/2 to 0. Figure 11.3 shows the behavior of components $p^{(j)}_{A,B,C}$ as functions of the time j, for the case $a = 1/4$. The concentration of A decreases as those of C and D, which coincide, increasing towards the equilibrium value.

Comparison with the experimental mean life will allow, in each case, to calibrate the value of the unit time used, that is, the "one-step" time for the corresponding process. The decreasing curve in the Figure represents $p^{(j)}_1 = (1-2a)^j$. In radioactive decay the mean life τ is defined by writing the concentration as a function of time as $c(t) = e^{-t/\tau}$. Putting, in terms of our discrete time "j", $p^{(j)}_1 = (M^j)_{11} = (1-2a)^j = e^{-j/\tau}$, we obtain the mean life $\tau = -\frac{1}{\ln(1-2a)}$.

To approach decays of type $A \to C + nE$, it is enough to take above $D = nE$. This amounts to change the matrix to

$$\mathbf{M} = \begin{pmatrix} 1-2a & 0 & 0 \\ a & 1-d & \frac{d}{n} \\ a & d & 1-\frac{d}{n} \end{pmatrix}$$

and use new distributions $\begin{pmatrix} p_A \\ p_C \\ p_E \end{pmatrix}$. Figure 11.4 shows the case $n = 2$.

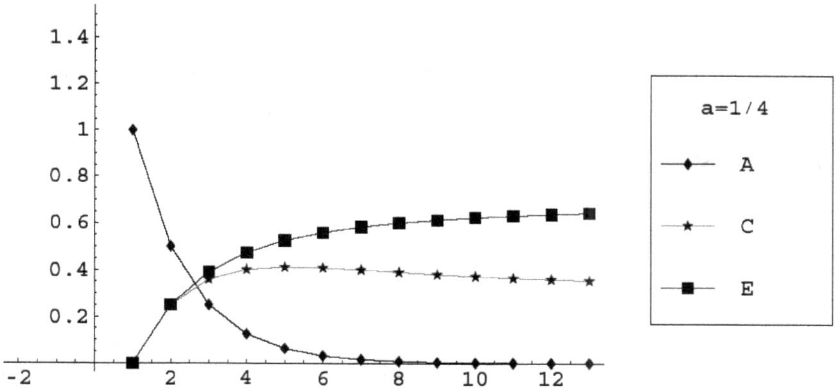

Fig. 11.4 *Decay* $A \to C + 2E$ $(a = 1/4)$.

To approach decays of type $A \to mB + nE$, it is enough to take further $C = mB$. This amounts to change the matrix to

$$\mathbf{M} = \begin{pmatrix} 1-2a & 0 & 0 \\ a & 1-\frac{d}{m} & \frac{d}{n} \\ a & \frac{d}{m} & 1-\frac{d}{n} \end{pmatrix}.$$

and use new distributions $\begin{pmatrix} p_A \\ p_B \\ p_E \end{pmatrix}$. Figure 11.5 shows the case $m = 3, n = 2$.

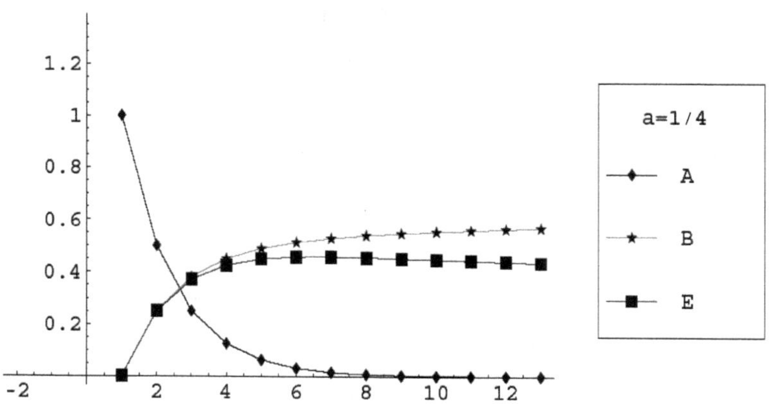

Fig. 11.5 *Decay* $A \to 3B + 2E$ $(a = 1/4)$.

11.3 Ternary decays

To consider decays of type $A \to B + C + D$, specially important in weak interactions, one more dimension is necessary. States will have the form $\begin{pmatrix} p_A \\ p_B \\ p_C \\ p_D \end{pmatrix}$. We start from the most general 4×4 stochastic matrix,

$$\begin{pmatrix} 1-a-b-c & d & g & r \\ a & 1-d-e-f & h & s \\ b & e & 1-g-h-j & t \\ c & f & j & 1-r-s-t \end{pmatrix},$$

and look for conditions to reduce the number of parameters. For example, A produces equal amounts of B, C and D, so that $a = b = c$. And the latter do not produce A, implying $d = g = r = 0$. Detailed balance with $(0, 1/3, 1/3, 1/3)^T$ gives $e+t = h+j$ and $e+f = h+s$. All these restrictions lead to

$$\begin{pmatrix} 1-3a & 0 & 0 & 0 \\ a & 1-e-f & h & e+f-h \\ a & e & 1-h-j & h+j-e \\ a & f & j & 1-f-j \end{pmatrix}, \tag{11.28}$$

still with 5 free parameters.

Up to this point final state interchanges are allowed. Forbidding them leaves only one free parameter:

$$\mathbf{M} = \begin{pmatrix} 1-3a & 0 & 0 & 0 \\ a & 1 & 0 & 0 \\ a & 0 & 1 & 0 \\ a & 0 & 0 & 1 \end{pmatrix}.$$

Figure 11.6 shows the quite natural evolution of the system with $a = 1/8$. Of course, the curves describing the growth of the populations B, C and D just superpose.

We have noticed in the previous section the irrelevance of the entries which are not in the first column. The same happens here: the decreasing curve represents $p_1^{(j)} = (1 - 3a)^j$, corresponding to the mean life $\tau = -\frac{1}{\ln(1-3a)}$.

11.4 Multiparticle decays

Applied to an initial state

$$\mathbf{p}^{(0)} = \begin{pmatrix} 1 \\ 0 \\ 0 \\ \cdots \\ \cdots \\ 0 \end{pmatrix}, \tag{11.29}$$

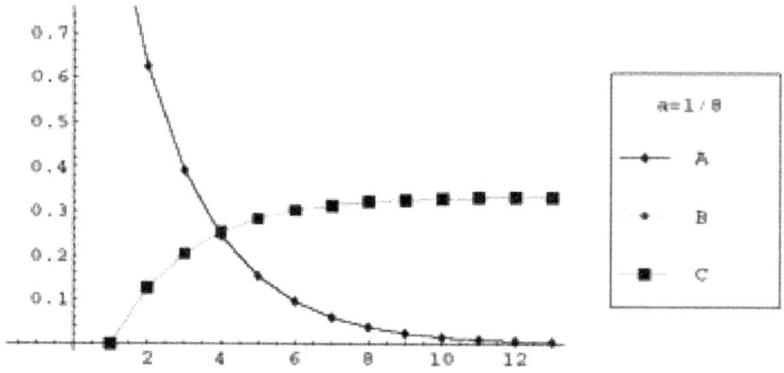

Fig. 11.6 *Decay $A \to B + C + D$. The curves for the products superpose ($a = 1/8$).*

any matrix gives just its own first column. Thus, the concentration of the decaying particle at the j-th step will be $(M^j)_{11}$, which depends only on first row and the first column of $\mathbf{M}$.

We might call "decay stochastic matrices" those whose first row is $(1 - [N-1]a, 0, 0, \dots, 0)$ and first column is $(1 - [N-1]a, a, a, \dots, a)^T$. This means a matrix like that of Eq.(11.28),

$$\mathbf{M} = \begin{pmatrix} 1-(N-1)a & 0 \quad 0 \quad 0 \quad \dots \quad 0 \\ \begin{matrix} a \\ a \\ a \\ \dots \\ a \end{matrix} & \begin{pmatrix} \text{any matrix} \\ \text{respecting} \\ \text{stochasticity} \\ \text{conditions} \end{pmatrix} \end{pmatrix} . \tag{11.30}$$

For a $N \times N$ decay matrix $\mathbf{M}$ acting on (11.29),

$$\mathbf{p}^{(j)} = \mathbf{M}^j \mathbf{p}^{(0)} = \begin{pmatrix} (1 - [N-1]a)^j \\ \frac{1}{2}[1 - (1 - [N-1]a)^j] \\ \dots \\ \dots \\ \frac{1}{2}[1 - (1 - [N-1]a)^j] \end{pmatrix} .$$

This corresponds to a mean life

$$\tau = - \frac{1}{\ln(1 - [N-1]a)} = - \frac{1}{\ln M_{11}} . \tag{11.31}$$

Of course, once $a = \frac{1-e^{-\frac{1}{\tau}}}{N-1}$, the matrix can be put in the more suggestive form

$$\mathbf{M} = \begin{pmatrix} e^{-\frac{1}{\tau}} & 0 \quad 0 \quad 0 \;\ldots\; 0 \\ \frac{1-e^{-\frac{1}{\tau}}}{N-1} & \\ \frac{1-e^{-\frac{1}{\tau}}}{N-1} & \begin{pmatrix} \text{any matrix} \\ \text{respecting} \\ \text{stochasticity} \\ \text{conditions} \end{pmatrix} \\ \frac{1-e^{-\frac{1}{\tau}}}{N-1} & \\ \cdots & \\ \frac{1-e^{-\frac{1}{\tau}}}{N-1} & \end{pmatrix}. \tag{11.32}$$

Decays are out-of-equilibrium processes. We use a matrix like (11.28), which was prepared to be in equilibrium with state $(0, 1/3, 1/3, 1/3)^T$, but apply it to (11.29). The examples above illustrate a very remarkable feature: in what concerns "dynamics", it is not really important which microstates are "reactants" and which are "products". The evolutive behavior in encoded in the matrix. The left-hand side of the reaction is given by the distribution taken as initial state. The distribution containing only the decaying particle has been taken in all examples above – the natural choice for a decay process. To show how reasonable the model is: the same matrix which, acting on $(1, 0, 0)$ gives the evolution depicted in Figure 11.3 will give, when applied to $(1/2, 1/2, 0)$, the evolution of Figure 11.7. Particles of type C are present from the start. With their production by the decaying A, their concentration increases at first. As the sum of all concentrations is fixed, the production of D is concomitantly inhibited if compared with that of Figure 11.3. This is a trivial example of blockage of a process by the excess of product ("final state blockage").

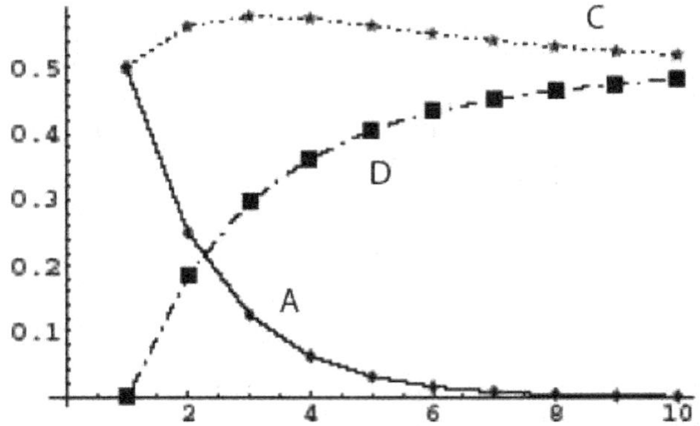

Fig. 11.7 *Decay of A into $C + D$, but starting in the presence of product C:* $\mathbf{p}^{(0)} = (1/2, 1/2, 0)^T$.

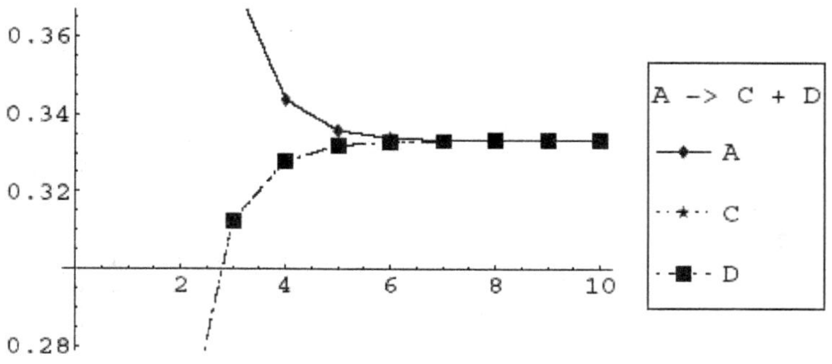

Fig. 11.8 *Ionization $A \to C + D$: start from a state containing only A. Lines for C and D actually coincide.*

We can proceed from decays to reactions by a simple modification of the matrices above. Take for example the case $A \longrightarrow C+D$. It was forced to be one-way process by the zero transitions in the matrix first row. Repeating the previous procedure, we start from the most general possible matrix and impose (i) that C and D contribute same value in their transitions to A; (ii) C and D regenerate themselves the same; (iii) detailed balance with the expected final state $(1/3, 1/3, 1/3)$. Instead of (11.33), we obtain

$$\mathbf{M} = \begin{pmatrix} 1-2a & a & a \\ a & 1-a-d & d \\ a & d & 1-a-d \end{pmatrix} . \tag{11.33}$$

Figure 11.8 shows the resul of applying it to the same initial state used above for the decay, $\mathbf{p}^{(0)} = (1, 0, 0)^T$, with parameters $a = 1/4$ and $d = 1/8$. We see C and D growing together towards equilibrium. This case is reminiscent, for example, of an ionization process like

$$H \longrightarrow H_+ + e^- \ ,$$

with H and H_+ representing neutral and ionized hydrogen, the latter supposed to be single. Numerical experiments indicate that the evolution pattern is rather insensible to the value of parameter d. Notice the rather cavalier approach of forgetting the dynamical details and even some reactants. A hot gas in equilibrium is composed of all the excited states of the atom or molecule plus photons, in different concentrations. It is as if a series of chemical reactions were simultaneously at work between different species, each species being one of the excited states. All of them are here summed up in H_+, and the photon in the more realistic

$$H + \gamma \longleftrightarrow H_+ + e^-$$

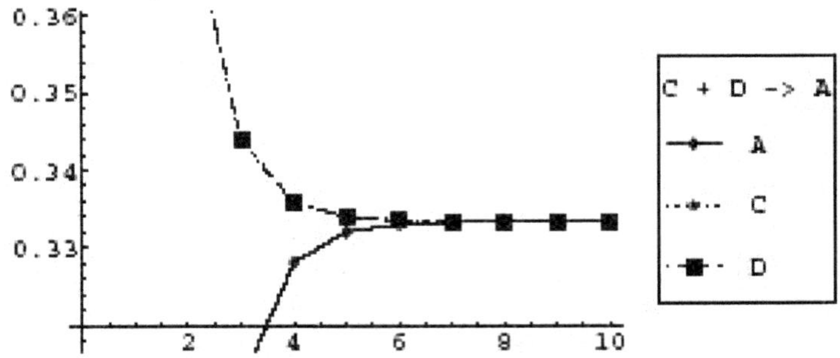

Fig. 11.9 *Formation $C + D \to A$: start from a state including only C and D, though in equal amounts.*

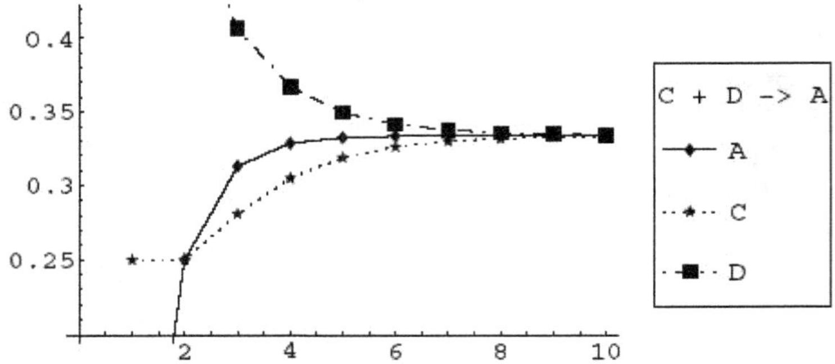

Fig. 11.10 *Formation: starting from C and D in different amounts.*

is incorporated in the "heat" of thermodynamical description. These abstractions lead, however, to very good results (such as Saha's formula) in the equilibrium case, and we feel authorized to proceed with our simple assumptions.

Instead of H decomposition, we can look at it formation (recombination $H_+ + e^- \longrightarrow H$) by taking as initial state H_+ and e^- in equal parts: $\mathbf{p}^{(0)} = (0, 1/2, 1/2)^T$. The process appears in Figure 11.9. Figure 11.10 shows the formation when C and D are introduced in different amounts, $\mathbf{p}^{(0)} = (0, 1/4, 3/4)^T$. As the D concentration is initially higher, C stays stationary for some time. of course, the system tends to the flat equilibrium distribution representing chemical equilibrium

$$H \longleftrightarrow H_+ + e^- \ .$$

11.5 Two-to-two reactions

Suppose molecules of two kinds, indicated by A and B, are introduced in an inert medium and assume that they are able to react so as to produce two other species C and D. This is represented in the standard way by

$$A + B \to C + D\ .$$

After some time there will be four species in the medium, with relative concentrations $p_A = \frac{N_A}{N}$, $p_B = \frac{N_B}{N}$, etc., which we shall put together as $p_i = \frac{N_i}{N}$. N is the total number of molecules of the four species and N_i is the number of molecules of species "i". If we take one molecule at random, the probability will be p_A of getting a molecule type A, p_C of getting a molecule type C, etc. The macroscopic state will be described by the distribution

$$\mathbf{p} = \begin{pmatrix} p_A \\ p_B \\ p_C \\ p_D \end{pmatrix},$$

with the microstates represented by the very same A, B, C, D. There will be other reactions, such as the "diagonal" $A+A \to A+A$, $C+C \to C+C$; the indifferent $A+B \to A+B$; the inverse $C+D \to A+B$. In equilibrium, this inverse reaction will proceed with the same velocity of the direct one, so that

$$A + B \leftrightarrow C + D\ .$$

By the very choice of coefficients, this means a stable distribution with equal concentrations of the four species. We shall suppose this chemical equilibrium is the final state: the flat distribution $p_i = \frac{1}{4}$.

Note. The discussion above suggests a backward procedure: represent any stochastic matrix through chemical reactions. Flux diagrams are preliminary to that enterprise. Take for instance matrix (7.1) of page 61,

$$\mathbf{U}'_{22} = \begin{pmatrix} 0 & 1 & 1/2 & 0 \\ 1 & 0 & 0 & 0 \\ 0 & 0 & 0 & 1 \\ 0 & 0 & 1/2 & 0 \end{pmatrix}, \qquad \begin{matrix} A & \longleftrightarrow & B \\ \uparrow & & \\ C & \longleftrightarrow & D \end{matrix} ,$$

whose assymptotic limit oscillates between the two matrices

$$\begin{pmatrix} 1 & 0 & 0 & 1 \\ 0 & 1 & 1 & 0 \\ 0 & 0 & 0 & 0 \\ 0 & 0 & 0 & 0 \end{pmatrix} \quad \begin{matrix} A & B \\ \uparrow & \uparrow \\ D & C \end{matrix} \quad \text{and} \quad \begin{pmatrix} 0 & 1 & 1 & 0 \\ 1 & 0 & 0 & 1 \\ 0 & 0 & 0 & 0 \\ 0 & 0 & 0 & 0 \end{pmatrix} \quad \begin{matrix} A & \longleftrightarrow & B \\ \uparrow & & \uparrow \\ C & & D \end{matrix} .$$

By the above procedure, matrix $\mathbf{U}'_{22}$ itself would come from the set of reactions

$$\begin{gathered} A \to B \\ B + C \to A \\ C \to D \\ D \to C \end{gathered}$$

which can actually be read in the flux diagram. The two limits represent the two reaction sets

$$\begin{matrix} A+D \to A \\ B+C \to B \end{matrix} \quad \text{and} \quad \begin{matrix} B+C \to A \\ A+D \to B \end{matrix} .$$

Notice that the reaction products oscillate between A and B. This means that the sistem tends to a situation in which the A and B concentrations are exchanged at each step: a childish toy model for the Belusov-Zhabotinsky reaction.[1]

We look now for a stochastic transition matrix describing the process. A first approach would be to modify as little as possible the matrix for the corresponding decay process. A very simple modification would be

$$\mathbf{M} = \begin{pmatrix} 1-2a & 0 & 2a & 0 \\ 0 & 1-2a & 2a & 0 \\ a & a & 1-4a & 2a \\ a & a & 0 & 1-2a \end{pmatrix}$$

This matrix can be used for many different cases, one for each initial state. Figure 11.11 shows A decaying into the other components. It is obtained by taking $\mathbf{p}^{(0)} = (1,0,0,0)^T$. Figure 11.12 shows the reaction $A+B \to C+D$, which comes up if $\mathbf{p}^{(0)} = (1/2,1/2,0,0)^T$. Figure 11.13 shows the converse reaction $C+D \to A+B$, which comes up if $\mathbf{p}^{(0)} = (0,0,1/2,1/2)^T$, and Figure 11.14 shows the A production $B+C+D \to A$, which comes up if $\mathbf{p}^{(0)} = (0,1/3,1/3,1/3)^T$. Notice that A increases from the beginning, but B is stable at first: it is not being consumed as, by the very structure of the matrix, only C produces A.

We could adopt a more conservative procedure in approaching the case $A+B \leftrightarrow C+D$, starting again from the most general form

$$\mathbf{M} = \begin{pmatrix} 1-a-b-c & d & g & r \\ a & 1-d-e-f & h & s \\ b & e & 1-g-h-j & t \\ c & f & j & 1-r-s-t \end{pmatrix}$$

of a 4×4 stochastic matrix. It has too many free parameters and some more assumptions are necessary. Let us assume detailed balancing: with a flat distribution, it imposes simply that the matrix be doubly stochastic. That

[1] See, for instance, reference [46].

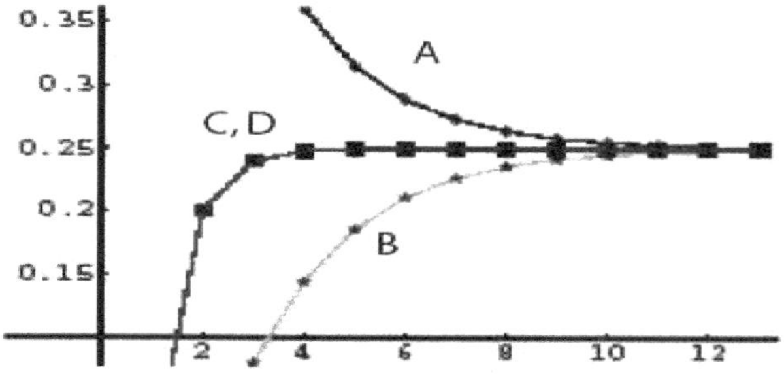

Fig. 11.11 *Decomposition* $A \to B + C + D$ $(a = 1/5)$.

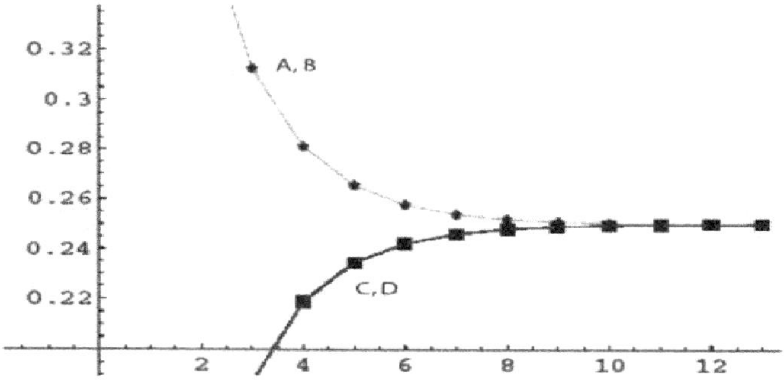

Fig. 11.12 *One-way reaction* $A + B \to C + D$. *To have it, start from* $\mathbf{p}^{(0)} = (1/2, 1/2, 0, 0)^T$.

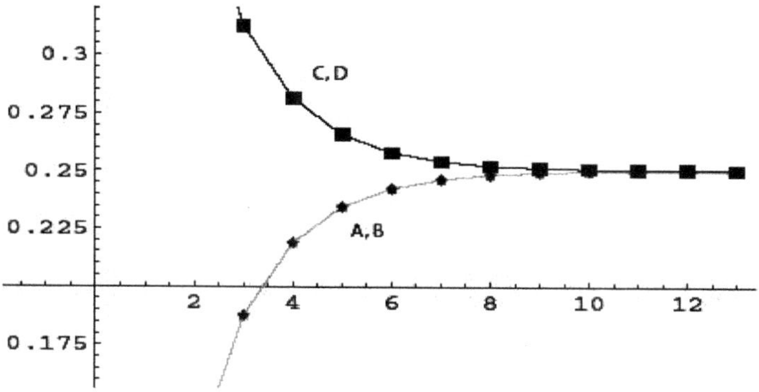

Fig. 11.13 *Converse reaction* $C + D \to A + B$. *It is enough to start from* $\mathbf{p}^{(0)} = (0, 0, 1/2, 1/2)^T$.

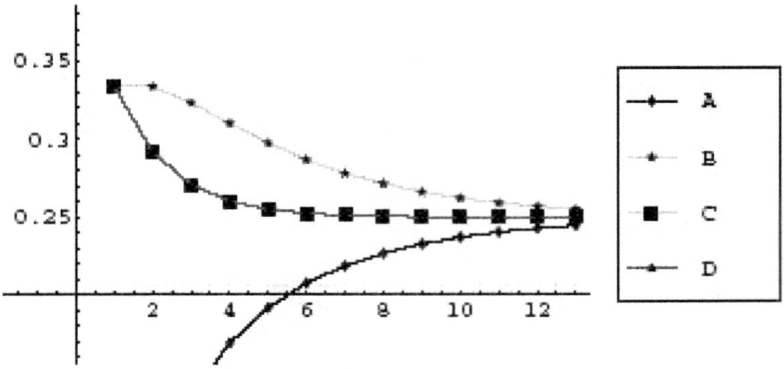

Fig. 11.14 *A production reaction:* $B + C + D \to A$.

leaves yet too many free parameters. We can impose that it be furthermore symmetric,

$$\mathbf{M} = \begin{pmatrix} 1-a-b-c & a & b & c \\ a & 1-a-e-f & e & f \\ b & e & 1-b-e-j & j \\ c & f & j & 1-c-f-j \end{pmatrix}.$$

Six parameters being still too many, we recall that a flat equilibrium distribution suggests a circulant. This would mean only two remaining free parameters:

$$\mathbf{M} = \begin{pmatrix} 1-2a-b & a & b & a \\ a & 1-2a-b & a & b \\ b & a & 1-2a-b & a \\ a & b & a & 1-2a-b \end{pmatrix}. \tag{11.34}$$

This is a degenerate matrix, with eigenvalues $(1, 1-4a, 1-2a-2b, 1-2a-2b)$. Notice that the simple imposition of circulant character would lead to

$$\mathbf{M} = \begin{pmatrix} 1-a-b-c & c & b & a \\ a & 1-a-b-c & c & b \\ b & a & 1-a-b-c & c \\ c & b & a & 1-a-b-c \end{pmatrix}, \tag{11.35}$$

which would be in equilibrium with the flat distribution, but not necessarily in detailed balancing with it. Notice also that the entry c above is related to the relationship between A and B, as is the entry a. This matrix is generically non-degenerate, the eigenvalues being $(1, 1-2a-2c, 1-a-2b-c-\sqrt{2ac-c^2-a^2}, 1-a-2b-c+\sqrt{2ac-c^2-a^2})$. The reaction, however, shows equal roles for both, so that the symmetry added by detailed balancing seems quite reasonable. Entry c is equally related to the relationship between C and D.

It is clear that the concentrations of C and D increase necessarily together, and those of A and B decrease the same. The symmetry of $\mathbf{M}$ enforces furthermore strict correspondences between A and D, as well as B and C.

Equation (11.34) is the same as

$$\mathbf{M} = a\mathbf{U} + b\mathbf{U}^2 + a\mathbf{U}^{-1} + (1 - 2a - b)\mathbf{I} .$$

$$\mathbf{M} = a(\mathbf{U} + \mathbf{U}^{-1} - 2\mathbf{I}) + b\mathbf{U}^2 + (1 - b)\mathbf{I} .$$

$$\mathbf{M} = \mathbf{I} + a\boldsymbol{\Delta} + b(\mathbf{U}^2 - \mathbf{I}) .$$

It is very curious the appearance of the Laplacian. Term $a\boldsymbol{\Delta}$ shows parameter a in the role of a diffusion coefficient, though here the "diffusion" takes place in the space of particle types. The presence of $\boldsymbol{\Delta}$ is due to the symmetry imposed on $\mathbf{M}$: for equation (11.35) we would have instead

$$\mathbf{M} = \mathbf{I} + a\mathbf{d}^\dagger + c\mathbf{d} + b(\mathbf{U}^2 - \mathbf{I}) .$$

Figure 11.15 shows the behavior presided by matrix (11.34), for a very small value of the diffusion coefficient a. The sense of the reaction is solely determined by the initial state taken, $\mathbf{p}^{(0)} = (1/2, 1/2, 0, 0)$. Waiving the condition of detailed balance and using matrix (11.35) gives more freedom of choice, and leads to the difficulties coming with it. The choice of parameters is very large, though the circulant character ensures the final flat equilibrium. Changing c leads to a whole zoo of possible behaviors. As long as c remains comparable to b the curves are gentle, though a separation of the constituents is to be remarked (Figures 11.16 and 11.17, in which the roles of b and c are exchanged).

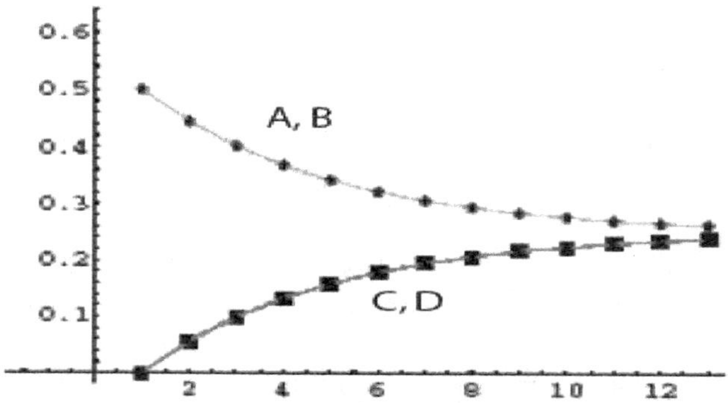

Fig. 11.15 $A + B \to C + D$ *(with* $a = 1/36$, $b = 3\ a$*). Behavior given by Eq.(11.34).*

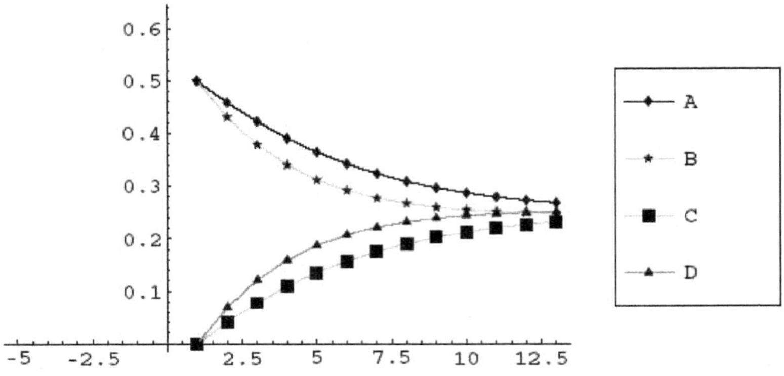

Fig. 11.16 *Case $a = 1/36$, $b = 2\ a$, $c = 3\ a$. Behavior given by Eq.(11.35).*

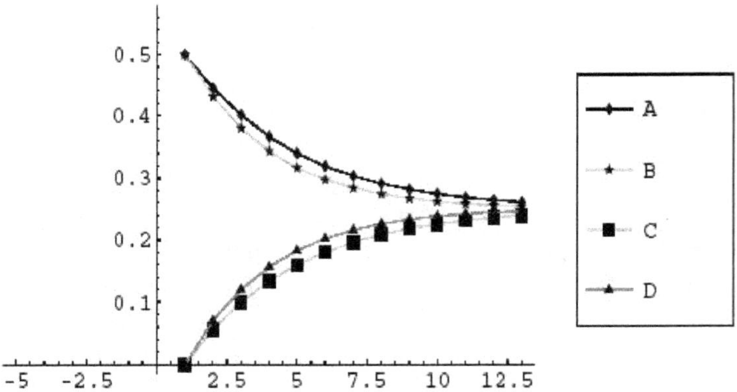

Fig. 11.17 *Case $a = 1/36$, $b = 3\ a$, $c = 2\ a$. Again Eq.(11.35), but with the roles of b and c exchanged with respect to previous Figure.*

A further repertory is given in Figures 11.18–11.21, in which parameter c is made to increase more and more. The concentration behavior changes quickly, becoming oscillatory at $c \approx 6a$. High c produces strong oscillations in both reactant and product concentrations. Comparison of Figures 11.20 and 11.21 shows that changing the ratio c/b causes only a small difference.

Perhaps the main point to be learned from these examples is the central role of the stochastic matrix. Given the "chemicals" involved, it is what we choose as initial state that determine which are the reactants and which are products.

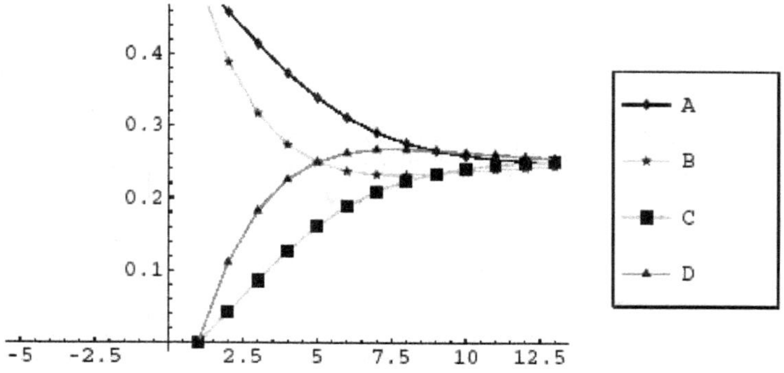

Fig. 11.18 *Case $a = 1/36$, $b = 2\ a$, $c = 6\ a$. Increasing c: behavior almost oscillatory when $c \approx 6a$.*

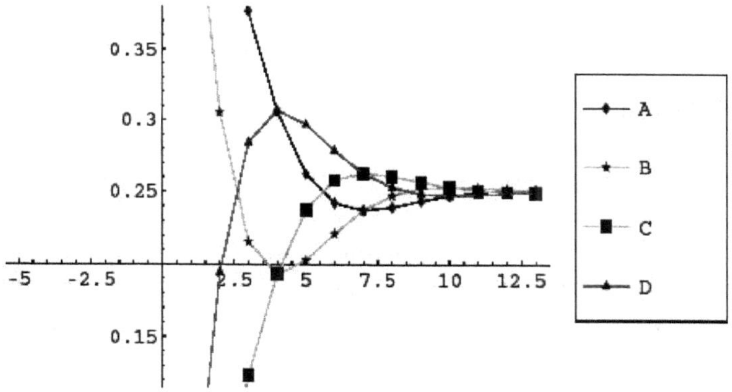

Fig. 11.19 *Case $a = 1/36$, $b = 2\ a$, $c = 12\ a$. Higher c's lead to clear oscillations in the concentrations.*

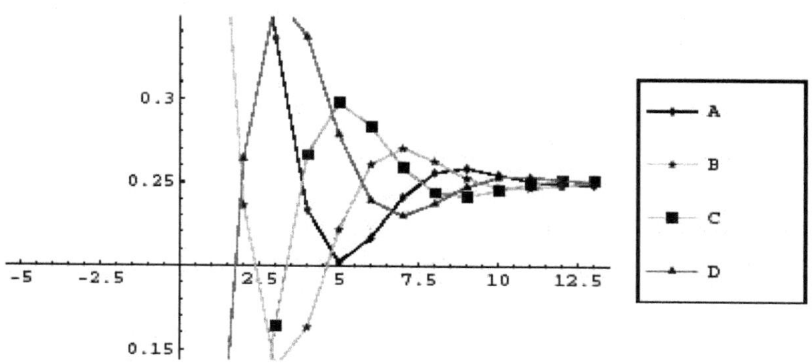

Fig. 11.20 *Case $a = 1/36$, $b = 2\ a$, $c = 18\ a$. High c produces strong oscillations in both reactant and product concentrations.*

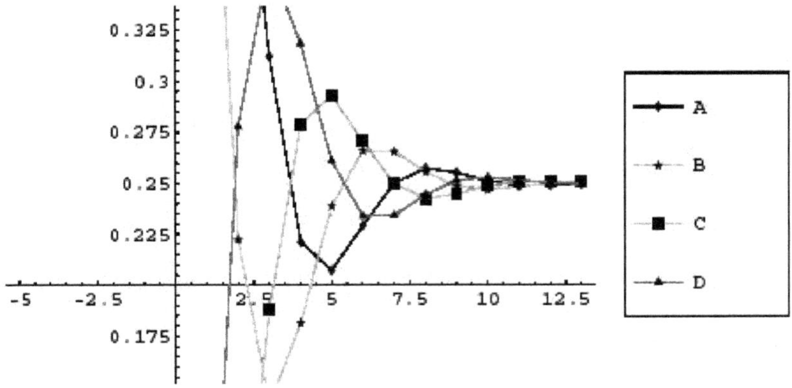

Fig. 11.21 *Case $a = 1/36$, $b = a$, $c = 18\ a$. Changing the ratio c/b causes only small differences.*

11.6 Relation to equilibrium thermodynamics

The above findings concern mainly out-of-equilibrium conditions, but their asymptotic limits should make connection with the usual results of equilibrium Statistical Mechanics on chemical reactions.[2] The latter could be used as a kind of "correspondence principle", hopefully providing some further insight. We shall take reactants/products $A, B, C, \ldots$ and make a few simplifying assumptions:

- in the stochastic side we shall suppose that the stochastic matrix $\mathbf{M}$ involved has a single eigenstate with eigenvalue $\lambda = 1$; this means that $M =$ (transient part) $+ Z_{\lambda=1}$.
  Each column of the projector $\mathbf{M}^{(\infty)} = Z_{\lambda=1}$ will be a copy of the state $p^{(\infty)}$ (Section 1.3.1).
- on the Statistical Mechanical side we suppose a non-relativistic classical ideal gas, in which case the relative equilibrium concentrations are given by $\frac{e^{\beta\mu_j}}{\lambda_j^3}$ for $j = A, B, C, \ldots$, where $\beta = 1/kT$ is the inverse temperature in terms of the Boltzmann constant k, μ_j is the chemical potential of species j and λ_k is the de Broglie wavelength of a particle j. For a particle of mass m_j, $\lambda_j = \frac{\hbar}{\sqrt{2\pi m_j kT}}$.

[2] See, for example, [47], or the excellent Chapter 9 of [33].

If $n = N/V = n_A + n_B + n_C$ is the total concentration, the probabilities we have been using above are $p_k^\infty = \frac{e^{\beta\mu_k}}{n\lambda_k^3}$. The asymptotic equilibrium state will in consequence have the form

$$p^\infty = \begin{pmatrix} \frac{e^{\beta\mu_A}}{n\lambda_A^3} \\ \frac{e^{\beta\mu_B}}{n\lambda_B^3} \\ \frac{e^{\beta\mu_C}}{n\lambda_C^3} \\ \cdots \\ \cdots \end{pmatrix}.$$

For the case $N = 3$, for example, the projector will be

$$Z_{\lambda=1} = \begin{pmatrix} \frac{e^{\beta\mu_A}}{n\lambda_A^3} & \frac{e^{\beta\mu_A}}{n\lambda_A^3} & \frac{e^{\beta\mu_A}}{n\lambda_A^3} \\ \frac{e^{\beta\mu_B}}{n\lambda_B^3} & \frac{e^{\beta\mu_B}}{n\lambda_B^3} & \frac{e^{\beta\mu_B}}{n\lambda_B^3} \\ \frac{e^{\beta\mu_C}}{n\lambda_C^3} & \frac{e^{\beta\mu_C}}{n\lambda_C^3} & \frac{e^{\beta\mu_C}}{n\lambda_C^3} \end{pmatrix}.$$

This is, in principle, all that can be obtained from *equilibrium* Thermodynamics. We can, of course, make additional assumptions. Detailed balancing, for instance, would impose some constraints.

Anyhow, there is one important point which this simple asymptotic "correspondence principle" brings to attention. The probabilities are dependent on three quite different families of parameters:

- individual characteristics of the species involved (such as the particle mass, which appears through the de Broglie wavelength);
- inter-particle properties, as their larger or smaller affinity, or tendency to couple to each other to engender other particles (the chemical potentials);
- collective properties, related to the system as a whole, or imposed from the exterior (as the temperature).

It is natural to expect that such asymptotic qualitative features be already present in the non-equilibrium, transient stage.

Chapter 12

Applications II: Genetics

One of the basic assumptions in Mendel's work is that every information concerning genetic heritage is transmitted by the sole parents to the child. Memory of previous generations is entirely contained in the parents' genetic charge, so that no further effect from the past is active. This is a typical discrete markovian process and means that that evolution is to be seen as a Markov chain. A tentative to describe it as a stochastic process is, however naïve, thereby justified. We shall view the whole process as a set of "chemical" reactions. On the same token, an approach to OAB blood types is presented.

12.1 Of Mendel, and garden peas

In his celebrated work [48] on garden peas, the father of Genetics examined mainly four characteristics of those carefully chosen plants, all of which presenting similar distributions: the color of the seeds albumen, which is yellow or green; the aspect of the seeds, rounded or wrinkled; the colors of the flowers, purple or white; and the very sizes of the plants, which can be tall or short. In each case, the first quality is dominant, the other recessive. We shall here consider the albumen colors, indicated Y (yellow) and g (green), and try to describe hybridization as the result of "chemical" reactions.

Some nomenclature seems necessary [9]. Peas (actually, their albumen) appear as yellow or green, but are genetically composite. They can actually be pure yellow (indicated YY), pure green (indicated gg) and also hybrids Yg. The external appearence, yellow and green are the phenotypes, while the internal YY, Yg and gg are the genotypes. The phenotypes are the only directly detectable forms. Yg appears as yellow because Y is dominant and g is recessive. A yellow pea is said to have two alleles (short for allelomorphs), YY and Yg. The real, basic microstates are the genotypes. We shall consequently treat with genotypes, not phenotypes. In our approach there will be the three microstates YY, Yg and gg, with probabilities p_{YY}, p_{Yg} and p_{gg}. The macrostate will be, at the n-th stage, a distribution

$$\mathbf{p}^{(n)} = \begin{pmatrix} p_{YY}^{(n)} \\ p_{Yg}^{(n)} \\ p_{gg}^{(n)} \end{pmatrix}.$$

12.1.1 *Merely counting*

Random mating is supposed throughout. That is to say, all matings are independent of genetic differences. Further, all individuals mate at the same time and are immediately replaced by their offspring. These are assumptions usually made in quantitative genetics, leading to distortions like the age problem, for which methods exist for ulterior corrections.[1]

Given the probability distribution

$$\mathbf{p} = \begin{pmatrix} p_{YY} \\ p_{Yg} \\ p_{gg} \end{pmatrix},$$

of the states in one generation, the next generation will have a probability distribution given by a stochastic matrix $\mathbf{M}$ applied to $\mathbf{p}$. With an exponent indicating the generation, so that $\mathbf{p}^{(0)}$ represents the probability distribution of a *bona fide* generation taken as initial, $\mathbf{p}^{(1)} = \mathbf{M}\mathbf{p}^{(0)}$, $\mathbf{p}^{(2)} = \mathbf{M}\mathbf{p}^{(1)} = \mathbf{M}^2\mathbf{p}^{(0)}$, $\mathbf{p}^{(3)} = \mathbf{M}\mathbf{p}^{(2)} = \mathbf{M}^3\mathbf{p}^{(0)}$, and so on. Therefore, the n-th generation will have a distribution $\mathbf{p}^{(n)} = \mathbf{M}^n\mathbf{p}^{(0)}$.

The rules governing parent-to-progeny transmission are alike to those of chemical reactions. If both parents are YY then the progeny is necessarily YY also: $YY + YY \Rightarrow YY$.

If the parent plants are YY and Yg, the progeny will have 50% of probability of being YY and 50% of probability of being Yg. Such mendelian rules give the probabilities of the two colors in the progeny given the parents types. What we need to construct the governing matrix is rather a different thing: the total probability to have a certain type in the progeny given a fixed type in the parents.

We shall encode the Mendel rules of inheritance in the language of reactions used in the previous chapter, but with one further consideration: account will be taken of the possible combinations. We suppose for example

[1] See for example reference [49], Chapter 6.

that, when YY meets Yg, the "atoms" Y and g are independent to combine at random. Thus, the reactions are

$$\begin{gathered} YY + YY \Rightarrow 4\, YY \\ YY + Yg \Rightarrow 2\, YY + 2\, Yg \\ YY + gg \Rightarrow 4\, Yg \\ Yg + Yg \Rightarrow YY + 2\, Yg + gg \\ Yg + gg \Rightarrow 2\, Yg + 2\, gg \\ gg + gg \Rightarrow 4\, gg\,. \end{gathered} \tag{12.1}$$

The first collumn in a stochastic matrix gives the contributios from the first-numbered microstate (here, YY) to each one of the others. The coefficients in the above reactions will be proportional to the corresponding value. Thus, from the first three reactions, YY will contribute with factors $4+2 = 6$ to YY itself and to Yg; and 0 to the others. The probabilities in the first column will be proportional to these numbers. The second column will be formed with the coefficients coming from the contributions of Yg and third with those of gg. In this way we build up a provisional matrix, with unnormalized columns:

$$\begin{matrix} 6 & 3 & 0 \\ 6 & 6 & 6 \\ 0 & 3 & 6\,. \end{matrix}$$

It is enough now to normalize the columns to get the stochastic matrix:

$$\mathbf{M} = \begin{pmatrix} 1/2 & 1/4 & 0 \\ 1/2 & 1/2 & 1/2 \\ 0 & 1/4 & 1/2 \end{pmatrix}, \qquad \underset{\text{flow diagram}}{\begin{matrix} YY & \Longleftrightarrow & Yg \\ & & \Updownarrow \\ & & gg \end{matrix}}\,. \tag{12.2}$$

This is our candidate governing matrix. It has eigenvalues $\lambda = 1$, 0 and 1/2, pointing to an equilibrium state which is the eigenvector for $\lambda = 1$,

$$\mathbf{p}^{(\infty)} = \begin{pmatrix} p^{(\infty)}_{YY} \\ p^{(\infty)}_{Yg} \\ p^{(\infty)}_{gg} \end{pmatrix} = \begin{pmatrix} 1/4 \\ 1/2 \\ 1/4 \end{pmatrix}. \tag{12.3}$$

A few comments:

(1) actually, $\mathbf{M}$ is in detailed balance with $\mathbf{p}^{(\infty)}$
(2) the flow diagram in (12.2), of course a reflex of the reaction table, shows that neither YY produce gg directly and not vice versa;
(3) the equilibrium proportion of phenotypes given by (12.3) is exactly that found by Mendel: yellow = YY + Yg : 3 :: green = gg: 1.

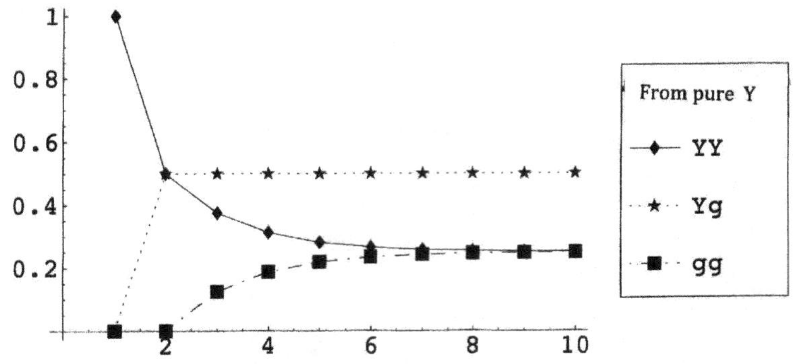

Fig. 12.1 *Starting from pure YY state.*

(4) imposing detailed balance of a general 3×3 stochastic matrix with Mendel's distribution (12.3) would still leave 3 undetermined parameters; additional conditions could be imposed by hand

(5) our contention is that Mendel ultimately *discovered* reactions (12.1), or the equivalent matrix (12.2).

We can proceed to examine the evolution processes given by (12.2). Of course, acting on an average stable population will cause no evolution at all. What happens if we start with a another state? Let us call $YY = (1,0,0)^T$ the pure Y phenotype state and, in the same line, $Yg = (0,1,0)^T$ and $g = (0,0,1)^T$.

Suppose we start with $\mathbf{p}^{(1)} = (1,0,0)^T$. Applying $\mathbf{M}$ successively we find the probability distributions of the ensuing generations. The result is given in Figure 12.1: we see the rapid decrease of YY, the equally quick increase of Yg and the slower increase of gg. Yg attains its equilibrium value in the second generation, but complete equilibrium only is manifest at the 8-th generation. This is of course a purely theoretical exercise, as in principle only the phenotypes can be prepared in experiments. The same happens with the evolution of a $\mathbf{p}^{(1)} = Yg = (0,1,0)^T$ shown in Figure 12.2. If we suppose yellow seeds to be be formed by YY and Yg in equal parts, the evolution is that of Figure 12.3. Finally, Figure 12.4 shows a development of a pure green state which is just the symmetric of that of a YY startoff. In all cases, a few generations suffice to near the equilibrium distribution.

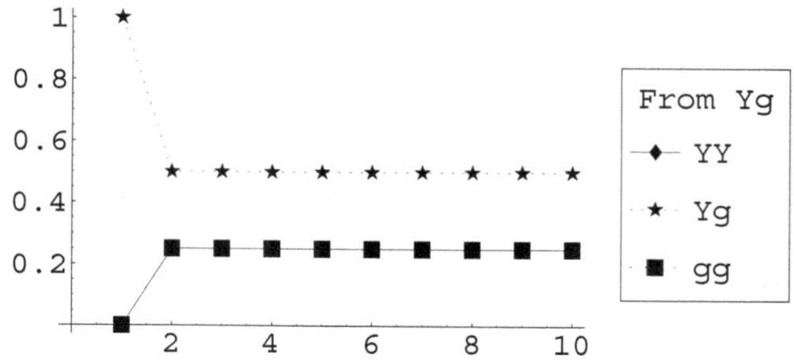

Fig. 12.2 *Starting from a hybrid Yg state. The lines for YY and gg coincide.*

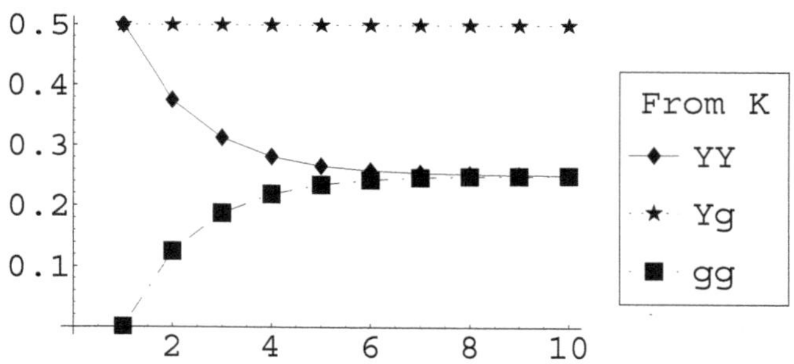

Fig. 12.3 *Starting from a yellow $\frac{YY+Yg}{2}$ state.*

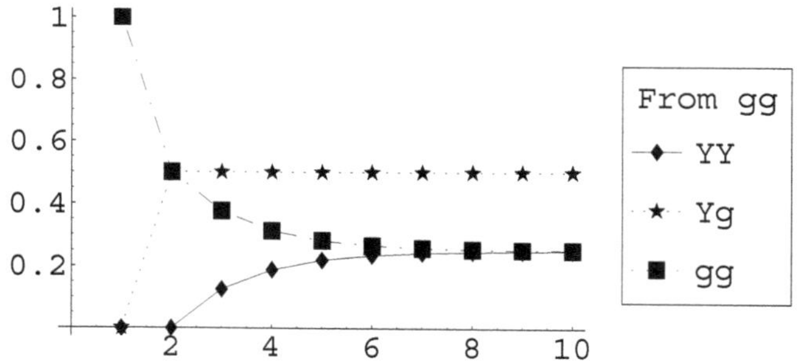

Fig. 12.4 *Starting from pure gg state.*

It may seem strange at first sight that a pure case, such as $gg = (0, 0, 1)^T$, produce other things. We should have in mind that, by the very nature of its phenomenological discovery, $\mathbf{M}$ actually pools the whole population data.

12.1.2 *An aside: valence-2 model*

The same results above would come from a model supposing the genotypes to be molecules of valence 2, or the system to be a graph with coordination number 2.

Indeed, lets us inspire ourselves in Kerner's approach to vitrification,[2] representing Y by $\circ-$, g by $\bullet-$, and their combinations by the schemes

$$YY = -\circ\cdots\circ- \, , \quad Yg = -\circ\cdots\bullet- \, , \quad gg = -\bullet\cdots\bullet- \, .$$

The medium ellipses have a meaning. Only the external valences are supposed to characterize the microstate, so that $YY = -\circ-\circ- = -\circ-\circ-\circ- = \ldots = -\circ-\bullet-\circ- = -\circ-\bullet-\circ-\bullet-\circ- = -\circ-(\text{anything})-\circ-$. Reactions (12.1) are then represented by

$$\begin{array}{c}
-\circ\cdots\circ- + -\circ\cdots\circ- \Rightarrow 4\ -\circ\cdots\circ- \\
-\circ\cdots\circ- + -\circ\cdots\bullet- \Rightarrow 2\ -\circ\cdots\circ- + 2\ -\circ\cdots\bullet- \\
-\circ\cdots\circ- + -\bullet\cdots\bullet- \Rightarrow 4\ -\circ\cdots\bullet- \\
-\circ\cdots\bullet- + -\circ\cdots\bullet- \Rightarrow -\circ\cdots\circ- + 2\ -\circ\cdots\bullet- + -\bullet\cdots\bullet- \\
-\circ\cdots\bullet- + -\bullet\cdots\bullet- \Rightarrow 2\ -\circ\cdots\bullet- + 2\ -\bullet\cdots\bullet- \\
-\bullet\cdots\bullet- + -\bullet\cdots\bullet- \Rightarrow 4\ -\bullet\cdots\bullet- \, .
\end{array} \qquad (12.4)$$

The numerical factors now come naturaly by counting the possible valence combinations. Notice the following: as only the valences are taken into account, the numbers will receive the same contributions when a $\circ-$ attaches itself to another $-\circ$ or to $-\bullet$ or when $\bullet-$ attaches to $-\bullet$. There is no difference in the "affinities" of the constituents, they all look the same for each one of them.

[2] Kerner's approach suppose an initial state in which nucleation has already taken place. Each constituent has most of its valences somehow "occupied" in the inner part of the agglomerate it belongs. Only the remaining, un-neutralized valences are presented to the exterior to determine the effective state. Our ellipses are remindful of those seed agglomerates.

12.1.3 *A "chemical" model*

The above approach seems rather naïve. It is the first thing to try and it is surprising that – in what concerns peas, at least – it does describe the facts of Nature. The beauty, the correctness and the *naïveté* are encoded in reactions (12.1). The main problem comes from our present-day molecular view of the process. The idea of *real* chemical reactions is not, nowadays, far-fetched. We know that molecular combinations do take place in the fertilization process. Nevertheless, macromolecules are involved, which combine in an extremely intricate way. Furthermore, even small molecules do not react in so simple a way. Mainly, there are chemical affinities between different molecules, larger and smaller. In the number factors shown in reactions (12.1), only valence combinations are taken into account. Consider for example the case

$$YY + Yg \Rightarrow 2\,YY + 2\,Yg$$
$$-\circ\cdots\circ - + - \circ\cdots\bullet - \Rightarrow 2\ \ -\circ\cdots\circ - + 2\ \ -\circ\cdots\bullet -.$$

The number 2 in 2 YY comes from pure counting: in $YY + Yg$, either the first or the second Y in YY couples to the Y in Yg. Concomitantly, the g can couple to either of the Y's in YY. If we think in terms of molecules, however, we would expect that Y couples to Y and to g with different *affinities.* In the case of real chemical reactions, these affinities are ultimately represented by the chemical potentials. Besides the pure counting factor 2 in 2 YY, a factor like $e^{\mu_{YY}/kT}$ would be expected, where kT is the absolute temperature multiplied by the Boltzmann constant, and the chemical potential μ_{YY} is an energy measuring the average attraction or repulsion between two "atoms" Y. The same would happen to the Yg term: it should have a factor like $e^{\mu_{Yg}/kT}$, with μ_{Yg} an energy measuring the average attraction or repulsion between "atoms" Y and g.

Of course, it would be also very naïve to describe the utterly complicated interactions between the real macromolecules involved in that simple way. We shall, however, do precisely that, looking for what would be a qualitative explanation of the fantastic success of reactions (12.1) in describing Nature. We shall have no place for detailed quantities like the temperature, but we shall retain a lesson from usual chemical reactionss. Whatever the affinities may be, they should depend both on the microscopic characteristics of the "atoms" (like the interaction energies between them) and on collective properties of the system as a whole (epitomized by the temperature).

We shall consider a very simple-minded approach, and *parametrize* the eventually missing affinities by factors $e^{\sigma_{YY}}$, $e^{\sigma_{Yg}}$ and $e^{\sigma_{gg}}$. Reactions (12.4) become

$$\begin{gathered}
-\circ\cdots\circ- + -\circ\cdots\circ- \Rightarrow 4\ e^{\sigma_{YY}}\ -\circ\cdots\circ- \\
-\circ\cdots\circ- + -\circ\cdots\bullet- \Rightarrow 2\ e^{\sigma_{YY}}\ -\circ\cdots\circ- + 2\ e^{\sigma_{Yg}}\ -\circ\cdots\bullet- \\
-\circ\cdots\circ- + -\bullet\cdots\bullet- \Rightarrow 4\ e^{\sigma_{Yg}}\ -\circ\cdots\bullet- \\
-\circ\cdots\bullet- + -\circ\cdots\bullet- \\
\Rightarrow e^{\sigma_{gg}}\ -\circ\cdots\circ- + 2\ e^{\sigma_{Yg}}\ -\circ\cdots\bullet- + e^{\sigma_{YY}}\ -\bullet\cdots\bullet- \\
-\circ\cdots\bullet- + -\bullet\cdots\bullet- \Rightarrow 2\ e^{\sigma_{gg}}\ -\circ\cdots\bullet- + 2\ e^{\sigma_{Yg}}\ -\bullet\cdots\bullet- \\
-\bullet\cdots\bullet- + -\bullet\cdots\bullet- \Rightarrow 4\ e^{\sigma_{gg}}\ -\bullet\cdots\bullet- \,.
\end{gathered} \tag{12.5}$$

Notice that the affinities must be inserted according to the realized coupling. In the one-but-last line, the term in $-\bullet\cdots\bullet-$ has the factor $e^{\sigma_{Yg}}$ because it comes from the a Yg coupling, $-\bullet\cdots\bullet- = -\bullet\cdots(\bullet-\circ)\cdots\bullet-$.

We can now repeat the procedure previously used to obtain the unnormalized matrix

$$\begin{pmatrix} 4\ e^{\sigma_{YY}} + 2\ e^{\sigma_{Yg}} & 2\ e^{\sigma_{Yg}} + e^{\sigma_{gg}} & 0 \\ 4\ e^{\sigma_{Yg}} + 2\ e^{\sigma_{YY}} & 2\ e^{\sigma_{YY}} + 2\ e^{\sigma_{gg}} + 2\ e^{\sigma_{Yg}} & 2\ e^{\sigma_{gg}} + 4\ e^{\sigma_{Yg}} \\ 0 & 2\ e^{\sigma_{Yg}} + e^{\sigma_{YY}} & 4\ e^{\sigma_{gg}} + 2\ e^{\sigma_{Yg}} \end{pmatrix}.$$

In the next step, column normalization, we find a novelty: the matrix will not depend on the $e^{\sigma_{XX}}$ themselves, but only on their differences. There is a freedom of choice, so that we take the differences with respect to σ_{Y_g}. more precisely, we introduce new parameters $A = e^{\sigma_{YY}-\sigma_{Yg}}$ and $B == e^{\sigma_{gg}-\sigma_{Yg}}$, which can be seen as relative affinities. The normalized matrix turns out to be

$$\begin{pmatrix} \frac{1+2A}{3+3A} & \frac{2+B}{6+3A+3B} & 0 \\ \frac{2+A}{3+3A} & \frac{2+2A+2B}{6+3A+3B} & \frac{2+B}{3+3B} \\ 0 & \frac{2+A}{6+3A+3B} & \frac{1+2B}{3+3B} \end{pmatrix}. \tag{12.6}$$

The eigenvalues are now $\lambda = 1$, $\frac{2(1+A+B)}{3(2+A+B)}$ and $\frac{AB-1}{3(1+A)(1+B)}$. The equilibrium eigenstate is

$$\mathbf{p}^{(\infty)} = \frac{1}{(4+A+B)[4+3B+A(3+2B)]} \begin{pmatrix} (1+A)(2+B)^2) \\ (2+A)(2+B)(2+A+B) \\ (2+A)^2\ (1+B)) \end{pmatrix}. \tag{12.7}$$

Suppose it so happens that the affinities are all equal, $\sigma_{YY} = \sigma_{Y_g} = \sigma_{gg}$. In that case the relative affinities are $A = B = 1$ and (12.5) reduces to (12.1), with its eigenvalues and asymptotic eigendistribution. From that we learn something: the success of the purely Mendelian approach tells us precisely that the last assumption is valid: at least in what peas are concerned, the affinities involved are all very close to each other. The two parameters of the model, the relative affinities, are ultimately irrelevant.

We have proceeded to all this seemingly useless discussion to call attention to a point. For other genetic systems there is no *a priori* reason for the relative affinities to be $= 1$ and, consequently, uneffective. They might, in principle, be very different from 1 and would, in that case, take into account the distinct tendencies of distinct macromolecules to couple to each other. We have said that, from chemical considerations, they would depend on individual aspects of each kind of molecules, but also on general characteristics of the system. What could be such "global" properties be? What would, for example, take the place of temperature in genetic systems? In kinetic gas theory, temperature is a measure of mobility. Peas are not remarked by their mobility, but humans are.

The fertilization process of human populations can depend on much more than their strictly genetic charge. Geography (food, climate-temperature-humidity, etc.), social tendencies (tabous, endogamy or exogamy habits, etc.), competition with other species (disease resistence) and even psychological ones (Oedipus complex).

In next section we consider a case for which the simple valence counting procedure fails completely, but the introduction of affinities can provide at least a means to parameterize the results: OAB blood types in humans.

12.2 OAB blood types

This is a case of genetics of human populations [49]. The phenotypes A, B and O are the only directly detectable forms. Type O is recessive with respect both to A and B: a person of type O is purely OO, but a type A can be either AO or AA; B can be either BO or BB. There is, furthermore, the composite type AB.

12.2.1 *Failure of the naïve approach*

The underlying picture in the process of fertilization is that of "micro-matings" between molecules – better, couplings between pieces of molecules. If we think of the differences in their constitutions (valence angles, for example), it is difficult to believe in the crude probabilities used below. Take for instance the case $OO + AO \Rightarrow 2OO + 2AO$.

Genes (we shall eventually use this rather misleading word) O and A will quite credibly exhibit distinct configurations, in number and geometry. There is no *a priori* reason to believe that O will couple with the same probability both to O itself and to A. We believe, however, that this is the first thing to try, as a first step into establishing a method. It will later be corrected by the use of affinities.

The states are, consequently, the probabilities of types OO, AO, AA, BO, BB and AB. We choose this ordering of states arbitrarily. The probability of type OO will be $p_1 = p_{OO}$, that of type AO will be $p_2 = p_{AO}$, etc., up to $p_6 = p_{AB}$. Each generation, we repeat, receives its genetic charge from the immediately previous generation. Given the probability distribution

$$\mathbf{p}^{(1)} = \begin{pmatrix} p_{OO} \\ p_{AO} \\ p_{AA} \\ p_{BO} \\ p_{BB} \\ p_{AB} \end{pmatrix} = \begin{pmatrix} p_1 \\ p_2 \\ p_3 \\ p_4 \\ p_5 \\ p_6 \end{pmatrix} \tag{12.8}$$

of the states in one generation, the next generation will have a probability distribution given by a stochastic matrix M applied to it. This matrix will have entries with a very clear meaning: $M_{12} = M_{(OO)(AO)}$, for example, is the probability to obtain OO from AO; $M_{46} = M_{(BO)(AB)}$, that to obtain BO from AB, etc.

The mendelian rules give the probabilities of the different blood-types in the progeny given the parents' types. If both parents are OO then the child is necessarily OO also. If the parents are OO and AO, the child will have 50% of probability of being OO and 50% of probability of being AO. They can be represented by the following set of reactions:

$$\begin{gathered} OO + OO \Rightarrow 4\ OO \\ OO + AO \Rightarrow 2\ OO + 2\ AO \\ OO + AA \Rightarrow 4\ AO \\ OO + BO \Rightarrow 2\ OO + 2\ BO \\ OO + BB \Rightarrow 4\ BO \\ OO + AB \Rightarrow 2\ AO + 2\ BO \end{gathered} \tag{12.9}$$

$$\begin{gathered} AO + AO \Rightarrow AA + 2\ AO + OO \\ AO + AA \Rightarrow 2\ AA + 2\ AO \\ AO + BO \Rightarrow AB + OO + BO + AO \\ AO + BB \Rightarrow 2\ AB + 2\ BO \\ AO + AB \Rightarrow AA + BO + AB + AO \end{gathered} \tag{12.10}$$

$$\begin{aligned} AA + AA &\Rightarrow 4\ AA \\ AA + BO &\Rightarrow 2\ AB + 2\ AO \\ AA + BB &\Rightarrow 4\ AB \\ AA + AB &\Rightarrow 2\ AA + 2\ AB \end{aligned} \tag{12.11}$$

$$\begin{aligned} BO + BO &\Rightarrow BB + 2\ BO + OO \\ BO + BB &\Rightarrow 2\ BB + 2\ BO \\ BO + AB &\Rightarrow AB + BO + BB + AO \\ \\ BB + BB &\Rightarrow 4\ BB \\ BB + AB &\Rightarrow 2\ BB + 2\ AB \end{aligned} \tag{12.12}$$

$$AB + AB \Rightarrow AA + BB + 2\ AB.$$

Counting leads to the unnormalized matrix

$$\begin{matrix} 8 & 4 & 0 & 4 & 0 & 0 \\ 8 & 8 & 8 & 4 & 0 & 4 \\ 0 & 4 & 8 & 0 & 0 & 4 \\ 8 & 4 & 0 & 8 & 8 & 4 \\ 0 & 0 & 0 & 4 & 8 & 4 \\ 0 & 4 & 8 & 4 & 8 & 8 \end{matrix}$$

which, after column normalization, gives the governing matrix

$$\mathbf{M} = \begin{pmatrix} 1/3 & 1/6 & 0 & 1/6 & 0 & 0 \\ 1/3 & 1/3 & 1/3 & 1/6 & 0 & 1/6 \\ 0 & 1/6 & 1/3 & 0 & 0 & 1/6 \\ 1/3 & 1/6 & 0 & 1/3 & 1/3 & 1/6 \\ 0 & 0 & 0 & 1/6 & 1/3 & 1/6 \\ 0 & 1/6 & 1/3 & 1/6 & 1/3 & 1/3 \end{pmatrix}. \tag{12.13}$$

The flux diagram is shown in Figure 12.5. The eigenvalues are $\lambda = 1$, 0, 0, 0, 1/2 and 1/2. Though degenerate, $\mathbf{M}$ is primitive (no extra root of modulus 1, regular (1 is the only maximal root), and totally regular (root 1 is single). As the diagonal terms are > 0, the expected return time is $= 1$ for all states. The states are positive-recurrent, so that so is the chain. Consequently, as it is also totally regular, the ergodic theorem applies.

The $\lambda = 1$ single equilibrium state is

$$\mathbf{p}^{(\infty)} = \begin{pmatrix} 1/9 \\ 2/9 \\ 1/9 \\ 2/9 \\ 1/9 \\ 2/9 \end{pmatrix}. \tag{12.14}$$

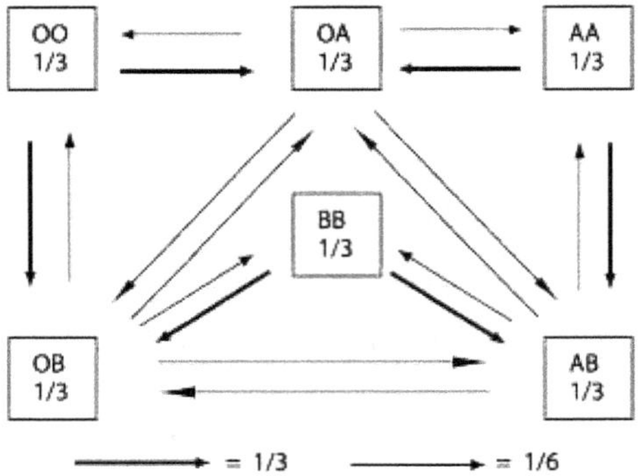

Fig. 12.5 *Flux diagram of matrix (12.13). Transition probabilities from one type to itself are shown in their corresponding boxes. Off-diagonal probabilities are indicated by the arrows, whose values appear in the lower part.*

The trouble is that this equilibrium is in clear contradiction with all known population data. It gives 11% for O blood type, which is everywhere majoritary. The other percentages, 33% for A, 33% for B and 11% for AB, also fall off the marks by far.

By the way, the existence of a single equilibrium distribution seems also to contradict known statistics.

The most puzzling aspect of the OAB data is that the percentages are very different for different populations around the world. The rough percentages for the phenotypes A, B and O are[3]

Region/Peoples		A	B	O
American natives		1.7	0.3	98
Australian natives		22	2	76
Africa		18	13	69
Europe		27	8	65
	Basques	23	2	75
	Italian	20	7	73
	English	25	8	67
East Asia		20	19	61

The high dispersion shown suggests strongly that the factors discussed at the end of the previous section are indeed important.

[3] See reference [50], mainly pages 13-15 and 105-106. The table is taken from that given at page 15.

It might happen that the process presided by matrix (12.13) were a very slow one. This is not so. If we take the extreme case of amerindians (with invented A and B subdivisions)

$$\mathbf{p}^{(1)} = \begin{pmatrix} 0.98 \\ 0.01 \\ 0.007 \\ 0.002 \\ 0.001 \\ 0 \end{pmatrix}. \tag{12.15}$$

our false equilibrium (12.15) would be attained in 20 generations. This is short if compared with the time amerindians are supposed to have been out of contact with other populations (at least 12 thousand years). The transient would be shorter for the other distributions. We are thus forced to add affinity factors. Due to the number and complexity of the possible environmental effects, they are to be seen as mere parameters hopefuly encapsulating a huge amount of information on endo/exogamic tendencies, resistence to diseases, presence/absence of disease agents in a region, etc., etc. The garden peas have been shown to be a two-parameter case.

Numerical tests can be easily performed to check trivial ideas. For example, we can imagine that endogamy would reinforce the diagonal entries in the governing matrix. By simply doubling the diagonal contributions in (12.13) (and renormalizing the columns), a new matrix is found which takes the pure O distribution $(1, 0, 0, 0, 0, 0)^T$ into the wrong equilibrium (12.15) in twenty-odd generations. It takes longer, but it happens notwithstanding. The point is that, at the sight of the above table, the equilibrium must be changed drastically, in a way that depends on the environmental effects.

12.2.2 *With external parameters*

Let us now address the task of introducing affinities, or "chemical potentials". The garden peas have been shown to be a two-parameter case. The present will be shown to represent a five-parameter problem. We start by simply inserting affinity factors in the most direct way. The reactions become

$$\begin{gathered}
OO + OO \Rightarrow 4\ e^{\mu_{OO}}\ OO \\
OO + AO \Rightarrow 2\ e^{\mu_{OO}}\ OO + 2\ e^{\mu_{AO}}\ AO \\
OO + AA \Rightarrow 4\ e^{\mu_{AO}}\ AO \\
OO + BO \Rightarrow 2\ e^{\mu_{OO}}\ OO + 2\ e^{\mu_{BO}}\ BO \\
OO + BB \Rightarrow 4\ e^{\mu_{BO}}\ BO \\
OO + AB \Rightarrow 2\ e^{\mu_{AO}}\ AO + 2\ e^{\mu_{BO}}\ BO
\end{gathered}$$

$$
\begin{gathered}
AO + AO \Rightarrow e^{\mu_{AA}}\ AA + 2\ e^{\mu_{AO}}\ AO + e^{\mu_{OO}}\ OO \\
AO + AA \Rightarrow 2\ e^{\mu_{AA}}\ AA + 2\ e^{\mu_{AO}}\ AO \\
AO + BO \Rightarrow e^{\mu_{AB}}\ AB + e^{\mu_{OO}}\ OO + e^{\mu_{BO}}\ BO + e^{\mu_{AO}}\ AO \\
AO + BB \Rightarrow 2\ e^{\mu_{AB}}\ AB + 2\ e^{\mu_{BO}}\ BO \\
AO + AB \Rightarrow e^{\mu_{AA}}\ AA + e^{\mu_{BO}}\ BO + e^{\mu_{AB}}\ AB + e^{\mu_{AO}}\ AO \\
\\
AA + AA \Rightarrow 4\ e^{\mu_{AA}}\ AA \\
AA + BO \Rightarrow 2\ e^{\mu_{AB}}\ AB + 2\ e^{\mu_{AO}}\ AOAA + BB \Rightarrow 4\ e^{\mu_{AB}}\ AB \\
AA + AB \Rightarrow 2\ e^{\mu_{AA}}\ AA + 2\ e^{\mu_{AB}}\ AB \\
\\
BO + BO \Rightarrow e^{\mu_{BB}}\ BB + 2\ e^{\mu_{BO}}\ BO + e^{\mu_{OO}}\ OO \\
BO + BB \Rightarrow 2\ e^{\mu_{BB}}\ BB + 2\ e^{\mu_{BO}}\ BO \\
BO + AB \Rightarrow e^{\mu_{AB}}\ AB + e^{\mu_{BO}}\ BO + e^{\mu_{BB}}\ BB + e^{\mu_{AO}}\ AO \\
\\
BB + BB \Rightarrow 4\ e^{\mu_{BB}}\ BB \\
BB + AB \Rightarrow 2\ e^{\mu_{BB}}\ BB + 2\ e^{\mu_{AB}}\ AB \\
\\
AB + AB \Rightarrow e^{\mu_{AA}}\ AA + e^{\mu_{BB}}\ BB + 2\ e^{\mu_{AB}}\ AB.
\end{gathered}
\tag{12.16}
$$

The provisional, unnormalized matrix is now

$$
\begin{matrix}
8e^{\mu_{OO}} & 4e^{\mu_{OO}} & 0 & 4e^{\mu_{OO}} & 0 & 0 \\
8e^{\mu_{AO}} & 8e^{\mu_{AO}} & 8e^{\mu_{AO}} & 4e^{\mu_{AO}} & 0 & 4e^{\mu_{AO}} \\
0 & 4e^{\mu_{AA}} & 8e^{\mu_{AA}} & 0 & 0 & 4e^{\mu_{AA}} \\
8e^{\mu_{BO}} & 4e^{\mu_{BO}} & 0 & 8e^{\mu_{BO}} & 8e^{\mu_{BO}} & 4e^{\mu_{BO}} \\
0 & 0 & 0 & 4e^{\mu_{BB}} & 8e^{\mu_{BB}} & 4e^{\mu_{BB}} \\
0 & 4e^{\mu_{AB}} & 8e^{\mu_{AB}} & 4e^{\mu_{AB}} & 8e^{\mu_{AB}} & 8e^{\mu_{AB}}
\end{matrix}
$$

Notice that the net result of introducing affinities in the reaction equations is to multiply each row by the corresponding self-affinity factor. In the procedure to normalize, we begin by dropping common factors in each column and multiplying by $e^{-\mu_{OO}}$:

$$
\begin{matrix}
1 & 1 & 0 & 1 & 0 & 0 \\
e^{\mu_{AO}-\mu_{OO}} & 2e^{\mu_{AO}-\mu_{OO}} & e^{\mu_{AO}-\mu_{OO}} & e^{\mu_{AO}-\mu_{OO}} & 0 & e^{\mu_{AO}-\mu_{OO}} \\
0 & e^{\mu_{AA}-\mu_{OO}} & e^{\mu_{AA}-\mu_{OO}} & 0 & 0 & e^{\mu_{AA}-\mu_{OO}} \\
e^{\mu_{BO}-\mu_{OO}} & e^{\mu_{BO}-\mu_{OO}} & 0 & 2e^{\mu_{BO}-\mu_{OO}} & e^{\mu_{BO}-\mu_{OO}} & e^{\mu_{BO}-\mu_{OO}} \\
0 & 0 & 0 & e^{\mu_{BB}-\mu_{OO}} & e^{\mu_{BB}-\mu_{OO}} & e^{\mu_{BB}-\mu_{OO}} \\
0 & e^{\mu_{AB}-\mu_{OO}} & e^{\mu_{AB}-\mu_{OO}} & e^{\mu_{AB}-\mu_{OO}} & e^{\mu_{AB}-\mu_{OO}} & 2e^{\mu_{AB}-\mu_{OO}}
\end{matrix}
$$

Introduce now the relative affinities

$$\alpha = e^{\mu_{AO}-\mu_{OO}}; \quad \alpha' = e^{\mu_{AA}-\mu_{OO}}\ ; \quad \beta = e^{\mu_{BO}-\mu_{OO}};$$
$$\beta' = e^{\mu_{BB}-\mu_{OO}}; \quad \gamma = e^{\mu_{AB}-\mu_{OO}}\ . \tag{12.17}$$

The matrix becomes

$$\begin{matrix} 1 & 1 & 0 & 1 & 0 & 0 \\ \alpha & 2\alpha & \alpha & \alpha & 0 & \alpha \\ 0 & \alpha' & \alpha' & 0 & 0 & \alpha' \\ \beta & \beta & 0 & 2\beta & \beta & \beta \\ 0 & 0 & 0 & \beta' & \beta' & \beta' \\ 0 & \gamma & \gamma & \gamma & \gamma & 2\gamma \end{matrix}\ .$$

The normalized version is then

$$\begin{pmatrix} \frac{1}{1+\alpha+\beta} & \frac{1}{1+2\alpha+\alpha'+\beta+\gamma} & 0 & \frac{1}{1+\alpha+2\beta+\beta'+\gamma} & 0 & 0 \\ \frac{\alpha}{1+\alpha+\beta} & \frac{2\alpha}{1+2\alpha+\alpha'+\beta+\gamma} & \frac{\alpha}{\alpha+\alpha'+\gamma} & \frac{\alpha}{1+\alpha+2\beta+\beta'+\gamma} & 0 & \frac{\alpha}{\alpha+\alpha'+\beta+\beta'+2\gamma} \\ 0 & \frac{\alpha'}{1+2\alpha+\alpha'+\beta+\gamma} & \frac{\alpha'}{\alpha+\alpha'+\gamma} & 0 & 0 & \frac{\alpha'}{\alpha+\alpha'+\beta+\beta'+2\gamma} \\ \frac{\beta}{1+\alpha+\beta} & \frac{\beta}{1+2\alpha+\alpha'+\beta+\gamma} & 0 & \frac{2\beta}{1+\alpha+2\beta+\beta'+\gamma} & \frac{\beta}{\beta+\beta'+\gamma} & \frac{\beta}{\alpha+\alpha'+\beta+\beta'+2\gamma} \\ 0 & 0 & 0 & \frac{\beta'}{1+\alpha+2\beta+\beta'+\gamma} & \frac{\beta'}{\beta+\beta'+\gamma} & \frac{\beta'}{\alpha+\alpha'+\beta+\beta'+2\gamma} \\ 0 & \frac{\gamma}{1+2\alpha+\alpha'+\beta+\gamma} & \frac{\gamma}{\alpha+\alpha'+\gamma} & \frac{\gamma}{1+\alpha+2\beta+\beta'+\gamma} & \frac{\gamma}{\beta+\beta'+\gamma} & \frac{2\gamma}{\alpha+\alpha'+\beta+\beta'+2\gamma} \end{pmatrix} \tag{12.18}$$

We fall back into the pure-counting case if $\alpha = \alpha' = \beta = \beta' = \gamma = 1$. The general equilibrium distribution is obtained by rather lengthy calculations. The result is

$$\mathbf{p}^{(\infty)} = \frac{1}{D} \begin{pmatrix} 1+\alpha+\beta \\ \alpha\big(1+2\alpha+\alpha'+\beta+\gamma\big) \\ \alpha'\big(\alpha+\alpha'+\gamma\big) \\ \beta\big(1+\alpha+2\beta+\beta'+\gamma\big) \\ \beta'\big(\beta+\beta'+\gamma\big) \\ \gamma\big(\alpha+\alpha'+\beta+\beta'+2\gamma\big) \end{pmatrix}, \tag{12.19}$$

where D is simply the summation of all the entries in the column-vector, $D = 1+2\,\alpha\,(1+\alpha+\alpha'+\beta+\gamma)+2\,\beta\,((1+\beta+\beta'+\gamma)+2\,\gamma\,(\alpha'+\beta'+\gamma)+\alpha'^2+\beta'^2$. It is then possible to examine what happens in the different cases of the table of page 166. It is necessary to resort to numerical methods, and we shall only consider some model cases.

To show what we can learn in what concerns the "chemical potentials", let us start with a simplification which is no far from the average of the table:

$$\mathbf{p}^{(\infty)} = \begin{pmatrix} 0.66 \\ 0.11 \\ 0.037 \\ 0.11 \\ 0.037 \\ 0.037 \end{pmatrix} . \tag{12.20}$$

It comes from the values $\alpha = 0.166, \beta = 0.166, \alpha' = 13.8333, \beta' = 13.8333, \gamma = 13.994$. Again from the definitions of Eq.(12.17), a more complex interplay of relationships shows up. A and B exhibit equally strong propensities to couple with other of the same type, as $e^{\mu_{AA}} = e^{\mu_{BB}} = 13.83\ e^{\mu_{OO}}$, and a still lightly stronger tendency to couble between themselves, as $e^{\mu_{AB}} = 13.994\ e^{\mu_{OO}}$. They aslso exhibit equal, much smaller, tendencies to couple with type O: $e^{\mu_{AO}} = e^{\mu_{BO}} = 0.166\ e^{\mu_{OO}}$.

There are many instructive points in these results. For instance, the distribution (12.20) shows equal probabilities for AA, BB and AB. Nevertheless, in order to keep on with an equal amount of AB, it is necessary that each A have a larger affinity to B than to A itself. Type O has a preference to couple to type O instead of couplint to A or B, This preference is, however, much smaller that those of A and B to couple within their own "families". All these things are, of course, contained in reactions (12.16), but we see the dominant role of affinities.

Let us list some sets of parameter values and the consequent equilibria:

close to european average:

$$\alpha = 0.10, \beta = 0.10, \alpha' = 0.48, \beta' = 0.17, \gamma = 0.02 \Rightarrow \begin{pmatrix} 0.637 \\ 0.095 \\ 0.152 \\ 0.079 \\ 0.026 \\ 0.009 \end{pmatrix}$$

close to basques:

$$\alpha = 0.07, \beta = 0.02, \alpha' = 0.45, \beta' = 0.02, \gamma = 0.001 \Rightarrow \begin{pmatrix} 0.747 \\ 0.077 \\ 0.159 \\ 0.155 \\ 0.0006 \\ 0.0004 \end{pmatrix}$$

close to amerindians I:

$$\alpha = 0.01, \beta = 0.001, \alpha' = 0.08, \beta' = 0.04, \gamma = 0.001 \Rightarrow \begin{pmatrix} 0.980 \\ 0.010 \\ 0.007 \\ 0.001 \\ 0.001 \\ 0.000 \end{pmatrix}$$

close to amerindians II:

$$\alpha = 0.009, \beta = 0.003, \alpha' = 0.08, \beta' = 0.00, \gamma = 0.009 \Rightarrow \begin{pmatrix} 0.9790 \\ 0.0096 \\ 0.0073 \\ 0.0029 \\ 0.0000 \\ 0.0009 \end{pmatrix}$$

The ideal situation would be to have the matrix parameters $\alpha, \alpha', \beta, \beta', \gamma$, and consequently, the relative "chemical potentials", for any given observed distribution. The analytical solution to this problem seems difficult (maybe impossible) to obtain, as it involves ultimately solving polynomials of order 5 or higher. It is necessary to resort to a numerical approach. Numerical exercises show some important points:

(1) as the last example illustrates, the set of parameters leading to a given observed distribution is not unique;
(2) small changes in the matrix parameters can produce large modifications in the equilibria;
(3) as a rule, smaller α's produce more OO; this is all natural, as they mean a large μ_{OO};
(4) equal parameters, $\alpha = \alpha' = \beta = \beta' = \gamma$, imply $AO = BO, BB = AA = AB/2$, whatever their common value;
    the pure-counting case if $\alpha = \alpha' = \beta = \beta' = \gamma = 1$.
(5) not for every "equilibrium" there exists a solution; this suggests restricted domains of possible equilibria.

12.2.3 *Reduced, realistic case*

For present-day data, the analysis above is by far too fine. Only results on the phenotypes O, A and B, that is, on OO, $AO + AA$ and $BO + BB$ are actually accessible. More detailed data, with separation of the alleles, will possibly be obtained ahead. Is it possible to reduce (not stochastic terminology!) our scheme so as to have only phenotype results? Well, we can easily add the components in the equilibrium distribution (12.19) to get the phenotypes probabilities:

$$\mathbf{p}^{(\infty)} = \frac{1}{D}\begin{pmatrix} 1+\alpha+\beta \\ \alpha(1+5\alpha+\beta+2\gamma) \\ \beta(1+\alpha+5\beta+2\gamma) \\ 2\gamma(\alpha+\beta+\gamma) \end{pmatrix}, \tag{12.21}$$

where D is now $D = 1+5\,\alpha^2+5\,\beta^2+2\,\gamma^2+2\,\alpha\,(1+\beta+2\,\gamma)+\beta\,(2+4\,\gamma)$. This corresponds to taking $\alpha' = \alpha$ (that is, $\mu_{AA} = \mu_{AO}$) and $\beta' = \beta$ (that is, $\mu_{BB} = \mu_{BO}$).

The corresponding reduced version of matrix (12.18) is obtained by first colapsing the AO, AA and BO, BB combined rows and columns in the non-normalized matrix and then normalizing again the columns to 1:

$$\begin{pmatrix} \frac{1}{1+\alpha+\beta} & \frac{1}{1+5\alpha+\beta+2\gamma} & \frac{1}{1+\alpha+5\beta+2\gamma} & 0 \\ \frac{\alpha}{1+\alpha+\beta} & \frac{5\alpha}{1+5\alpha+\beta+2\gamma} & \frac{\alpha}{1+\alpha+5\beta+2\gamma} & \frac{\alpha}{1+\beta+\gamma} \\ \frac{\beta}{1+\alpha+\beta} & \frac{\beta}{1+5\alpha+\beta+2\gamma} & \frac{5\beta}{1+\alpha+5\beta+2\gamma} & \frac{\beta}{1+\beta+\gamma} \\ 0 & \frac{2\gamma}{1+5\alpha+\beta+2\gamma} & \frac{2\gamma}{1+\alpha+5\beta+2\gamma} & \frac{\gamma}{1+\beta+\gamma} \end{pmatrix} \tag{12.22}$$

It is now possible, by imposing equality between the above equilibrium distribution and the data for different populations, to obtain the corresponding relative affinities. As examples:

european average:

$$\begin{matrix} \alpha{=}0.24125, \quad \beta{=}0.0956844, \quad \gamma{=}0.0 \\ \mu_{OO}{=}\mu_{AO}{+}1.42, \quad \mu_{OO}{=}\mu_{BO}{+}2.35 \end{matrix} \Rightarrow \begin{pmatrix} 0.65 \\ 0.27 \\ 0.08 \\ 0. \end{pmatrix}$$

basques:

$$\begin{matrix} \alpha{=}0.18897, \quad \beta{=}0.024662, \quad \gamma{=}0.0 \\ \mu_{OO}{=}\mu_{AO}{+}1.67, \quad \mu_{OO}{=}\mu_{BO}{+}3.70 \end{matrix} \Rightarrow \begin{pmatrix} 0.75 \\ 0.23 \\ 0.02 \\ 0. \end{pmatrix}$$

east Asia:

$$\begin{matrix} \alpha{=}0.2065, \quad \beta{=}0.198889146, \quad \gamma{=}0.0 \\ \mu_{OO}{=}\mu_{AO}{+}1.58, \quad \mu_{OO}{=}\mu_{BO}{+}1.61 \end{matrix} \Rightarrow \begin{pmatrix} 0.61 \\ 0.20 \\ 0.19 \\ 0. \end{pmatrix}$$

amerindians:

$$\begin{matrix} \alpha{=}0.0163038, \quad \beta{=}0.00302531, \quad \gamma{=}0.0 \\ \mu_{OO}{=}\mu_{AO}{+}4.12, \quad \mu_{OO}{=}\mu_{BO}{+}5.80 \end{matrix} \Rightarrow \begin{pmatrix} 0.980 \\ 0.017 \\ 0.003 \\ 0.0 \end{pmatrix}$$

In all these cases, there are many mathematical solutions for the equations turning up, but only one acceptable (with α, β, γ real positive). Notice that $\gamma = 0$ signals $\mu_{OO} >> \mu_{AB}$.

Chapter 13

Discrete quantum mechanics

The real Mechanics of Nature is Quantum Mechanics, and classical structures should come out as survivals of those quantal characteristics which are not completely "erased" in the process of taking the semiclassical limit. It is consequently amazing that precisely those quantal structures backing the fundamental Hamiltonian formalism of Classical Mechanics, in particular the symplectic structure, be poorly known. We shall now see how spaces of matrices can be endowed with a differential geometry. And how, applied to those matrices which are basic to Quantum Mechanics, this differential structure leads to a quantum symplectic space, from which the classical phase space inherits its makeup.

13.1 Introduction

Consider the classical phase space $\mathbf{E}^{2n}$ of some simple mechanical system with generalized coordinates $q = (q^1, q^2, \ldots, q^n)$ and conjugate momenta $p = (p_1, p_2, \ldots, p_n)$. A mechanical system is called "simple" precisely when its phase space is topologically trivial, as the Euclidean $2n$-dimensional space $\mathbf{E}^{2n}$. Notice that $\mathbf{E}^{2n}$, the simplest example of phase space, is enough to model any case in which the configuration space is a vector space. The dynamic quantities are functions $F(q,p)$, $G(q,p)$, etc., defined on $\mathbf{E}^{2n}$, and constitute an associative algebra with the usual pointwise product, $(F \cdot G)(x) = F(x)\, G(x)$, as operation. Given any associative algebra, one may get a Lie algebra with the commutator as operation. Of course, due to its commutativity, the classical Lie algebra of dynamic functions coming from the pointwise product is trivial. In Classical Mechanics, it is the peculiar noncommutative Lie algebra defined by the Poisson bracket which is physically significant. This is a rather strange situation from the mathematical point of view, as natural brackets are those coming as commutators in associative algebras and the Poisson bracket does not come that way. We know, however, that a powerful geometric background, the Hamiltonian (or symplectic) structure, lies behind the Poisson bracket, giving to its algebra a meaningful and deep content.

In Quantum Mechanics, the product in the algebra of dynamic functions (that is, the operators) is noncommutative and the consequent commutator Lie algebra plays a fundamental role. Nevertheless, despite the foresight of Dirac who, in his seminal paper [52], calls commutators "quantum derivations", the well-known noncommutativity of Quantum Mechanics has much more of algebra than of geometry. The difference rests, of course, in the absence of specifically geometric structures in the algebra of operators, such as differential forms, connections, metrics and the like — in a word, in the absence of a *differential* geometry.

Commentary 176.1. ► In this respect much knowledge has come up from those quantization procedures related to geometric quantization. For instance, there does exist a certain connection underlying prequantization, but the whole procedure does not lead to full quantization [53] and its study is still in progress. ◄

The recent developments in noncommutative geometry [54] have led to renewed attempts to make explicit the quantum symplectic structure. Manin's "quantum space" [55], for example, is a step towards it. We intend now to examine the question of quantum differential geometry and try to unravel the problem of the symplectic structure. We have seen in our previous discussion that circulant matrices are ultimately defined on the discrete circle Z_N. If we take Z_N as the configuration space, then its Fourier-dual, the "momentum" space, will be another copy of Z_N. Phase space will consequently be a lattice torus $Z_N \otimes Z_N$. We shall see that the continuum classical limit of the lattice $2N$-torus is precisely the above $\mathbf{E}^{2n}$ and, strong of this foreknowledge, shall restrict ourselves to that specially simple case, in which everything seems to work quite well. It is always good to have such a fair, straight working case in sight before one proceeds to more involved problems.

Let us summarize in crude words the main gate into noncommutative geometry [54], [56], [57]. The complex- or real-valued functions defined on a differentiable manifold M constitute an associative algebra $C(M)$ with the pointwise product. This algebra contains all the information about M, in particular everything concerning its topology and differentiable structure. The differentiable structure, for instance, is expressed in terms of vector fields, which are derivatives acting on $C(M)$.

On usual manifolds (point manifolds, as the phase space above), this algebra is commutative. The passage into noncommutative geometry is made by working out everything on $C(M)$ *as if* the product were noncommutative, while retaining associativity. In the case of $\mathbf{E}^{2n}$, this would mean that $F \cdot G$ is transformed into some non-abelian product $F \circ G$, with a new operation "$\circ$" (of which a particular realization will be shown later on). A manifold is essentially a space on which coordinates (an ordered set of real, commutative point functions into the real line) can be defined. When we think of $C(M)$ as noncommutative also the coordinates, like the other functions, become noncommutative. The resulting geometry of M will thereby "become" noncommutative.

Differentials come up in a very simple way through the (no more trivial) commutator. Associativity of the product $F \circ G$ implies the Jacobi identity (i.e., the very character of Lie algebra) and this makes of the commutator $[F, G] = F \circ G - G \circ F$, with fixed F, a derivative "with respect to F". Indeed, the Jacobi identity

$$[F, [G, H]] = [[F, G], H] + [G, [F, H]]$$

is just the Leibniz rule for the "product" defined by the operation $[,]$. The product "$\circ$" related to quantization is the so-called "star-product" [58], and the quantum-classical relationship is better seen in the Weyl–Wigner picture of Quantum Mechanics. In that picture, quantum operators are obtained from classical dynamic functions via the Weyl prescription and the quantum formalism is expressed in terms of Wigner functions. These are "c-number" functions, which however multiply each other via the star product instead of the pointwise product. It will be seen that this procedure is in one-to-one correspondence with the operator point of view in which, instead of the function algebra, a matrix algebra is at work.

This chapter, devoted to Discrete Quantum Mechanics, is deeply correlated to the one which follows, in which the symplectic aspects are discussed. We intend first to establish a general notion of the Weyl prescription, by recognizing as such the expressions of quantum operators in a certain well-chosen basis of unitary operators. Then, with that notion in mind, we shall examine the differential geometry of the space of quantum operators, with central interest in its symplectic structure. The lattice torus will be adopted as a privileged example. It exhibits the basic ideas in a simple way and, in the continuum classical limit, tends just to the phase space $\mathbf{E}^{2n}$. While from time to time making contact with the continuum limit, we shall actually (in Section 13.2) use the discrete Weyl–Wigner procedure inherent to the lattice, which avoids most of the difficulties involving integration measures and reduces quantum operators to finite matrices. This means that what will be obtained is actually a differential geometry on a space of matrices. The Hamiltonian quantum structure turns up in a direct way and it is an easy task to compare their characteristics with those of its classical correspondent.

The "well-chosen" unitary basis above alluded to has been introduced by Schwinger to describe Weyl's version of the Heisenberg group. In that basis, operators appear as the quantum versions of classical dynamic quantities, whenever the latter exist. This is to say that this basis provides a direct realization of the correspondence principle. It would be of great interest to examine in this formalism the announced breakdown of the principle [59] for physically realistic potentials with attractive tails of the form $\sim -C/r^n$, with $n \geq 3$. The Weyl–Wigner transformations will help bridging the gap between classical differential geometric concepts, of which the symplectic form is an example, and their quantum counterparts. The operators belonging to Schwinger's basis are labeled by a double-integer index whose values span just a lattice torus, which we call quantum phase space (QPS hereafter). The coefficients in the operator expansions are functions defined on QPS. The role and basic characteristics of this space are well-known for the simplest cases [60].

Two ingredients will be essential to establish a differential geometry on the space of quantum operators and/or on the space of Wigner functions. The first will be the Weyl–Wigner transformation. After introducing the necessary notation, we shall list the main properties of Schwinger's basis and recall some general facts on QPS. Operator expansions in that basis will then be related to Weyl–Wigner transformations (in Section 13.3). The notions of twisted convolutions and products will be introduced with the aim of making clear the interpretation of the Weyl prescription as a Fourier operator expansion. Although we avoid as a rule the rather involved mathematics which stays behind the whole subject, a few comments of a more formal character are made after that, the main intention being to explain why our case is so simple that full quantization can be achieved. Quantum groups are then briefly discussed, as well as their relationship to braiding and to the Yang–Baxter equations. They appear as algebras of matrices whose entries are themselves matrices (Section 13.4). Imprimitive matrices of index h, for example, can be written as $h \times h$ matrices of matrices (see Comment 44.1). The second ingredient will be the differential geometry of matrix space, examined in the next chapter (Section 14.1).

The structure corresponding to the quantum commutator in a c-number formulation is not the Poisson bracket, but the Moyal bracket [61]. We find this bracket in Section 14.2, as a natural feature of matrix differential geometry. It is also shown the perfect isomorphism between this operator approach and the formalism using Wigner functions with the star product. We look for braiding behind the formalism and exhibit (Section 14.3) non-trivial solutions for the braid (or Yang–Baxter) equations. The existence of these solutions is related to a property specific to the quantum case: unlike its classical analogue, which is responsible for the Poisson structure, the inverse quantum symplectic matrix is not an antisymmetric matrix.

13.2 Weyl–Heisenberg groups

We have, in the previous chapter, found the operators $\mathbf{U}$ and $\mathbf{V}$ as basic entities in the formalism of circulant matrices. We shall now retrieve them as the building bricks of general operators in Weyl's approach to Quantum Mechanics. We shall use Dirac's bracket notation in what follows. Thus, $|v_k\rangle$ will indicate the $\mathbf{V}$ eigenkets: $\mathbf{V}|v_k\rangle = \omega^k|v_k\rangle$. This corresponds to the canonical linear basis $\{\Delta_{rs}\} = \{|v_r\rangle\langle v_s|\}$ described in Section 9.1. These eigenkets exhibit clearly the action of the shift operator as $\mathbf{U}|v_k\rangle = |v_{k+1}\rangle$. The remarkable fact is that $\mathbf{V}$, on its side, acts as a shift operator on the $\mathbf{U}$ eigenkets: $\mathbf{V}|u_k\rangle = |u_{k-1}\rangle$. This duality between $\mathbf{U}$ and $\mathbf{V}$ leads to a deep and rich structure standing behind the Quantum Mechanics of simple systems. This structure is better seen in a simplified, discrete, finite-matrix version, the so-called Discrete Quantum Mechanics.[1] In addressing this topic we shall find it convenient to repeat some of the previous results in an explicitly quantum-mechanical context.

13.2.1 *Weyl's operators*

Weyl realizations [63] of the Heisenberg group are built up in terms of two conjugate unitary operators $\mathbf{U}$ and $\mathbf{V}$ satisfying the basic relation $\mathbf{VU} = \omega\,\mathbf{UV}$, where ω is a complex number. Finite dimensional representations are obtained by taking for $\mathbf{U}$ and $\mathbf{V}$ matrices $N \times N$ such that $\mathbf{U}^N = \mathbf{I}$, $\mathbf{V}^N = \mathbf{I}$. There is one realization for each integer N. It follows by taking determinants in the basic relation that $\omega = e^{i(2\pi/N)}$, so that we are just in the case of equation (9.37).

The group elements are then products of powers of the type $\omega^p\,\mathbf{U}^m\mathbf{V}^n$, which are related to the Heisenberg group defining-relations in terms of triples,

$$(m, n, p) * (r, s, q) = (m + r, n + s, p + q + \tfrac{1}{2}[nr - ms]) . \tag{13.1}$$

[1] See [62] for an appraisal.

These relations are satisfied if we put $(m, n, p) = \omega^p \omega^{mn/2}\, \mathbf{U}^m \mathbf{V}^n$. Such monomials in **U** and **V** constitute (see Section 9.1) a complete basis for all quantum operators related to the physical system. This is to say that a quantum operator belongs to the group ring in some representation of the Heisenberg group.

In a basis for the space of states given by orthonormalized kets $|v_k\rangle$, with k an integer and with the cyclic condition $|v_{k+N}\rangle \equiv |v_k\rangle$, **U** can be *defined* by

$$\mathbf{U}|v_k\rangle = |v_{k+1}\rangle \ , \tag{13.2}$$

which says that

$$\mathbf{U}^m = \sum_{k=1}^{N} |v_{k+m}\rangle\langle v_k| \ . \tag{13.3}$$

The cyclic condition imposed on the kets implies $\mathbf{U}^N = 1$. **U**'s eigenvalues $u_k = \exp[i(2\pi/N)k] = \omega^k$ correspond to eigenkets $|u_k\rangle$, with $|u_{k+N}\rangle = |u_k\rangle$ and fixed by

$$\mathbf{U}|u_k\rangle = u_k\ |u_k\rangle \ . \tag{13.4}$$

The operator **V** is *defined* by

$$\mathbf{V}|u_k\rangle = |u_{k-1}\rangle \ , \tag{13.5}$$

so that

$$\mathbf{V}^n = \sum_{k=1}^{N} |u_{k-n}\rangle\langle u_k| \ . \tag{13.6}$$

Then also $\mathbf{V}^N = 1$ and the **V** eigenvalues are $v_k = \omega^k$. The **V** eigenkets $|v_k\rangle$, such that

$$\mathbf{V}|v_k\rangle = v_k|v_k\rangle \ , \tag{13.7}$$

are just those we have started with. A direct calculation shows that the basic relation, or its enlarged form

$$\mathbf{V}^n\mathbf{U}^m = \omega^{mn}\mathbf{U}^m\mathbf{V}^n \ , \tag{13.8}$$

is implied by the ket cyclic conditions. All numbers k, m, n, etc., are defined mod(N).

13.2.2 *The Schwinger basis*

Expression (13.8) is invariant under the simultaneous changes $\mathbf{U} \to \mathbf{V}$, $\mathbf{V} \to \mathbf{U}^{-1}$, $m \to n$, $n \to -m$. This symmetry leaves also invariant the Schwinger operators

$$\mathbf{S}_{(m,n)} = e^{i\frac{\pi}{N}mn}\mathbf{U}^m\mathbf{V}^n = \omega^{\frac{mn}{2}}\ \mathbf{U}^m\mathbf{V}^n\ , \tag{13.9}$$

which are unitary. It is indeed immediate to verify that

$$\mathbf{S}_{(0,0)} = \mathbf{I}\ ,\ \mathbf{S}^{-1}_{(m,n)} = \mathbf{S}^{\dagger}_{(m,n)} = \mathbf{S}_{(-m,-n)}\ .$$

The set $\{\mathbf{S}_{(m,n)}\}$ constitutes [64] a complete orthonormal (in a natural metric given below) matrix basis to whose special virtues we shall repeatedly refer. Besides those given above, it has many other beautiful properties:

- associativity:
$$\left(\mathbf{S}_{(m,n)}\mathbf{S}_{(r,s)}\right)\ \mathbf{S}_{(k,l)} = \mathbf{S}_{(m,n)}\ \left(\mathbf{S}_{(r,s)}\mathbf{S}_{(k,l)}\right)\ ; \tag{13.10}$$
- quasi-periodicity: $\mathbf{S}_{(N,p)} = (-)^p\mathbf{S}_{(0,p)}$ and $\mathbf{S}(p,N) = (-)^p\mathbf{S}(p,0)$;
- two-foldedness: $\mathbf{S}_{(N,N)} = (-)^N\mathbf{S}_{(0,0)}$ (more on this property may be seen in [65]);
- powers: $\mathbf{S}^p_{(m,n)} = \mathbf{S}_{(pm,pn)}$;
- tr $\mathbf{S}_{(m,n)} = N\delta_{n0}\delta_{m0}$.

The entries are $\left(\mathbf{S}_{(m,n)}\right)_{rs} = \delta_{r-s,m}\omega^{n(r-m/2)}$ in the canonical base, in which $\mathbf{V}$ is diagonal. Using $\mathbf{V}$-kets, $\mathbf{S}_{(m,n)}$ is written

$$\mathbf{S}_{(m,n)} = \sum_k e^{i(\pi/N)(2k+m)n}|v_{k+m}\rangle\langle v_k|\ . \tag{13.11}$$

The explicit expression has the general form

$$\mathbf{S}_{(m,n)} = \omega^{mn/2}\ \times$$
$$\begin{pmatrix}
0 & 0 & 0 & \cdots & 0 & 0 & 0 & 0 & \omega^{n(N-m+1)} & 0 & 0 & \cdots & 0 \\
0 & 0 & 0 & \cdots & 0 & 0 & 0 & 0 & 0 & \omega^{n(N-m+2)} & 0 & \cdots & 0 \\
0 & 0 & 0 & \cdots & 0 & 0 & 0 & 0 & 0 & 0 & \omega^{n(N-m+3)} & \cdots & 0 \\
\cdots & \cdots & \cdots & \cdots & \cdots & \cdots & \cdots & \cdots & \cdots & \cdots & \cdots & \cdots & \cdots \\
0 & 0 & \cdots & \cdots & 0 & 0 & 0 & 0 & 0 & 0 & \cdots & \cdots & \omega^{nN} \\
\omega^n & 0 & \cdots & \cdots & 0 & & & & 0 & 0 & \cdots & 0 \cdots & 0 \\
0 & \omega^{2n} & 0 & \cdots & 0 & & & & 0 & 0 & \cdots & 0 \cdots & 0 \\
0 & 0 & \omega^{3n} & \cdots & 0 & & & & 0 & 0 & \cdots & 0 \cdots & 0 \\
\cdots & \cdots & \cdots & \cdots & \cdots & & & & \cdots & \cdots & \cdots & \cdots \cdots & \cdots \\
0 & 0 & \cdots & \cdots & \omega^{n(N-m)} & & & & 0 & \cdots & \cdots & 0 \cdots & 0
\end{pmatrix}. \tag{13.12}$$

The matrix, without the overall factor $\omega^{mn/2}$, comes from $\mathbf{U}$-shifting $\mathbf{V}^n$ for m times.

A general matrix $\mathbf{A}$ can now be written as

$$\mathbf{A} = \tfrac{1}{N} \sum_{(m,n)} A_{(m,n)} \, \mathbf{S}_{(m,n)} \ , \tag{13.13}$$

with as coefficients the complex numbers

$$A_{(m,n)} = tr[\mathbf{S}^{\dagger}_{(m,n)} \mathbf{A}] \ . \tag{13.14}$$

Summing up, the canonical linear basis and the Schwinger basis are related by

$$\Delta_{ab} = \tfrac{1}{N} \sum_{n=1}^{N} \omega^{-n\frac{a+b}{2}} \, \mathbf{S}_{(a-b,n)} \ ; \ \ \mathbf{S}_{(m,n)} = \sum_{i=1}^{N} \omega^{n[i-m/2]} \Delta_{i,i-m} \ . \tag{13.15}$$

As to the components in both bases, we can compare

$$\mathbf{A} = \sum_{a,b} A_{ab} \, \Delta_{ab} = \tfrac{1}{N} \sum_{a,b,n=1}^{N} A_{ab} \, \omega^{-\frac{n}{2}(a+b)} \, \mathbf{S}_{(a-b,n)}$$

with (13.13) to find

$$A_{(i,j)} = \omega^{-\frac{ij}{2}} \sum_{s} A_{s+i,s} \, \omega^{-js} \ .$$

Each dynamic quantity will be an operator $\mathbf{A}$ belonging to the group algebra, and will be written in Schwinger's basis as in (13.13), with coefficients given in (13.14). A few other properties are easily checked:

$$tr\mathbf{A} = A_{(0,0)} \ ; \ A_{(m,n)} = \left[\mathbf{S}^{\dagger}_{(m,n)} \mathbf{A}\right]_{(0,0)} \ ; \tag{13.16}$$

$$tr\mathbf{A}^{\dagger}\mathbf{A} = \sum_{(m,n)} |A_{(m,n)}|^2 \ . \tag{13.17}$$

The trace actually defines an internal product on the operator algebra by

$$\langle \mathbf{A}, \mathbf{B} \rangle = tr\left(\mathbf{B}^{\dagger}\mathbf{A}\right) \ . \tag{13.18}$$

A trivial metric is thereby defined,

$$g_{(m,n),(r,s)} = \tfrac{1}{N} \, tr\left[\mathbf{S}^{\dagger}_{(m,n)} \mathbf{S}_{(r,s)}\right] = \delta^{(2)}_{(m,n),(r,s)} = \delta_{mr} \, \delta_{ns} \ . \tag{13.19}$$

The basis $\{\mathbf{S}_{(m,n)}\}$ is, by definition, orthonormal in this metric. The set of operators constitutes thus a metric algebra with unity. The double indices are a token of phase space, as will become clear when we discuss the classical limit.

We shall most of the time use an obvious compact notation in terms of 2-vectors, with $\mathbf{m} = (m_1, m_2)$, $\mathbf{n} = (n_1, n_2)$, $\mathbf{r} = (r_1, r_2)$, $\mathbf{0} = (0, 0)$, $-\mathbf{m} = (-m_1, -m_2)$, $\mathbf{m} + \mathbf{r} = (m_1 + r_1, m_2 + r_2)$, $\mathbf{m} \times \mathbf{r} = (m_1 r_2 - m_2 r_1)$, $\mathbf{m} \cdot \mathbf{r} = (m_1 r_1 + m_2 r_2)$, etc. The basis members are then written

$$\mathbf{S_m} = e^{i(\pi/N)m_1 m_2} \mathbf{U}^{m_1} \mathbf{V}^{m_2} \tag{13.20}$$

and the above properties become $\mathbf{S_0} \equiv \mathbf{I}$, $\mathbf{S}_\mathbf{m}^{-1} = \mathbf{S}_\mathbf{m}^\dagger = \mathbf{S}_{-\mathbf{m}}$, etc. We introduce also a double delta $\delta_{\mathbf{m},\mathbf{n}}$, in terms of which the metric becomes simply

$$g_{\mathbf{m},\mathbf{n}} = \tfrac{1}{N}\, tr\left[\mathbf{S}_\mathbf{m}^\dagger \mathbf{S_n}\right] = \delta^{(2)}_{\mathbf{m},\mathbf{n}} = \delta_{\mathbf{m},\mathbf{n}} \, . \tag{13.21}$$

An arbitrary matrix has now the form

$$\mathbf{A} = \tfrac{1}{N} \sum_{\mathbf{m}} A^{\mathbf{m}}\, \mathbf{S_m} \, , \tag{13.22}$$

with

$$A^{\mathbf{m}} = tr\left[\mathbf{S}_\mathbf{m}^\dagger \mathbf{A}\right] . \tag{13.23}$$

The symbol $\sum_{\mathbf{m}}$ means, of course, summation over all the distinct pairs $\mathbf{m} = (m_1, m_2)$. We shall see later that, in the continuum limit of the present formalism, each component $A^{\mathbf{m}}$ will tend to a Wigner density related to the operator $\mathbf{A}$. The above expansion of $\mathbf{A}$ is actually a discrete version of the Weyl prescription.

Another form of Eq.(13.11) is

$$\mathbf{S_m}\, |v_k\rangle = e^{i\alpha_1(k;\mathbf{m})} |v_{k+m_1}\rangle \, , \tag{13.24}$$

where we have introduced the phase

$$\alpha_1(k; \mathbf{m}) = \tfrac{\pi}{N}\, (2k + m_1)\, m_2 \, . \tag{13.25}$$

For each value of N, the $\mathbf{S_m}$'s realize a representation of the Heisenberg group in the form of a projective representation (or abelian extension) of the double cyclic group $Z_N \otimes Z_N$. Their product actually satisfies

$$\mathbf{S_r}\, \mathbf{S_m} = e^{2i\alpha_2(\mathbf{m},\mathbf{r})}\, \mathbf{S_m}\, \mathbf{S_r} = e^{i\alpha_2(\mathbf{m},\mathbf{r})}\, \mathbf{S_{m+r}} \tag{13.26}$$

with

$$\alpha_2(\mathbf{m}, \mathbf{r}) = \tfrac{\pi}{N}\, (m_1 r_2 - m_2 r_1) = \tfrac{\pi}{N}\, (\mathbf{m} \times \mathbf{r}) \, . \tag{13.27}$$

Unlike α_1, the phase α_2 is independent of the state label. There are some properties of consequence in future calculations:

- for simple index summation, the Kronecker expression (9.6);
- for double indices,

$$\frac{1}{N^2} \sum_{\mathbf{m}} e^{2i\alpha_2(\mathbf{m},\mathbf{r}-\mathbf{s})} = \delta_{\mathbf{r},\mathbf{s}} \; ; \tag{13.28}$$

- finally, the operator identity

$$\frac{1}{N} \sum_{\mathbf{m}} \mathbf{S_m} \; \mathbf{A} \; \mathbf{S_m^\dagger} = (tr\mathbf{A}) \; \mathbf{I} \; . \tag{13.29}$$

The operators **U** and **V**, first introduced by Weyl, were largely used by Schwinger who, in a series of major papers [64], stressed both their role as purveyors of a complete basis and the optimal uncertainty provided by their duality. The basic facts of Quantum Mechanics can be expressed in this formalism.

Commentary 184.1. ► A point of particular interest is the following: when N is a prime number, the pair (**U**, **V**) describes one degree of freedom taking on N possible values. When N is not prime, it is a product of prime numbers and the basis $\{\mathbf{S_m}\}$ factorizes correspondingly into a product of independent sub-bases, one for each prime factor, that is, one for each degree of freedom. This fact, reminiscent of the fast Fourier procedure, leads to a classification of the quantum degrees of freedom in terms of prime decompositions of integers. If we want to work simultaneously with two or more degrees of freedom of a physical system, we must use for N a well-chosen non-prime value, but it is, in general, more convenient to analyze the system into its independent degrees and examine each one at a time. In any case, we must make more precise the condition for the validity of expression (13.22): it holds for any operator which is dependent on those degrees of freedom which are under consideration (and, evidently, of the corresponding conjugate momenta). ◄

The choice of the most suitable base depends, of course, on the problem in view. The canonical linear base is the simplest for general purposes. The set of products $\{\mathbf{U}^m\mathbf{V}^n\}$ constitutes another base. From this set have sprung the bases of interest to Quantum Mechanics. Bases formed by hermitian operators are more closely related to observables, and have been studied in detail [51], [66]. The unitary basis $\{\mathbf{S_m}\}$ is of particular interest because of the remarkable properties listed above. It is immediate from (13.12) that the operators $\mathbf{S_m}$ reduce to the Pauli matrices for $N = 2$ (see Comment 193.1 below). For $N \geq 2$, they are those generalizations of Pauli matrices providing the finest grading of the linear complex Lie algebra $gl(N, \mathbb{C})$. Furthermore, we shall see that this basis is directly related, in the semiclassical limit, to Fourier and Weyl–Wigner transformations.

Commentary 184.2. ► Gradings of Lie algebras are physically important because they establish the existence of preferred basis admitting additive quantum numbers [67]. $\{\mathbf{S_m}\}$ is a preferred basis in this sense. ◄

The lattice torus spanned by the labels (m, n) has been called *quantum phase space* (QPS). Because of their two-foldedness property, the operators $\mathbf{S_m}$ constitute a "double covering modulus N" of the torus. It has been shown in [60] how closed paths on QPS lead to open paths in operator space and how this fact is related to noncommutativity. The labels m and n may be seen as coordinates (mod N). Or else, we may think of $\mathbf{U}^m$ and $\mathbf{V}^n$ as global "coordinates", with values on the abelian group $Z_N \otimes Z_N$. In this case, due to the projective character of the representation, such coordinates will appear as noncommutative "point functions".

13.2.3 *Continuum limit and interpretation*

The usual continuum limit corresponds to taking to infinity both the torus radii, making simultaneously

- $N \to \infty$;
- each number of form $\sqrt{2\pi/N}m \to$ some real constant a;
- $\sqrt{2\pi/N}\sum_k \to \int da$;
- the area $(2\pi/N)$ of each elementary lattice plaquette on QPS $\to 0$. The resulting expressions are dependent on Planck's constant and the classical limit is, of course, that for $\hbar \to 0$.

The standard case of the position and momentum operators $\mathbf{q}$, $\mathbf{p}$ turns up when we choose $\mathbf{U}^m = e^{i\sqrt{2\pi/N}m\mathbf{q}}$ and $\mathbf{V}^n = e^{i\sqrt{2\pi/N}n\mathbf{p}}$. In the continuum limit, as $\sqrt{2\pi/N}m$ and $\sqrt{2\pi/N}n$ tend to constants a_1, a_2, etc., the eigenvalues of $\mathbf{q}$ and $\mathbf{p}$, which are of the form $\sqrt{2\pi/N}k$, tend to numbers q, p. This is the case of quantization on $\mathbf{E}^{2n}$. For each pair of conjugate variables, say, $(\mathbf{q}_x, \mathbf{p}_x)$, $\mathbf{S}_{(m,0)} \to \mathbf{S}_{(a_1,0)} = \mathbf{U}^{a_1} = e^{ia_1\mathbf{q}_x}$, $\mathbf{S}_{(0,n)} \to \mathbf{S}_{(0,a_2)} = \mathbf{V}^{a_2} = e^{ia_2\mathbf{p}_x}$ and $\mathbf{S_m} \to \mathbf{S_a} = e^{i(a_1\mathbf{q}_x + a_2\mathbf{p}_x)} = e^{-ia_1a_2/2}e^{ia_1\mathbf{q}_x}e^{ia_2\mathbf{p}_x} = e^{-ia_1a_2/2}\mathbf{S}_{(a_1,0)}\mathbf{S}_{(0,a_2)}$. Summing up, in the continuum limit,

$$\mathbf{S}_{(m,n)} \to \mathbf{S}_{(a_1,a_2)} = e^{i(a_1\mathbf{q}+a_2\mathbf{p})} \; . \tag{13.30}$$

Commentary 185.1. ▶ Matrices of type $\mathbf{A} = \int dp\, a(p)\, e^{ip\mathbf{q}}$ are continuum circulants: they have entries $< p'|\mathbf{A}|p'' > = a(p'' - p')$. ◀

Commentary 185.2. ▶ Notice that, in the finite case, the determinant of $\mathbf{U}$ has different signs for even and odd orders: det $\mathbf{U}$ = det $\mathbf{V} = (-)^{N-1}$, foreboding an ambiguity in the $N \to \infty$ limit. In the continuous case, there is indeed a duplication of the degrees of freedom [68]. ◀

Note. Take

$$\mathbf{U} = e^{i\sqrt{\frac{2\pi}{N}}\mathbf{q}} \,.$$

In the circulant base, $\mathbf{q} = \sum_{n=1}^{N} q_n \mathbf{U}^n$, with components

$$q_n = \sqrt{\tfrac{2\pi}{N}} \sum_{k=1}^{N} \tfrac{k}{N} e^{-i\frac{2\pi}{N}nk} = \begin{cases} i\,\sqrt{\frac{2\pi}{N}} \frac{e^{-i\frac{\pi}{N}n}}{2\sin\frac{\pi}{N}n} & \text{for } \; n \neq N \\ \frac{(N+1)}{2}\,\sqrt{\frac{2\pi}{N}} \text{ for } \; n = N \,. \end{cases}$$

This can be formally written as

$$q_n = \sqrt{\tfrac{2\pi}{N}} \left\{ \delta_{nN} \frac{(N+1)}{2} + (1 - \delta_{nN}) \frac{1}{1 - e^{i\frac{2\pi}{N}n}} \right\}$$

$$\therefore \; \mathbf{q} = \sqrt{\tfrac{2\pi}{N}} \left\{ \frac{(N+1)}{2} \mathbf{I} + \sum_{n=1}^{N-1} \mathbf{U}^n \frac{1}{1 - e^{i\frac{2\pi}{N}n}} \right\} .$$

It comes from $\ln \mathbf{U} = \sum_{k=1}^{N} (\ln \lambda_k) \mathbf{Z}_k[\mathbf{U}]$, which gives

$$\mathbf{q} = \sqrt{\tfrac{2\pi}{N}} \sum_{n=1}^{N} \sum_{k=1}^{N} \tfrac{k}{N} e^{-i\frac{2\pi}{N}kn} \mathbf{U}^n \,,$$

and from the results of Note (10.4.1). The entries are, as due for a circulant, $q_{rs} = q_{r-s}$. For $N = 4$, for example,

$$\mathbf{q} = \sqrt{\frac{\pi}{8}} \begin{pmatrix} 5 & (-i-1) & -1 & (i-1) \\ (i-1) & 5 & (-i-1) & -1 \\ -1 & (i-1) & 5 & (-i-1) \\ (-i-1) & -1 & (i-1) & 5 \end{pmatrix} . \tag{13.31}$$

This suggests a clue for the "incompleteness" of Quantum Mechanics, advocated by Einstein in his celebrated debate with Bohr. Position is given by the matrix $\mathbf{q}$. The coordinates of a matrix are all its entries. In realistic cases, the position $\mathbf{q}$ is an infinite matrix. Even if we were somehow able to go beyond the probabilistic interpretation, infinite measurements would be necessary to completely fix the position.

If we take the Fourier transform $\tilde{\mathbf{q}} = \mathbf{F}\mathbf{q}\mathbf{F}^{-1}$ of this matrix (see Note 9.2), we find the matrix $\mathbf{p}$ with entries $p_{rs} = \sqrt{\frac{2\pi}{N}} r \delta_{rs}$, which shows that $\mathbf{V} = e^{i\sqrt{\frac{2\pi}{N}}\mathbf{p}}$ and, as expected, $\mathbf{p} = \sum_{n=1}^{N} q_n \mathbf{V}^n$ with the same components. The same would come from comparison with the eigenvalues of the discrete Laplacian $\Delta = \mathbf{U} + \mathbf{U}^{-1} - 2\mathbf{I} = -\,4\sin^2 \sqrt{\frac{\pi}{2N}}\mathbf{q}$. The entries of the commutator $[\mathbf{q}, \mathbf{p}]$ show its circulant character:

$$[\mathbf{q}, \mathbf{p}]_{rs} = \sqrt{\tfrac{2\pi}{N}} (s - r) q_{rs} = \frac{2\pi}{N} \frac{s - r}{1 - e^{i\frac{2\pi}{N}(r-s)}} \,.$$

The phases (13.25) and (13.27) are the result of the action of algebraic cochains on the group elements: $\alpha_1(k; \mathbf{m}) = \alpha_1(k; \mathbf{S_m})$ and $\alpha_2(\mathbf{m}, \mathbf{n}) = \alpha_2(k; \mathbf{S_m}, \mathbf{S_n})$. On such cochains is defined a derivative d which satisfies $d^2 \equiv 0$ and defines a cohomology analogous to that of differential forms [see Eq.(14.5) below]. This cohomology gives information on the projective representation involved. Thus, if α_1 is exact, i.e., $\alpha_1 = d\alpha_0$ for some 0-cochain α_0, then α_1 may be eliminated by adding a phase α_0 to the wave functions. When α_2 is exact, that is, when $\alpha_2 = d\beta_1$ for some 1-cochain β_1, then α_2 can be eliminated by absorbing its "integral" β_1 as a phase in the operators, which thereby appear as "gaugefied", or state-dependent. It turns out actually [60] that $\alpha_2 = d\alpha_1$ in the present case, so that $d\alpha_2 \equiv 0$. This means that the 2-cochain α_2 is a cocycle, a condition which is equivalent to the associativity shown in Eq.(13.10). The 2-cochain α_2 being a cocycle means also that it might be gauged out by modifying the operators $\mathbf{S_m}$. We shall not do it because some important aspects are more easily seen when α_2 is explicitly present, but the whole formalism keeps in consequence some analogy to gauge theories. We may call α_2 the "fundamental cocycle", because it is the analogue on QPS of the symplectic form in Classical Mechanics, and actually tends to it in the classical limit. It defines a "pre-symplectic" structure on QPS, which is at the origin of the classical Hamiltonian structure. The cochain α_1, on the other hand, has a role analogous to Liouville's canonical form. When there exists a limiting classical system, such cochain properties will have classical counterparts.

If we calculate $\mathbf{S}_{(a_1,0)}\ \mathbf{S}_{(0,a_2)} = e^{i\alpha_2[(0,a_2),(a_1,0)]}\ \mathbf{S}_{(a_1,a_2)}$ from the very definition of α_2, we find $\alpha_2[(0,a_2),(a_1,0)] = -(1/2)a_1a_2$. We see that, in the classical limit, α_2 measures the area in phase space related to the operators appearing in its arguments. The fundamental relation (9.37) is a very particular case of (13.26), which shows how the shift-matrix $\mathbf{U} = \mathbf{S}_{(1,0)}$ and its diagonalized $\mathbf{V} = \mathbf{S}_{(0,1)}$ determine a minimal cell on phase space.

As $\mathbf{U}$ determines $\mathbf{V}$ and all which follows, we can say that the matrix $\mathbf{U}$ is the cornerstone of all the process of quantization on phase space in the simple (q, p) case. For more involved phase spaces, the Heisenberg group must be replaced by other groups and the treatment becomes much more complicated. Nevertheless, a general quantization principle based on Fourier transformations can be retained [69].

We have been using $\hbar = 1$, which can be corrected when necessary by simple dimensional analysis. While a commutator has no dimension by itself, the Poisson bracket has dimension [action^{-1}]. The case above has an especially simple classical limit, as higher order terms vanish in the Weyl–Wigner transformation. Only the Poisson bracket remains: we find that $\alpha_2[(a_1,0),(0,a_2)] = \alpha_2(e^{ia_1\mathbf{q}}, e^{ia_2\mathbf{p}}) = -(\hbar/2)a_1a_2\{q,p\}= -(\hbar/2)a_1a_2$. Notice that the operators appear in the exponents. The reason will become clear later on, where we shall see that the symplectic structure on the operator algebra is in reality related to the sine of α_2. But we can learn already from this example that, as soon as we leave the group to consider general operators, our interest becomes focused on the operator space itself: we go from the group elements to members of the algebra they generate. In the classical Hamiltonian formalism, we may start with phase space as defined (say) by the (q,p) pair, but canonical invariance implies that any other pair of conjugate dynamic functions $F(q,p)$, $G(q,p)$ can be used as well. And, indeed, only through the use of general dynamic functions do we arrive at the complete picture. Here we shall be led to work on the operator algebra. It will be on this algebra that the quantum symplectic structure will find its general expression.

The phase chosen in (13.20) is at the origin of the nice properties of the basis $\{\mathbf{S_m}\}$. The special form (13.30) of the continuum limit exhibits the main reason for that choice: it gives to the basis $\{\mathbf{S_m}\}$ the role of a Fourier basis and makes of it the fundamental object in Weyl–Wigner transformations.

13.3 Weyl–Wigner transformations

The dynamical quantities of Classical Mechanics are usual, commutative functions on phase space. We say that they are c-number (classical number) dynamical quantities. The dynamical quantities of Quantum Mechanics, on the other hand, are operators. The Weyl prescription is a Fourier-based procedure to raise the ambiguities arising in the transcription of classical into quantum quantities. It provides the quantum operator in correspondence with a certain classical quantity, the Wigner density. It is possible to represent the operators as functions on phase space, the Wigner functions. These are Fourier transforms of the Wigner densities. The Weyl–Wigner transformations provide a to and fro relationship between the Wigner functions and the quantum operators. The Wigner functions multiply each other via a special operation, the star product, so that their algebra differs from that of usual functions. We can say that the dynamical quantities are the same in Classical and in Quantum Mechanics, only (the product defining) their algebra being different. Working with Wigner functions is doing matrix algebra with c-number functions.

The operator expansions (13.22) are discrete versions of the Weyl prescription giving the quantum operator in correspondence with a classical dynamic quantity. Let us only recall how the prescription[2] works for the $\mathbf{E}^{2n}$ coordinate-momentum case. The *Wigner functions* $A_W(q,p)$ are written as Fourier transforms [70], [71] of certain *Wigner densities* $A(a,b)$:

$$A_W(q,p) = F[A] = \int da\ db\ e^{i(aq+bp)} A(a,b)\ . \tag{13.32}$$

Then the Weyl operator $\mathbf{A}(\mathbf{q},\ \mathbf{p})$, function of operators $\mathbf{q}$ and $\mathbf{p}$ which corresponds to A_W, will be

$$\mathbf{A}(\mathbf{q},\mathbf{p}) = \int da\ db\ e^{i(a\mathbf{q}+b\mathbf{p})}\ A(a,b)\ . \tag{13.33}$$

We may denote by $\mathbf{F}$ this operator Fourier transform, so that

$$A_W = F\left[\mathbf{F}^{-1}\left[\mathbf{A}\right]\right]\ ; \mathbf{A} = \mathbf{F}\left[F^{-1}\left[A_W\right]\right]\ . \tag{13.34}$$

The Wigner functions are not classical objects, but c-number representatives of quantum quantities. They are quantum reconstructions of the classical functions obtained from the most elementary products of its basic variables. The products $\mathbf{qp}$ and $\mathbf{pq}$, for instance, become different. The function obtained at the end of this reconstruction, including all the powers of $\hbar$ turning up in the process, will tend to the classical quantities in correspondence when $\hbar \to 0$. Only with these details in mind can we really state (as we have done in the epigraph to this section) that the dynamical functions are the same in the classical and the quantum cases. On their side, the Wigner densities $A(a,b) = F^{-1}[A_W] = \mathbf{F}^{-1}[\mathbf{A}]$ include usually Dirac deltas and their derivatives. Because of (13.30), Eq.(13.33) can be written as

$$\mathbf{A}(\mathbf{q},\mathbf{p}) = \int da\ db\ \mathbf{S}(a,b)\ A(a,b)\ , \tag{13.35}$$

of which (13.22) is a discrete version with the coefficients $A^{\mathbf{m}}$ as Wigner densities. This also shows how quantization casts its roots in Fourier analysis: an operator is given as an operator Fourier expansion.

13.3.1 *Products and commutators*

The product of two operators has the expression

$$\begin{aligned}\mathbf{AB} &= \frac{1}{N^2}\sum_{\mathbf{m}}\sum_{\mathbf{n}} A^{\mathbf{m}} B^{\mathbf{n}} e^{i\alpha_2(\mathbf{n},\mathbf{m})}\ \mathbf{S}_{\mathbf{m}+\mathbf{n}} \\ &= \tfrac{1}{N}\sum_{\mathbf{p}}\left[\tfrac{1}{N}\sum_{\mathbf{m}} A^{\mathbf{m}} B^{\mathbf{p}-\mathbf{m}} e^{i\alpha_2(\mathbf{p},\mathbf{m})}\right]\ \mathbf{S}_{\mathbf{p}}\ ,\end{aligned} \tag{13.36}$$

[2] See for example [51].

from which the Wigner densities

$$(\mathbf{AB})^{\mathbf{p}} = tr\left[\mathbf{S}_{\mathbf{p}}^{\dagger}\mathbf{AB}\right] = \tfrac{1}{N}\sum_{\mathbf{m}} A^{\mathbf{m}} B^{\mathbf{p}-\mathbf{m}} e^{i\alpha_2(\mathbf{p},\mathbf{m})} \tag{13.37}$$

are directly read. Expressions of this type, called "twisted convolutions", will be discussed later. The commutators

$$[\mathbf{S}_{\mathbf{n}}, \mathbf{S}_{\mathbf{m}}] = 2i \sin\left[\alpha_2(\mathbf{m}, \mathbf{n})\right] \mathbf{S}_{\mathbf{m}+\mathbf{n}} \tag{13.38}$$

lead, for two general operators, to

$$[\mathbf{A}, \mathbf{B}] = \tfrac{1}{N^2} \sum_{\mathbf{p}} \sum_{\mathbf{m}} A^{\mathbf{m}} B^{\mathbf{p}-\mathbf{m}}\ 2i \sin\left[\alpha_2(\mathbf{p}, \mathbf{m})\right] \mathbf{S}_{\mathbf{p}}\ . \tag{13.39}$$

This is the Weyl prescription for the commutator, and the expressions

$$[\mathbf{A}, \mathbf{B}]^{\mathbf{p}} = tr\left(\mathbf{S}_{\mathbf{p}}^{\dagger}\left[\mathbf{A}, \mathbf{B}\right]\right) = \tfrac{1}{N} \sum_{\mathbf{m}} A^{\mathbf{m}} B^{\mathbf{p}-\mathbf{m}}\ 2i \sin\left[\alpha_2(\mathbf{p}, \mathbf{m})\right] \tag{13.40}$$

will be the densities of the discrete version of the Moyal bracket. It is obvious that α_2, which had previously marked the projective character of the representation, appears as the source of noncommutativity in (13.39).

For the sake of completeness and to allow immediate comparison with the classical case, we write down also the continuum versions, for once reintroducing Planck's constant wherever due:

$$[\mathbf{S}_{\mathbf{a}}, \mathbf{S}_{\mathbf{b}}] = 2i \sin\left[-\hbar\, \frac{\mathbf{a}\times\mathbf{b}}{2}\right] \mathbf{S}_{\mathbf{a}+\mathbf{b}}\ ; \tag{13.41}$$

$$[\mathbf{A}, \mathbf{B}]^{\mathbf{p}} = \frac{2i}{\hbar} \int d\mathbf{a} \int d\mathbf{b}\ A(\mathbf{b})\ B(\mathbf{a}-\mathbf{b}) \sin\left[-\hbar\, \frac{\mathbf{a}\times\mathbf{b}}{2}\right] \mathbf{S}_{\mathbf{a}}\ . \tag{13.42}$$

Summing up, the Schwinger symmetrized basis appears as a (discrete finite at first, or continuum infinite in the limit) operator Fourier basis. In the expansion (13.22), for example, the operator $\mathbf{A}$ is just the operator Fourier transform of the Wigner density $A^{\mathbf{m}}$.

Commentary 190.1. ► An example of Wigner density: the j-th power of Weyl's shift operator has density $(\mathbf{U}^j)^{\mathbf{m}} = N\delta_j^{m_1}\delta_0^{m_2}$. Then, $(\mathbf{U}^j)_W(r,s) = e^{i(2\pi/N)jr}$, independent of s and coherent with the previously indicated continuum limit:

$$(\mathbf{U}^j)_W(r,s) = e^{i\sqrt{2\pi/N}j\sqrt{2\pi/N}r} \to e^{i\sqrt{2\pi/N}jq} \to e^{iaq}\ .$$

◄

13.3.2 *On convolutions, straight and twisted*

The star product (or twisted product) is introduced in a simple way through the notion of twisted convolution.[3] Let us again consider the phase space $\mathbf{E}^{2n}$, using now the notation $x = (x_1, x_2, \ldots, x_n, x'_1, x'_2, \ldots, x'_n)$, $y = (y_1, y_2, \ldots, y_n, y'_1, y'_2, \ldots, y'_n)$, etc. for its points. The complex functions f, g, etc., defined on $\mathbf{E}^{2n}$, constitute a commutative algebra with the usual pointwise product. A product of two functions is the Fourier transform of their convolution. More precisely, if $F[f]$ and $F[g]$ are the Fourier transforms of f and g, the pointwise product is $f \cdot g = F^{-1}[F[f] * F[g]]$. By the way, to the usual convolution of two functions,

$$(f * g)(x) = \int f(y) g(x - y) dy \ ,$$

will clearly correspond the expression $\frac{1}{N} \sum_{\mathbf{m}} A^{\mathbf{m}} B^{\mathbf{p}-\mathbf{m}}$ in the discrete case.

We can realize the passage from the commutative pointwise-product algebra $C^\infty(\mathbf{E}^{2n})$ of complex differentiable functions on $\mathbf{E}^{2n}$ to a noncommutative-product algebra in the following way. The twisted convolution of index c is defined by

$$(f *_c g)(x) = \int e^{ic/2(x \times y)} f(y) g(x - y) dy \ ,$$

where $(x \times y)_i = x_i y'_i - x'_i y_i$. The corresponding expression in our case,

$$\frac{1}{N} \sum_{\mathbf{m}} A^{\mathbf{m}} B^{\mathbf{p}-\mathbf{m}} e^{i(c/2)p \times m} \ ,$$

shows that $(\mathbf{AB})^{\mathbf{p}}$ given by (13.37) is just a twisted convolution whose index is the area $2\pi/N$ of an elementary lattice plaquette. Now, the twisted product "$\circ$" stands to the twisted convolution as the pointwise product stands to the usual convolution. It is defined as an inverse Fourier transform of the twisted convolution of Fourier transforms:

$$f \circ g = F^{-1} \left[F[f] *_c F[g] \right] .$$

It defines on $C^\infty(\mathbf{E}^{2n})$ a deformation of the algebra defined by the pointwise product, precisely the new noncommutative algebra we are looking for.

We have up to now avoided giving the explicit form of the Wigner functions, or the discrete version of (13.32). The reason is that Fourier transforms are expansions in terms of irreducible unitary representations, and in our case we should use representations which are not unitary, but projective. In order to have a Weyl realization of the Heisenberg group,

[3] See [72] and references therein.

it is necessary to perform an extension of the group $Z_N \otimes Z_N$. The truly unitary representations would in reality be related to $Z_N \otimes Z_N$ and not to the Heisenberg group. We are now in a position to circumvent this difficulty. We start by learning how to read the coefficients, or Wigner densities, and by establishing as a rule that

Wigner densities never simply multiply, but always convolute in the twisted way given by (13.37).

Once this is kept in mind, we may use a unitary representation and write

$$A_W(\mathbf{r}) = \tfrac{1}{N}\sum_{\mathbf{m}} A^{\mathbf{m}} e^{i(2\pi/N)\mathbf{m}\cdot\mathbf{r}} . \tag{13.43}$$

This expression, we repeat, only makes sense if the rule (13.37) for twist-convolving the coefficients is used every time some multiplication is performed. $A_W(\mathbf{r})$ with $\mathbf{r} = (r_1, r_2) \in Z_N \otimes Z_N$ may be seen as a function on QPS (in reality, on its Fourier dual). We

- either use operators (13.22) with the coefficients $A^{\mathbf{m}}$ taken as functions belonging to the usual commutative algebra of pointwise products and usual convolutions,
- or use the Wigner representation (13.43) with $A^{\mathbf{m}}$ belonging to the twisted algebra.

This point is of extreme importance, for it allows us to keep using usual Fourier transformations, while taking noncommutativity into account exclusively in the coefficients. We shall below [after Eq.(13.45)] give an alternative which is formally simpler.

By using (9.6), (13.43) gives immediately

$$A^{\mathbf{m}} = \tfrac{1}{N}\sum_{\mathbf{r}} e^{-i(2\pi/N)(\mathbf{m}\cdot\mathbf{r})} A_W(\mathbf{r}) .$$

It becomes also immediate to verify that the Wigner function corresponding to an hermitian operator $\mathbf{A} = \frac{1}{N}\sum_{\mathbf{m}} A^{\mathbf{m}}\,\mathbf{S}_{\mathbf{m}} = \frac{1}{N}\sum_{\mathbf{m}} A^{(-\mathbf{m})*}\,\mathbf{S}_{\mathbf{m}}$ is a real function.

Looking at (13.36) we recognize that $\mathbf{AB} = \mathbf{F}[F^{-1}[A_W] *_c F^{-1}[B_W]]$. It follows that the density is $\mathbf{F}^{-1}[\mathbf{AB}] = [F^{-1}[A_W] *_c F^{-1}[B_W]]$ and, consequently,

$$(AB)_W = F\left[F^{-1}[A_W] *_c F^{-1}[B_W]\right] = A_W \circ B_W ,$$

a twisted product in which the phase α_2 provides the twisting, i.e., the deformation of the function algebra. As $\alpha_2 \neq 0$, the "classical" dynamic quantities A_W, B_W in correspondence with quantum dynamic quantities multiply each other no more by the pointwise product, but by the noncommutative twisted product. If α_2 vanished, twisted convolution would turn into simple convolution and twisted product would reduce to pointwise product. But this is never the case, and twisting is an inevitable mark of quantum behavior.

Commentary 193.1. ▶ Let us consider an example in the case $N = 2$. It is immediate to identify one of the usual conventions for the Pauli matrices,

$$\sigma_1 = \mathbf{S}_{10} = \mathbf{U} = \left(\begin{smallmatrix} 0 & 1 \\ 1 & 0 \end{smallmatrix}\right); \ \sigma_2 = \mathbf{S}_{11} = \left(\begin{smallmatrix} 0 & i \\ -i & 0 \end{smallmatrix}\right); \ \sigma_3 = \mathbf{S}_{01} = \mathbf{V} = \left(\begin{smallmatrix} -1 & 0 \\ 0 & 1 \end{smallmatrix}\right).$$

We obtain for the Wigner densities expressions like $\sigma_1 = \frac{1}{2}\sum_{\mathbf{m}} (2\delta_1^{m_1}\delta_0^{m_2})\,\mathbf{S}_{\mathbf{m}}$, and the Wigner functions as their Fourier transforms, for example $(\sigma_1)_W(\mathbf{r}) = \frac{1}{2}\sum_{\mathbf{m}} (2\delta_1^{m_1}\delta_0^{m_2})$ $e^{i(2\pi/N)\mathbf{m}\cdot\mathbf{r}}$. The results found are

$$(\sigma_1)_W(r_1, r_2) = e^{i\pi r_1}\ , \quad (\sigma_2)_W(r_1, r_2) = e^{i\pi(r_1+r_2)} \ \text{ and } \ (\sigma_3)_W(r_1, r_2) = e^{i\pi r_2}\ .$$

The twisted convolutions are of the type

$$(\sigma_3 *_c \sigma_1)(i_1, i_2) =$$
$$\frac{1}{2}\sum_{\mathbf{m},\mathbf{n}} (2\delta_0^{m_1}\delta_1^{m_2})(2\delta_1^{i_1-m_1}\delta_0^{i_1-m_2})e^{i(\pi/2)(i_1m_2-i_2m_1)} = 2\delta_1^{i_1}\delta_1^{i_2}\,e^{-i(\pi/2)}\ .$$

The twisted product comes out then as

$$((\sigma_3)_W \circ (\sigma_1)_W)\,(r_1, r_2) = i\ e^{i\pi(r_1+r_2)} = i\ (\sigma_2)_W(r_1, r_2) = ((\sigma_3\sigma_1)_W)\,(r_1, r_2)\ .$$

This example illustrates the extreme simplicity which comes from discreteness. It also exhibits clearly the role of $Z_N \otimes Z_N$. But, above all, it shows how the noncommutativity is taken into account by the twisting. ◀

Again a review of the main points. The geometry of a manifold M is encoded in the properties of the algebra $C(M)$ of functions defined on M. The very definition of a differentiable manifold, for example, is based on the differentiability of some special functions, the coordinate transformations. The subset $C^0(M)$ of continuous functions will contain information on general topological properties, but no information on the differentiable structure. All such information is, nevertheless, contained in the set $C^\infty(M)$ of indefinitely differentiable functions. Such function spaces are usually taken as algebras with the operation of pointwise product. Mathematically, they are simple cases of $*$-algebras. We have seen the emergence of noncommutative geometry related to QPS, coming precisely from the fact that, in order to keep the quantum information, the functions representing quantum quantities — the Wigner functions — do not belong to an algebra built up with this simple commutative product, but require a twisted product. We have passed into a noncommutative $*$-algebra. We have said at the beginning of this chapter that, in a noncommutative geometry, coordinates also become noncommutative. After all we have seen, the pair of coordinates q, p in Quantum Mechanics is a paradigmatic example: they must multiply each other via the star product and their commutator coincides with that of the operators $\mathbf{q}, \mathbf{p}$.

13.3.3 *The c-number realization*

Let us go back to (13.43). It represents a Fourier expansion of a function defined on the lattice torus, in terms of the basis $\{\phi_{\mathbf{m}}(\mathbf{r}) = e^{i(2\pi/N)\mathbf{m}\cdot\mathbf{r}}\}$. As it is always the case with Fourier transformations, this is actually a basis of characters [73] of the group $Z_N \otimes Z_N$. Characters are unitary representations of the group into the unit circle, $f: G \to S^1$. The Wigner functions $(\mathbf{S}_{\mathbf{m}})_W$ corresponding to the basis operators will be simply

$$(\mathbf{S}_{\mathbf{m}})_W(\mathbf{r}) = \phi_{\mathbf{m}}(\mathbf{r}) \ . \tag{13.44}$$

Thus, in the Weyl prescription, the Wigner functions corresponding to the basic Schwinger operators are just the Fourier basic functions. This is one more quality of the symmetrized Schwinger basis. The correspondence between functions and operators is then complete.

We can translate the operator properties into the language of Wigner functions. We shall only make explicit the main results. With the help of $\frac{1}{N^2}\sum_{\mathbf{r}} \phi^*_{\mathbf{m}}(\mathbf{r})\phi_{\mathbf{n}}(\mathbf{r}) = \delta_{\mathbf{m},\mathbf{n}}$, the properties involving the trace come out directly, like for instance

$$A^{\mathbf{m}} = \frac{1}{N}\sum_{\mathbf{r}} \phi^*_{\mathbf{m}}(\mathbf{r}) A_W(\mathbf{r}) \ .$$

One finds the twisted product $\left((\mathbf{S}_{\mathbf{r}})_W \circ (\mathbf{S}_{\mathbf{u}})_W\right)(\mathbf{q}) = e^{i\alpha_2(\mathbf{u},\mathbf{r})}(\mathbf{S}_{\mathbf{r}+\mathbf{u}})_W(\mathbf{q})$, leading to the important result

$$\phi_{\mathbf{r}}(\mathbf{q}) \circ \phi_{\mathbf{u}}(\mathbf{q}) = (\mathbf{S}_{\mathbf{r}})_W(\mathbf{q}) \circ (\mathbf{S}_{\mathbf{u}})_W(\mathbf{q}) = e^{i\alpha_2(\mathbf{r},\mathbf{u})}\ \phi_{\mathbf{r}+\mathbf{u}}(\mathbf{q}) \ . \tag{13.45}$$

Notice that the characters, initially providing a unitary representation, give now a projective one. And this leads to another point of view concerning the use of the star product. Instead of convoluting the coefficients, we may write $A_W(\mathbf{r}) = \frac{1}{N}\sum_{\mathbf{m}} A^{\mathbf{m}}\ \phi_{\mathbf{m}}(\mathbf{r})$ and use directly (13.45). With the twisted product as the operation in the function algebra, the measure of its noncommutativity gives automatically the Moyal bracket for the Wigner functions,

$$[\phi_{\mathbf{r}}, \phi_{\mathbf{m}}]_\circ(\mathbf{q}) = [\phi_{\mathbf{r}} \circ \phi_{\mathbf{m}} - \phi_{\mathbf{m}} \circ \phi_{\mathbf{r}}](\mathbf{q}) = 2i \sin[\alpha_2(\mathbf{m},\mathbf{r})]\phi_{\mathbf{r}+\mathbf{m}}(\mathbf{q}) \ , \tag{13.46}$$

which is the Weyl–Wigner version of (13.38). The counterpart of (13.36) will be

$$(A_W \circ B_W)(\mathbf{q}) = \sum_{\mathbf{p}} (AB)^{\mathbf{p}} \, \phi_{\mathbf{p}}(\mathbf{q}) \, , \tag{13.47}$$

with the coefficients given by (13.37). Finally, (13.39) is translated into

$$\begin{aligned} [A_W, B_W]_\circ (\mathbf{q}) &= \tfrac{1}{N^2} \sum_{\mathbf{m}} \sum_{\mathbf{r}} A^{\mathbf{m}} B^{\mathbf{r}} \; 2i \sin[\alpha_2(\mathbf{r}, \mathbf{m})] \; \phi_{\mathbf{r}+\mathbf{m}}(\mathbf{q}) \\ &= \tfrac{1}{N^2} \sum_{\mathbf{p}} \sum_{\mathbf{m}} A^{\mathbf{m}} B^{\mathbf{p}-\mathbf{m}} \; 2i \sin[\alpha_2(\mathbf{p}, \mathbf{m})] \; \phi_{\mathbf{p}}(\mathbf{q}) \, . \end{aligned} \tag{13.48}$$

We see that the parallelism is complete. Quantization can be seen

(i) either as a passage from classical dynamic quantities into operators,
(ii) or as a deformation in the algebra of the coefficients,
(iii) or still as a deformation in the algebra of the basic functions, from pointwise product to twisted product.

It will be seen in the following that the parallelism goes over to the Hamiltonian structure.

13.4 Braiding and quantum groups

General spaces allowing to and fro Fourier transformations are quantum groups, or better, Hopf algebras. Such algebras show a natural tendency to have a braided underlying structure.

It is convenient to make here a parenthesis and digress briefly on some mathematical topics. The discussion, though not quite indispensable, will help us to get a better understanding both of what we have been doing and of the forthcoming results. In particular, it will bring to light the main reasons for the particular simplicity of our case.

13.4.1 *Why Hopf algebras*

One of the most fascinating developments of the last decades has been the discovery of the relation of Hopf algebras (or bialgebras, or quantum groups) to the Yang–Baxter equation and, consequently, to braid groups.[4]

[4] Readable appraisals are given in [74] and [75]. On braid groups, some more intuition may come from the Statistical Mechanics they define (see Section 18.4).

The simplest example of bialgebra in Physics appears in the sum of two angular momenta, usually written $J = J_1 + J_2$. In reality, this operator acts on direct-sum kets $|Y> = |Y_1> \oplus |Y_2>$, which shows that it should actually be written in the form $J = J_1 \otimes I + I \otimes J_2$. An object of this form belongs to a Hopf algebra. Further investigation has uncovered the relationship of Hopf algebras to an impressive number of seemingly disparate topics. It has established the existence of common points in as distinct subjects as statistical lattice models, integrability of nonlinear equations, infinite algebras, the inverse scattering method, topology of low-dimensional manifolds, high-temperature superconductivity, non-abelian harmonic analysis, conformal models in field theory, noncommutative geometry, quantum Hall effect, etc. An idea of this broadness of scope comes from recalling that C. N. Yang has found the basic equations as integrability conditions for the Schrödinger equation with a special potential, while R. J. Baxter has uncovered them as identities involving symmetries in statistic mechanical lattice models.[5] We shall, of course, ignore almost everything of all that and concentrate in a few topics of immediate interest to our theme.

There are many ways to approach Hopf algebras [78] besides the purely algebraic one. Among physicists, the most usual method involves group deformations [79], but we may also generalize the method used to introduce the classical groups as those transformation groups which preserve given sesquilinear forms [80]. Or still make an entrance through free calculus, as in knot theory [81], where they make a brief act of presence in terms of Alexander matrices, but are immediately abelianized in the calculations leading to invariant polynomials. The main point is that the entries of the usual matrices appearing in "classical" subjects become noncommutative, although subject to certain rules.

In very rough lines, what happens is the following. The entries t^i_j of usual matrices consist of real or complex numbers and consequently commute with each other: $t^i_j \, t^m_n = t^m_n \, t^i_j$. Matrices can constitute Lie groups, that is, smooth manifolds. To each group element corresponds a matrix. Each point on a manifold, each matrix in the case, is entitled to have coordinates, a set of real numbers describing it. Thus, each matrix will have its coordinates, which are just the entries t^m_n if they are real and, if they are complex, their real and imaginary parts. If now we take the very entries of the matrices as noncommutative, the new "coordinates" will be of a new kind. Noncommutative geometry comes to the fore. This noncommutativ-

[5] A general reference is [76].

ity is represented in the form

$$R^{rs}_{im}\, t^i_j\, t^m_n = R^{pq}_{jn}\, t^r_p\, t^s_q\,,$$

where the R^{ij}_{kl}'s are complex coefficients. Some restrictions, in the form of constraints on these coefficients, are imposed to ensure a minimum of "respectability" to the new algebraic structure turning up. In particular, the imposition of associativity leads to the Yang–Baxter equation, whose form and relationship to braid groups are given in Section 14.3. The resulting new structure is a Hopf algebra. Thus, "quantum groups" is a generic name for some sets of matrices whose entries are themselves noncommutative. They are not groups at all, but structures generalizing them. In our case, we shall find "hypermatrices", matrices whose entries are themselves matrices.

We should say a few words, in the same naïve style, on another approach [82], which is the nearest to our purposes. It emphasizes the role of Fourier transformations and is concerned with harmonic analysis on groups. It takes as starting algebras the spaces of functions on groups. Recall classical Fourier transformations on the line or on Euclidean 3-space: they establish a duality between the space of functions on the original space and the space on which the Fourier transforms are defined. The latter is the Fourier dual of the original space. The original space is actually a translation group T and its dual $\hat{T}$ is the space of (equivalence classes of) unitary irreducible representations of T. In the classical, abelian case, the dual $\hat{T}$ is another group. It happens that, when the original group G is locally compact commutative and compact, or discrete, the dual set $\hat{G}$ is a locally compact commutative group, which is furthermore respectively discrete or compact (just in that order). This is the "Pontryagin duality". Quantization in a box (q-space compact) leads to discrete momenta. Actually, the transformations take place between the respective group rings, to which the representations can be extended. Important examples of Fourier-dual spaces are: the group $\mathbb{R}$ of the real numbers, dual to itself (continuum q, idem p); the circle S^1 and the group $\mathbb{Z}$ of integers, dual to each other (angle ϕ, angular momentum component J_z any integer); and, especially important for our purposes, the cyclic group Z_N which is, like $\mathbb{R}$, self-dual. This is a first reason for the simplicity of our case: we remain in the same Z_N while going to and fro by Fourier transformations.

A great change comes up when the original group is noncommutative: the corresponding dual space is no more a group. Much is known on the case in which G is compact. The dual $\hat{G}$ is then a category, that of the finite dimensional representations of G, that is, a category of vector spaces (better, modules). This new duality, between a group and a category, is called Tanaka–Krein duality. The reason for the special simplicity of abelian groups is that their unitary irreducible representations have dimension one and the tensor product of two such representations is another one-dimensional representation. Each such representation may be considered simply as a complex function f,

$$f : G \to \mathbf{C} \ , \ g \to f(g) \ , \ \text{ with } \ f(g_1 g_2) = f(g_1) f(g_2) \ .$$

Recall that characters are representations of this kind, $f : G \to S^1$. Tensor product is then reduced to the simple pointwise product of functions and the set of inequivalent unitary irreducible representations is itself a group. This is the property which does not generalize to the noncommutative case.

The good question is then the following: is it possible to enlarge the notion of group to another object, so that its dual come to be an object of the same kind? The complete answer has been found in the case of finite groups: the more general objects required are precisely Hopf algebras. So, quantum groups are those generalizations of groups allowing for a "good" notion of Fourier duality.

We have in the previous section prepared ourselves, through the resource of absorbing the noncommutativity in the coefficients, to perform the Fourier transformation starting from the commutative group $Z_N \otimes Z_N$. The Heisenberg group is non-abelian, but with the twisted-convolution prescription we have been able to circumvent this problem. The algebra of functions on $Z_N \otimes Z_N$ with convolution given by "$*_c$" is Fourier-transformed into the algebra of functions on $Z_N \otimes Z_N$ with multiplication "$\circ$". It becomes possible to work as if the domain space were in both cases commutative groups. The prescription we have used is actually a trick allowing the use of the simpler formalism of Pontryagin duality, instead of that related to the Tanaka–Krein duality. Or, if we prefer, allowing to avoid the explicit use of Hopf algebras. This is another reason for the simplicity of our case.

13.4.2 *Infinite algebras*

In the limit $N \to \infty$, we touch the subject of infinite algebras.[6] The $\{\mathbf{S_m}\}$ basis has been a source of ideas in model building. It was extensively used by Floratos, first in the study of the (discretized) torus membrane, then to obtain special representations of the $GL_q(2)$ quantum group and finally

[6] See for instance [83] and references therein.

to relate Manin quantum space to Quantum Mechanics [84]. The double-graded algebra (13.38), with its "fermionic" anticommutator counterpart

$$[\mathbf{S_r}, \mathbf{S_m}]_+ = 2\ \cos[\alpha_2(k; \mathbf{S_m}, \mathbf{S_r})]\ \mathbf{S_{m+r}}\ ,$$

have been used [85] in the $N \to \infty$ case as a supersymmetric example of Kac–Moody type algebras. But the main analogues to our case are the so-called algebras of area-preserving transformation generators [86].

Let us recall the main points for the special case of a phase space. We start by supposing that a set $\{\phi_\mathbf{m}(\mathbf{q})\}$ of basic functions exists, in terms of which every function on the space may be expanded. Given the Poisson bracket $\{,\}_P$, the basis functions will establish an algebra $\{\phi_\mathbf{m}, \phi_\mathbf{n}\}_P = \sum_\mathbf{r} c^\mathbf{r}{}_{\mathbf{mn}}\ \phi_\mathbf{r}$. In Hamiltonian theory, each function $F = \frac{1}{N}\sum_\mathbf{m} F_\mathbf{m}\ \phi_\mathbf{m}$ on phase space will define a Hamiltonian field X_F which is such that $X_F G = \{F, G\}_P$ for all G. With the notation $\mathbf{x} = (x_1, x_2, \ldots, x_n, x'_1, x'_2, \ldots, x'_n)$ for the coordinates on phase space, the operator

$$X_F = \sum_k (\partial_k F)\,\partial_{k'} - (\partial_{k'} F)\,\partial_k = (\vec{\partial}\ F)\times \vec{\partial}$$

is the field generating the canonical transformations whose generating function is F. We may thus attach to the basis $\{\phi_\mathbf{m}(\mathbf{r})\}$ a basis $\{X_\mathbf{m}\}$ of Hamiltonian fields defined by $X_\mathbf{m} G = \{\phi_\mathbf{m}, G\}_P$ for all G. An algebra of differential operators turns out, with

$$[X_\mathbf{m}, X_\mathbf{n}] = \sum_\mathbf{r} c^\mathbf{r}{}_{\mathbf{mn}} X_\mathbf{r}\ .$$

Notice that $X_\mathbf{m}\phi_\mathbf{n} = \{\phi_\mathbf{m}, \phi_\mathbf{n}\}_P = \sum_\mathbf{r} c^\mathbf{r}{}_{\mathbf{mn}}\ \phi_\mathbf{r}$. The $X_\mathbf{m}$'s will generate canonical transformations, preserving the area (or volume) by Liouville theorem.

The formalism is actually more general, as such area-preserving transformations may be defined also on manifolds which are not phase spaces. According to Fletcher [83], every infinite double-indexed Lie algebra can be put into the Moyal form (or its particular case, the Poisson form) *in some conveniently chosen function basis.* Actually, it was shown that any (infinite double-indexed) bracket algebra may be transformed into Moyal's, provided only that it satisfies Jacobi identities.

The system of basic functions $\{\phi_\mathbf{m}(\mathbf{x})\}$ depends, of course, on which phase space is considered. Proposed basis are, for instance: $\{\phi_\mathbf{m}(\mathbf{x}) = x_1^{m_1+1} x_2^{m_2+1}\}$ for the plane and $\{\phi_\mathbf{m}(\mathbf{x}) = x_1^{m_1-1} e^{i m_2 x_2}\}$ for the cylinder. In reality, such basis are chosen so as to provide infinite algebras of interest in other contexts. But the question which remains is: when expanding general functions in terms of such basis, how do we calculate the coefficients? In order to have well-defined situations, functions should constitute an inner-product space, with an integration measure provided. Notice further that, in the regular representation needed, also the coefficients represent the group. The basis proposed above are local basis and do not allow expansions of functions on the plane and the cylinder, but only on limited domains of them. However, different domains mean actually different spaces. It is enough to recall one true basis for the plane, the quickly converging Hermite functions, to realize that the behavior at infinity must be accounted for in some way. Furthermore, for quantization we need unitary characters and polynomials are not enough. Abraham and Marsden [53], for instance, have considered the Lie algebra of real-valued polynomials on $\mathbf{E}^{2n}$, but showed soon after that no full quantization is possible. Non-unitary characters[7] may be very helpful for other ends, such as analytic continuations to make contact with Laplace transform, but quantization is related to true Fourier expansions, summations over (non equivalent) unitary irreducible representations. Thus, in what concerns quantization, the whole issue is not clear and deserves further study.

All this leads to the third reason for the simplicity of our case: our basis $\{\phi_{\mathbf{m}}(\mathbf{r}) = e^{i(2\pi/N)\mathbf{m}\cdot\mathbf{r}}\}$ for the torus, as well as its limit, do respect the unitarity requirement, and that is why we shall be able to obtain a full quantization. In what follows we shall do something similar to the infinite algebra procedure, but finding directly the Moyal bracket

$$\{f, g\}_{Moyal} = \lim_{x \to x'} \left[\frac{1}{\sigma} \sin\left(\sigma \; \vec{\partial}_x \times \vec{\partial}_{x'}\right) \; f(x) \; g(x') \right],$$

(σ is an elementary area element on x-space) and with full justification coming from operator space via Weyl–Wigner transformations. Let us say once more that ours is the simplest case and that quantization on non-trivial spaces is a vast and very difficult problem [87].

Finally, we are here committed to Weyl prescription. There are other prescriptions [71], leading to other infinite algebras and to other kinds of quantization, which we shall ignore.

[7] See [73], Chapter 25.

Chapter 14

Quantum symplectic structure

14.1 Matrix differential geometry

Differentiable manifolds have a large amount of structure, implicitly assumed every time we take a derivative or a differential. They are topological manifolds, on which coordinates are defined; and through coordinates, differentiability of functions can be introduced. Once the entries are taken as coordinates, matrices can be seen as points on a topological manifold. Differentiability can then be introduced, but carefully: unlike those of usual manifolds, these points are noncommutative. This hard algebraic nut is best cracked through algebraic methods.

The basic properties of classical phase space are of differential nature. The Hamilton equations, for example, are differential equations describing the time behavior of the degrees of freedom and their conjugate momenta. They are particular cases of the Liouville equation, which describes the time evolution of *any* dynamical quantity $F(q,p)$ and has the form

$$X_H F = \{F, H\}_P \ . \tag{14.1}$$

The left-hand side shows, applied to $F(q,p)$, the time evolution operator X_H, that is, the vector field which generates those infinitesimal canonical transformations whose generating function is the Hamiltonian $H(q,p)$. The right-hand side shows the Poisson bracket. The latter is also of differential character. It actually represents an action of the symplectic form Ω, an exterior 2-form which is the fundamental object of any phase space. We recall that a p-form is an antisymmetric tensor field of order p. The bracket is given by that 2-form acting on two vector fields,

$$\{H, F\}_P = \Omega(X_H, X_F) \ ,$$

where X_F is the vector field which generates those infinitesimal canonical transformations whose generating function is F. The symplectic form Ω is invariant under the action of X_H, a property spelled out by the vanishing of the Lie derivative L_{X_H} with respect to X_F when applied to Ω:

$$L_{X_H}\Omega = 0 \ .$$

The Lie derivative L_X measures the infinitesimal change of any tensor when submitted to a transformation whose generator is X. When applied to a function F (a 0-th order tensor), it reduces to $X(F)$, and this is what actually happens in the left-hand side of the Liouville equation. It is another character of the differential structure. When acting on differential forms, it can be written

$$L_X = d \circ i_X + i_X \circ d \ . \tag{14.2}$$

In this expression, d is the exterior derivative; i_X is the interior product [defined by Eq.(14.6) below], actually another kind of derivative; and $\circ$ here means simply the composition of two operations. The interior product reduces the degree of a form. Thus, $i_{X_F}\Omega$ is a 1-form, and the relation between a dynamical function F and its Hamiltonian vector field is just given by

$$i_{X_F}\Omega = dF \ .$$

It may happen that a 1-form σ exists, the Liouville canonical form, such that $\Omega = -d\sigma$. If that form is preserved ($L_{X_F}\sigma = 0$), then also Ω is preserved, F is a generating function of a canonical transformation, and actually $F = \sigma(X_F)$.

Concerning bases for vectors and covectors (1-forms): the most usual are $\{\frac{\partial}{\partial q^i}, \frac{\partial}{\partial p_i}\}$ for vectors (used in the usual expression for the Poisson bracket), and its dual base $\{dq^i, dp_i\}$ for covectors. These bases are global for $\mathbf{E}^{2n}$. For a general phase space, they are only locally defined, with distinct systems of coordinates $\{q^i, p_i\}$ in different domains of the space. By the way, the Darboux theorem says that coordinates exist around each point which allow one to write $\Omega = \sum_i dq^i \wedge dp_i$. Notice, however, that any set $\{e_n\}$ of linearly independent derivatives (vector fields), together with their covector duals $\{e^m \mid e^m(e_n) = \delta^m_n\}$ can be used as a base. Unlike the previous case, the base members do not commute in general. The commutators $[e_r, e_s]$ are vectors themselves, and can be written as $c^k{}_{rs}e_k$. The components $c^k{}_{rs}$ are the base structure coefficients. For the dual base holds the equation

$$de^k = -\ \tfrac{1}{2}\ c^k{}_{rs}e^r \wedge e^s \ . \tag{14.3}$$

We shall call this the "Maurer–Cartan equation", though the name is usually reserved to its appearance in Lie algebras.

The base, and its dual, are characterized by the "canonical form" $\theta = \sum_k e_k \, e^k$, which sums up the above equations as $d\theta + \theta \wedge \theta = 0$.

We shall not at all proceed to a discussion of the symplectic (or Hamiltonian) structure of Classical Mechanics.[1] The lines above are only intended to recall the main points and to impress the idea that what is fundamental is a *differential* structure. We shall eventually add a few more classical results in the quantum treatment below, where they come to be of interest.

What we shall proceed to do is to look for a Hamiltonian structure in quantum phase space. The quantum analogue to classical phase space is, we have seen, a matrix space. In the case of the lattice torus, a space of finite matrices. The first thing we need is, consequently, a differential geometry on finite-matrix space. Such a geometry has been constructed [88] by Dubois-Violette, Kerner and Madore (DKM hereafter). Besides other important characteristics, these authors have found a natural symplectic structure on differential matrix space, leading to a naturally defined "Poisson bracket". What follows in this section is a simplified *resumé* of the DKM results.

We shall need some basic facts of algebra [91]. Given the set M_N of $N \times N$ matrices of complex elements, the set $End(M_N)$ of all the endomorphisms (that is, linear operators) $M_N \to M_N$ is an associative algebra. The set of its commutators, indicated by $[End(M_N)]$, is a Lie algebra. Any endomorphism $D : M_N \to M_N$ for which holds the Leibnitz rule

$$D(\mathbf{AB}) = (D\mathbf{A})\mathbf{B} + \mathbf{A}(D\mathbf{B})$$

is a derivation. The Lie algebra $[End(M_N)]$ contains $D(M_N)$, the subspace of all the derivations of M_N, as a Lie subalgebra. Each matrix $\mathbf{A} \in M_N$ defines a derivation $ad(\mathbf{A}) = ad_{\mathbf{A}}$ by the adjoint action,

$$ad_{\mathbf{A}}(\mathbf{B}) = [\mathbf{A}, \mathbf{B}] \ \forall \ \mathbf{B} \in M_N \ .$$

Conversely, every derivative on M_N is of the form $ad_{\mathbf{A}}$ for some $\mathbf{A} \in M_N$.

We state these facts to emphasize the distinction between an algebra and its derivative algebra. Perhaps the example which best illustrates this point comes precisely from Classical Mechanics, where the derivative corresponding to a member of the algebra of the dynamic quantities, say $G(q,p)$, is the corresponding field X_G, whose effect on another quantity F is given by the "commutator" involved, the Poisson bracket: $X_G(F) = \{F, G\}_P$.

[1] The reader is referred to the classics: Arnold [15] and Abraham–Marsden [53]. For our purposes, the shorter introduction given by Kirillov [82] or the section Phys.1 of reference [89] are enough.

The Liouville equation (14.1) is just such an expression, with the Hamiltonian H in the role of G. The field X_G is, we repeat, the infinitesimal generator of the canonical transformation whose generating function is G.

On a differentiable manifold V, differential forms map fields into the space $C^\infty(V)$ of differentiable functions on V [89]. $C^\infty(V)$ is, with the operation of pointwise multiplication, a commutative algebra. We shall find out a noncommutative geometry by replacing functions by matrices, the space $C^\infty(V)$ by M_N.

Let us then introduce differentials in matrix space. Differential forms will become M_N-valued and some new properties will come forth.

A set of unitary $N \times N$ matrices $\mathbf{E}_k$, with

$$[\mathbf{E}_j, \mathbf{E}_k] = \sum_i C^i{}_{jk}\mathbf{E}_i \ ,$$

provides a basis for the algebra M_N, and the derivative algebra will have a basis $\{e_k\}$, with $e_k = ad_{\mathbf{E}_k}$. A general derivative X will be written $X = X^k e_k$ in this basis. In a way similar to usual differential forms, a whole graded algebra $\Omega = \oplus_p \Omega^p(M_N)$ of matrix-valued forms is given in the following way [90]:

(i) $\Omega^0(M_N)$ is identified to M_N itself. This is the matrix analogue to the space $C^\infty(V)$ of functions (0-forms) on the manifold V.

(ii) Just as the differential of a function f acts on a field X according to $df(X) = X(f)$, the space $\Omega^1(M_N)$ of 1-forms is introduced once a differential operator d is defined by

$$d\mathbf{M}(X) = X(\mathbf{M}) = ad_{\mathbf{X}}\mathbf{M} = [\mathbf{X}, \mathbf{M}] \ . \tag{14.4}$$

(iii) For higher orders, $d : \Omega^p \to \Omega^{p+1}$ is given by the alternating expression

$$\begin{aligned}(p+1)(d\Xi)(e_1, e_2, \ldots, e_{p+1}) &= e_1\Xi(e_2, e_3, \ldots, e_{p+1}) \\ &\quad - e_2\Xi(e_1, e_3, \ldots, e_{p+1}) + \cdots + (-)^p e_{p+1}\Xi(e_1, e_2, \ldots, e_p) \\ &\quad - \Xi([e_1, e_2], e_3, \ldots, e_{p+1}) + \Xi([e_1, e_3], e_2, \ldots, e_{p+1}) \\ &\quad - \Xi([e_1, e_4], e_2, \ldots, e_{p+1}) \\ &\quad + \ldots + (-)^p\Xi([e_1, e_{p+1}], e_2, \ldots, e_p) + \Xi(e_1, [e_2, e_3], \ldots, e_{p+1}) \\ &\quad + \cdots + (-)^p\Xi(e_1, e_2, e_3, \ldots, [e_p, e_{p+1}]) \ .\end{aligned} \tag{14.5}$$

This "d" is a well known derivative used in the study of Lie algebra cohomology.[2] Each expression like $e_1\Xi(e_2, e_3, \ldots, e_{p+1})$ means here that e_1 is to operate, as in (14.4), on the result of applying Ξ to $(e_2, e_3, \ldots, e_{p+1})$. Thus, $e_1\Xi(e_2, e_3, \ldots, e_{p+1}) = [\mathbf{E}_1, \Xi(e_2, e_3, \ldots, e_{p+1})]$. It comes out from the definition that $d^2 \equiv 0$. For example,

$$\begin{aligned}(d^2\mathbf{M})(e_1, e_2) &= e_1(d\mathbf{M}(e_2)) - e_2(d\mathbf{M}(e_1)) - d\mathbf{M}([e_1, e_2]) \\ &= e_1(e_2(\mathbf{M})) - e_2(e_1(\mathbf{M})) - [e_1, e_2](\mathbf{M}) = 0 \ .\end{aligned}$$

Notice however that, once functions are replaced by matrices,

$$(d\mathbf{MN})(X) = d\mathbf{M}(X)\mathbf{N} = [X, \mathbf{M}]\mathbf{N} \neq (\mathbf{N}d\mathbf{M})(X) = \mathbf{N}[X, \mathbf{M}] \ ,$$

that is, $d\mathbf{MN} \neq \mathbf{N}d\mathbf{M}$. Quite different, of course, from the trivial property $dfg = gdf$ of commutative geometry.

An interior product is provided by an antiderivation $i_{\mathbf{X}}$ corresponding to each derivation $\mathbf{X}$ of the algebra. In complete formal analogy to usual differential forms, it is fixed by the requirement that the condition

$$(i_{\mathbf{X}}\Xi)(\mathbf{X}_1, \mathbf{X}_2, \ldots, \mathbf{X}_{p-1}) = \Xi(\mathbf{X}, \mathbf{X}_1, \ldots, \mathbf{X}_{p-1}) \qquad (14.6)$$

hold for all sets $\{\mathbf{X}_1, \mathbf{X}_2, \ldots, \mathbf{X}_{p-1}\}$ of derivatives. Then $L_{\mathbf{X}} = d \circ i_{\mathbf{X}} + i_{\mathbf{X}} \circ d$ is a derivation, the matrix analogue to (14.2). Properties similar to those valid in differential calculus, such as

$$d \circ L_{\mathbf{X}} = L_{\mathbf{X}} \circ d, \ \{i_{\mathbf{X}_1}, i_{\mathbf{X}_2}\} = i_{\mathbf{X}_1} \circ i_{\mathbf{X}_2} + i_{\mathbf{X}_2} \circ i_{\mathbf{X}_1} = 0, [L_{\mathbf{X}_1}, L_{\mathbf{X}_2}] = L_{[\mathbf{X}_1, \mathbf{X}_2]} \ ,$$

etc., can be shown to be true also here. It is immediate that

$$d\mathbf{E}_k(e_j) = e_j(\mathbf{E}_k) = [\mathbf{E}_j, \mathbf{E}_k] = \textstyle\sum_i C^i{}_{jk}\mathbf{E}_i \ .$$

The forms $d\mathbf{E}_k$ constitute a basis for $\Omega^1(M_N)$. Higher-order multilinear mappings would constitute the analogous to higher order forms, but there is an important difference. Unlike the usual case $df \wedge dg = -\, dg \wedge df$ and due to matrix noncommutativity, $d\mathbf{M}d\mathbf{N} \neq -d\mathbf{N}d\mathbf{M}$ in general. This makes the use of basis $\{d\mathbf{E}_k\}$ a rather cumbersome affair. A far more convenient basis is formed by those more abstract forms θ^j which are dual to the e_j's,

$$\theta^j(e_k) = \delta^j_k \ . \qquad (14.7)$$

In this basis,

$$d\mathbf{E}_k = \textstyle\sum_{i,j} C^i{}_{jk} \, \mathbf{E}_i \, \theta^j \ . \qquad (14.8)$$

Now the θ^j's, unlike the $d\mathbf{E}_k$'s, are found to satisfy

$$\theta^j \, \theta^i = - \, \theta^i \, \theta^j \ , \qquad (14.9)$$

[2] A short introduction to the subject can be found in [92].

and furthermore

$$d\theta^i = -\ \tfrac{1}{2} \sum_{j,k} C^i{}_{jk}\ \theta^j\ \theta^k\ , \tag{14.10}$$

a formula analogous to the Maurer–Cartan equations (14.3) of the classical case. Going further with the analogy and introducing the "canonical form"

$$\theta = \mathbf{E}_j\ \theta^j\ , \tag{14.11}$$

DKM have shown that the 2-form

$$\Omega = d\theta \tag{14.12}$$

provides a natural symplectic structure on the matrix space, defining consequently a Poisson bracket.

In Classical Mechanics, the Poisson bracket of two dynamic functions F and G equals the symplectic form applied to the corresponding fields. Actually, there is a list of equalities in the case:

$$\{F, G\} = \Omega(X_F, X_G) = -X_F(G) = X_G(F) = dF(X_G) = -dG(X_F)\ .$$

On matrix space, to each $\mathbf{A} \in M_N$ will correspond a Hamiltonian field $X_\mathbf{A} \in D(M_N)$ which is such that $\Omega(X_\mathbf{A}, Y) = Y(\mathbf{A}) = d\mathbf{A}(Y)$ for all $Y \in D(M_N)$. The matrix Poisson bracket is defined by

$$\{\mathbf{A}, \mathbf{B}\}_M = -\Omega(X_\mathbf{A}, X_\mathbf{B})\ . \tag{14.13}$$

Due to linearity, it is enough to examine the relations for the basis members. Thus, $\{\mathbf{E}_i, \mathbf{E}_j\}_M = -\Omega(e_i, e_j) = d\mathbf{E}_j(e_i) = e_i(\mathbf{E}_j) = [\mathbf{E}_i, \mathbf{E}_j]$ and in consequence $\{\mathbf{A}, \mathbf{B}\}_M = [\mathbf{A}, \mathbf{B}]$. We shall find it convenient, for reasons of hermiticity, to define the derivatives as $ad_{i\mathbf{X}}\mathbf{M} = i\ [\mathbf{X}, \mathbf{M}]$ in the following. With this convention,

$$\{\mathbf{A}, \mathbf{B}\}_M = i[\mathbf{A}, \mathbf{B}]\ . \tag{14.14}$$

Thus, the Poisson bracket coming out from the natural Hamiltonian structure on a matrix manifold is just the commutator. We shall in the following show that, despite some peculiarities, once the formalism is applied to the quantum operator algebra, this symplectic structure is exactly that provided by the cocycle α_2, and the Poisson bracket is directly related to the Moyal bracket.

Many more things may be introduced on matrix space, such as integration, coderivative and the Laplacian, connections and their curvatures. Also general theorems, like the Hodge decomposition, have been proved by DKM. We shall not need them here.

14.2 The symplectic form

A phase space is, to begin with, a differential manifold. Its special character comes from the presence of a 2-form, the symplectic form, which can be seen as an antisymmetric matrix. This matrix is nondegenerate, that is, invertible. The inverse to the symplectic form is an antisymmetric matrix which defines the Poisson structure. Using the differential structure above introduced on matrix space, it is possible to define a quantum version of the classical symplectic form. It appears as a hypermatrix, a matrix of matrices. It has an inverse, but there comes an important difference with respect to the classical case: this inverse is not antisymmetric.

In line with the DKM formalism of the previous section, take now the $\mathbf{S_m}$'s as a basis for the algebra M_N, and consider the related derivative algebra generated by the operators $e_\mathbf{m} = ad(i\mathbf{S_m}) = ad_{i\mathbf{S_m}}$ (for all $\mathbf{S_m} \neq \mathbf{S_0}$). It is an essential point that, for all that concerns the derivative fields, the vanishing index $\mathbf{0} = (0,0)$ is excluded. Using the Jacobi identity, we find that, for all $\mathbf{S_k}$, $[e_\mathbf{r}, e_\mathbf{m}](\mathbf{S_k}) = 2\sin[\alpha_2(\mathbf{r},\mathbf{m})]e_{\mathbf{m+r}}(\mathbf{S_k})$, so that the general commutators are

$$[e_\mathbf{r}, e_\mathbf{m}] = 2\sin[\alpha_2(\mathbf{r},\mathbf{m})]\ e_{\mathbf{m+r}}\ . \tag{14.15}$$

The structure coefficients

$$C^\mathbf{p}{}_\mathbf{rm} = 2\sin[\alpha_2(\mathbf{r},\mathbf{m})]\ \delta^\mathbf{p}_{\mathbf{m+r}} \tag{14.16}$$

(where $\delta^\mathbf{p}_{\mathbf{m+r}} = \delta^{p_1}_{m_1+r_1}\delta^{p_2}_{m_2+r_2}$) have the symmetry $C^\mathbf{p}{}_\mathbf{rm} = C^\mathbf{r}{}_{-\mathbf{m},\mathbf{p}}$ and define a Cartan–Killing metric,

$$k_\mathbf{rs} = -\ \tfrac{1}{2N^2}\sum_\mathbf{m}\sum_\mathbf{p} C^\mathbf{p}{}_\mathbf{rm}\ C^\mathbf{m}{}_\mathbf{sp} = \delta_{\mathbf{r},-\mathbf{s}}\ . \tag{14.17}$$

Commentary 207.1. ► The structure coefficients $C^c{}_{ab}$ define the Killing bilinear form $\gamma_{ab} = C^c{}_{ad}C^d{}_{bc}$. If for all coefficients the condition $C_{dab} = \gamma_{dc}C^c{}_{ab} = C_{bda}$ is satisfied, then the algebra is semi-simple and γ_{ab} is the Cartan metric (invertible nondegenerate bilinear form) on the group manifold (see Section 16.5). Only a true metric, not any bilinear form, can properly lower and raise indices. ◄

This metric has an obvious relation to the metric (13.21): $k_{\mathbf{r},-\mathbf{s}} = \delta_{\mathbf{r},\mathbf{s}} = \frac{1}{N}\ tr(\mathbf{S_r}\mathbf{S_s^\dagger}) = g_\mathbf{rs}$. It is important to keep in mind that all numbers, also in the Kronecker deltas, are defined mod N. And that the vanishing index $\mathbf{0} = (0,0)$ is excluded. For this reason we must use an adaptation of (13.28) in the above calculations: the summation, excluding the $(0,0)$ contributions, becomes

$$\sum_{\mathbf{m}\neq\mathbf{0}} e^{2i\alpha_2(\mathbf{m},\mathbf{r}-\mathbf{s})} = N^2\delta_{\mathbf{r},\mathbf{s}} - 1\ . \tag{14.18}$$

The metric $k_{\mathbf{rs}}$ establishes a relationship with the hermitian conjugates, which furthermore allows an improvement in notation. In fact, we verify that, defining $\mathbf{S}^{\mathbf{m}} = k^{\mathbf{mn}}\mathbf{S}_{\mathbf{n}}$, such "covariant" versions of the "contravariant" $\mathbf{S}_{\mathbf{m}}$'s are just their hermitian conjugates:

$$\mathbf{S}^{\mathbf{m}} = k^{\mathbf{mn}}\, \mathbf{S}_{\mathbf{n}} = \mathbf{S}_{-\mathbf{m}} = \mathbf{S}^{\dagger}_{\mathbf{m}}\ .$$

Thus, (13.23) is also $A^{\mathbf{m}} = \mathrm{tr}\,[\mathbf{S}^{\mathbf{m}}\mathbf{A}]$.

We introduce now a basis of 1-forms $\theta^{\mathbf{r}}$ dual to the $e_{\mathbf{r}}$,

$$\theta^{\mathbf{m}}(e_{\mathbf{r}}) = \delta^{\mathbf{m}}_{\mathbf{r}}\ \mathbf{I}$$

which satisfy

$$\theta^{\mathbf{m}}\ \theta^{\mathbf{r}} = -\ \theta^{\mathbf{r}}\ \theta^{\mathbf{m}}\ . \tag{14.19}$$

It is immediately seen that

$$d\mathbf{S}_{\mathbf{m}}(e_{\mathbf{r}}) = e_{\mathbf{r}}(\mathbf{S}_{\mathbf{m}}) = i\ [\mathbf{S}_{\mathbf{r}}, \mathbf{S}_{\mathbf{m}}]$$

$$= \sum_{\mathbf{p}} C^{\mathbf{p}}{}_{\mathbf{rm}}\ \mathbf{S}_{\mathbf{p}} = 2\sin[\alpha_2(\mathbf{r},\mathbf{m})]\ \mathbf{S}_{\mathbf{m}+\mathbf{r}}\ ,$$

so that

$$d\mathbf{S}_{\mathbf{m}} = \sum_{\mathbf{p},\mathbf{r}} \mathbf{S}_{\mathbf{p}}\ C^{\mathbf{p}}{}_{\mathbf{rm}}\theta^{\mathbf{r}} = \sum_{\mathbf{r}} 2\sin[\alpha_2(\mathbf{r},\mathbf{m})]\ \mathbf{S}_{\mathbf{m}+\mathbf{r}}\ \theta^{\mathbf{r}}\ . \tag{14.20}$$

Calculating $d\theta^{\mathbf{p}}(e_{\mathbf{m}}, e_{\mathbf{r}})$ by (14.5), we find

$$d\theta^{\mathbf{p}} = -\ \tfrac{1}{2}\ \textstyle\sum_{\mathbf{r},\mathbf{m}} C^{\mathbf{p}}{}_{\mathbf{rm}}\theta^{\mathbf{r}}\theta^{\mathbf{m}} = \sum_{\mathbf{m}} \sin[\alpha_2(\mathbf{m},\mathbf{p})]\theta^{\mathbf{m}}\theta^{\mathbf{p}-\mathbf{m}}\ . \tag{14.21}$$

We define now the matrix-valued 1-form

$$\theta = \sum_{\mathbf{p}} \mathbf{S}_{\mathbf{p}}\ \theta^{\mathbf{p}} \tag{14.22}$$

and the symplectic 2-form

$$\Omega = d\,\theta\ . \tag{14.23}$$

We see the "canonical form" θ in the character of (minus) the classical Liouville form. It follows that

$$\Omega = \tfrac{1}{2}\ \textstyle\sum_{\mathbf{p},\mathbf{m},\mathbf{r}} \mathbf{S}_{\mathbf{p}}\ C^{\mathbf{p}}{}_{\mathbf{mr}}\ \theta^{\mathbf{m}}\ \theta^{\mathbf{r}} = \sum_{\mathbf{p},\mathbf{m}} \sin[\alpha_2(\mathbf{m},\mathbf{p})]\ \mathbf{S}_{\mathbf{p}}\ \theta^{\mathbf{m}}\ \theta^{\mathbf{p}-\mathbf{m}}\ ,$$

or equivalently

$$\Omega = \sum_{\mathbf{r},\mathbf{m}} \sin[\alpha_2(\mathbf{r},\mathbf{m})]\ \mathbf{S}_{\mathbf{r}+\mathbf{m}}\ \theta^{\mathbf{r}}\ \theta^{\mathbf{m}}\ , \tag{14.24}$$

or still

$$\Omega = \tfrac{1}{2}\textstyle\sum_{\mathbf{r},\mathbf{m}} \Omega_{\mathbf{rm}}\ \theta^{\mathbf{r}}\ \theta^{\mathbf{m}}\ . \tag{14.25}$$

Here we have isolated the components

$$\Omega_{\mathbf{rm}} = \sum_{\mathbf{p}} C^{\mathbf{p}}{}_{\mathbf{rm}} \mathbf{S_p} = i\, [\mathbf{S_r}, \mathbf{S_m}] \ , \tag{14.26}$$

in terms of which many interesting expressions can be written:

$$e_{\mathbf{m}}(\mathbf{A}) = \tfrac{1}{N} \sum_{\mathbf{r}} \Omega_{\mathbf{mr}}\, A^{\mathbf{r}} \ ; \tag{14.27}$$

$$[\mathbf{A}, \mathbf{B}] = -\ \tfrac{i}{N^2} \sum_{\mathbf{r},\mathbf{m}} A^{\mathbf{r}}\, \Omega_{\mathbf{rm}}\, B^{\mathbf{m}} \ ; \tag{14.28}$$

$$d\mathbf{S_m} = \sum_{\mathbf{r}} \Omega_{\mathbf{rm}}\, \theta^{\mathbf{r}} \ ; \tag{14.29}$$

$$d\mathbf{A} = \tfrac{1}{N} \sum_{\mathbf{r},\mathbf{m}} A^{\mathbf{r}}\, \Omega_{\mathbf{mr}}\, \theta^{\mathbf{m}} \ , \tag{14.30}$$

and so on.

We can now form a matrix with entries $\Omega_{\mathbf{rm}}$, $\hat{\Omega} = (\Omega_{\mathbf{rm}})$, which will be the analogue of the classical symplectic matrix. $\hat{\Omega}$ will be a hypermatrix, a matrix whose entries are themselves matrices, and its presence is an intimation of a Hopf algebra.

Commentary 209.1. ► We have met hypermatrices before. As said in Comment 44.1, an imprimitive matrix of index h is similar to the hypermatrix (6.5), whose h-th power is matrix-diagonal. Imprimitive matrices suggest, by the way, a generalization of the shift operator U to spaces of hypermatrices. ◄

The hypermatrix $\hat{\Omega}$ has actually an inverse, defined as a hypermatrix $\hat{r}$ with the properties $\hat{r}\,\hat{\Omega} = \hat{\Omega}\,\hat{r} = \hat{I}$. Using (14.18), it is found that

$$r^{\mathbf{ij}} = \tfrac{i}{N^2}\, \mathbf{S^j}\, \mathbf{S^i} = \tfrac{i}{N^2}\, e^{i\alpha_2(\mathbf{i},\mathbf{j})}\, \mathbf{S^{i+j}} = \tfrac{i}{N^2}\, e^{i(\pi/N)\mathbf{i}\times\mathbf{j}}\, \mathbf{S^{i+j}} \tag{14.31}$$

leads indeed to

$$\sum_{\mathbf{j}} r^{\mathbf{ij}}\, \Omega_{\mathbf{jm}} = \sum_{\mathbf{j}} \Omega_{\mathbf{mj}}\, r^{\mathbf{ji}} = \delta^{\mathbf{i}}_{\mathbf{m}}\, \mathbf{I} \ .$$

We insist that all indices are $\neq \mathbf{0}$. The components of the hypermatrix $\hat{r}$ turn up in formulas like

$$\theta^{\mathbf{j}} = -\tfrac{1}{N} \sum_{\mathbf{m}} r^{\mathbf{jm}}\, d\mathbf{S_m} = \tfrac{1}{N} \sum_{\mathbf{m}} d\mathbf{S_m}\, r^{\mathbf{mj}} \ . \tag{14.32}$$

The classical symplectic matrix is an even-order antisymmetric matrix whose inverse is another antisymmetric matrix. This is not the case for our quantum symplectic hypermatrix: $r^{\mathbf{ij}}$ is not antisymmetric. The definition (14.31) tells us that this anomaly is related to the fundamental cocycle:

$$r^{\mathbf{ij}} = e^{2i\alpha_2(\mathbf{i},\mathbf{j})}\, r^{\mathbf{ji}} \; . \tag{14.33}$$

Equation (14.26) gives the antisymmetric part:

$$r^{\mathbf{ji}} - r^{\mathbf{ij}} = \tfrac{1}{N^2}\, \Omega^{\mathbf{ij}} = \tfrac{1}{N^2}\, k^{\mathbf{im}}\, k^{\mathbf{jn}}\, \Omega_{\mathbf{mn}} \; . \tag{14.34}$$

Given any field $X = \frac{1}{N}\sum_{\mathbf{m}} X^{\mathbf{m}}\, e_{\mathbf{m}}$, the interior product $i_X\theta$ will be $= \theta(X) = \frac{1}{N}\sum_{\mathbf{m}} \mathbf{S_m}\ \ \theta^{\mathbf{m}}(X) = \frac{1}{N}\sum_{\mathbf{m}} \mathbf{S_m}\ X^{\mathbf{m}}$. We verify easily that $di_X\theta = -i_X\Omega$. Thus, the Lie derivative of θ vanishes for all X. Then $L_X\Omega = L_X \circ d\theta = d \circ LX\theta = 0$, and consequently, in classical terminology, all fields X are strictly Hamiltonian [82]. On quantum phase space, all transformations are canonical, all are symmetries of the symplectic form.

The operators of the form $\theta(X) = \frac{1}{N}\sum_{\mathbf{m}} \mathbf{S_m}\ X^{\mathbf{m}}$ act as the generating functions: $d[\theta(X)](Y) = \Omega(X,Y) = -d[\theta(Y)](X)$ for all X, Y. The fundamental cocycle α_2 vanishes under the action of the Lie derivative defined by any transformation. It has, thus, the same behavior of the symplectic form under canonical transformations. Conversely, to operators $\mathbf{A} = \frac{1}{N}\sum_{\mathbf{m}} A^{\mathbf{m}}\mathbf{S_m}$ and $\mathbf{B} = \frac{1}{N}\sum B^{\mathbf{r}}\mathbf{S_r}$, will correspond fields $X_{\mathbf{A}} = -\frac{i}{N}\sum_{\mathbf{m}} A^{\mathbf{m}} e_{\mathbf{m}}$ and $X_{\mathbf{B}} = -\frac{i}{N}\sum_{\mathbf{r}} B^{\mathbf{r}} e_{\mathbf{r}}$ [the factors $(-i)$ being necessary if the coefficients are to remain the same as in (13.40)]. The corresponding "Poisson bracket" is defined by (14.13) and, in consequence,

$$\{\mathbf{A},\mathbf{B}\}_M = \tfrac{2}{N^2} \sum_{\mathbf{p},\mathbf{m}} A^{\mathbf{m}} B^{\mathbf{p}-\mathbf{m}} \sin[\alpha_2(\mathbf{m},\mathbf{p})]\ \mathbf{S_p} \; . \tag{14.35}$$

A glimpse at (13.39) corroborates (14.14),

$$\{\mathbf{A},\mathbf{B}\}_M = i\ [\mathbf{A},\mathbf{B}] \; , \tag{14.36}$$

and we can retrieve the (discrete version of the) Moyal bracket components in (14.35). We see thus that the DKM symplectic form gives directly the Moyal bracket. Concerning the relationship between α_2 and Ω, we might say that α_2 stands to Ω as the Poisson bracket stands to Moyal's. As a last point, it is easy to verify that $[e_{\mathbf{m}}, e_{\mathbf{r}}](\mathbf{S_p}) = d\{\mathbf{S_m},\mathbf{S_r}\}_M(e_{\mathbf{p}}) = e_{\mathbf{p}}(\{\mathbf{S_m},\mathbf{S_r}\}_M)$, from which it follows that

$$[X_{\mathbf{A}}, X_{\mathbf{B}}](C) = d\{\mathbf{A},\mathbf{B}\}_M(X_C) \tag{14.37}$$

for all operators $\mathbf{A}$, $\mathbf{B}$, $\mathbf{C}$. This means that the matrix Poisson bracket of two operators works as the generating function of the commutator of the respective derivative fields. In reality, the classical formula

$$\{\mathbf{A},\mathbf{B}\} = \Omega(X_{\mathbf{A}}, X_{\mathbf{B}}) = e_{\mathbf{i}}(\mathbf{B})\ r^{\mathbf{ij}}\ e_{\mathbf{j}}(\mathbf{A}) \tag{14.38}$$

keeps holding here. We must only be attentive to the fact that, as each factor in the right-hand side is now a matrix, the expression does not hold for every order of the factors. The order written above is just the good one.

Equation (14.36) does not come as a real surprise. It has a clear analogue in Classical Mechanics: suppose the Liouville canonical form σ has vanishing Lie derivatives with respect to all members of a field basis $\{e_k\}$, such that $[e_i, e_j] = \sum_k c^k{}_{ij} e_k$. In that case, the fields have global generating functions $\sigma(e_k)$ and their Poisson algebra just mimic the field algebra, $\{\sigma(e_i), \sigma(e_j)\}_P = \sum_k c^k{}_{ij} \sigma(e_k)$. However, this is not always the case in Classical Mechanics. For classical systems in general, a Hamiltonian field is only *locally* related to a generating function. We have found that all fields are *strictly* Hamiltonian in our special quantum case. Furthermore, in a classical system the commutator of two fields is not necessarily generated by the Poisson bracket of the respective generating functions [15]. The difficulty comes, in the classical case, from the necessity of global fields. That requires the phase space to be a parallelizable manifold, such as a Lie group. Here, everything works quite well, because we do have a group for phase space and, furthermore, the semiclassical continuum limit is purely euclidean.

Let us now quickly examine the Weyl–Wigner counterparts of all that. With the star product, the functions constitute a unit algebra, as $\phi_{\mathbf{0}}(\mathbf{p}) = \phi_{(0,0)}(\mathbf{p}) = 1$; $\phi_{\mathbf{r}}(\mathbf{p}) \circ \phi_{\mathbf{0}}(\mathbf{p}) = \phi_{\mathbf{r}}(\mathbf{p})$, etc. We could obtain the whole structure without further resource to the operator version, through the use of the derivatives like $X_{\mathbf{m}}\phi_{\mathbf{r}}(\mathbf{q}) = \{\phi_{\mathbf{m}}(\mathbf{p}), \phi_{\mathbf{r}}(\mathbf{p})\}_{Moyal}$ and their dual forms $\alpha^{\mathbf{s}}$ such that $\alpha^{\mathbf{s}}(X_{\mathbf{r}}) = \delta^{\mathbf{s}}_{\mathbf{r}}$. We shall, however, prefer to use the parallelism of both formalisms. The Weyl–Wigner versions of the symplectic form entries are

$$\omega_{\mathbf{ij}}(\mathbf{q}) = (\Omega_{\mathbf{ij}})_W(\mathbf{q}) = \textstyle\sum_{\mathbf{p}} C^{\mathbf{p}}{}_{\mathbf{ij}}\ \phi_{\mathbf{p}}(\mathbf{q})$$
$$= X_{\mathbf{i}}\phi_{\mathbf{j}}(\mathbf{q}) = \{\phi_{\mathbf{i}}(\mathbf{p}), \phi_{\mathbf{j}}(\mathbf{p})\}_{Moyal}\ . \tag{14.39}$$

Those of the inverse are

$$\rho^{\mathbf{ij}}(\mathbf{q}) = (r^{\mathbf{ij}})_W(\mathbf{q}) = \tfrac{i}{N^2} e^{i\alpha_2(\mathbf{i},\mathbf{j})} \phi^*_{\mathbf{i}+\mathbf{j}}(\mathbf{q})\ . \tag{14.40}$$

Good use of (14.18) leads to the expected result

$$\sum_{\mathbf{j}} \rho^{\mathbf{ij}}(\mathbf{q}) \circ \omega_{\mathbf{jk}}(\mathbf{q}) = \delta^{\mathbf{i}}_{\mathbf{k}}\ . \tag{14.41}$$

Equally expected, we find

$$d\phi_{\mathbf{m}}(\mathbf{p})(X_{\mathbf{r}}) = X_{\mathbf{r}}\phi_{\mathbf{m}}(\mathbf{p}) = \{\phi_{\mathbf{r}}(\mathbf{p}), \phi_{\mathbf{m}}(\mathbf{p})\}_{Moyal} = \omega_{\mathbf{rm}}(\mathbf{p})\ , \tag{14.42}$$

so that

$$d\phi_{\mathbf{m}}(\mathbf{p}) = \sum_{\mathbf{r}} \{\phi_{\mathbf{r}}(\mathbf{p}), \phi_{\mathbf{m}}(\mathbf{p})\}_{Moyal} \alpha^{\mathbf{r}} = \sum_{\mathbf{r}} \omega_{\mathbf{rm}}(\mathbf{p}) \alpha^{\mathbf{r}} . \tag{14.43}$$

On the other hand, we have $\alpha^{\mathbf{r}}(\mathbf{p}) = -\sum_{\mathbf{m}} \rho^{\mathbf{rm}}(\mathbf{p})\, d\phi_{\mathbf{m}}(\mathbf{p})$ and

$$\sum_{\mathbf{r},\mathbf{m}} \omega_{\mathbf{rm}}(\mathbf{p})\, \alpha^{\mathbf{r}}(\mathbf{p})\, \alpha^{\mathbf{m}}(\mathbf{p}) = \sum_{\mathbf{m}} d\phi_{\mathbf{m}}(\mathbf{p})\, \alpha^{\mathbf{m}}(\mathbf{p})$$
$$= -\sum_{\mathbf{m},\mathbf{k}} d\phi_{\mathbf{m}}(\mathbf{p})\, \rho^{\mathbf{mk}}(\mathbf{p})\, d\phi_{\mathbf{k}}(\mathbf{p}) .$$

It follows then from

$$0 = d^2\phi_{\mathbf{m}}(\mathbf{p}) = \sum_{\mathbf{i},\mathbf{r}} X_{\mathbf{i}} X_{\mathbf{r}} \phi_{\mathbf{m}}(\mathbf{p}) \alpha^{\mathbf{i}}(\mathbf{p}) \alpha^{\mathbf{r}}(\mathbf{p}) + \sum_{\mathbf{r}} [X_{\mathbf{r}} \phi_{\mathbf{m}}(\mathbf{p})] d\alpha^{\mathbf{r}}(\mathbf{p})$$

that

$$d\alpha^{\mathbf{r}}(\mathbf{p}) = -\tfrac{1}{2} \textstyle\sum_{\mathbf{i},\mathbf{j}} C^{\mathbf{r}}{}_{\mathbf{ij}}\, \alpha^{\mathbf{i}}(\mathbf{p})\, \alpha^{\mathbf{j}}(\mathbf{p}) , \tag{14.44}$$

and

$$\sum_{\mathbf{m}} d(\phi_{\mathbf{m}}(\mathbf{p})\, \alpha^{\mathbf{m}}(\mathbf{p})) = \tfrac{1}{2} \textstyle\sum_{\mathbf{i},\mathbf{j}} \omega_{\mathbf{ij}}\, \alpha^{\mathbf{i}}(\mathbf{p})\, \alpha^{\mathbf{j}}(\mathbf{p}) . \tag{14.45}$$

These are the Weyl–Wigner versions of (14.25) and (14.23). There is a complete isomorphism of differential algebras, a direct correspondence between the operator algebra and the noncommutative algebra of Wigner functions. We may use either the hypermatrices with matrix entries or usual matrices with entries in the "star-product algebra". The statement that the dynamic quantities are the same in classical and in Quantum Mechanics, only (the product defining) their algebra being different, is thereby vindicated.

14.3 The quantum fabric

The above formalism leads to non-trivial solutions of the Yang–Baxter equations. These equations are equivalent to the relations defining braid groups. We shall start with some general comments on the latter. Next, it is shown how the Yang–Baxter equations are related to them and to associativity. Finally, the mentioned solutions are exhibited. They are specifically quantal — related to the non–antisymmetric character of the inverse quantum symplectic form — and suggest the existence of some kind of weaving pattern in the background of quantum phase space.

14.3.1 *Braid groups*

The name is not at all gratuitous. Braid groups, which generalize the permutation groups, describe real braids. They can actually be used to classify woven and plaited materials. Hair braids, for example: there will be one element of some braid group for each hair curling.

Let us consider, for illustration, an exchange of particles. Suppose that we look at a plane projection of the exchange, and that a rubber strand links the initial to the final position of each particle. When two of them are exchanged, their strands have two possible ways of going around each other: the first string passes over or below the second. An experiment with two strings will show that these two ways are inverse to each other: performing the exchange one way, then the other, brings back the initial situation. A member of the group represents one of them, its inverse represents the other. The two-strands case is described by the smallest non-trivial braid group, denoted B_2, which has only one generator σ_1. There are only two ways of braiding once, corresponding to σ_1 and σ_1^{-1}. However, even in this simplest of all cases, the group is infinite. We can twine one of the strings around the other n times in the same way (which corresponds to σ_1^n or σ_1^{-n}, depending on the way we have chosen). This, for any value of n. It follows that the group has infinite members. We have been forgetting the identity, which describes no-braiding and which, by the way, is the only element of the one-strand group B_1. And the name "generator" is due to the fact that any other group element can be obtained as a power of σ_1. We can then add a third particle and consider the case of 3 strands. The corresponding group B_3 will have two generators σ_1 and σ_2, the latter describing the exchange of the second and the third particles, just as σ_1 describes the exchange of the first and the second. An exchange of the first and the third particles is not independent: it can be obtained by a composition of the previous generators. And any other group element is some product of powers of the two generators. Things become more interesting when 4 particles are present. There is a third generator σ_3, responsible for exchanging once the third and the fourth particles. If we perform now experiments with 4 strings, we discover a few remarkable facts, rather difficult to put into words, but very simply described by algebra: $\sigma_1\sigma_2\sigma_1 = \sigma_2\sigma_1\sigma_2$ and $\sigma_2\sigma_3\sigma_2 = \sigma_3\sigma_2\sigma_3$, but $\sigma_1\sigma_3 = \sigma_3\sigma_1$. Only nearest-neighbors have non-trivial relationships.

These results can be extended to any number of strands, and have been formalized by Artin.[3] For the N-strand braid group B_N, there exists a basis $\{\sigma_j\}$ of $(N-1)$ generators obeying the relations

$$\sigma_i\sigma_{i+1}\sigma_i = \sigma_{i+1}\sigma_i\sigma_{i+1} \quad \text{for} \quad i = 1, 2, \ldots, N-1 ;$$

$$\sigma_i\sigma_j = \sigma_j\sigma_i \quad \text{for} \quad |i-j| \geq 2 . \tag{14.46}$$

The symmetric group S_N is the special case coming up when all these σ_i's satisfy the additional conditions $(\sigma_i)^2 = \mathbf{I}$. This would mean that each generator is equal to its own inverse, that is, that the difference between the two ways of curling two strands is not taken into account.

Let us here remark that particle exchange, if ruled by B_N, will have a quite different character from exchange taking place in usual gases, which is governed by S_N. More will be said in Section 18.4 on braiding, on the effect of such exchanges on the particle wave functions and on the consequent Statistical Mechanics.

Braid groups are effectively present, as previously said, whenever some braiding process is at work. Every hair braid and every tapestry, tissue or material formed by woven strings, is accurately described by a member of some braid group. Braid groups classify weaving patterns. And, conversely, finding solutions of the above equations in a formalism signals the presence of some kind of weaving. As announced in Section 13.4, the Yang–Baxter equation has turned up in many fields of interest. We shall now show that that equation is equivalent to the above braid equations. Then, we shall see that the condition really meant by them is associativity. Finally, we shall exhibit a non-trivial solution involving the structure of quantum phase space.

14.3.2 *The Yang–Baxter equation*

We have been meeting matrices the entries of which are themselves $N \times N$ matrices. The natural abodes of such hypermatrices are direct product spaces. Recall that, in direct product index notation, the product $\mathbf{A} \otimes \mathbf{B}$ of two matrices has entries $\langle ij|\mathbf{A} \otimes \mathbf{B}|mn\rangle = \langle i|\mathbf{A}|m\rangle\langle j|\mathbf{B}|n\rangle$, the direct product of three matrices will have elements $\langle ijk|\mathbf{A} \otimes \mathbf{B} \otimes \mathbf{C}|mnr\rangle = \langle i|\mathbf{A}|m\rangle\langle j|\mathbf{B}|n\rangle\langle k|\mathbf{C}|r\rangle$, and so on. If $\mathbf{R} = \mathbf{A} \otimes \mathbf{B}$, and $\mathbf{E}$ is the $N \times N$ identity matrix, we can use the suggestive compact writing

$$\mathbf{R}_{12} = \mathbf{A} \otimes \mathbf{B} \otimes \mathbf{E}, \ \mathbf{R}_{13} = \mathbf{A} \otimes \mathbf{E} \otimes \mathbf{B}, \ \mathbf{R}_{23} = \mathbf{E} \otimes \mathbf{A} \otimes \mathbf{B}, \ \text{etc.}$$

[3] The classic on the subject is Birman [135]. For a resumé, see [93].

Matrix elements will be written $\langle ij|\mathbf{R}|mn\rangle = R^{ij}{}_{mn}$ and, more important for what follows, $\langle ijr|\mathbf{R}_{12}|mns\rangle = \langle ijr|\mathbf{R}\otimes\mathbf{E}|mns\rangle = \delta^r_s R^{ij}{}_{mn}$. This notation extends to the case in which the pieces of $\mathbf{R}$ are not contiguous — for example, $\langle ijr|\mathbf{R}_{13}|mns\rangle = \delta^j_n R^{ir}{}_{ms}$.

Let us find a direct product representation of the braid equations. It is enough to look for the braid group generators σ_1 and σ_2 in the forms $\sigma_1 = \mathbf{B}_{12}$ and $\sigma_2 = \mathbf{B}_{23}$, $\mathbf{B}$ being some direct product. In that case,

$$\langle kji|\sigma_1\sigma_2\sigma_1|mnr\rangle = \langle kji|\mathbf{B}_{12}|abs\rangle\langle abs|\mathbf{B}_{23}|ucx\rangle\langle ucx|\mathbf{B}_{12}|mnr\rangle$$

$$= \mathbf{B}^{kj}{}_{ab}\mathbf{B}^{bi}{}_{cr}\mathbf{B}^{ac}{}_{mn}\ ;$$

on the other hand,

$$\langle kji|\sigma_2\sigma_1\sigma_2|mnr\rangle = \langle kji|\mathbf{B}_{23}|abs\rangle\langle abs|\mathbf{B}_{12}|ucd\rangle\langle ucd|\mathbf{B}_{23}|mnr\rangle$$

$$= B^{ji}{}_{bd}B^{kb}{}_{mc}B^{cd}{}_{nr}\ .$$

The braid equation $\sigma_1\sigma_2\sigma_1 = \sigma_2\sigma_1\sigma_2$, or

$$\mathbf{B}_{12}\mathbf{B}_{23}\mathbf{B}_{12} = \mathbf{B}_{23}\mathbf{B}_{12}\mathbf{B}_{23}\ , \tag{14.47}$$

becomes

$$B^{kj}{}_{ab}B^{bi}{}_{cr}B^{ac}{}_{mn} = B^{ji}{}_{ca}B^{kc}{}_{mb}B^{ba}{}_{nr}\ . \tag{14.48}$$

Given $\mathbf{B}$ satisfying this equation, the matrix $\mathbf{R} = \mathbf{PB}$, defined by the index permutation $\mathbf{P}$ fixed by $R^{ij}{}_{mn} = B^{ji}{}_{mn}$, will satisfy

$$R^{jk}{}_{ab}R^{ib}{}_{cr}R^{ca}{}_{mn} = R^{ij}{}_{ca}R^{ck}{}_{mb}R^{ab}{}_{nr}\ . \tag{14.49}$$

This is exactly the Yang–Baxter equation, whose compact version is

$$\mathbf{R}_{23}\mathbf{R}_{13}\mathbf{R}_{12} = \mathbf{R}_{12}\mathbf{R}_{13}\mathbf{R}_{23}\ . \tag{14.50}$$

14.3.3 *Quantum tapestry*

There is a direct realization of the above relations in terms of the metric (14.17). Indeed, $R^{\mathbf{ij}}{}_{\mathbf{mn}} = (\delta^{\mathbf{i}}_{\mathbf{n}}\delta^{\mathbf{j}}_{\mathbf{m}} + ak^{\mathbf{ij}}k_{\mathbf{mn}})$ is a solution for the constant a in some range. This is a rather trivial case. A quantum group can be introduced in a way analogous to a classical group, as that Hopf algebra of hypermatrix transformations preserving a given bilinear form. This $\mathbf{R}$-matrix is just that related to the preservation of the metric $k_{\mathbf{mn}}$ [80].

We shall, however, show the presence of another kind of solution, which is non-trivial and which is due to the multiplicity hidden in (13.26). When we write $\mathbf{S_r S_m} = e^{2i\alpha_2(\mathbf{m},\mathbf{r})}\mathbf{S_m S_r}$, we are not actually telling the whole story. This equation gives only the commutation condition for two fixed matrices $\mathbf{S_r}$ and $\mathbf{S_m}$. But $\mathbf{S_r S_m}$ is actually related by phases to any other product of pairs of basic matrices for which the sum of indices is $\mathbf{r} + \mathbf{m}$.

Let us define the set $\mathbf{B} = \{B^{\mathbf{ij}}{}_{\mathbf{kl}}\}$ of commutation coefficients between the members of the algebra basis by:

$$\mathbf{S^i S^j} = \sum_{\mathbf{m,n}} B^{\mathbf{ij}}{}_{\mathbf{mn}} \mathbf{S^m S^n} . \tag{14.51}$$

Then the general solution is

$$B^{\mathbf{ij}}{}_{\mathbf{mn}} = \delta^{\mathbf{i+j}}_{\mathbf{m+n}} e^{i[\alpha_2(\mathbf{m,n}) - \alpha_2(\mathbf{i,j})]} . \tag{14.52}$$

We may check directly that this in effect respects Eq.(14.48), but we may also use another way which, though lengthier, shows clearly how representations of the braid groups come up very simply from any associative algebra.

We shall here follow a simple method leading directly to the braid equations for the Schwinger basis, but it will be clear that the procedure is valid for more general cases. The only specificity will be that all indices are double.

Let us impose coherence of the commutation coefficients with associativity and compare two different ways of bracketing the $\mathbf{S^i}$'s.

- a first series of bracketings is

$$\mathbf{S^k S^j S^i} = (\mathbf{S^k S^j})\mathbf{S^i} = B^{\mathbf{kj}}{}_{\mathbf{ab}} \mathbf{S^a S^b S^i} = B^{\mathbf{kj}}{}_{\mathbf{ab}} \mathbf{S^a} (\mathbf{S^b S^i})$$

$$= B^{\mathbf{kj}}{}_{\mathbf{ab}} B^{\mathbf{bi}}{}_{\mathbf{cr}} \mathbf{S^a S^c S^r} = B^{\mathbf{kj}}{}_{\mathbf{ab}} B^{\mathbf{bi}}{}_{\mathbf{cr}} (\mathbf{S^a S^c}) \mathbf{S^r}$$

$$= B^{\mathbf{kj}}{}_{\mathbf{ab}} B^{\mathbf{bi}}{}_{\mathbf{cr}} B^{\mathbf{ac}}{}_{\mathbf{mn}} \ \mathbf{S^m S^n S^r} \ ;$$

- another order is

$$\mathbf{S^k S^j S^i} = \mathbf{S^k}(\mathbf{S^j S^i}) = B^{\mathbf{ji}}{}_{\mathbf{ca}} \ \mathbf{S^k S^c S^a} = B^{\mathbf{ji}}{}_{\mathbf{ca}} \ (\mathbf{S^k S^c})\mathbf{S^a}$$

$$= B^{\mathbf{ji}}{}_{\mathbf{ca}} B^{\mathbf{kc}}{}_{\mathbf{mb}} \ \mathbf{S^m S^b S^a} = B^{\mathbf{ji}}{}_{\mathbf{ca}} B^{\mathbf{kc}}{}_{\mathbf{mb}} \ \mathbf{S^m}(\mathbf{S^b S^a})$$

$$= B^{\mathbf{ji}}{}_{\mathbf{ca}} B^{\mathbf{kc}}{}_{\mathbf{mb}} B^{\mathbf{ba}}{}_{\mathbf{nr}} \ \mathbf{S^m S^n S^r} \ ;$$

- consequently,

$$B^{\mathbf{kj}}{}_{\mathbf{ab}} B^{\mathbf{bi}}{}_{\mathbf{cr}} B^{\mathbf{ac}}{}_{\mathbf{mn}} = B^{\mathbf{ji}}{}_{\mathbf{ca}} B^{\mathbf{kc}}{}_{\mathbf{mb}} B^{\mathbf{ba}}{}_{\mathbf{nr}}$$

  appears as an acceptable condition. This is just the braid equation (14.47), though here with double indices.

Notice to begin with the role of $\mathbf{B}$ as a kind of "grandmatrix", representing the exchange of matrices, with each $\mathbf{S}$ related to one strand of the braid. More interesting is the fact that the subindices in the equation obtained,

$$\mathbf{B}_{12}\mathbf{B}_{23}\mathbf{B}_{12} = \mathbf{B}_{23}\mathbf{B}_{12}\mathbf{B}_{23} \ ,$$

reflect just the order used in the bracketing. We have indeed, to obtain the left-hand side, taken first the 1st and the 2nd of the $\mathbf{S}$'s (which finally yield $\mathbf{B}_{12}$), then the 2nd and the 3rd (to obtain $\mathbf{B}_{23}$), finally again the 1st and the 2nd of the $\mathbf{S}$'s (to get the final $\mathbf{B}_{12}$). And the same is true of the right-hand side.

Equation (14.47) is a presentation (see Comment 312.2) of the 3rd braid group B_3, corresponding to only three strands. For higher number of strands, we may consider higher order direct products and obtain elements of higher order braid groups. Adjacent bracketings lead to expressions of the type $\sigma_i\sigma_{i+1}\sigma_i = \sigma_{i+1}\sigma_i\sigma_{i+1}$ above. Noncontiguous bracketings will then lead to exchanges in which the $\mathbf{S}^{\mathbf{j}}$'s ignore each other, and thus to expressions of the type $\sigma_k\sigma_i = \sigma_i\sigma_k$ for $|i-k| \geq 2$. Untouched $\mathbf{S}^{\mathbf{j}}$'s correspond to non-exchanging strands.

An inspection of the matrix elements of (14.52) reveals that: (i) in the diagonal, $m = i$ and $n = j$, we have just the identity; (ii) the crossed case $m = j$; $n = i$, with

$$B^{\mathbf{ij}}{}_{\mathbf{kl}} = \delta^{\mathbf{j}}_{\mathbf{k}}\delta^{\mathbf{i}}_{\mathbf{l}} e^{2i\alpha_2(\mathbf{j},\mathbf{i})} = \delta^{\mathbf{j}}_{\mathbf{k}}\delta^{\mathbf{i}}_{\mathbf{l}} e^{2i\alpha_2(\mathbf{k},\mathbf{l})} \ , \tag{14.53}$$

corresponds to (13.26). This case is trivial, because $\mathbf{B}^2 = \mathbf{I}$ and it is not the braid group but the symmetric group which is involved. The other cases are non-trivial and come up from the mentioned fact that the product $\mathbf{S}^{\mathbf{i}}\mathbf{S}^{\mathbf{j}}$ is, up to the compensating phases of (14.52), equal to any other product $\mathbf{S}^{\mathbf{m}}\mathbf{S}^{\mathbf{n}}$, with the only proviso that $\mathbf{i}+\mathbf{j} = \mathbf{m}+\mathbf{n}$.

The existence of non-trivial matrices $\mathbf{B}$ is directly related to the fact that $r^{\mathbf{ij}}$ is not simply antisymmetric. Indeed, (14.51) is equivalent to $r^{\mathbf{ji}} = \sum_{\mathbf{m},\mathbf{n}} B^{\mathbf{ij}}{}_{\mathbf{mn}} r^{\mathbf{nm}}$. If the matrix $\mathbf{B}$ were a negative-diagonal hypermatrix, then $\mathbf{B}^2 = \mathbf{I}$ and only the symmetric group would be present. Braiding as found above is consequently a specifically quantum effect.

We have seen that the Yang–Baxter equation itself comes from associativity. Quantum Mechanics parts from Classical Mechanics by the noncommutativity of its dynamical quantities. Their algebras are different in that aspect. The departure is not, however, as thorough as it could be. It is irrelevant to speak of associativity of an abelian algebra, but still it is a fundamental algebraic property which is present in the classical algebra. And the Quantum algebra breaks commutativity while preserving associativity. The braiding disclosed is a consequence of that preservation.

The matrices **B** are very large. Because indices are double, the matrices are $N^4 \times N^4$, too large for direct manipulations. For $N = 2$, **B** is a 16×16 matrix; for $N = 3$, it is a 81×81 matrix, etc.

We have started with QPS as we would have started with the simple pair (q, p) in Classical Mechanics, and finished by finding a symplectic structure in the very algebras of general operators and/or Wigner functions. The latter turn up directly as quantum versions of the algebra of general dynamic functions. Quantization appears directly as a deformation of the algebra of classical observables, as a manifestation of the noncommutative character of the geometry of physical observables. Perhaps the most striking aspect of the formalism above is precisely this directness, an obvious consequence of working at the same time with Weyl–Wigner transformations and with matrices. The expression of each operator is automatically a discrete and finite version of the Weyl prescription relating classical dynamic quantities to their quantum representatives through their Fourier components.

The differential geometric treatment provides a clear distinction between the operators and the corresponding derivative fields (the true "quantum derivations"). The DKM natural symplectic structure on the differential matrix space is related to the Moyal bracket, thereby vindicating those authors' assertion that "Quantum Mechanics is noncommutative symplectic geometry". Structures of Classical Mechanics come in principle from quantum mechanical structures, although something may be lost or distorted in the limiting process. From all that was said here, the classical symplectic structure stems in the long run from the cocycle α_2 and is, as a consequence, a relic signal of the basic noncommutative character of quantum mechanical geometry. The mathematically "strange" Poisson bracket, "strange" because it did not come from an associative product, comes out now as a limiting case of the quantum bracket, thus as part of the quantum heritage of Classical Mechanics. The existence of non-trivial solutions of the Yang–Baxter equation signals the possible presence of braiding in QPS and, so to speak, behind the scene which classical phase space, as its limit, represents.

PART 3
BELL MATRICES

Chapter 15

Bell polynomials

15.1 An organizing tool

One of the main tasks of Mathematical Physics is to provide new tools, adapted to the new resources continuously turning up. The recent advances in computer algebraic manipulations have brought to the forefront a renewed interest in closed expressions. We present now an organizing tool, whose main quality is precisely to provide closed expressions for a variety of mathematical and physical results, ranging from the characteristic classes of Differential Geometry to the virial series in the theory of real gases. This organizing tool is the multi-variable polynomial introduced by Eric Temple Bell in the 1930s.[1]

Bell polynomials allow to represent a function f by a matrix $\mathbf{B}[f]$, in such a way that the operation of function composition is transformed into matrix product. They thereby linearize the operation of function composition. For example, function inversion, when well-grounded, is transformed into matrix inversion. The successive iterations of a fixed function are transformed into powers of matrices. This leads to a convenient approach to the map iterations used in the study of dynamical systems. A matrix can have any complex number as a power, in particular any real number. This leads to a notion of continuous iteration which preserves the meaning of the word "iteration".

The multiple variables involved are the Taylor coefficients of the formal series. Bell polynomials are, consequently, "functional" objects, involved exclusively with the functional behavior of the function f and not at all with

[1] As already said in the Preface 2, there seems to be no standard nomenclature here. Authors frequently call "Carleman matrices" those we are calling "Bell matrices", and also exchange both by their transposes.

its arguments and values. Focussing on purely functional characteristics helps clarifying which aspects are essential.

Closed relationships between determinants and traces of powers of general matrices turn up, which relate symmetric functions of various kinds. These lead to compact formulae for Lie algebra invariants, as well as to the above mentioned results on characteristic classes and real gases. Also some recursion formulae emerge for wave function normalization and phase space volumes, and for the canonical partition and correlation functions. A closed expression for Mayer's virial coefficients is obtained, as well as a trivial demonstration of the Matsubara theorem concerning graphs.

Of some topics are given only very short indications, suggesting further investigation. These are the cases, for example, of the relationships to Padé approximants (Comment 270.1) and to fractional calculus (Section 15.5.3). In the same vein is the discussion on the existence of a basic, underlying matrix behind the theory of real gases, and its possible unitarity in the thermodynamic limit (Section 18.5.2), as well as the relationship of these questions with the Fredholm theory (18.6).

The reader will have noticed our cavalier usage of the word "function": we have been (and shall be) using it for "formal series".

General references on the polynomials are Riordan [94], [95] and Comtet [96]. We are mainly indebted to the latter, from which we have taken the bulk of the mathematics and much of the notation. A very useful online resource is Weisstein's site [100].

15.2 Definition and elementary properties

Bell polynomials are introduced in the most traditional way, through their appearance in the Faà di Bruno formula. Their main formal properties are developed by analytical methods, but with a bias toward combinatorial properties.

Consider two Taylor series,

$$f(u) = \sum_{j=1}^{\infty} \frac{f_j}{j!} u^j \quad \text{and} \quad g(x) = \sum_{k=1}^{\infty} \frac{g_k}{k!} x^k \, . \tag{15.1}$$

Their convergence will be of no concern to us, so that they are taken as purely formal series. They have been chosen to have vanishing independent terms, with $f(0) = 0$ and $g(0) = 0$. We shall be interested in their composition $F(x) = [f \circ g](x) = f[g(x)]$. It will be a series

$$F(x) = \sum_{n=1}^{\infty} \frac{x^n}{n!} F_n[f;g] \ , \tag{15.2}$$

whose Taylor coefficients will have the form

$$F_n[f;g] = \sum_{k=1}^{n} \mathbb{B}_{nk}(g_1, g_2, \ldots, g_{n-k+1}) \ f_k \ . \tag{15.3}$$

The coefficients $\mathbb{B}_{nk}$ are certain polynomials in the Taylor coefficients g_i, precisely the multi-variable Bell polynomials we shall be interested in. Such polynomials appear in the multinomial theorem, which says that

$$\frac{1}{m!} \left[\sum_{j=1}^{\infty} \frac{g_j}{j!} \ x^j \right]^m = \sum_{n=m}^{\infty} \frac{x^n}{n!} \ \mathbb{B}_{nm}(g_1, g_2, \ldots, g_{n-m+1}) \tag{15.4}$$

and is a particular case of (15.2), with $f_k = \delta_{km}$. This relationship provides the easiest access to most properties of the polynomials. It also conveys their real meaning as Taylor coefficients themselves:

$$\mathbb{B}_{nk}(g_1, g_2, \ldots, g_{n-k+1}) = \frac{1}{k!} \left[\frac{d^n}{dx^n} \left(\sum_{j=1}^{\infty} \frac{g_j}{j!} \ x^j \right)^k \right]_{x=0} \ . \tag{15.5}$$

The detailed expression is rather involved:

$$\mathbb{B}_{nk}(g_1, g_2, \ldots, g_{n-k+1}) = \sum_{\{\nu_i\}}{}^{''} \frac{n!}{\prod_{j=1}^{n} [\nu_j!(j!)^{\nu_j}]} \ g_1^{\nu_1} g_2^{\nu_2} \cdots g_{n-k+1}^{\nu_{n-k+1}} \ . \tag{15.6}$$

The double prime indicates a double constraint: the summation takes place only on those sets $\{\nu_j\}$ of non-negative integers satisfying simultaneously the two conditions

$$\sum_{j=1}^{n} j \ \nu_j = n \ ; \tag{15.7}$$

$$\sum_{j=1}^{n} \nu_j = k \ . \tag{15.8}$$

Expression (15.7) represents a partition of the number n, and (15.8) says that the number of summands is k. Thus, the summation takes place on all the partitions of n with k summands.

Commentary 224.1. ▶ **Partitions**[2] A partition of the positive integer n is a set of positive integers λ_k ("parts") with $\lambda_1 \geq \lambda_2 \geq \lambda_3 \geq \ldots \geq \lambda_n$ such that $\sum_{k=1}^{n} \lambda_k = n$. Such sets $\{\lambda_k\}$ are in one-to-one correspondence with the sets $\{\nu_j\}$ above. Given $\{\nu_j\}$, it is enough to define the λ_k's by $\lambda_k = \sum_{j=k}^{n} \nu_j$. Conversely, $\nu_i = \lambda_i - \lambda_{i+1}$. By (15.7) and (15.8), ν_j counts how many times number j appears as a part, and k is the total number of summands in a given partition. It is convenient to establish by convention that the first prime in summation (15.6) refers to condition (15.7), so that Σ' means a summation over the partitions of n. The second prime adds condition (15.8), which is waived if we add a summation over k. In consequence, $\Sigma' = \sum_{k=1}^{n} \Sigma''$. How many partitions are there for a given number n? The number $p(n)$ of partitions of n is the number of distinct ways n can be obtained as a sum of positive integers. It can be found from the generating function

$$1 + \sum_{n=1}^{\infty} p(n)\, x^n = \left(\prod_{k=1}^{\infty} (1 - x^k) \right)^{-1} . \tag{15.9}$$

◀

Commentary 224.2. ▶ From (15.7) and (15.8), the factor $n!/\prod_j [j!]^{\nu_j}$ in (15.6) counts the number of ways n identical particles may be separated into aggregates, ν_j being the number of aggregates composed of j particles. Further division by $\prod_j \nu_j!$ means that we consider as indistinguishable those aggregates which have the same size. ◀

Equation (15.3), giving the successive derivatives of a composite function (that is, of a function applied to a function), is the *Faà di Bruno formula*, whose complete expression is

$$f[g(x)] = \sum_{n=1}^{\infty} \frac{x^n}{n!} \sum_{k=1}^{n} \mathbb{B}_{nk}(g_1, g_2, \ldots, g_{n-k+1})\, f_k \, . \tag{15.10}$$

It is immediate to obtain the double generating function

$$e^{ug(x)} - 1 = \sum_{n=1}^{\infty} \frac{x^n}{n!} \sum_{k=1}^{n} u^k\, \mathbb{B}_{nk}(g_1, g_2, \ldots, g_{n-k+1}) \, . \tag{15.11}$$

The expression

$$\mathbb{Y}_n[u; g] = \sum_{k=1}^{n} u^k\, \mathbb{B}_{nk}(g_1, g_2, \ldots, g_{n-k+1}) \tag{15.12}$$

is frequently called the *complete Bell polynomial* and the $\mathbb{B}_{nk}$'s are, in this terminology, the *partial Bell polynomials*.

We have been showing the arguments g_k explicitly, while emphasizing a property: given n and k, $\mathbb{B}_{nk}$ depends only on the Taylor coefficients up to order $n - k + 1$. However, depending on the circumstances, it may be more convenient to use one or another of the alternative notations

$$\mathbb{B}_{nk}(g_1, g_2, \ldots, g_{n-k+1}) = \mathbb{B}_{nk}[g(x)] = \mathbb{B}_{nk}[g] = \mathbb{B}_{nk}\{g_j\} \, . \tag{15.13}$$

[2] A commendable account of the general theory of partitions and its ramifications is [97].

In the last expression, $\{g_j\}$ represents the coefficient-list of the series g, with g_j a typical element.

In the alphabet terminology, the particular Bell polynomial $\mathbb{B}_{nk}[g]$ is a function on the alphabet $\mathbf{g}$ with $(n-k+1)$ letters. It will take into consideration only a finite number of terms in the case of an infinite series.

$\mathbb{B}_{nk}\{g_j\}$ is a polynomial of total order k in its $(n-k+1)$ variables, with integer positive coefficients. This knowledge allows a quick calculation of some particular cases. Thus, for $k=n$, there is only the first variable g_1. As $\nu_{k\neq 1}=0$, the only solution of conditions (15.7) and (15.8) is $\nu_1=n$. Consequently, (15.6) gives

$$\mathbb{B}_{nn}[g] = g_1^n \,. \tag{15.14}$$

If $k=n-1$, there will be only g_1 and g_2 as variables: $\nu_{k\neq 1,2}=0$. The only solution is that with $\nu_1 = n-2$ and $\nu_2=1$, leading to

$$\mathbb{B}_{n,n-1}[g] = \binom{n}{2}\, g_1^{n-2} g_2 \,. \tag{15.15}$$

For notational convenience, we shall establish that

$$\mathbb{B}_{n0}[g] = \delta_{n0} \tag{15.16}$$

for any g. With this convention (adopted everywhere in this text), (15.11) can be put into the form

$$e^{ug(x)} = \sum_{n=0}^{\infty} \frac{x^n}{n!}\, \mathbb{Y}_n[u;g] = \sum_{n=0}^{\infty} \frac{x^n}{n!} \sum_{j=0}^{n} u^j\, \mathbb{B}_{nj}[g] \,. \tag{15.17}$$

Commentary 225.1. ► Wherefrom comes another expression for $\mathbb{Y}_n[u;g]$, this time in terms of "sandwiched" differentials: $\mathbb{Y}_n[u;g] = e^{-ug(x)}\, D_{x^n}\, e^{ug(x)}$, with $D_{t^n} f(t) = \left[\frac{d^n f(t)}{dt^n}\right]_{t=0}$. The coefficient F_n in (15.2) can be formally obtained as $\mathbb{Y}_n[f;g(x)]$, with the formal substitution rule $f^k \Rightarrow f_k$. ◄

Commentary 225.2. ► Notice that $e^{h(x)} = \sum_{n=0}^{\infty} \frac{x^n}{n!} \sum_{j=0}^{n} \mathbb{B}_{nj}[h]$ implies

$$f(x) = \sum_{n=0}^{\infty} \frac{x^n}{n!} \sum_{j=0}^{n} \mathbb{B}_{nj}[\ln f] \,.$$

◄

Commentary 225.3. ► For finite calculations, it is sometimes of interest to introduce truncated series, defining polynomial functions as the truncated exponential $\exp_{(r)}(x) = \sum_{n=0}^{r} \frac{x^n}{n!}$, the truncated logarithm $\ln_{(r)}(1+x) = \sum_{n=1}^{r} \frac{(-)^{n-1} x^n}{n}$, etc. ◄

Commentary 225.4. ► Our convention, using Taylor series with inverse factorials, corresponds to what Combinatorics experts call "exponential" Bell polynomials. An alternative would be to start with series of the form $f(u) = \sum_j f_j u^j$ instead of those of (15.1). This would lead to different numerical factors almost everywhere. The choice made here seems to be the simplest and most useful. The factorial in the denominator is frequently responsible for the series convergence (see the next Comment and, for an example, Comment (231.1)). ◄

Commentary 225.5. ► The connection between series with and without factorials in the denominators is given by a Borel transform [101]. The Borel transform of a general series $g(z) = \sum_{n=0}^{\infty} \frac{g_n}{n!} z^n$ is the series $F(z) = \sum_{n=0}^{\infty} g_n z^n$, which can be obtained as $F(z) = \int_0^\infty dte^{-t} g(tz)$ by using the integral $\int_0^\infty dte^{-t}t^n = n!$. We write, in this case, $\mathcal{B}[g](z) = F(z)$ and, of course, $\mathbb{B}_{nk}[\mathcal{B}[g]] = \mathbb{B}_{nk}\{j!\, g_j\}$.

The function $\phi(t,z) = g(tz)$ is the "Borel associate" of $F(z)$. If $\phi(t,z)$ is convergent, $F(z)$ is "Borel-summable". With such enhanced convergence, it is possible to give a sense to otherwise meaningless series. A famous case is that of the series $1-1+1-1+1\ldots.$ This is $F(1)$, with $F(z) = \sum_{n=0}^{\infty}(-)^n z^n$. The associate is $\phi(t,z) = e^{-tz}$, corresponding to $F(z) = 1/(1+z)$, so that $F(1) = 1/2$. Formally, $\mathcal{B}[e^{-u}](z) = \frac{1}{1+z}$. ◄

Commentary 226.1. ► Borel transformations are, like the combings of Comment 88.0, manipulations of formal series, transformations leading to quite different "functions". Notice that utmost attention must be paid to the detailed form of the argument. Consider an example involving modified Bessel functions of I-type,

$$I_j(z) = \sum_{p=0}^{\infty} \frac{1}{p!(p+j)!} \left(\frac{z}{2}\right)^{2p+j} .$$

The particular case $I_0(x) = \sum_{q=0}^{\infty} \frac{1}{(q!)^2}(x/2)^{2q}$, in which the variable appears quadratically, is such that $\int_0^\infty due^{-u} I_0(ux) = \sum_{q=0}^{\infty} \frac{(2q)!}{(q!)^2}(\frac{x}{2})^{2q}$. This means simply that $\mathcal{B}[I_0](z) = \frac{1}{\sqrt{1-z^2}}$. ◄

Commentary 226.2. ► **Borel and Laplace transformations** Let us recall the expressions of the Laplace transform

$$\hat{g}(\beta) = \int_0^\infty dte^{-\beta t} g(t) \tag{15.18}$$

and its inverse (Bromwich integral)

$$\frac{1}{2\pi i}\int_{c-i\infty}^{c+i\infty} d\beta e^{\beta t}\hat{g}(\beta) = \begin{cases} g(t) & \text{for } t>0 \\ 0 & \text{for } t<0\ . \end{cases} \tag{15.19}$$

We have seen in Comment 225.5 how a Borel transformation takes off the inverse factorials of a series $g(x) = \sum_{n=0}^{\infty} g_n \frac{x^n}{n!}$:

$$\mathcal{B} : g_n \to n! g_n \ , \ \mathcal{B}[g](z) = \sum_{n=0}^{\infty} g_n x^n \ .$$

The relation to the Laplace transform is given by

$$\mathcal{B}[g](z) = \int_0^\infty due^{-u}\, g(uz) = \frac{1}{z}\int_0^\infty dte^{-t/z}\, g(t) \doteq \frac{1}{z}\,\hat{g}(\frac{1}{z}) \ .$$

◄

The results (15.14) and (15.15) have been obtained from the detailed expression (15.6). The best source of particular expressions is, however, the multinomial theorem. Some properties coming immediately from (15.4) are:

$$\mathbb{B}_{n1}[g] = g_n \tag{15.20}$$

$$\mathbb{B}_{n2}[g] = \frac{1}{2}\sum_{j=1}^{n-1} \binom{n}{j} g_j \, g_{n-j} \tag{15.21}$$

$$\mathbb{B}_{nk}[ag(x)] = \mathbb{B}_{nk}(ag_1, ag_2, \ldots, ag_{n-k+1}) = a^k \, \mathbb{B}_{nk}[g(x)] \tag{15.22}$$

$$\mathbb{B}_{nk}[g(bx)] = \mathbb{B}_{nk}(b^1 g_1, b^2 g_2, \ldots, b^{n-k+1} g_{n-k+1}) = b^n \, \mathbb{B}_{nk}[g(x)] \tag{15.23}$$

where a and b are constants. These relations are of great practical importance, and will be repeatedly used throughout this text.

Suppose, for example, that we fall upon a summation of the form showed in the right-hand side of (15.17). It is a double summation, requiring the knowledge of all the $\mathbb{B}_{nk}$'s. Equation (15.20) tells us that this is not actually so. The summation is $\exp\left\{u \sum_{n\geq 1} \mathbb{B}_{n1}[g] \, \frac{x^n}{n!}\right\}$, so that only the $\mathbb{B}_{nk}$'s with $k = 1$ are necessary. This seemingly innocent remark has some vital consequences. We shall see that the grand canonical partition function of Statistical Mechanics (Section 18.3) has just that form. In the diagrammatic approach to real gases the index k in $\mathbb{B}_{nk}$ will count the number of disconnected pieces a diagram involving n particles is made of. A diagram with only one piece ($k = 1$) is a connected diagram, so that the summation, originally over all diagrams, reduces to (the exponential of) that of the connected graphs (see Section 18.3.2).

Equation (15.22) says that, when the function f is multiplied by a constant a, the Bell polynomial $\mathbb{B}_{nk}$ gets a factor a^k. Due to this property, the complete Bell polynomial (15.12) can be alternatively written as

$$\mathbb{Y}_n[u; g] = \sum_{j=1}^{n} \mathbb{B}_{rj}[ug] = \sum_{j=1}^{n} \mathbb{B}_{rj}(ug_1, ug_2, \ldots, ug_{n-j+1}) \, . \tag{15.24}$$

Another relationship coming from (15.4) is

$$\mathbb{B}_{nk}[f(x) + g(x)] = \sum_{j=0}^{n} \sum_{m=0}^{k} \binom{n}{j} \mathbb{B}_{jm}[f(x)] \, \mathbb{B}_{n-j,k-m}[g(x)] \, . \tag{15.25}$$

A particular case of this double weighted convolution is

$$\mathbb{B}_{nk}[2f(x)] = \sum_{j=0}^{n} \sum_{m=0}^{k} \binom{n}{j} \mathbb{B}_{jm}[f(x)] \, \mathbb{B}_{n-j,k-m}[f(x)] \, , \tag{15.26}$$

which is also $2^k \mathbb{B}_{nk}[f]$ by (15.22).

Commentary 227.1. ► For a finite sum of functions: take first the simple expression

$$\frac{1}{k!}\left[\sum_{j=1}^{N} g_j\right]^k = \sum_{n=k}^{kN} \frac{1}{n!}\, \mathbb{B}_{nk}\{j! g_j\} = \sum_{n=k}^{kN} \sum_{\{\nu_i\}}{}'' \prod_{j=1}^{n} \frac{g_j^{\nu_j}}{\nu_j!} ,$$

valid for any g_j's. If now each of them is a function, $g_j = f_j(x)$,

$$\frac{1}{k!}\left[\sum_{j=1}^{R} f_j(x)\right]^k = \sum_{n=k}^{\infty} \frac{1}{n!}\, \mathbb{B}_{nk}\{j! f_j(x)\} = \sum_{n=k}^{kR} \sum_{\{\nu_i\}}{}'' \prod_{j=1}^{n} \frac{[f_j(x)]^{\nu_j}}{\nu_j!} ,$$

in which expression $\frac{[f_j(x)]^{\nu_j}}{\nu_j!} = \sum_{\mu_j=\nu_j}^{\infty} \frac{x^{\mu_j}}{\mu_j!} \mathbb{B}_{\mu_j \nu_j}[f_j(x)]$ can be used again for each factor. ◄

Concerning (15.23), it says that the Bell polynomial $\mathbb{B}_{nk}$ acquires a factor b^n when the argument of f is multiplied by a constant b. This result comes from a simple change of variable $x \to bx$ in the multinomial theorem. In particular, when b is only a sign, $b = -1$, we have

$$\mathbb{B}_{nk}\{(-)^j g_j\} = (-)^n\, \mathbb{B}_{nk}[g] \, . \tag{15.27}$$

A case we shall often meet, combining (15.22) and (15.23), is

$$\mathbb{B}_{nk}\{(-)^{j-1} g_j\} = (-)^{n-k}\, \mathbb{B}_{nk}[g] \, . \tag{15.28}$$

If all the letters g_{k+p} beyond the k-th are zero, then $\mathbb{B}_{Nm}(g_1, g_2, \ldots, g_k,$ $0, 0, \ldots, 0)$ is the Bell polynomial of a polynomial of degree k in the independent variable,

$$\mathbb{B}_{Nm}(g_1, g_2, \ldots, g_k, 0, 0, \ldots, 0) = \mathbb{B}_{Nm}\left[\sum_{j=1}^{k} \frac{g_j}{j!} x^j\right] , \tag{15.29}$$

and is nonvanishing for N in the range $km \geq N \geq m$.

There are many usual notations for the identity function $g(x) = Id(x) = e(x) = x$, which we shall use rather indifferently. The last formula provides

$$\mathbb{B}_{Nm}[e(x)] = \mathbb{B}_{Nm}(1, 0, \ldots, 0) = \delta_{Nm} \, . \tag{15.30}$$

15.2.1 *Formal examples*

The series in (15.1) have been chosen to have vanishing constant terms. Let us further suppose that the first-order coefficient is nonvanishing. Series of this type are invertible, and actually constitute a group under the composition operation

$$(f \circ g)(x) = f\,[g(x)] \, .$$

The identity series e such that $e(t) = t$ plays the role of a neutral element and each series g possesses an inverse $g^{<-1>}$, with

$$g^{<-1>} \circ g = g \circ g^{<-1>} = e .$$

As composition is associative, all the group conditions are satisfied.

A bit more exactly: Given $g(x) = g_0 + \sum_{k=1}^{\infty} g_k \frac{x^k}{k!}$ with $g_1 \neq 0$, there exists a series $f(u) = f_0 + \sum_{j=1}^{\infty} f_j \frac{u^j}{j!}$, unique in a small enough neighbourhood of $u = 0$ and $x = 0$, which is such that $f[g(x)] = x$ and $g[f(u)] = u$. In that case f and g are inverse to each other and $f_1 = 1/g_1$.

Bell polynomials have special significance for series without independent term. This means that, in the notation above, we shall either consider the case $g_0 = 0$, $f_0 = 0$ or work with the series $g(x) - g_0$, $f(u) - f_0$.

It will be convenient to use the Γ function to extend to integer negative, real or complex numbers some binomial notations which are usual for integers, such as

$$x! = \Gamma(x+1) \; ; \quad \binom{x}{k} = \frac{x!}{k!(x-k)!} \; ; \quad \binom{-k}{n} = \frac{(-)^n (k+n-1)!}{(k-1)!\; n!} \; . \tag{15.31}$$

The simplest case of inversion is the series

$$f(u) = e^u - 1 = \sum_{j=1}^{\infty} \frac{1}{j!} \, u^j \; , \tag{15.32}$$

whose inverse is

$$g(x) = \ln(1+x) = \sum_{j=1}^{\infty} \frac{(-)^{j-1}}{j} \, x^j \; . \tag{15.33}$$

These series are related to the Stirling numbers. The generating functions for the Stirling numbers of the first kind $s_k^{(j)}$ are, for $k \geq 1$,

$$x(x-1)(x-2)...(x-k+1) = \binom{x}{k} k! = \sum_{j=0}^{k} s_k^{(j)} x^j \tag{15.34}$$

and

$$\frac{1}{k!} \left[\ln(1+x)\right]^k = \sum_{n=k}^{\infty} \frac{x^n}{n!} \, s_n^{(k)} \; . \tag{15.35}$$

We use $s_n^{(0)} = \delta_n^0$ by convention. The multinomial theorem (15.4) provides a prompt identification:

$$s_n^{(k)} = \mathbb{B}_{nk}[\ln(1+x)] = \mathbb{B}_{nk}(0!, -1!, 2!, -3!, \ldots) = \mathbb{B}_{nk}\{(-)^{j-1}(j-1)!\} \; . \tag{15.36}$$

Properties (15.22) and (15.23) give

$$\mathbb{B}_{nk}(0!, 1!, 2!, 3!, \ldots) = \mathbb{B}_{nk}\{(j-1)!\} = |s_n^{(k)}| = (-)^{n+k}\, s_n^{(k)}\,, \qquad (15.37)$$

which is the modulus of the first-kind Stirling number.

Commentary 230.0. ▶ These $s_n^{(k)}$ have alternating signs. We find indeed from Eq.(15.17) that $\sum_{k=1}^{n} s_n^{(k)} = \delta_n^1$. That same equation, combined with (15.34), leads to Newton's formula in the form $(1+x)^u = \sum_{n=0}^{\infty} \frac{x^n}{n!} \sum_{j=0}^{n} s_n^{(j)} u^j$. From (15.37) it follows that

$$|s_n^{(k)}| = {\sum_{\{\nu_i\}}}'' \frac{n!}{\prod_{j=1}^{n} [\nu_j!\, j^{\nu_j}]}\,.$$

See Comment 231.1 below for one more result. ◀

The generating functions for the Stirling numbers of the second kind $S_n^{(j)}$ are, for $n \geq 1$,

$$x^n = \sum_{j=1}^{n} S_n^{(j)} x(x-1)(x-2)\ldots(x-j+1) = \sum_{j=1}^{n} S_n^{(j)}\, j! \binom{x}{j} \qquad (15.38)$$

and

$$\frac{1}{k!}\left(e^u - 1\right)^k = \sum_{n=k}^{\infty} \frac{u^n}{n!}\, S_n^{(k)}\,, \qquad (15.39)$$

($S_0^{(0)} = 1$ by convention) from which

$$\mathbb{B}_{nk}[e^u - 1] = \mathbb{B}_{nk}(1, 1, 1, \ldots, 1) = \mathbb{B}_{nk}\{1\} = S_n^{(k)}\,. \qquad (15.40)$$

It follows from (15.6) that

$$S_n^{(k)} = {\sum_{\{\nu_i\}}}'' \frac{n!}{\prod_{j=1}^{n} [\nu_j!(j!)^{\nu_j}]}\,, \qquad (15.41)$$

with the same conditions (15.7) and (15.8).

Commentary 230.1. ▶ The modulus $(-)^{n+k} s_n^{(k)}$ is the number of those permutations of n letters which have k cycles (see Comment 312.1). $S_n^{(k)}$ is the number of ways of partitioning a set of n members into k non-empty subsets. ◀

Commentary 230.2. ▶ Let us see how the double convolution formula (15.25) can be used to produce a combinatorial identity. Consider the difference between $f(x) = e^{\alpha x} - 1$ and $g(x) = e^{\beta x} - 1$. The convolution formula is, in that case,

$$\mathbb{B}_{nk}[e^{\alpha x} - e^{\beta x}] = \sum_{j=0}^{n} \sum_{m=0}^{k} (-)^{k-m} \binom{n}{j}\, \mathbb{B}_{jm}\left[e^{\alpha x} - 1\right]\, \mathbb{B}_{n-j,k-m}\left[e^{\beta x} - 1\right].$$

Using (15.23) and (15.40), we find that $\mathbb{B}_{jm}[e^{\alpha x}-1] = \mathbb{B}_{jm}\{\alpha^r\} = \alpha^j \mathbb{B}_{jm}\{1\} = \alpha^j S_j^{(m)}$ and $\mathbb{B}_{n-j,k-m}[e^{\beta x}-1] = \beta^{n-j} S_{n-j}^{(k-m)}$. Consequently,

$$\mathbb{B}_{nk}[e^{\alpha x} - e^{\beta x}] = \sum_{j=0}^{n} \sum_{m=0}^{k} (-)^{k-m} \binom{n}{j} \alpha^j \beta^{n-j} S_j^{(m)} \, S_{n-j}^{(k-m)} .$$

If now we choose $\beta = \alpha$, the left-hand side vanishes. Dropping an overall factor α^n, we arrive at

$$\sum_{j=0}^{n} \sum_{m=0}^{k} (-)^{k-m} \binom{n}{j} S_j^{(m)} \, S_{n-j}^{(k-m)} = 0 . \tag{15.42}$$

This is, of course, a particular case of the identity

$$\sum_{j=0}^{n} \sum_{m=0}^{k} \binom{n}{j} (-)^{k-m} \, \mathbb{B}_{jm}[f(x)] \, \mathbb{B}_{n-j,k-m}[f(x)] = 0 , \tag{15.43}$$

which is (15.25) when $g = -f$. A variant of this exercise is applied in Comment 237.2 to obtain the Bell polynomials of the Fibonacci series. ◀

Commentary 231.1. ▶ **The Bell number** From (15.39),

$$e^{t[e^u-1]} = 1 + \sum_{n=1}^{\infty} \frac{u^n}{n!} \sum_{k=1}^{n} t^k S_n^{(k)} .$$

When $t = 1$, we have the generating function

$$e^{[e^u-1]} = 1 + \sum_{n=1}^{\infty} \frac{u^n}{n!} \, \omega(n) , \tag{15.44}$$

where the number

$$\omega(n) = \sum_{k=1}^{n} S_n^{(k)} \tag{15.45}$$

is the Bell number, or exponential number. As a consequence of Comment 230.1, $\omega(n)$ represents the total number of ways of partitioning a set $\hat{n}$ of n members into non-empty subsets. The members of the set are supposed distinguishable, so that $\omega(n)$ is also the number of equivalence relations that can be defined on $\hat{n}$. For coherence we put $\omega(0) = 1$. A useful recursion relation follows from taking derivatives in (15.44):

$$\omega(n+1) = \sum_{k=0}^{n} \binom{n}{k} \, \omega(k) . \tag{15.46}$$

This will follow also from Eq.(15.107) below. Expanding the left-hand side of (15.44) and comparing the coefficients, the Dobiński formula is found:

$$\omega(n) = \sum_{k=1}^{n} \mathbb{B}_{nk}\{1\} = e^{-1} \sum_{k=0}^{\infty} \frac{k^n}{k!} . \tag{15.47}$$

This last series converges due to the denominator $k!$. Putting $x = -1$ in (15.34), we find

$$\sum_{m=0}^{N} (-)^{N-m} s_N^{(m)} = N!$$

As a consequence,

$$\sum_{m=1}^{N} \mathbb{B}_{Nm}\{(j-1)!\} = \sum_{m=1}^{N} (-)^{N-m} s_N^{(m)} = \sum_{m=1}^{N} |s_N^{(m)}| = N!$$

This is consistent with the interpretation of the modulus given in Comment 230.1. From Comment 230.0 and the above result follows

$$\sum_{\{\nu_i\}}{}' \frac{1}{\prod_{j=1}^{n} [\nu_j! \, j^{\nu_j}]} = 1 \; .$$

As will be seen in some more detail in Comment 256.1, a hallmark property of the Stirling numbers [146] of the second kind $S_n^{(k)}$ is the expression

$$\left(x \, \frac{d}{dx} \right)^n f(x) = \sum_{k=1}^{n} S_n^{(k)} x^k \frac{d^k}{dx^k} f(x) \tag{15.48}$$

which gives, when applied to the exponential,

$$\left(x \, \frac{d}{dx} \right)^n e^x = \sum_{j=0}^{\infty} \frac{j^n}{j!} \, x^j = e^x \sum_{k=1}^{n} S_n^{(k)} x^k \; . \tag{15.49}$$

This leads, for $x = 1$, again to the Dobiński formula

$$\sum_{j=0}^{\infty} \frac{j^n}{j!} = e \sum_{k=1}^{n} S_n^{(k)} = e \, \omega(n) \; . \tag{15.50}$$

◀

Commentary 232.1. ▶ Contracting the generating functions for first and second kind respectively with Stirling numbers of the second and first kinds, we find

$$x^m = m! \sum_{k=m}^{\infty} \frac{(e^x - 1)^k}{k!} \, s_k^{(m)} = m! \sum_{k=m}^{\infty} S_k^{(m)} \frac{[\ln(1+x)]^k}{k!} \; .$$

Suppose we want to write a function $g(x) = \sum_{m=1}^{\infty} g_m \frac{x^m}{m!}$ not in terms of the variable x, but of its exponential. Then,

$$g(x) = \sum_{k=1}^{\infty} G_k \frac{(e^x - 1)^k}{k!} = \sum_{k=1}^{\infty} \frac{(e^x - 1)^k}{k!} \sum_{m=1}^{k} s_k^{(m)} g_m$$

$$\therefore \; G_k = \sum_{m=1}^{k} s_k^{(m)} g_m \; .$$

A transformation of variables appears as a transformation of coefficients. In the same way,

$$g(x) = \sum_{k=1}^{\infty} \gamma_k \frac{[\ln(1+x)]^k}{k!} = \sum_{k=1}^{\infty} \frac{[\ln(1+x)]^k}{k!} \sum_{m=1}^{k} S_k^{(m)} g_m$$

$$\therefore \; \gamma_k = \sum_{m=1}^{k} S_k^{(m)} g_m \; .$$

If all $g_m = 1$, $\gamma_k = \omega(k)$ (the k-th Bell number) and we find an expansion of function $e^x - 1$ in terms of its very inverse:

$$e^x - 1 = \sum_{k=1}^{\infty} \omega(k) \frac{[\ln(1+x)]^k}{k!} ,$$

which can be easily tested by using the Dobiński formula. Generalizing: given any two functions $g(x) = \sum_{k=1}^{\infty} g_k \frac{x^k}{k!}$ and $f(x) = \sum_{j=1}^{\infty} f_j \frac{x^j}{j!}$, we have the rather trivial identities

$$x^j = j! \sum_{k=j}^{\infty} \frac{1}{k!} [g(x)]^k \, \mathbb{B}_{kj}[g^{<-1>}]$$

$$f(x) = \sum_{k=1}^{\infty} \frac{[g(x)]^k}{k!} \sum_{j=1}^{\infty} \mathbb{B}_{kj}[g^{<-1>}] f_j = \sum_{k=1}^{\infty} \frac{[g(x)]^k}{k!} \mathbb{B}_{k1}[f \circ g^{<-1>}]$$

$$f(x) = \sum_{k=1}^{\infty} \frac{\left[f^{<-1>}(x)\right]^k}{k!} \mathbb{B}_{k1}[f^{<2>}] .$$

◄

Commentary 233.1. ► An example of continuum comb (see Comment 88.0) is the "Dirac comb" probability distribution

$$P(x) = e^{-1} \sum_{k=1}^{\infty} \frac{\delta(x-k)}{k!} ,$$

which leads to averages like

$$\bar{f} = < f > = \int_0^{\infty} dx f(x) P(x) = e^{-1} \sum_{k=1}^{\infty} \frac{f(k)}{k!} .$$

The Dobiński formula says that the Bell numbers are momenta of that comb distribution:

$$\omega(n) = < x^n > = \int_0^{\infty} dx \, x^n \, P(x) .$$

◄

Commentary 233.2. ► Concerning combs (see Comment 88.0), notice that we can use the multinomial theorem in each term to find

$$\sum_{p=1}^{\infty} g_{pN} \frac{x^{pN}}{(pN)!} = \sum_{p=1}^{\infty} g_{pN} \sum_{j=pN}^{\infty} \frac{x^j}{j!} \mathbb{B}_{j,pN}[f(x) = x] = \sum_{p=1}^{\infty} g_{pN} \sum_{j=pN}^{\infty} \frac{x^j}{j!} \delta_{j,pN} .$$

If we like complications, we can further use the Borel transform of Comment 225.5 to get

$$\mathcal{B}\left[\mathrm{Comb}_N[g]\right](x) = \int_0^{\infty} du \, e^{-u} \sum_{p=1}^{\infty} g_{pN} \frac{x^{pN} u^{pN}}{(pN)!} = \sum_{p=1}^{\infty} g_{pN} x^{pN} .$$

In particular,

$$\mathcal{B}\left[\mathrm{Comb}_N[\exp(x) - 1]\right] = \sum_{p=1}^{\infty} x^{pN} = \frac{x^N}{1 - x^N} .$$

It is not unnatural that such a summation on the multiples of N turn up in the generating function for the number d_N of its divisors, which is

$$\sum_{N=1}^{\infty} d_N x^N = \sum_{N=1}^{\infty} \frac{x^N}{1-x^N} = \sum_{N=1}^{\infty} \mathcal{B}\left[\mathrm{Comb}_N[\exp(x)-1]\right] .$$

◀

Other examples and special cases come from comparing series. Suppose r to be a positive integer. Then,

$$(1+u)^r - 1 = \sum_{k=1}^{r} \frac{r!}{(r-k)!} \frac{u^k}{k!} ,$$

so that

$$\mathbb{B}_{nk}[(1+u)^r - 1] = \mathbb{B}_{nk}\left\{\frac{r!}{(r-j)!}\right\} . \tag{15.51}$$

By direct computation of the Taylor coefficients, we find

$$(1+u)^{-r} - 1 = \sum_{n=1}^{\infty} \frac{u^n}{n!} (-)^n \frac{(r+n-1)!}{(r-1)!} ,$$

and consequently

$$\mathbb{B}_{nk}[(1+u)^{-r} - 1] = \mathbb{B}_{nk}\left\{(-)^j \frac{(r+j-1)!}{(r-1)!}\right\} ; \tag{15.52}$$

$$\mathbb{B}_{n1}[(1+u)^{-r} - 1] = (-)^n \frac{(r+n-1)!}{(r-1)!} . \tag{15.53}$$

Let us insist on one aspect of the last notation introduced in (15.13) that can perhaps lead to confusion. In Eq.(15.52), the index "j" appears on the right-hand side and not on the left-hand side. In the curly-bracket notation $\mathbb{B}\{g_j\}$, the argument simply exhibits the typical Taylor coefficient g_j of the series $g(x)$. That is the meaning of the "free" index "j".

Commentary 234.1. ▶ A formal trick generalizing that of Comment 225.1 is of frequent use [94], [95]. It uses operator derivatives and their functions. For example, define the operator D_t as that operator whose powers, acting on a function $f = \sum_{n=1}^{\infty} f_n \frac{t^n}{n!}$, give the respective Taylor coefficients: $D_t^n f(t) = \left[\frac{d^n f(t)}{dt^n}\right]_{t=0} = f_n$. Then,

$$e^{\alpha D_t} f(t) = \sum_{r=0}^{\infty} \frac{\alpha^r}{r!} D_t^r f(t) = f(\alpha) .$$

Take now the function

$$f(t) = e^{gV(t)} - 1 = \sum_{N=1}^{\infty} \frac{t^N}{N!} \mathbb{Y}_N[g; V]$$

and apply the second order exponential operator:

$$e^{\alpha D_t^2}\left[e^{gV(t)}-1\right]=\sum_{n=1}^{\infty}\frac{\alpha^n}{n!}\,\mathbb{Y}_{2n}[g;V]\,.$$

The above expression appears in Quantum Field Theory, as the generating function of all the vacuum Feynman diagrams with line amplitude 2α, coupling constant g and field potential V. Bender, Brody and Meister have used conveniently chosen field theories to obtain combinatorial results [145] on graphs. ◀

Commentary 235.1. ▶ Reciprocal series like $[f(x)]^{-n}$ will turn up in the manipulations leading to the Lagrange formula for the inverse of a series (Section 15.4). If u is a function, the Faà di Bruno formula will give

$$\mathbb{B}_{r1}[(1+g(x))^{-n}-1]=\sum_{s=1}^{r}\mathbb{B}_{rs}[g]\,\mathbb{B}_{s1}[(1+u)^{-n}-1]$$
$$=\sum_{s=1}^{r}(-)^s\frac{(n+s-1)!}{(n-1)!}\,\mathbb{B}_{rs}[g]\,.$$

Therefore,

$$(1+g(x))^{-n}-1=\sum_{m=1}^{\infty}\frac{x^m}{m!}\sum_{j=1}^{m}(-)^j\frac{(n+j-1)!}{(n-1)!}\,\mathbb{B}_{mj}[g]\,. \tag{15.54}$$

We can use the above result to calculate

$$\frac{1}{[f(x)]^n}=\frac{1}{\left[f_1x+\sum_{m=2}^{\infty}\frac{x^m}{m!}f_m\right]^n}=\frac{1}{(f_1x)^n\,\left[1+\sum_{m=2}^{\infty}\frac{x^{m-1}}{(m-1)!}\frac{f_m}{mf_1}\right]^n}$$
$$=\frac{1}{(f_1x)^n\,\left[1+\sum_{m=1}^{\infty}\frac{f_{m+1}}{(m+1)f_1}\,\frac{x^m}{m!}\right]^n}\,,$$

so that finally

$$\frac{1}{[f(x)]^n}=\frac{1}{(f_1x)^n}\left[1+\sum_{l=1}^{\infty}\frac{x^l}{l!}\sum_{j=1}^{l}(-)^j\frac{(n+j-1)!}{(n-1)!}\,\mathbb{B}_{lj}\left(\frac{f_2}{2f_1},\frac{f_3}{3f_1},\frac{f_4}{4f_1},\dots\right)\right]. \tag{15.55}$$

We shall also retain for later use the expression

$$\left[\frac{f_1x}{f(x)}\right]^n-1=\sum_{l=1}^{\infty}\frac{x^l}{l!}\sum_{j=1}^{l}(-)^j\frac{(n+j-1)!}{(n-1)!}\,\mathbb{B}_{lj}\left(\frac{f_2}{2f_1},\frac{f_3}{3f_1},\frac{f_4}{4f_1},\dots\right). \tag{15.56}$$

◀

Also from a direct calculation of the Taylor coefficients, we find

$$\ln[1+g(x)]=\sum_{n=1}^{\infty}\frac{x^n}{n!}\sum_{k=1}^{n}(-)^{k-1}(k-1)!\,\mathbb{B}_{nk}[g]\,. \tag{15.57}$$

The main source of particular relations is the manipulation of generating functions. For instance, comparing

$$g(x)\,e^{ug(x)}=\sum_{n=1}^{\infty}\frac{x^n}{n!}\sum_{r=0}^{n-1}\binom{n}{r}g_{n-r}\sum_{s=0}^{r}u^s\,\mathbb{B}_{rs}[g] \tag{15.58}$$

with the derivative of (15.11) with respect to u, we find

$$m\,\mathbb{B}_{nm}[g] = \sum_{k=m-1}^{n-1} \binom{n}{k} g_{n-k}\,\mathbb{B}_{k,m-1}[g]\ . \qquad (15.59)$$

This formula [see more about it in Eq.(15.127) and below] can be used to obtain recursively another expression for the Bell polynomials,

$$m\,\mathbb{B}_{k_0 m}[g] = \sum_{k_1=m-1}^{k_0-1} \sum_{k_2=m-1}^{k_1-1} \sum_{k_3=m-1}^{k_2-1} \cdots \sum_{k_{m-1}=1}^{k_m} \frac{k_0!}{k_{m-1}!} \prod_{j=1}^{m-1} \frac{g_{k_{j-1}-k_j}}{(k_{j-1}-k_j)!}\, g_{k_{m-1}}\ . \qquad (15.60)$$

Special series can lead to formulae involving particular Bell polynomials. One example is

$$\mathbb{B}_{nk}[\sin x] = \mathbb{B}_{nk}(1,0,-1,0,1,0,-1,0,\ldots)\ .$$

The beautiful expression

$$\exp\left\{\sum_{k\geq 1} k^{k-1}\frac{x^k}{k!}\right\} = \sum_{n\geq 0}(n+1)^{n-1}\frac{x^n}{n!} \qquad (15.61)$$

gives

$$\sum_{j=1}^{n} \mathbb{B}_{nj}\{k^{k-1}\} = (n+1)^{n-1}\ . \qquad (15.62)$$

Commentary 236.1. ▶ Call $g(x) = \sum_{k\geq 1} k^{k-1}\frac{x^k}{k!}$. Then the above quoted expression says that

$$\exp\{g(x)\} = \sum_{n\geq 0}(n+1)^{n-1}\frac{x^n}{n!} = \sum_{n\geq 0}(n+1)^n\frac{x^n}{(n+1)!}$$
$$= \frac{1}{x}\sum_{n\geq 0}(n+1)^n\frac{x^{n+1}}{(n+1)!} = \frac{1}{x}\,g(x)\ . \qquad (15.63)$$

Consequently, $g(x)$ is the solution of equation $u = g(u)\,e^{-g(u)}$. This is a very special case of Lambert's W-function (see Comment 248.2 below). ◀

Commentary 236.2. ▶ **Lah numbers** The expression

$$\frac{1}{k!}\left(\frac{t}{1-t} = \sum_{j=1}^{\infty} t^j\right)^k = \sum_{n=k}^{\infty}\frac{t^n}{n!}\,\mathbb{B}_{nk}\{j!\} \qquad (15.64)$$

leads, after use of

$$(1-t)^{-m} = \sum_{r=0}^{\infty}\binom{m+r-1}{r} t^r\ , \qquad (15.65)$$

to the *Lah numbers*

$$\mathbb{B}_{nk}\{j!\} = \mathbb{B}_{nk}(1!,2!,3!,\ldots) = \binom{n-1}{k-1}\frac{n!}{k!} = \binom{n}{k}\frac{(n-1)!}{(k-1)!}\ . \qquad (15.66)$$

◀

Commentary 237.1. ▶ **Idempotent numbers** The elements of the Bell matrix of function $f(x) = xe^x = \sum_{k=1}^{\infty} k \frac{x^k}{k!}$ are the idempotent numbers:

$$\mathbb{B}_{nk}[xe^x] = \mathbb{B}_{nk}(1, 2, \ldots, n-k+1) = \mathbb{B}_{nk}\{j\} = \binom{n}{k} k^{n-k} . \tag{15.67}$$

This comes from applying $e^{ug(x)} - 1 = \sum_{n=1}^{\infty} \frac{x^n}{n!} \sum_{k=1}^{n} u^k \, \mathbb{B}_{nk}(g_1, g_2, \ldots, g_{n-k+1})$ to $f(x) = xe^x$. It is enough to compare

$$e^{uxe^x} - 1 = \sum_{n=1}^{\infty} \frac{x^n}{n!} \sum_{k=1}^{n} u^k \, \mathbb{B}_{nk}(1, 2, \ldots, n-k+1)$$

to the direct expansion

$$e^{uxe^x} - 1 = \sum_{k=1}^{\infty} \frac{u^k x^k e^{kx}}{k!} = \sum_{k=1}^{\infty} \frac{u^k x^k}{k!} \sum_{j=0}^{\infty} \frac{(kx)^j}{j!} = \sum_{k=1}^{\infty} \sum_{j=0}^{\infty} \frac{u^k}{k!} \frac{x^{k+j} k^j}{j!}$$
$$= \sum_{k=1}^{\infty} \sum_{n=k}^{\infty} \frac{u^k}{k!} \frac{x^n k^{n-k}}{(n-k)!} = \sum_{n=1}^{\infty} \frac{x^n}{n!} \sum_{k=1}^{n} u^k \binom{n}{k} k^{n-k} . \tag{15.68}$$

We shall later find the inverse to the matrix of idempotent numbers, which is the Bell matrix of Lambert's W-function (Comment 248.2). ◀

Commentary 237.2. ▶ **Fibonacci numbers** Let us use the Fibonacci numbers as an excuse for an exercise using many of the above results. The Fibonacci sequence (1, 1, 2, 3, 5, 8, 13, 21, 34, 55, ...) is formed by integers F_n such that $F_n = F_{n-1} + F_{n-2}$, initialized by the conditions $F_1 = F_2 = 1$. They have the generating function

$$F(z) = \sum_{n=1}^{\infty} F_n z^n = \frac{z}{1 - z - z^2} .$$

One of their many fascinating characteristics is a "miraculous" relationship with the Golden Ratio $\alpha = \frac{1+\sqrt{5}}{2}$. In effect, for every n,

$$F_n = \frac{1}{\sqrt{5}} \left[\alpha^n - \frac{1}{(-\alpha)^n} \right]$$

(see more on Fibonacci numbers in Comment 288.2). The Borel associate of $F(z)$ (see Comment 225.5) is the function

$$g(u) = \sum_{n=1}^{\infty} F_n \frac{u^n}{n!} = \frac{1}{\sqrt{5}} \left[e^{\alpha u} - e^{-u/\alpha} \right] .$$

Of course, $\mathbb{B}_{nk}\{r! \; F_r\} = \mathbb{B}_{nk}[F(t)]$. We can try to avoid the factorials in the arguments, by finding $\mathbb{B}_{nk}\{F_r\} = \mathbb{B}_{nk}[g(u)]$ through the double convolution formula (15.25), which is

$$\mathbb{B}_{nk}[e^{\alpha u} - e^{-u/\alpha}] = \sum_{j=0}^{n} \sum_{m=0}^{k} (-)^{k-m} \binom{n}{j} \, \mathbb{B}_{jm}\left[e^{\alpha u} - 1\right] \, \mathbb{B}_{n-j,k-m} \left[e^{-u/\alpha} - 1 \right]$$

for the present case. What follows now is a continuation of Comment 230.2, with $\beta = -1/\alpha$. There, we had found $\mathbb{B}_{jm}[e^{\alpha u} - 1] = \alpha^j S_j^{(m)}$. The same procedure leads to

$\mathbb{B}_{n-j,k-m}[e^{-u/\alpha}-1]=(-1/\alpha)^{n-j}S_{n-j}^{(k-m)}$. The factors of $1/\sqrt{5}$ are taken into account by using Eq.(15.22). We arrive thus at a rather involved expression:

$$\mathbb{B}_{nk}\{F_r\}=\frac{1}{(\sqrt{5})^k}\sum_{j=0}^{n}\sum_{m=0}^{k}(-)^{k-m}\binom{n}{j}(-)^{n-j}\alpha^{2j-n}\;S_j^{(m)}S_{n-j}^{(k-m)}\;.$$

The $\sinh(x)=\frac{1}{2}\left[e^x-e^{-x}\right]$ series would give a similar result, with $\alpha\Rightarrow 1$ and $\sqrt{5}\Rightarrow 2$:

$$\mathbb{B}_{nk}[\sinh(x)]=\frac{1}{2^k}\sum_{j=0}^{n}\sum_{m=0}^{k}(-)^{k-m}\binom{n}{j}(-)^{n-j}\;S_j^{(m)}S_{n-j}^{(k-m)}\;,$$

or

$$\mathbb{B}_{nk}(1,0,1,0,1,0,1,0,\ldots)=\frac{1}{2^k}\sum_{j=0}^{n}\binom{n}{j}(-)^{n-j}\sum_{m=0}^{k}(-)^{k-m}\;S_j^{(m)}S_{n-j}^{(k-m)}\;.$$

◄

Commentary 238.1. ► Some particular combinations receive special names. Thus, the *logarithmic polynomials* $L_n[g]$ are those appearing in (15.57),

$$L_n[g]=L_n\{g_j\}=\sum_{k=1}^{n}(-)^{k-1}(k-1)!\;\mathbb{B}_{nk}[g]\;, \tag{15.69}$$

so that

$$\ln[1+g(x)]=\sum_{n=1}^{\infty}\frac{x^n}{n!}\;L_n[g]\;. \tag{15.70}$$

We may for example check that $L_n[e(x)=x]=(-)^{n-1}(n-1)!$. ◄

Commentary 238.2. ► The *power polynomials* extend the use of Bell polynomials to series with an independent term. They are defined through

$$\left(1+\sum_{j=1}^{\infty}\frac{g_j}{j!}\;t^j\right)^k=1+\sum_{n=1}^{\infty}\frac{t^n}{n!}\;P_n^{(k)}[g]\;, \tag{15.71}$$

so that

$$P_n^{(k)}[g]=\sum_{j=1}^{n}\frac{k!}{(k-j)!}\;\mathbb{B}_{nj}[g]\;. \tag{15.72}$$

This holds for any k. For k integer, we can find that

$$P_n^{(k)}[g]=\binom{n+k}{k}^{-1}\mathbb{B}_{n+k,k}(1,2g_1,3g_2,\ldots)\;.$$

Other relationships can be established, such as

$$k!\;\mathbb{B}_{nk}[g]=\sum_{r=0}^{k}\binom{k}{r}(-)^{k-r}P_n^{(r)}[g]\;; \tag{15.73}$$

$$L_n[g]=\mathbb{B}_{n1}\left[\ln\{1+g(x)\}\right]=\sum_{k=1}^{n}\frac{(-)^{k-1}}{k}\;P_n^{(k)}[g]\;. \tag{15.74}$$

◄

15.3 The matrix representation

Bell polynomials can be considered as matrix entries. The resulting matrices provide a linear realization of the group of invertible series and preserve the group properties order-by-order.

Taking a series $g(x)$ up to order N, the polynomials $\mathbb{B}_{nk}[g]$ constitute a (lower-) triangular $N \times N$ matrix $\mathbb{B}[g]$, with n as row index and k as column index. The general aspect of such a *Bell matrix* is

$$\mathbb{B}[g] = \begin{pmatrix} g_1 & 0 & 0 & 0 & \cdots\cdots & 0 \\ g_2 & g_1^2 & 0 & 0 & \cdots\cdots & 0 \\ g_3 & 3g_1g_2 & g_1^3 & 0 & \cdots\cdots & 0 \\ g_4 & 4g_1g_3 + 3g_2^2 & 6g_1^2g_2 & g_1^4 & \cdots\cdots & 0 \\ \cdots & \cdots & \cdots & \cdots & \cdots\cdots & 0 \\ g_N & \cdots & \cdots & \cdots & \cdots\cdots & g_1^N \end{pmatrix}. \tag{15.75}$$

For the case $N = 5$, for example, the exact matrix is

$$\mathbb{B}[g] = \begin{pmatrix} g_1 & 0 & 0 & 0 & 0 \\ g_2 & g_1^2 & 0 & 0 & 0 \\ g_3 & 3g_1g_2 & g_1^3 & 0 & 0 \\ g_4 & 4g_1g_3 + 3g_2^2 & 6g_1^2g_2 & g_1^4 & 0 \\ g_5 & 10g_2g_3 + 5g_1g_4 & 15g_1g_2^2 + 10g_1^2g_3 & 10g_1^3g_2 & g_1^5 \end{pmatrix}. \tag{15.76}$$

When $N \to \infty$, the Bell matrix $\mathbb{B}[g]$ is clearly an overcomplete description of g, whose Taylor coefficients occupy its first column. Associated to the multinomial theorem, Eq.(15.20) tells us that the first column is formed by the Taylor coefficients of $g(x)$; the second column, by the Taylor coefficients of $\frac{1}{2!}g^2(x)$; the third column, by those of $\frac{1}{3!}g^3(x)$; the entries in the j-column are the Taylor coefficients of $\frac{1}{j!}[g(x)]^j$. Consequently,

$$\mathbb{B}_{rs}[g] = \left(\frac{1}{s!}[g(x)]^s\right)_r = \tfrac{1}{s!}\,\mathbb{B}_{r1}[g^s]\,. \tag{15.77}$$

No confusion should be made between the g^s appearing here, which is the s-th power of the series g, and the s-th iterate $g^{<s>}$ of g.

Commentary 239.1. ▶ The above expressions provide the simplest way to obtain $\mathbb{B}_{rs}[g]$ in computers: (i) calculate the s-th power of $g(x)$; (ii) take its r-th Taylor coefficient; (iii) divide by s! An application is the closed form

$$S_n^{(k)} = \frac{1}{k!}\sum_{j=0}^{k}(-)^{k-j}\binom{k}{j}j^n$$

for the Stirling numbers of the second kind. It comes simply from $S_n^{(k)} = \frac{1}{k!}\,\mathbb{B}_{n1}[(e^u-1)^k]$ by expanding the function $(e^u - 1)^k$. Another concerns the so-called logistic map is defined by $g(x) = ax(1-x)$ for $0 \le x \le 1$, with a a real parameter in the range $0 \le a \le 4$. From the above formulas it is trivial to get $\mathbb{B}_{rs}[ax(1-x)] = \frac{(-)^{r-s} r! a^s}{(r-s)!(2s-r)!}$. ◄

Commentary 240.1. ► Commutators of Bell matrices are related to the functions by

$$e^{g[f(x)]} - e^{f[g(x)]} = \sum_{n=0}^{\infty} \frac{x^n}{n!} \sum_{j=0}^{n} \left[\mathbb{B}[g], \mathbb{B}[f]\right]_{nj} \ .$$

Bell matrices by themselves do not constitute a closed algebra with the operation defined by the commutator because, though $[\mathbb{B}[g], \mathbb{B}[f]] = \mathbb{B}[\ln(e^{g[f(x)]} - e^{f[g(x)]})]$, the function is not of the necessary type: $\ln(e^{g[f(x)]} - e^{f[g(x)]}) \to -\infty$ when $x \to 0$. ◄

Commentary 240.2. ► Traces and determinants of $\mathbb{B}[g]$ depend only on the first coefficient of g. For the $N \times N$ case and $g_1 \neq 1$, det $\mathbb{B}[g] = g_1^{N(N+1)/2}$ and $tr_N \mathbb{B}[g] = \frac{g_1^{(N+1)} - 1}{g_1 - 1} - 1$. For the infinite matrix, under the condition that $|g_1| < 1$, $tr_\infty \mathbb{B}[g] = \frac{g_1}{1-g_1}$. If $g_1 = 1$, $tr_N \mathbb{B}[g] = N$ and det $\mathbb{B}[g] = 1$. ◄

Invertible triangular matrices constitute a group, of which the set of Bell matrices $\{\mathbb{B}[g] = (\mathbb{B}_{nk}[g])\}$, for all the invertible series g, constitutes a subgroup. Hereby comes the most fascinating property of Bell polynomials, which justifies the use of this overcomplete description of series: the matrices $\mathbb{B}[g]$, with the operation of matrix product, provide a representation of the series composition group:

$$\mathbb{B}[g]\ \mathbb{B}[f] \ = \ \mathbb{B}[f \circ g] \ . \tag{15.78}$$

This property comes easily by using (15.4) twice, as

$$\frac{1}{k!}\left(f\left[g(t)\right]\right)^k = \frac{1}{k!}\left(\sum_{j=1}^{\infty} \frac{f_j}{j!}\, g^j\right)^k = \sum_{n=k}^{\infty} \frac{g(t)^n}{n!}\ \mathbb{B}_{nk}[f]$$

$$= \sum_{n=k}^{\infty} \sum_{j=n}^{\infty} \frac{t^j}{j!}\ \mathbb{B}_{jn}[g]\ \mathbb{B}_{nk}[f] = \sum_{j=k}^{\infty} \frac{t^j}{j!} \sum_{n=k}^{j} \mathbb{B}_{jn}[g]\ \mathbb{B}_{nk}[f] \ ,$$

from which $\mathbb{B}_{jk}[f(g(t))] = \sum_{n=k}^{j} \mathbb{B}_{jn}[g]\ \mathbb{B}_{nk}[f]$, which is just (15.78). Associativity can be verified by a tedious but straightforward computation. The identity function $e(t) = t$ provides the group identity element. It has $\mathbb{B}_{nk}[e]$ given by (15.30), which is $\mathbb{B}[e] = \mathbf{I}$.

Commentary 240.3. ► What we have above is in reality an anti-representation (or right-action) because of the inverse order shown in (15.78), but this is by no means a problem. We could alternatively use higher-triangular matrices and attribute to g the transpose $\mathbf{B}^T[g]$, which would lead to a usual (left-action) representation. ◄

Summing up, infinite Bell matrices constitute a linear representation of the group of invertible formal series without independent terms. If we consider only the first N rows and columns, we have an approximation of the series by the polynomial formed by its first N terms. The important point is that the group properties hold at each order N.

Bell matrices are not normal matrices (see Section 1.4). This means that there exists no unitary matrix $\mathbf{U}$ such that $\mathbf{U}\mathbb{B}\mathbf{U}^{-1}$ is diagonal. We know of course their eigenvalues, but the eigenvectors do not necessarily constitute a complete orthogonal system. There is another point we want to make. Take for example the cases $N = 2$ and $N = 3$. For

$$\mathbb{B} = \begin{pmatrix} g_1 & 0 \\ g_2 & g_1^2 \end{pmatrix},$$

the eigenvalues g_1 and g_1^2 will have respectively the eigenvectors

$$v_1 = \begin{pmatrix} \frac{g_1(1-g_1)}{g_2} \\ 1 \end{pmatrix}, \quad v_2 = \begin{pmatrix} 0 \\ 1 \end{pmatrix}.$$

For

$$\mathbb{B}[g] = \begin{pmatrix} g_1 & 0 & 0 \\ g_2 & g_1^2 & 0 \\ g_3 & 3g_1g_2 & g_1^3 \end{pmatrix}, \tag{15.79}$$

the eigenvectors corresponding to the eigenvalues g_1, g_1^2 and g_1^3 are

$$v_1 = \begin{pmatrix} \frac{g_1(g_1-1)^2(1+g_1)}{3g_2^2+g_3-g_1g_3} \\ -\frac{g_2(g_1^2-1)}{3g_2^2+g_3-g_1g_3} \\ 1 \end{pmatrix}, \quad v_2 = \begin{pmatrix} 0 \\ \frac{g_1(1-g_1)}{3g_2} \\ 1 \end{pmatrix}, \quad v_3 = \begin{pmatrix} 0 \\ 0 \\ 1 \end{pmatrix}.$$

We have said that smaller Bell matrices are contained in larger ones, and are approximations to them. What we want to stress here is that the eigenvectors, as seen in the examples above, do not have this property.

Commentary 241.1. ▶ **Wronski formula** Concerning products of formal series, another matrix procedure is well known [102]. Given a series $P_a(x) = a_0 + a_1x + a_2x^2 + a_3x^3 + \ldots$, to it corresponds the semi-circulant matrix

$$\mathbf{A} = \begin{pmatrix} a_0 & 0 & 0 & 0 & \ldots \\ a_1 & a_0 & 0 & 0 & \ldots \\ a_2 & a_1 & a_0 & 0 & \ldots \\ a_3 & a_2 & a_1 & a_0 & \ldots \\ \ldots & \ldots & \ldots & \ldots & \ldots \end{pmatrix}.$$

To another series $P_b(x) = b_0 + b_1x + b_2x^2 + b_3x^3 + \ldots$ will correspond the matrix

$$\mathbf{B} = \begin{pmatrix} b_0 & 0 & 0 & 0 & \ldots \\ b_1 & b_0 & 0 & 0 & \ldots \\ b_2 & b_1 & b_0 & 0 & \ldots \\ b_3 & b_2 & b_1 & b_0 & \ldots \\ \ldots & \ldots & \ldots & \ldots & \ldots \end{pmatrix} .$$

Given now the two series $P_a(x)$ and $P_b(x)$, the matrix corresponding to the pointwise product $P_a(x)P_b(x)$ is just the matrix product $\mathbf{AB}$ (they commute!). Under the condition $a_0 \neq 0$, there exists a reciprocal series $P_a^{-1}(x)$, which is the inverse of $P_a(x)$ by the product. Moreover, the matrix corresponding to the reciprocal $P_a^{-1}(x)$ is just the inverse matrix $\mathbf{A}^{-1}$. This leads to the Wronski formula for the coefficients of the reciprocal. If we write $P_a^{-1}(x) = b_0 + b_1x + b_2x^2 + b_3x^3 + \ldots$ then

$$b_n = \frac{(-)^n}{a_0^{n+1}} \begin{vmatrix} a_1 & a_0 & 0 & 0 & \ldots & 0 & 0 \\ a_2 & a_1 & a_0 & 0 & \ldots & 0 & 0 \\ a_3 & a_2 & a_1 & a_0 & \ldots & 0 & 0 \\ \ldots & \ldots & \ldots & \ldots & \ldots & \ldots & 0 \\ a_{n-1} & \ldots & \ldots & \ldots & \ldots & 0 & a_0 \\ a_n & a_{n-1} & a_{n-2} & a_{n-3} & \ldots & a_2 & a_1 \end{vmatrix} .$$

Take $a_0 = 0$ in the matrix $\mathbf{A}$ above. By the multinomial theorem, $m!\ \mathbb{B}_{nm}\{j!\ a_j\} = n!\ A^m_{n1}$. ◀

15.3.1 *An important inversion formula*

We shall frequently meet expressions of the form

$$F_n = \mathbb{Y}_n[1; g] = \sum_{j=1}^{n} \mathbb{B}_{nj}(g_1, g_2, \ldots, g_{n-j+1}) .$$

And we shall as frequently need its inverse, the expression of the g_k's in terms of the F_n's. To obtain that inverse, let us go back to (15.17), put $u = 1$ and write

$$F(x) = \exp\left[\sum_{k=1}^{\infty} \frac{g_k}{k!} x^k\right] - 1 = \sum_{n=1}^{\infty} \frac{x^n}{n!} \sum_{j=1}^{n} \mathbb{B}_{nj}(g_1, g_2, \ldots, g_{n-j+1}) .$$

Let us also introduce $f(z) = e^z - 1$, so that $f^{<-1>}(w) = \ln(1+w)$ and $F(x) = f[g(x)]$. From (15.36), $f_j^{<-1>} = s_j^{(1)} = (-)^{j-1}(j-1)!$. Notice then that

$$\frac{1}{m!}[F(x)]^m = \sum_{N=m}^{\infty} \frac{x^N}{N!} \sum_{r=1}^{N} \mathbb{B}_{nr}(F_1, F_2, \ldots, F_{n-r+1}) ,$$

from which

$$\mathbb{B}_{Nm}[F] = \mathbb{B}_{Nm}[f \circ g] = \sum_{j=m}^{N} \mathbb{B}_{Nj}[g]\ \mathbb{B}_{jm}[f] = \mathbb{B}_{Nm}(F_1, F_2, \ldots, F_{n-r+1}) .$$

From what we have seen in the previous section, $\mathbb{B}[g] = \mathbb{B}[F]\,\mathbb{B}[f^{<-1>}]$. In consequence,

$$g_n = \mathbb{B}_{n1}[g]$$
$$= \sum_{j=1}^{n} \mathbb{B}_{nj}[F]\,\mathbb{B}_{j1}[f^{<-1>}] = \sum_{j=1}^{n} \mathbb{B}_{nj}[F]\, s_j^{(1)} = \sum_{j=1}^{n} (-)^{j-1}(j-1)!\,\mathbb{B}_{nj}[F]\,.$$

These formulas are so important that we shall rewrite them together and repeat the result:

the expressions

$$F_n = \sum_{j=1}^{n} \mathbb{B}_{nj}(g_1, g_2, \ldots, g_{n-j+1}) \tag{15.80}$$

and

$$g_n = \sum_{j=1}^{n} (-)^{j-1}(j-1)!\,\mathbb{B}_{nj}(F_1, F_2, \ldots, F_{n-j+1}) \tag{15.81}$$

are inverse to each other. They only spell the fact that $F(u) = e^u - 1$ is inverse to $g(v) = \ln(1+v)$. Notice that $g_n = L_n[g]$ (see Comment 238.1). We shall find that these formulas provide, for example,

- relationships between the determinant of a matrix $\mathbf{M}$ and the traces of the powers of $\mathbf{M}$ (Section 16.1);
- relationships between elementary and power symmetric functions (Section 16.2);
- the Viète formulas relating the coefficients of a polynomial and the powers of its roots (Section 16.3);
- relations between cohomology classes and characters on a differentiable manifold (Section 16.5.1);
- relations between the canonical partition function and the virial coefficients for a real gas (Section 18.2).

15.4 The Lagrange inversion formula

Given a series $g(x) = \sum_{j\geq 1} g_j \frac{x^j}{j!}$ with $g_1 \neq 0$, the Lagrange formula provides the coefficients f_k of the inverse series $f(y) = \sum_{k\geq 1} f_k \frac{y^k}{k!}$, which is such that $f[g(z)] = g[f(z)] = z$. This famous result is here given a combinatorial proof.

15.4.1 *Inverting Bell matrices*

Consider the matrices formed with Stirling numbers of first and second kind, say up to order 4:

$$s = (s_n^{(k)}) = \begin{pmatrix} 1 & 0 & 0 & 0 \\ -1 & 1 & 0 & 0 \\ 2 & -3 & 1 & 0 \\ -6 & 11 & -6 & 1 \end{pmatrix} ; \quad S = (S_n^{(k)}) = \begin{pmatrix} 1 & 0 & 0 & 0 \\ 1 & 1 & 0 & 0 \\ 1 & 3 & 1 & 0 \\ 1 & 7 & 6 & 1 \end{pmatrix} .$$

They are inverse to each other. This property holds for matrices of any order: the Stirling numbers satisfy

$$\sum_{k=m}^{n} s_n^{(k)} \, S_k^{(m)} = \sum_{k=m}^{n} S_n^{(k)} \, s_k^{(m)} = \delta_n^m \, . \tag{15.82}$$

We have seen that they are Bell matrices, (15.36) and (15.40), corresponding to functions inverse to each other. This is the best example of the general property

$$\mathbb{B}[g] \, \mathbb{B}[g^{<-1>}] = I \, . \tag{15.83}$$

From this, it is evident that, given the series g, the inverse series $g^{<-1>}$ can be obtained from

$$\mathbb{B}[g^{<-1>}] = \mathbb{B}^{-1}[g] \tag{15.84}$$

by matrix inversion. For example, the inverse of (15.79) is

$$\mathbb{B}^{-1}[g] = \begin{pmatrix} \frac{1}{g_1} & 0 & 0 \\ -\frac{g_2}{g_1^3} & g_1^{-2} & 0 \\ \frac{3g_1g_2^2 - g_1^2 g_3}{g_1^6} & -\frac{3g_2}{g_1^4} & g_1^{-3} \end{pmatrix} . \tag{15.85}$$

The first three Taylor coefficients of $g^{<-1>}$ can be read in the first column. Thus, any algorithm for matrix inversion is an algorithm for series inversion.

Commentary 244.1. ► The Lah numbers $\mathbb{B}_{nk}\{j!\}$ (Comment 236.2) appear as the entries of the Bell matrix for the function $f(x) = \frac{x}{1-x}$. As the inverse to f is $f^{<-1>}(t) = \frac{t}{1+t}$, the inverse matrix will have entries $\mathbb{B}^{-1}_{nk}\{j!\} = \mathbb{B}^{-1}_{nk}\left[\frac{x}{1-x}\right] = \mathbb{B}_{nk}\left[\frac{t}{1+t}\right]$ $= \mathbb{B}_{nk}\{(-)^{j-1}j!\} = (-)^{n-k}\mathbb{B}_{nk}\{j!\}$. ◄

This method to obtain the inverse of a series becomes, however, unpractical when the matrices are of large orders. It would be better to have a general, closed expression. The celebrated Lagrange formula, which we now proceed to find, is precisely a closed expression for the series $f(x)$ which is the inverse of a given $g(x)$,

$$f(x) = g^{<-1>}(x) \ .$$

Let us beforehand make a short parenthesis on the Leibniz rule, which will be necessary.

15.4.2 *The Leibniz formula*

The Leibniz formula provides the n-th derivative of the pointwise product $F(t)G(t)$ of two functions. It follows directly from (15.4) that

$$\mathbb{B}_{nk}[F(t)\ G(t)] = \begin{cases} k! \sum_{j=k}^{n-k} \binom{n}{j}\ \mathbb{B}_{jk}[F]\ \mathbb{B}_{n-j,k}[G] \ \text{ if } \ n \geq 2k \\ 0 \text{ for } \ k \leq n < 2k \ . \end{cases} \tag{15.86}$$

The Leibniz rule for $n \geq 2$ is the particular case

$$\mathbb{B}_{n1}[F(t)G(t)] = \sum_{j=1}^{n-1} \binom{n}{j} \mathbb{B}_{j1}[F]\ \mathbb{B}_{n-j,1}[G] = \sum_{j=1}^{n-1} \binom{n}{j} F_j\ G_{n-j} \ . \tag{15.87}$$

It acquires its simplest, binomial-like expression when we recall the convention $\mathbb{B}_{n0} = \delta_{n,0}$ and add the two values $j = 0, n$ to the above summation:

$$[F(t)G(t)]_n = \sum_{j=0}^{n} \binom{n}{j} F_j G_{n-j} \ . \tag{15.88}$$

15.4.3 *The inverse series*

Let us go back to Eqs.(15.2) and (15.3) for $F(t) = [f \circ g](t) = f[g(t)]$. As $f_k = \mathbb{B}_{k1}[f]$, we have of course

$$F_n[f;g] = \sum_{k=1}^{n} \mathbb{B}_{nk}[g] f_k = \sum_{k=1}^{n} \mathbb{B}_{nk}[g]\ \mathbb{B}_{k1}[f] = \mathbb{B}_{n1}[f \circ g] \ . \tag{15.89}$$

Thus, series f will be the inverse of series g if

$$\sum_{k=1}^{n} \mathbb{B}_{nk}[g] f_k = \delta_{n1} \ . \tag{15.90}$$

For $n = 1$, this gives $f_1 = 1/g_1$. For $n \geq 2$, use of (15.14) leads to a recursion formula

$$f_n = -\frac{1}{g_1^n} \sum_{k=1}^{n-1} f_k\ \mathbb{B}_{nk}[g] \ . \tag{15.91}$$

This leads to a step-by-step computation of the inverse series [102].

A determinant form comes out if we solve (15.90) by Cramer's rule:

$$f_n = \frac{(-)^{n-1}}{g_1^{n(n+1)/2}} \begin{vmatrix} \mathbb{B}_{21}[g] & \mathbb{B}_{22}[g] & 0 & 0 \ldots & 0 \\ \mathbb{B}_{31}[g] & \mathbb{B}_{32}[g] & \mathbb{B}_{33}[g] & 0 \ldots & 0 \\ \cdots & \cdots & \cdots & \cdots\cdots & \cdots \\ \cdots & \cdots & \cdots & \cdots\cdots & 0 \\ \mathbb{B}_{n-1,1}[g] & \mathbb{B}_{n-1,2}[g] & \cdots & \cdots\cdots & \mathbb{B}_{n-1,n-1}[g] \\ \mathbb{B}_{n1}[g] & \mathbb{B}_{n2}[g] & \cdots & \cdots\cdots & \mathbb{B}_{n,n-1}[g] \end{vmatrix} . \tag{15.92}$$

Indeed, f_2 comes when the determinant reduces simply to $\mathbb{B}_{21}[g] = g_2$; for $n = 3$, only the upper two rows appear and $f_3 = g_1^{-6} \begin{vmatrix} g_2 & g_1^2 \\ g_3 & 3g_1g_2 \end{vmatrix}$. And so on. Formula (15.91) turns up when the determinant in (15.92) is expanded along the last row.

Let us see how the Lagrange formula can be obtained in our language. Analytical methods provide a more elegant approach to the problem [101], [102], [96], [104]. Our method, however, besides being constructive and elementary [105], [106], gives not only the inverse series, but directly its Bell polynomials.

We shall actually look for a solution of the recursion relation (15.91).

- Let us start by repeating that the first coefficient is $f_1 = g_1^{<-1>} = 1/g_1$. Once this is fixed, we look for f_n for $n \geq 2$.
- We shall need to calculate $[f(x)]^{-n} f' = \frac{1}{1-n} \frac{d}{dx} f^{1-n}$.
- Equation (15.55) can be used to give

  $$[f(x)]^{1-n} = \tfrac{1}{(f_1 x)^{n-1}}$$
  $$+ \frac{1}{f_1^{n-1}} \sum_{l=1}^{\infty} \frac{x^{l-n+1}}{l!} \sum_{j=1}^{l} (-)^j \frac{(n+j-2)!}{(n-2)!} \mathbb{B}_{lj} \left(\frac{f_2}{2f_1}, \frac{f_3}{3f_1}, \frac{f_4}{4f_1}, \ldots \right).$$

- Taking the derivative and multiplying by $f_1^n x^{n+1}$, we find

  $$f_1^n x^{n+1} \, [f(x)]^{-n} f' = f_1 \, x$$
  $$+ f_1 \sum_{l=1}^{\infty} \frac{x^{l+1}}{l!} \frac{(l-n+1)}{1-n} \sum_{j=1}^{l} (-)^j \frac{(n+j-2)!}{(n-2)!} \mathbb{B}_{lj} \left\{ \frac{f_{k+1}}{(k+1)f_1} \right\}.$$

- We observe that, in this expression, the n-th coefficient is zero:

  $$\mathbb{B}_{n1}[x f' (f_1 x)^n \, [f(x)]^{-n}] = 0 \, .$$

- We observe also that

  $$\mathbb{B}_{n1}[x f'] = n f_n \, .$$

  Though the derived series $g'(t) = \sum_{j=0}^{\infty} \frac{g_{j+1}}{j!} t^j$ does not belong to the group, $g'(t) - g_1$ does, and $\mathbb{B}[g'(t) - g_1] = \mathbb{B}\{g_{j+1}\}$.

- In consequence,

$$\mathbb{B}_{n1}[xf'((f_1x)^n\,[f(x)]^{-n}-1)]$$
$$=\mathbb{B}_{n1}[xf'((f_1x)^n\,[f(x)]^{-n})]-\mathbb{B}_{n1}[xf']=-nf_n\,.$$

- But $\mathbb{B}_{n1}[xf'((f_1x)^n\,[f(x)]^{-n}-1)]$, as a product, can also be evaluated by using Leibniz formula (15.87):

$$\mathbb{B}_{n1}[xf'((f_1x)^n\,[f(x)]^{-n}-1)]$$
$$=\sum_{j=1}^{n-1}\binom{n}{j}\mathbb{B}_{j1}[xf']\;\mathbb{B}_{n-j,1}[(f_1x)^n\,[f(x)]^{-n}-1]$$
$$=\sum_{j=1}^{n-1}\binom{n}{j}jf_j\;\mathbb{B}_{n-j,1}[(f_1x)^n\,[f(x)]^{-n}-1]\,.$$

- Comparison with the previous expression gives

$$f_n=-\sum_{j=1}^{n-1}\binom{n-1}{j-1}\;\mathbb{B}_{n-j,1}[(f_1x)^n\,[f(x)]^{-n}-1]f_j\;,$$

which is (15.91) with

$$\mathbb{B}_{n,j}[g]=g_1^n\binom{n-1}{j-1}\;\mathbb{B}_{n-j,1}\left[\left(\frac{f_1x}{f(x)}\right)^n-1\right]. \tag{15.93}$$

We can further use (15.56) to arrive at

$$\mathbb{B}_{n,j}[g(x)]=\frac{1}{f_1^n}\;\binom{n-1}{j-1}\;\sum_{l=1}^{n-j}(-)^l\frac{(n+l-1)!}{(n-1)!}\;\mathbb{B}_{n-j,l}\left\{\frac{f_{k+1}}{(k+1)f_1}\right\}. \tag{15.94}$$

The last two formulae show the relationship between a series $g=f^{<-1>}$ and its inverse $f=g^{<-1>}$. We can, of course, interchange g and f at will.

- The identity $(f_1x)^n\,[f(x)]^{-n}-1=\exp\left(-n\ln(f/f_1x)\right)-1$ leads to

$$\mathbb{B}_{n-j,1}[(f_1x)^n\,[f(x)]^{-n}-1]=\sum_{l=1}^{n-j}(-)^l n^l\;\mathbb{B}_{n-j,1}\left[\ln\left(f(u)/f_1u\right)\right].$$

Given all the above relations, the result can be put into several forms:

$$g_{n\geq 2}^{<-1>}=\frac{1}{g_1^n}\sum_{k=1}^{n-1}(-)^k\frac{(n+k-1)!}{(n-1)!}\;\mathbb{B}_{n-1,k}\left(\frac{g_2}{2g_1},\frac{g_3}{3g_1},\frac{g_4}{4g_1},\dots\right); \tag{15.95}$$

$$g_{n\geq 2}^{<-1>}=\frac{1}{g_1}\sum_{k=1}^{n-1}(-)^k\frac{(n+k-1)!}{(n-1)!}\;\mathbb{B}_{n-1,k}\left\{\frac{g_{j+1}}{(j+1)g_1^{j+1}}\right\}. \tag{15.96}$$

If we recall convention (15.16) then all cases, including $g_1^{<-1>} = 1/g_1$, are put together in

$$g_n^{<-1>} = \frac{1}{g_1^n} \sum_{k=0}^{n-1} (-)^k n^k \, \mathbb{B}_{n-1,k} \left[\ln \frac{g(t)}{g_1 t} \right] . \qquad (15.97)$$

The Bell polynomials (15.94) of the inverse can be written in a compact form:

$$\mathbb{B}_{nm}[g^{<-1>}] = \frac{1}{g_1^n} \binom{n-1}{m-1} \sum_{k=0}^{n-m} \mathbb{B}_{n-m,k} \left[-n \ln \left(\frac{g(t)}{g_1 t} \right) \right] . \qquad (15.98)$$

Equation (15.93) provides what are perhaps the most practical versions:

$$g_n^{<-1>} = \frac{1}{g_1^n} \left[\left(\frac{g_1 x}{g(x)} \right)^n - 1 \right]_{n-1} ; \qquad (15.99)$$

$$g^{<-1>}(u) = \sum_{n=1}^{\infty} \frac{u^n}{n! g_1^n} \left[\left(\frac{g_1 x}{g(x)} \right)^n - 1 \right]_{n-1} . \qquad (15.100)$$

Other expressions for the inverse series will be given later [see Eqs.(16.49), (17.49) and Comment 295.1].

Summing up: given a series $g(x) = \sum_{j \geq 1} g_j \frac{x^j}{j!}$ with $g_1 \neq 0$, its inverse series will be

$$f(y) = \sum_{n=1}^{\infty} \frac{y^n}{n!} \frac{1}{g_1^n} \sum_{k=0}^{n-1} \mathbb{B}_{n-1,k} \left[\ln \left(\frac{x}{g(x)} \right)^n \right] . \qquad (15.101)$$

Commentary 248.1. ► It is a simple exercise to use the above formulas to show that an inverse to the logistic map (see Comment 239.1) $g(x) = ax(1-x)$ is $f(y) = \frac{1}{2}[1 - \sqrt{1 - 4y/a}]$. ◄

Commentary 248.2. ► **Lambert's W-function** The Lambert W-function (also called Ω-function) is the inverse to function $f(x) = xe^x$. We have in Comment 237.1 shown that the Bell matrix $\mathbb{B}[f(x) = xe^x]$ has the idempotent numbers for entries. Function $W(u) = f^{<-1>}(u)$ is defined by

$$x = W(xe^x) ; \;\; W(u) e^{W(u)} = u \, .$$

It is a multisheeted function, actually with a sheet structure similar to that of the logarithm function (wherefrom its alternative name "logarithm-product"). Its first principal is given by the formal series

$$W(u) = \sum_{n=1}^{\infty} (-)^{n-1} n^{n-1} \frac{u^n}{n!} \qquad (u < e^{-1}) \, , \qquad (15.102)$$

whose numerical convergence, given by the radius of convergence $e^{-1} = 0.367879...$ in the parenthesis, is very limited.[3] We have anyhow the formal identity

$$x = W(xe^x) = \sum_{n=1}^{\infty} (-)^{n-1} n^{n-1} \frac{x^n e^{nx}}{n!} \ , \ \forall \, x \, .$$

The Bell matrix will have entries

$$\mathbb{B}_{rk}[W(u)] = \mathbb{B}_{rk}\{(-)^{n-1} n^{n-1}\} = \binom{r-1}{k-1} (-r)^{r-k} \, . \tag{15.103}$$

That would mean that

$$\sum_{k=r}^{n} (-)^{k-r} k^{n-r} \binom{n}{k} \binom{k-1}{r-1} = \delta_{rn}$$

is an identity. To find (15.103), we look for the inverse g to $f(x) = xe^x = \sum_{k=1}^{\infty} k \frac{x^k}{k!}$. Notice that $f_1 = 1, f_k = k$; then the Lagrange formula (15.94) for the Bell matrix entries gives

$$\begin{aligned}
\mathbb{B}_{n,j}[W = f^{<-1>}] &= \frac{1}{f_1^n} \binom{n-1}{j-1} \sum_{l=1}^{n-j} (-)^l \frac{(n+l-1)!}{(n-1)!} \, \mathbb{B}_{n-j,l} \left\{ \frac{f_{k+1}}{(k+1) f_1} \right\} \\
&= \binom{n-1}{j-1} \sum_{l=1}^{n-j} (-)^l \frac{(n+l-1)!}{(n-1)!} \, \mathbb{B}_{n-j,l}\, \{1\} = \binom{n-1}{j-1} \sum_{l=1}^{n-j} (-)^l l! \, \frac{(n+l-1)!}{(n-1)! l!} \, S_{n-j}^{(l)} \\
&= \binom{n-1}{j-1} \sum_{l=1}^{n-j} (-)^l \, (n-1+l)(n-1+l-1)(n-1+l-2)...(n+1) n \, S_{n-j}^{(l)} \\
&= \binom{n-1}{j-1} (-n)^{n-j} \, .
\end{aligned} \tag{15.104}$$

as announced. As another exercise, we can find series (15.102) by using the standard Lagrange formula (15.96):

$$\begin{aligned}
\mathbb{B}_{n,1}[W] &= \sum_{l=1}^{n-1} (-)^l \, \frac{(n-1+l)!}{(n-1)!} \, S_{n-1}^{(l)} \\
&= \sum_{l=1}^{n-1} (-)^l \, (n-1+l)(n-1+l-1)(n-1+l-2)...(n+1) n \, S_{n-1}^{(l)} \\
&= \sum_{l=1}^{n-1} (-n+1-l)(-n+2-l)(-n+3-l)...(-n-1)(-n) \, S_{n-1}^{(l)} \\
&= (-n)^{n-1} = (-)^{n-1} n^{n-1} \, .
\end{aligned} \tag{15.105}$$

Consequently, $f^{<-1>}(u) = \sum_{n=1}^{\infty} (-)^{n-1} n^{n-1} \frac{u^n}{n!} = - \sum_{n=1}^{\infty} n^{n-1} \frac{(-u)^n}{n!}$.

[3] See Eric W. Weisstein. "Lambert W-Function." From MathWorld – A Wolfram Web Resource. http://mathworld.wolfram.com/LambertW-Function.html. And further Keith Briggs, "W-ology, or Some Exactly Solvable Growth Models", http://www.btexact.com/people/briggsk2/W-ology.html (Keith Briggs, Department of Plant Sciences, University of Cambridge, Downing St., Cambridge CB2 3EA kmb28@cam.ac.uk.).

The function $g(u)$, solution of $u = g(u)\, e^{-\, g(u)}$, which turned up in Comment 236.1, is simply $g(u) = -\, W(-u)$. Equation (15.61) is

$$e^{-W(-x)} = \frac{W(-x)}{(-x)} \ .$$

Some properties of Lambert's W-function:

- $xy = W(xye^{xy}) = W(xe^{x})\, W(ye^{y});\quad W(e) = 1 = W(ze^{z})\, W(\frac{1}{z} e^{\frac{1}{z}});$
- identity $z = W(ze^{z})$ and its consequence $z^{\gamma} = W(z^{\gamma} e^{z^{\gamma}})$ entail
$$W^{\gamma}(ze^{z}) = W(z^{\gamma} e^{z^{\gamma}}) \ ;$$
- putting $z = \frac{\beta}{\alpha} x$ in identity $z = W(ze^{z})$ shows that the inverse to $u(x) = x^{\alpha} e^{\beta x}$ is $\frac{\alpha}{\beta} W(\frac{\beta}{\alpha} u^{1/\alpha})$ and the trivial result $x = \frac{\alpha}{\beta} W(\frac{\beta}{\alpha} x e^{\frac{\beta}{\alpha} x})$ is recovered; the inverse to $u(x) = K x^{\alpha} e^{\beta x}$ is $\frac{\alpha}{\beta} W\left[\frac{\beta}{\alpha} (\frac{u}{K})^{1/\alpha}\right]$;
- scaling law: if $x \to Ax$, $W(Axe^{Ax}) = AW(xe^{x})$;
- the differential equation:
$$u\, W'(u)\, (1 + W(u)) = W(u) \ ;$$
- as found above, $\mathbb{B}_{rk}[W(u)] = \mathbb{B}_{rk}\{(-)^{n-1} n^{n-1}\} = \binom{k-1}{r-1} (-k)^{r-k}$.

◄

Commentary 250.1. ► **Power-tower function** Use identity $W[f(x)] e^{W[f(x)]} = f(x)$ and its consequence $W[f(x)] = f(x) e^{-\, W[f(x)]}$ for $f(x) = -\ln x$. Then $\frac{W[-\ln x]}{(-\ln x)} = e^{-W[-\ln x]}$, so that $t(x) := \frac{W[-\ln x]}{(-\ln x)}$ satisfies $t(x) = e^{(\ln x) t(x)} = x^{t(x)} = x^{x^{t(x)}} = \ldots$. The resulting infinite exponentiation of a variable is its so-called power-tower "function", sometimes indicated $t(x) = x \uparrow\uparrow \infty$:

$$t(x) = -\frac{W(-\ln x)}{\ln x} = x \uparrow\uparrow \infty = x^{x^{x^{\cdot^{\cdot^{\cdot}}}}} \ .$$

The finite exponentiates have different behaviors at the origin: even order cases $x, x^{x^{x}}, \ldots$ vanish at $x = 0$, while odd order cases $x^{x}, x^{x^{x^{x}}}, \ldots = 1$ at $x = 0$. ◄

15.5 Developments

Bell polynomials, as a rule, provide mere information rearrangements. A formula with this character is shown, which will find many applications in the following, even eventually leading to real recursion relations. Connections with orthogonal polynomials and with derivatives and/or integrations of fractional order are exhibited. Derivatives of Bell polynomials with respect to their arguments are given simple expressions.

15.5.1 *A useful pseudo-recursion*

We may now find a kind of recursion relation for summations of Bell polynomials (that is, for complete Bell polynomials). Take again (15.11), in the particular case $f(u) = e^u - 1$, $F(t) = e^{g(t)} - 1$. Clearly $Y_n(f; g)$ is

$$F_n(e; g) = \frac{d^n}{dt^n}\left[e^{g(t)} - 1\right]_{t=0} = \frac{d^{n-1}}{dt^{n-1}}\left[g'(t)e^{g(t)}\right]_{t=0} ;$$

$g'(t)$ is the series

$$g'(t) = \sum_{j=0}^{\infty} \frac{g_{j+1}}{j!} t^j , \tag{15.106}$$

which does not belong to our group. We can, however, use the full Leibniz rule (15.86) to get

$$F_n = \left[g'(t)\ e^{g(t)}\right]_{n-1} = \sum_{j=0}^{n-1} \binom{n-1}{j} g'_j \left[e^{g(t)}\right]_{n-1-j}$$

$$= \sum_{j=0}^{n-1} \binom{n-1}{j} g_{j+1} \left[\frac{d^{n-1-j}}{dt^{n-1-j}}\ e^{g(t)}\right]_{t=0} = g_n + \sum_{j=0}^{n-2} \binom{n-1}{j} g_{j+1} F_{n-1-j} .$$

From this result, we easily obtain an expression for the summation of all the entries of the n-th line of a Bell matrix:

$$\sum_{k=1}^{n} \mathbb{B}_{nk}[g] = \sum_{p=0}^{n-1} \binom{n-1}{p} g_{n-p} \sum_{m=0}^{p} \mathbb{B}_{pm}[g] = \frac{1}{n} \sum_{j=1}^{n} \binom{n}{j} j g_j \sum_{r=0}^{n-j} \mathbb{B}_{n-j,r}[g] . \tag{15.107}$$

This expression is not really a recursion relation, but an information rearrangement, a relation between different entries of the Bell matrix which is useful in certain cases. For example, when g_j's are known, as in the case of Eq.(15.46) above, and in the virial series for bosons and fermions to be seen later. It will find applications in the discussion of symmetric functions [around Eq.(16.17)] and in the virial approach to real gases [below Eq.(18.53)]. For the time being, let us only notice that (15.107) can be used to give an analogous relationship for the complete Bell polynomials (15.12):

$$\mathbb{Y}_n[u; g] = \sum_{j=1}^{n} \mathbb{B}_{nj}(ug_1, ug_2, \ldots, ug_{n-j+1}) = \sum_{p=0}^{n-1} \binom{n-1}{p} g_{n-p} \sum_{m=0}^{p} \mathbb{B}_{pm}[ug]$$

$$= \sum_{p=0}^{n-1} \binom{n-1}{p}\ g_{n-p} \sum_{m=0}^{p} u^m\ \mathbb{B}_{pm}[g] = \sum_{p=0}^{n-1} \binom{n-1}{p}\ g_{n-p} \mathbb{Y}_p[u; g] . \tag{15.108}$$

We recall that the convention (15.16) is systematically adopted. By the way, once the matrix representation has been introduced, each complete Bell polynomial $\mathbb{Y}_n[u;g]$ acquires the status of a "horizontal" (row-along) generating function for the polynomials $\mathbb{B}_{nj}[g]$.

Commentary 252.1. ▶ **Solving a convolution equation** Let be given a finite sequence (or alphabet) $\mathbf{c} = (c_1, c_2, c_3, ..., c_n)$. Then equation

$$F_n = \sum_{k=1}^{n} c_k F_{n-k} = \sum_{p=0}^{n-1} c_{n-p} F_p$$

has for solution the alphabet $F_n = \frac{1}{n!}\sum_{j=1}^{n} \mathbb{B}_{nj}\{(l-1)!\, n\, c_l\}$. To show it, it is enough to introduce $F_n = \mathbb{Y}_n[u;g]/n!$ and $g_k = n\,(k-1)!\, c_k$ in the last expression of (15.108),

$$\mathbb{Y}_n[u;g] = \sum_{j=1}^{n} u^j \mathbb{B}_{nj}[g] = \sum_{p=0}^{n-1} \binom{n-1}{p} g_{n-p} \mathbb{Y}_p[u;g] \ .$$

It follows in effect that

$$\frac{\mathbb{Y}_n[u;g]}{n!} = \sum_{p=0}^{n-1} \frac{g_{n-p}}{(n-1-p)!\, n} \frac{\mathbb{Y}_p[u;g]}{p!} = \sum_{k=1}^{n} \frac{g_k}{n\,(k-1)!} \frac{\mathbb{Y}_{n-k}[u;g]}{(n-k)!}$$

$$\therefore \ F_n = \sum_{k=1}^{n} \frac{g_k}{n\,(k-1)!} F_{n-k} \ .$$

Notice: $g(t) = n \sum_{k=1}^{\infty} c_k \frac{t^k}{k}$, and u must be chosen $u = 1$. The closed solution is then

$$F_n = \frac{1}{n!}\mathbb{Y}_n[1; \{(l-1)!\, n\, c_l\}] = \frac{1}{n!}\sum_{j=1}^{n} \mathbb{B}_{nj}\{(l-1)!\, n\, c_l\} \ .$$

For an example in Statistical Mechanics, including a clue towards a determinantal form of the above result, see Comment 323.1. ◀

15.5.2 *Relations to orthogonal polynomials*

In some cases an easy relationship can be obtained with the orthogonal polynomials of Mathematical Physics. For example, looking at the generating function for the Hermite polynomials $H_n(x)$,

$$e^{-t^2+2tx} - 1 = \sum_{n=1}^{\infty} \frac{t^n}{n!} H_n(x) \ , \tag{15.109}$$

in the light of (15.11) and (15.12), we find

$$H_n(x) = \mathbb{Y}_n[1; 2tx - t^2] = \sum_{j=1}^{n} \mathbb{B}_{nj}[2tx - t^2] = \sum_{j=1}^{n} \mathbb{B}_{nj}(2x, -2, 0, 0, \ldots, 0) \ . \tag{15.110}$$

The lower order examples $H_0(x) = 1$; $H_1(x) = 2x$; $H_2(x) = 4x^2 - 2$; $H_3(x) = 8x^3 - 12x$, etc., follow immediately by inspecting (15.75). When identifying the coefficients, one must keep in mind the "canonical", exponential definition of series coefficients we are adopting since equations (15.1). The Hermite polynomials are, thus, special kinds of complete Bell polynomials (notice that we are using the conventions of [34]; there are other conventions in use, including some factors of $\sqrt{2}$). Use of (15.108) gives the well-known recursion relation

$$H_n(x) = 2xH_{n-1}(x) - 2(n-1)H_{n-2}(x) .$$

In other cases, the orthogonal polynomials appear as arguments of the Bell polynomials. Thus, the generating function for the Legendre polynomials $P_n(x)$,

$$\frac{1}{\sqrt{1-2tx+t^2}} = \sum_{n=0}^{\infty} P_n(x)\ t^n , \tag{15.111}$$

leads immediately to

$$\exp\left[\frac{1}{\sqrt{1-2tx+t^2}}\right] - 1 = \sum_{N=1}^{\infty} \frac{t^N}{N!} \sum_{m=1}^{N} \mathbb{B}_{Nm}\{j!\ P_j(x)\} . \tag{15.112}$$

Legendre polynomials are particular cases of the Gegenbauer polynomials $C_N^{\alpha}(x)$ generated as

$$(1-2tx+t^2)^{-\alpha} = \sum_{N=0}^{\infty} \frac{t^n}{N!}\ C_N^{\alpha}(x) . \tag{15.113}$$

As

$$(1-2tx+t^2)^{-\alpha} = \sum_{N=0}^{\infty} \frac{t^N}{N!} \sum_{m=[N/2]}^{N} \frac{(-)^m\ \Gamma(1-\alpha)}{(1-\alpha-m)!}\ \mathbb{B}_{Nm}(2x, -2, 0, 0, \ldots, 0) ,$$

we obtain

$$C_N^{\alpha}(x) = \sum_{m=[N/2]}^{N} \frac{(-)^m\ \Gamma(1-\alpha)}{(1-\alpha-m)!}\ \mathbb{B}_{Nm}(2x, -2, 0, 0, \ldots, 0) . \tag{15.114}$$

For the particular value $\alpha = 1/2$,

$$\begin{aligned} P_N(x) &= \tfrac{1}{N!} \sum_{m=[N/2]}^{N} \frac{1}{2^m\ (2m-1)!}\ \mathbb{B}_{Nm}(2x, -2, 0, 0, \ldots, 0) \\ &= \tfrac{1}{N!} \sum_{m=[N/2]}^{N} \frac{1}{2^m\ (2m-1)!}\ \mathbb{B}_{Nm}[2x\ t - t^2] . \end{aligned} \tag{15.115}$$

The symbol $[N/2]$ represents the integer which is in the set $\{N/2, (N+1)/2\}$. Notice that the Bell polynomial $\mathbb{B}_{Nm}(2x, -2, 0, 0, \ldots, 0)$ appears in all the cases above, an obvious consequence of the "seed" expression $(2x\ t - t^2)$ in all the generating functions.

Let us go back to (15.78). An obvious consequence is that function iteration is transformed into simple matrix power:

$$\mathbb{B}[g \circ g] = \mathbb{B}[g]\ \mathbb{B}[g] \ = \ \mathbb{B}^2[g]\ ;\ \mathbb{B}[g^{<n>}] = \mathbb{B}^n[g] . \tag{15.116}$$

We shall discuss iterates at length in Section 17.2. Here we only present a beautiful relationship involving the Chebyshev polynomials $T_n(x)$ of the first kind. These polynomials can be introduced through the generating function

$$\frac{1-t^2}{1-2tx+t^2}+1=2\sum_{n=0}^{\infty}T_n(x)\,t^n\ ,\quad \text{or}\quad \frac{xt-t^2}{1-2tx+t^2}=\sum_{n=1}^{\infty}T_n(x)t^n\ , \tag{15.117}$$

from which the first examples can be easily obtained: $T_0(x) = 1$; $T_1(x) = x$; $T_2(x) = 2x^2 - 1$; $T_3(x) = 4x^3 - 3x$; $T_4(x) = 8x^4 - 8x^2 + 1$, etc. But there is another, equivalent definition:

$$T_n(x) = \cos\left[n \arccos x\right] . \tag{15.118}$$

This expression has the form $T_n(x) = F^{<-1>}[n\ F(x)]$, from which it follows that

$$T_m[T_n(x)] = [T_m \circ T_n](x) = T_{mn}(x) = T_n[T_m(x)] . \tag{15.119}$$

This means that Chebyshev polynomials transform composition into index-product and vice versa. In particular,

$$T_n^{<m>}(x) = [\underbrace{T_n \circ T_n \circ T_n \ldots \circ T_n}_{m\ times}](x) = T_{n^m}(x) . \tag{15.120}$$

As can be seen from the first examples given above, only those of odd order have well-defined Bell polynomials. For the other, care must be taken to extract the values at the origin. The last expression implies, provided the involved Bell polynomials are defined, the following relations:

$$\mathbb{B}[T_n^{<m>}] = \mathbb{B}[\underbrace{T_n \circ T_n \circ T_n \ldots \circ T_n}_{m\ times}] = \mathbb{B}^m[T_n] . \tag{15.121}$$

Sequences $\{f_n\}$ of functions of the form $f_n(x) = F^{<-1>}[n\ F(x)]$ are of clear interest to iterations, as $f_n[f_n(x)] = F^{<-1>} \circ n\ F \circ F^{<-1>}[n\ F(x)] = F^{<-1>} \circ [n^2\ F(x)] = f_{n^2}(x)$. They are involved with the equation

$$F \circ f_n = n\ F\ ,$$

which is a particular form of the Schröder functional equation. When Bell polynomials can be applied, the expression above leads to

$$\sum_s \mathbb{B}_{rs}[f_n]\ \mathbb{B}_{s1}[F] = \sum_s \mathbb{B}_{rs}[f_n]\ F_s = n\ F_r\ ,$$

clearly related to an eigenvector of the Bell matrix. This will be discussed in Section 17.2.3.

15.5.3 *Differintegration*

Derivatives and integrals to arbitrary orders are the subject of a rather neglected chapter of classical mathematics, to which researchers sometimes arrive when approaching particular problems. It has occupied the minds of people like Leibniz, Euler, Lagrange, Fourier and Abel, but seems to have remained unsystematic up to the works of Liouville, Riemann and Holmgren in the course of the 19th century. A thorough treatment is given in [107] and an application to fractals in [108]. There are many expressions for the derivative or integral of order q. Only to give the gist of the thing, let us recall two of the most suggestive expressions defining differintegrals. One is Liouville's definition of the fractional derivative (of order q) of a series $f(x) = \sum_j a_j e^{jx}$, which is

$$\frac{d^q}{dx^q} f(x) = \sum_j a_j e^{jx} j^q \ .$$

The other will be Riemann's definition of a fractional integration. In present-day notation, it would read:

$$\frac{d^{-r}}{dx^{-r}} f(x) = \frac{1}{\Gamma(r)} \int_0^x (x-y)^{r-1} f(y) dy \ .$$

We shall here only quote a formula which involves Bell polynomials:

$$\frac{d^q}{dx^q} f[g(x)] = \frac{x^{-q}}{\Gamma(1-q)} f[g(x)] + \sum_{n=1}^{\infty} \binom{q}{n} \frac{x^{n-q}}{\Gamma(n+1-q)} \sum_{m=1}^{n} \mathbb{B}_{nm}[g] \ f_m \ . \tag{15.122}$$

This expression is a certain generalization of the Faà di Bruno formula (15.3) to any complex order of derivation q. The binomial is defined as $\binom{q}{n} = q!/[n!(q-n)!]$ and, as previously said [see (15.31)], the factorial of a general complex number is defined through Γ functions.

15.5.4 *Homogeneity degree*

The derivative of a Bell polynomial with respect to its particular argument g_i is

$$\frac{\partial}{\partial g_i} \mathbb{B}_{nk}[g] = \binom{n}{i} \mathbb{B}_{n-i,k-1}[g] \ . \tag{15.123}$$

This expression is obtained by applying $\frac{\partial}{\partial g_i}$ to both sides of (15.4). Comparing with (15.59), written in the form

$$m\,\mathbb{B}_{nm}[g] = \sum_{j=1}^{n-m+1} \binom{n}{j} g_j\,\mathbb{B}_{n-j,m-1}[g]\ , \tag{15.124}$$

we find

$$m\,\mathbb{B}_{nm}[g] = \sum_{j=1}^{n-m+1} g_j \frac{\partial}{\partial g_j}\,\mathbb{B}_{n,m}[g]\ , \tag{15.125}$$

which is to be expected in view of (15.6) and (15.8). Recall that, for fixed n and m, $\mathbb{B}_{n,m}[g]$ depends only on g_j for $j = 1, 2, \ldots, n-m+1$. We can define the operator

$$D = \sum_{j=1}^{N} g_j \frac{\partial}{\partial g_j} \tag{15.126}$$

and write

$$D\,\mathbb{B}_{n,m}[g] = \sum_{j=1}^{N} g_j \frac{\partial}{\partial g_j}\,\mathbb{B}_{n,m}[g] = m\,\mathbb{B}_{n,m}[g]\ . \tag{15.127}$$

D measures the total degree of homogeneity of the polynomial $\mathbb{B}_{n,m}[g]$ in the variables g_j.

We shall see later some physical appearances of this degree in the theory of gases, as the number of clusters in an N-particle gas (Section 18.2), or as the average size of clusters (Section 18.5.1), which is related to the pressure.

Commentary 256.1. ▶ Notice that

$$\left(x\frac{d}{dx}\right)^n \sum_{r=1}^{\infty} \frac{f_r}{r!} x^r = \sum_{r=1}^{\infty} \frac{f_r}{r!} x^r r^n = \sum_{r=1}^{\infty} \frac{f_r}{r!} x^r \sum_{k=1}^{n} S_n^{(k)} r(r-1)(r-2)...(r-k+1)$$

by using (13.36) for r^n; then,

$$= \sum_{k=1}^{n} S_n^{(k)} \sum_{r=1}^{\infty} \frac{f_r}{r!} x^r r(r-1)(r-2)...(r-k+1) = \sum_{k=1}^{n} S_n{}^{(k)} x^k \left(\frac{d}{dx}\right)^k f(x)\ .$$

Therefore,

$$\left(x\frac{d}{dx}\right)^n f(x) = \sum_{k=1}^{n} S_n{}^{(k)} x^k \left(\frac{d}{dx}\right)^k f(x) \tag{15.128}$$

Examples:

$$\left(x\frac{d}{dx}\right)^n x^k = k^n x^k; \quad \left(x\frac{d}{dx}\right)^n e^x = \sum_{k=0}^{\infty} k^n \frac{x^k}{k!}\ ; \quad \left(\frac{d}{dx}\right)^n e^x = e^x\ ;$$

$$\therefore\ \sum_{k=0}^{\infty} k^n \frac{x^k}{k!} = \sum_{k=1}^{n} S_n{}^{(k)} x^k e^x\ .$$

In the form

$$e^{-x}\sum_{k=0}^{\infty}\frac{k^n}{k!}x^k = \sum_{k=1}^{n} S_n{}^{(k)} x^k ,$$

this gives a relationship of Stirling numbers to the Poisson distribution which, for $x = 1$, reduces to the Dobiński formula (15.50).

Formula (15.128) is related to the purely algebraic normal ordering problem in Quantum Field Theory [109]: it is enough to put $a = \frac{d}{dx}$ for the annihilation operator and $a^\dagger = x$ for the creator to see that

$$\left(a^\dagger a\right)^n = \sum_{k=1}^{n} S_n{}^{(k)}(a^\dagger)^k a^k .$$

We can use Rodrigues' formula for the Hermite polynomials, $H_n(x) = (-)^n e^{x^2}\frac{d^n}{dx^n}e^{-x^2}$ to find

$$e^{x^2}\left(x\frac{d}{dx}\right)^n e^{-x^2} = \sum_{k=1}^{n} S_n{}^{(k)} x^k e^{x^2}\left(\frac{d}{dx}\right)^k e^{-x^2} = \sum_{k=1}^{n} S_n{}^{(k)}(-x)^k H_k(x) .$$

◀

Commentary 257.1. ▶ **Using the circulant shifting** We have said (and, in a few cases, seen) that (15.4) is the most convenient formula to obtain properties of Bell polynomials. In order to get the polynomials themselves, many of the above expressions are used, mainly (15.5) and (15.59). We have seen a practical rule in Comment 239.1. A particular Bell polynomial $\mathbb{B}_{nm}[g]$ depends only on the coefficients $g_1, g_2, \ldots, g_{n-m+1}$. If the task is to calculate it isolately, it is enough to consider the series to order $(n-m+1)$:

$$g(x) = \sum_{j=1}^{n-m+1}\frac{g_j}{j!}x^j . \tag{15.129}$$

The shift circulant operator $\mathbf{U}$ of Chapter 9 leads to some properties of interest in manipulations. As examples,

$$\begin{aligned}(\mathbf{U}^j\mathbb{B})_{nm} &= \mathbb{B}_{n-j,m} ;\\ (\mathbb{B}\;\mathbf{U}^j)_{nm} &= \mathbb{B}_{n,m+j} ;\\ (\mathbf{U}^{n-m}\;\mathbb{B})_{nm} &= \mathbb{B}_{m,m} = g_1^m ;\\ (\mathbf{U}^{n-r}\;\mathbb{B}\;\mathbf{U}^{1-m})_{nm} &= \mathbb{B}_{r1} = g_r .\end{aligned}$$

And we can rewrite (15.123) in the form

$$\frac{\partial}{\partial g_i}\,\mathbb{B}_{nk}[g] = \binom{n}{i}(\mathbf{U}^i\;\mathbb{B}\;\mathbf{U}^{-1})_{nk} . \tag{15.125}$$

◀

Chapter 16

Determinants and traces

The determinant of a matrix $\mathbf{A}$ can be written as a Bell polynomial in the traces of powers of $\mathbf{A}$. This leads to many important classical relationships: between different kinds of symmetric functions, between the coefficients and the roots of a polynomial, between the invariants of a Lie algebra, between the characteristic classes and characters of a differentiable manifold.

16.1 Introduction

The relationships between the determinant, the principal minors and the traces of powers of a given matrix $\mathbf{A}$ are reviewed.

The main clue leading to the relations between the determinant, the principal minors and the traces of powers of a matrix is the formal identity

$$\det\ \mathbf{X} = \exp\{tr[\ln \mathbf{X}]\}\ . \tag{16.1}$$

Suppose $\mathbf{A}$ to be an $N \times N$ matrix. All the relations of interest for us are summed up in the general formula

$$\det\left[\mathbf{I}\ \lambda + z\ \mathbf{A}\right] = \sum_{j=0}^{N} \frac{z^j\ \lambda^{N-j}}{j!} \sum_{m=0}^{j} \mathbb{B}_{jm}\{(-)^{k-1}(k-1)!\ tr(\mathbf{A}^k)\}\ . \tag{16.2}$$

This is obtained by taking $\mathbf{X} = \mathbf{I}\ \lambda + z\ \mathbf{A}$ in (16.1) and using $\ln(1+x) = \sum_{j\geq 1} \frac{(-)^{j-1}}{j} x^j$. The main steps are:

$$\det[\mathbf{I}\ \lambda + z\ \mathbf{A}] = \lambda^N \det[\mathbf{I} + \tfrac{z}{\lambda}\ \mathbf{A}]$$

$$= \lambda^N \exp\left[tr \ln\left(\mathbf{I} + \frac{z}{\lambda}\ \mathbf{A}\right)\right] = \lambda^N \exp\left[\sum_{k\geq 1} \frac{(-)^{k-1}}{k} \left(\frac{z}{\lambda}\right)^k\ tr\mathbf{A}^k\right]$$

$$= \lambda^N \left\{ 1 + \sum_{m=1}^{\infty} \frac{1}{m!} \left[\sum_{k \geq 1} \frac{(-)^{k-1}}{k} \left(\frac{z}{\lambda}\right)^k \; tr\mathbf{A}^k \right]^m \right\}$$

$$= \lambda^N \left\{ 1 + \sum_{m=1}^{\infty} \sum_{j=m}^{\infty} \left(\frac{z}{\lambda}\right)^j \frac{1}{j!} \, \mathbb{B}_{jm}\{(-)^{k-1}(k-1)! \; tr\mathbf{A}^k\} \right\}$$

$$= \lambda^N \left\{ 1 + \sum_{j=1}^{\infty} \left(\frac{z}{\lambda}\right)^j \frac{1}{j!} \sum_{m=1}^{j} \mathbb{B}_{jm}\{(-)^{k-1}(k-1)! \; tr\mathbf{A}^k\} \right\},$$

which is (16.2). This two-variable polynomial, which we shall denote

$$\Delta_{\mathbf{A}}(\lambda, z) = \det \left[\mathbf{I} \, \lambda + z \, \mathbf{A} \right], \tag{16.3}$$

will be ubiquitous in what follows. As it is, $\Delta_{\mathbf{A}}(\lambda, z)$ is the characteristic polynomial of the matrix $(-z \; \mathbf{A})$. The characteristic polynomial of $\mathbf{A}$ will be $\Delta_{\mathbf{A}}(\lambda) = \Delta_{\mathbf{A}}(\lambda, -1)$ and $\Delta_{\mathbf{A}}(1, z)$ will come up as the generating function of the symmetric products of the eigenvalues of $\mathbf{A}$.

Commentary 260.1. ► There are many useful relations involving these determinants, such as $\Delta_{\mathbf{A}}(\lambda, -1) = \lambda^N (1, 1/\lambda)$; $\Delta_{\mathbf{A}}(z, \lambda) = (\det \mathbf{A}) \; \Delta_{\mathbf{A}^{-1}}(\lambda, z)$ when $\mathbf{A}^{-1}$ exists, etc. ◄

A relation of particular beauty comes out when we put $\lambda = 0$, $z = 1$:

$$\det \mathbf{A} = \tfrac{1}{N!} \sum_{m=0}^{N} \mathbb{B}_{Nm}\{(-)^{k-1}(k-1)! \; tr(\mathbf{A}^k)\} \; , \tag{16.4}$$

[see Eq.(16.20) below] whose inverse, as a consequence of the relation between Eqs.(15.80) and (15.81), is

$$tr(\mathbf{A}^n) = \tfrac{(-)^{n-1}}{(n-1)!} \sum_{k=1}^{n} (-)^{k-1}(k-1)! \; \mathbb{B}_{nk}\{j! \; \det{}_j \mathbf{A}\} \; . \tag{16.5}$$

In this expression,

$$\det{}_j \mathbf{A} = \tfrac{1}{j!} \sum_{m=0}^{j} \mathbb{B}_{jm}\{(-)^{k-1}(k-1)! \; tr(\mathbf{A}^k)\} \; . \tag{16.6}$$

The meaning of $\det_j \mathbf{A}$ will become clear in the discussion of symmetric functions: it is the sum of all principal minors of order j of $\mathbf{A}$ (see

Section 16.2, page 265). Such sums of minors have a strong presence in Fredholm theory (see Section 18.6), in which they appear under another form:

$$\det_j \mathbf{A} = \frac{1}{j!} \sum_{s_1=1}^{N} \sum_{s_2=1}^{N} \cdots \sum_{s_j=1}^{N} \begin{vmatrix} A_{s_1 s_1} & A_{s_1 s_2} & \cdots & A_{s_1 s_j} \\ A_{s_2 s_1} & A_{s_2 s_2} & \cdots & A_{s_2 s_j} \\ \cdot & \cdot & \cdots & \cdot \\ \cdot & \cdot & \cdots & \cdot \\ A_{s_j s_1} & A_{s_j s_2} & \cdots & A_{s_j s_j} \end{vmatrix} . \tag{16.7}$$

Notice that $\mathbf{A}$ is an $N \times N$ matrix and that the summations are over all the values $1, 2, \ldots, N$, but the summands are $j \times j$ determinants.

Note. In terms of such minors, Eq.(16.2) takes the form

$$\det [\mathbf{I}\, \lambda + z\, \mathbf{A}] = \sum_{j=0}^{N} z^j\, \lambda^{N-j} \det_j \mathbf{A}\ .$$

Commentary 261.1. ▶ Some trivial relations: take $\mathbf{A}^T$ the transpose of $\mathbf{A}$; $\mathbf{A}^*$, the matrix whose entries are the complex-conjugate of those of $\mathbf{A}$; $\mathbf{A}^\dagger = (\mathbf{A}^T)^*$; and $\mathbf{A}^{-1}$ the inverse of $\mathbf{A}$ (if existent). Then,

$$\Delta_{\mathbf{A}}(\lambda, z) = \Delta_{\mathbf{A}^T}(\lambda, z)\ ; [\Delta_{\mathbf{A}}(\lambda, z)]^* = \Delta_{\mathbf{A}^*}(\lambda^*, z^*)\ ;$$
$$\Delta_{\mathbf{A}^*}(\lambda, z) = \Delta_{\mathbf{A}^\dagger}(\lambda, z); [\Delta_{\mathbf{A}^*}(\lambda, z)]^* = \Delta_{\mathbf{A}^\dagger}(\lambda^*, z^*)\ ;$$
$$\Delta_{\mathbf{A}^{-1}}(\lambda, z) = \Delta_{\mathbf{A}}(z, \lambda)/(\det \mathbf{A})\ .$$

◀

The above relations between determinant and traces of powers may seem trivial, and are rather immediate for diagonalizable (that is, normal) matrices. Nevertheless, as presented above, only the spectrum has been used and they are valid for non-diagonalizable (not normal) matrices as well. This is the case, for example, for the Bell matrices themselves, whose case will be examined in Section 17.2.

16.2 Symmetric functions

If the matrix is diagonal, the relationships previously found reduce to those involving the main kinds of symmetric functions. These include the well-known Viète formulae relating the roots and the coefficients of a polynomial. We begin with the symmetric functions because, besides their intrinsic interest, they help visualizing the more general relations.

A symmetric function in the N variables $x_1, x_2, x_3, \ldots, x_N$ is a polynomial $P(x_1, x_2, x_3, \ldots, x_N)$ which is invariant under any permutation of the

x_k's [110], [96], [111]. That is to say that a symmetric function is a polynomial $F(x)$ giving the same value for any word formed with the alphabet $\mathbf{x}$.

Commentary 262.1. ► We are considering, of course, commutative alphabets: in a product, the letters x_k can exchange positions at will. As in current speech, noncommutative alphabets appear in the study of word groups. Symmetric functions on noncommutative alphabets constitute a recent field of research [112], [113], mainly related to quantum groups. Such studies have called attention to the interest of having closed forms like those given below, even for the classical case. ◄

Of particular interest are the monomial symmetric functions of type

$$F(x) = \sum_{\substack{\{a_1,a_2,a_3,\ldots\},\\ \{i_1,i_2,i_3,\ldots,i_N\}}} x_{i_1}^{a_1}\, x_{i_2}^{a_2}\, x_{i_3}^{a_3} \ldots x_{i_r}^{a_r} ,$$

of which, as will be seen later, the partition functions for ideal and real gases are examples. A few families of such functions are specially important. We shall be interested in three of them. Before addressing them, a *caveat* is in order: the generating functions are given in their truncated version (in the sense of Comment 225.3).

(1) The j-th *elementary* symmetric functions, $e_j[x]$ = sum of all distinct words of j factors that can be formed with the N variables,

$$e_j[\mathbf{x}] = \Sigma^{(N)} x_1 x_2 \ldots x_j .$$

Some examples are:

$$e_1[\mathbf{x}] = x_1 + x_2 + x_3 + \ldots + x_N ;$$

$$e_2[\mathbf{x}] = x_1x_2 + x_1x_3 + \ldots + x_1x_N + x_2x_3 + x_2x_4 + \ldots + \ldots + x_{N-1}x_N ;$$

$$\ldots$$

$$e_N[\mathbf{x}] = x_1x_2x_3 \ldots x_{N-1}x_N .$$

When the x_j's are the eigenvalues of an $N \times N$ matrix $\mathbf{A}$, clearly $e_1[\mathbf{x}] = \operatorname{tr} \mathbf{A}$ and $e_N[\mathbf{x}] = \det \mathbf{A}$. These symmetric functions have the generating function

$$\Sigma[\mathbf{x}; t] = \sum_{n=0}^{N} e_n[\mathbf{x}] t^n = \prod_{j=1}^{N} (1 + x_j t) , \tag{16.8}$$

with $e_0[x] = 1$ by convention. The *fundamental theorem* of the symmetric functions says that the e_k's are algebraically independent and any symmetric function on $\{x_1, x_2, x_3, \ldots, x_N\}$ can be written in terms of the e_k's. In this sense, $\{e_k\}$ constitutes a basis. For $N \to \infty$, this means that, up to an overall constant, any formal series $f(t)$ can be seen as the generating function $\Sigma[\mathbf{x}; t]$ with some convenient choice of alphabet

$$\mathbf{x} = \{x_1, x_2, x_3, \ldots\} .$$

(2) The *complete homogeneous* symmetric functions $h_k[\mathbf{x}]$, given by

$$H\,[\mathbf{x}; t] = \sum_{j=0}^{N} h_j[\mathbf{x}] t^j = \prod_{j=1}^{N} (1 - x_j t)^{-1} = \frac{1}{\Sigma[\mathbf{x}; -t]} . \tag{16.9}$$

By convention, $h_0[x] = 1$. The above h_j's differ from the e_j's because they include repetitions. For $N = 3$, the first examples are

$$h_1[x] = x_1 + x_2 + x_3 ;$$

$$h_2[x] = x_1x_2 + x_1x_3 + x_2x_3 + x_1^2 + x_2^2 + x_3^2 ;$$

$$h_3[x] =$$

$$x_1x_2x_3 + x_1^2x_2 + x_1^2x_3 + x_2^2x_1 + x_2^2x_3 + x_3^2x_2 + x_3^2x_1 + x_1^3 + x_2^3 + x_3^3 .$$

Commentary 263.1. ► If we recall that the ideal grand canonical partition functions (see Section 18.3) are written, in terms of the fugacity $z = e^{\beta\mu}$, as $\Xi^{(F)} = \prod_k (1 + z e^{-\beta \epsilon_k})$ for fermions and $\Xi^{(B)}(z) = \prod_k (1 - z e^{-\beta \epsilon_k})^{-1}$ for bosons, we recognize, with $x_k = e^{-\beta \epsilon_k}$, that $\Xi^{(F)}(z) = \Sigma[\mathbf{x}; z]$ and $\Xi^{(B)}(z) = H[\mathbf{x}; z]$. The above relation between Σ and H turns up as the well-known property $\Xi^{(B)}(z) = \frac{1}{\Xi^{(F)}(-z)}$. See Comment 335.1. ◄

(3) The k-th *power-sum* symmetric functions,

$$p_k[\mathbf{x}] = \sum_{j=1}^{N} x_j^k = x_1^k + x_2^k + x_3^k + \ldots + x_N^k . \tag{16.10}$$

When the x_j's are the eigenvalues of a matrix $\mathbf{A}$, $p_k[\mathbf{x}]$ is clearly the trace of $\mathbf{A}^k$. We shall adopt the conventions $p_0[\mathbf{x}] = N$ and $p_{(k>N)}[\mathbf{x}] = 0$. It is easily found that the $p_k[\mathbf{x}]$'s have the generating function

$$P[\mathbf{x}; z] = \sum_{j=1}^{N} p_j[\mathbf{x}] z^{j-1} = \sum_{j=1}^{N} x_j (1 - x_j z)^{-1} . \tag{16.11}$$

This is the same as

$$P[\mathbf{x}; z] = \frac{d}{dz} \ln H[\mathbf{x}; z] = - \frac{d}{dz} \ln \Sigma[\mathbf{x}; -z] . \tag{16.12}$$

There is another generating function,

$$p[\mathbf{x},z]=\sum_{j=1}^{N}e^{zx_j}=\sum_{n=0}^{N}\frac{1}{n!}\sum_{j=1}^{N}(x_j\;z)^n=\sum_{n=0}^{N}\frac{1}{n!}\;p_n\;[\mathbf{x}]\;z^n\;, \qquad (16.13)$$

which is related to $[N+zP[\mathbf{x};z]]$ by a Borel transform (see Comment 225.5).

Commentary 264.1. ▶ A few simple relations: (i) $\frac{d}{dx_j}p_k = kx_j^{k-1}$; (ii) $\sum_{j=1}^{N} x_j \frac{d}{dx_j}p_k = kp_k$; (iii) $\frac{1}{m!}(p_k)^m=\sum_{r=m}^{Nm}\frac{1}{r!}\mathbb{B}_{rm}\{j!x_j^k\}$. ◀

Commentary 264.2. ▶ In the same statistical-mechanical vein of Comment 263.1: $P[x;z]$ corresponds to $N(z)/z$. ◀

A first relationship between the e_k's and the p_k's can be found by the following steps:

$$\sum_{n=0}^{N}e_n[\mathbf{x}]t^n=\prod_{i=1}^{N}(1+x_it)=e^{\{\sum_{i=1}^{N}\ln[1+tx_i]\}}=\sum_{j=0}^{\infty}\frac{1}{j!}\left[\sum_{i=1}^{N}\ln(1+tx_i)\right]^j$$

$$=\sum_{j=0}^{\infty}\frac{1}{j!}\left[\sum_{k=1}^{\infty}\frac{(-)^{k-1}t^k}{k}\sum_{i=1}^{N}x_i^k\right]^j=\sum_{j=0}^{\infty}\frac{1}{j!}\left[\sum_{k=1}^{\infty}\frac{(-)^{k-1}t^k}{k!}(k-1)!\;p_k[\mathbf{x}]\right]^j$$

$$=\sum_{n=0}^{N}\frac{t^n}{n!}\sum_{j=0}^{n}\mathbb{B}_{nj}\{(-)^{k-1}(k-1)!\;p_k[\mathbf{x}]\}\;.$$

Consequently,

$$e_n[\mathbf{x}]=\frac{1}{n!}\sum_{j=1}^{n}\mathbb{B}_{nj}\{(-)^{k-1}(k-1)!\;p_k[\mathbf{x}]\}\;. \qquad (16.14)$$

Note. Many results on symmetric functions can be obtained from properties of Bell polynomials. Two examples from the last formula above:

- $e_n[\mathbf{1}]=e_n[\{1,1,1,...,1\}]=\binom{N}{n}$ follows from

$$e_n[1,1,1,...,1]=\frac{1}{n!}\sum_{j=1}^{n}\mathbb{B}_{nj}\{(-)^{k-1}(k-1)!\;p_k[\mathbf{1}]\}$$

$$=\frac{1}{n!}\sum_{j=1}^{n}\mathbb{B}_{nj}\{(-)^{k-1}(k-1)!\,N\}=\frac{1}{n!}\sum_{j=1}^{n}N^j s_n^{(j)}=\binom{N}{n}\;,$$

  where we have used Eqs.(15.34) and (15.22).
- $e_j[a\mathbf{x}]=a^j\,e_j[\mathbf{x}]$ follows from Eq.(15.23).

The inverse relation is found by noticing that

$$\ln \Sigma[\mathbf{x}, t] = \ln\left(1 + \sum_{j=1}^{N} e_j[\mathbf{x}]t^j\right) = \sum_{k=1}^{\infty} \frac{(-)^{k-1}(k-1)!}{k!}\left[\sum_{j=1}^{N} e_j[\mathbf{x}]t^j\right]^k$$

$$= \sum_{k=1}^{\infty}(-)^{k-1}(k-1)! \sum_{n=k}^{\infty} \frac{t^n}{n!}\, \mathbb{B}_{nk}\{j!e_j[\mathbf{x}]\}$$

$$= \sum_{n=1}^{\infty} \frac{t^n}{n!} \sum_{k=1}^{n}(-)^{k-1}(k-1)!\, \mathbb{B}_{nk}\{j!e_j[\mathbf{x}]\} ;$$

but then

$$P[\mathbf{x}, t] = -\frac{d}{dt} \ln \Sigma[\mathbf{x}, -t]$$

$$= \sum_{n=1}^{\infty} \frac{(-)^{n-1}t^{n-1}}{(n-1)!} \sum_{k=1}^{n}(-)^{k-1}(k-1)!\, \mathbb{B}_{nk}\{j!e_j[\mathbf{x}]\} ,$$

and consequently

$$p_k[\mathbf{x}] = \frac{(-)^{k-1}}{(k-1)!} \sum_{j=1}^{k}(-)^{j-1}(j-1)!\, \mathbb{B}_{kj}\{n!e_n[\mathbf{x}]\} \;\; (\text{for } \; k \geq 1) . \qquad (16.15)$$

If we take

$$\ln \Sigma[\mathbf{x}, t] = \ln\{1 + \sum_{j=1}^{N} e_j[\mathbf{x}]t^j\}$$

$$= \sum_{k=1}^{\infty} \frac{t^k}{k!} \sum_{j=1}^{k}(-)^{j-1}(j-1)!\, \mathbb{B}_{kj}\{n!e_n[\mathbf{x}]\} = -\sum_{k=1}^{\infty} \frac{(-t)^k}{k} p_k[\mathbf{x}] ,$$

we find (16.12) again, though now in the form

$$\ln \Sigma[\mathbf{x}, t] \doteq \int_0^t dt P[\mathbf{x}, -t] .$$

Suppose for a moment that the alphabet $\mathbf{x}$ is formed by the eigenvalues of a diagonal matrix $\mathbf{A}$. Then clearly $p_r[\mathbf{x}] = tr\mathbf{A}^r$ and $e_k[\mathbf{x}] =$ sum of all the k-order principal minors of $\mathbf{A}$. This is related to what was said about minors below Eq.(16.5).

Other relationships can be obtained by manipulations analogous to those above. Some examples are:

$$\ln H[\mathbf{x},t] = \int_0^t dt P[\mathbf{x},t] \ ;$$

$$h_n[\mathbf{x}] = \frac{1}{n!}\sum_{j=1}^{n} \mathbb{B}_{nj}\{(k-1)!p_k[\mathbf{x}]\} \ ;$$

$$nh_n[\mathbf{x}] = \sum_{j=1}^{n} p_j h_{n-j} \ .$$

Using the pseudo-recursion (15.107) in (16.14), we find some important relations between the symmetric functions:

$$e_n = \frac{1}{n}\sum_{j=0}^{n-1} (-)^{n-j-1} p_{n-j} e_j = \frac{1}{n}\sum_{k=1}^{n} (-)^{k-1} p_k e_{n-k} \ , \qquad (16.16)$$

which is the same as $p_n = \sum_{k=0}^{n-1} (-)^{n-k-1} p_k e_{n-k} = \sum_{r=1}^{n} (-)^{r-1} p_{n-r} e_r$, and can be put into the form

$$\sum_{j=0}^{n} (-)^{n-j} p_{n-j} e_j = \sum_{i=1}^{n}\sum_{j=0}^{n} (-)^{n-j} x_i^{n-j} e_j = 0 \ . \qquad (16.17)$$

Commentary 266.1. ▶ In the case of the root-alphabet of a polynomial, this is the so-called "Newton formula" and has a "finer" version, as it holds term-by-term. In effect, each term $\sum_{j=0}^{n}(-)^{n-j} x_i^{n-j} e_j = 0$ says simply that x_i is a root. In Section 16.4, these expressions will appear related to the characteristic polynomial of a matrix with eigenvalues x_i. ◀

Commentary 266.2. ▶ The relations above hold for a fixed alphabet of N members. We can also find relations between distinct alphabets, with successive values of N. Adjoining a letter x_{N+1} to the alphabet $\mathbf{x}$ (with, by convention, $e_{-1} = 0$ and $e_{N+1}^{(N)} = 0$), we find from the generating functions that

$$e_k^{(N+1)} = e_k^{(N)} + x_{N+1}\, e_{k-1}^{(N)} \ .$$

This means also that $e_k^{(N)}$ can be obtained from $e_k^{(N+1)}$ by simply putting $x_{N+1} = 0$. ◀

An interesting relation is

$$\sum_{m=1}^{N} \mathbb{B}_{Nm}\{(-)^{j-1}(j-1)!a_j\} = \begin{vmatrix} a_1 & 1 & 0 & 0 & . & . & 0 \\ a_2 & a_1 & 2 & 0 & . & . & 0 \\ a_3 & a_2 & a_1 & 3 & . & . & 0 \\ . & . & . & . & . & . & 0 \\ . & . & . & . & . & a_1 & \text{N-1} \\ a_N & a_{N-1} & . & . & . & . & a_1 \end{vmatrix} \ , \qquad (16.18)$$

which can be checked from (16.2). It is actually an identity, and can be used, together with $g_j = (-)^{j-1}(j-1)!a_j$, to recast many of the previous formulas, down from (15.4), into determinant form. The general aspect would be

$$\sum_{m=1}^{N} \mathbb{B}_{Nm}\{g_j\} = \begin{vmatrix} g_1 & 1 & 0 & 0 & . & . & 0 \\ -g_2 & g_1 & 2 & 0 & . & . & 0 \\ g_3/2! & -g_2 & g_1 & 3 & . & . & 0 \\ . & . & . & . & . & . & . \\ . & . & . & . & . & g_1 & \text{N-1} \\ (-)^{N-1}\frac{g_N}{(N-1)!} & (-)^N\frac{g_{N-1}}{(N-2)!} & . & . & . & . & g_1 \end{vmatrix} . \quad (16.19)$$

The relation (16.4), for example, takes on the curious form

$$\det \mathbf{A} = \frac{1}{N!} \begin{vmatrix} tr\mathbf{A} & 1 & 0 & 0 & . & . & 0 \\ tr\mathbf{A}^2 & tr\mathbf{A} & 2 & 0 & . & . & 0 \\ tr\mathbf{A}^3 & tr\mathbf{A}^2 & tr\mathbf{A} & 3 & . & . & 0 \\ . & . & . & . & . & . & 0 \\ . & . & . & . & . & tr\mathbf{A} & \text{N-1} \\ tr\mathbf{A}^N & tr\mathbf{A}^{N-1} & . & . & . & tr\mathbf{A}^2 & tr\mathbf{A} \end{vmatrix} . \quad (16.20)$$

Another particular case involves the symmetric functions:

$$n!e_n = \begin{vmatrix} p_1 & 1 & 0 & 0 & . & . & 0 \\ p_2 & p_1 & 2 & 0 & . & . & 0 \\ p_3 & p_2 & p_1 & 3 & . & . & 0 \\ . & . & . & . & . & . & . \\ . & . & . & . & . & p_1 & \text{n-1} \\ p_n & p_{n-1} & . & . & . & . & p_1 \end{vmatrix} . \quad (16.21)$$

This expression appears in Statistical Mechanics, with e_n for the canonical partition function and p_n related to the clusters integrals (see Eq.(18.54) below).

We have said that, by the fundamental theorem, the elementary functions e_k constitute a basis for the symmetric functions. The power-sums p_k constitute another basis, as the Jacobian between them is generically nonvanishing.

An *alternating function* on the alphabet $\mathbf{x} = \{x_1, x_2, x_3, \ldots, x_N\}$ is a polynomial $F(x_1, x_2, x_3, \ldots, x_N)$ which is transformed into $-F(x_1, x_2, x_3, \ldots, x_N)$ by any odd permutation of the x_k's. The difference product

$$p(\mathbf{x}) = \prod_{1 \leq i < j \leq N} (x_j - x_i)$$

is called the *simplest alternating polynomial* of the alphabet $\mathbf{x}$. It vanishes, of course, when at least two of the letters are equal. Its square $D(\mathbf{x}) = [p(\mathbf{x})]^2$ is a symmetric function and can consequently be expressed in terms of the e_k's. An equivalent form of $p(\mathbf{x})$ is the Vandermonde determinant

$$p(\mathbf{x}) = \begin{vmatrix} 1 & 1 & 1 & 1\,. & 1 \\ x_1 & x_2 & x_3 & .\,. & x_N \\ x_1^2 & x_2^2 & x_3^2 & .\,. & x_N^2 \\ . & . & . & .\,. & . \\ x_1^{N-1} & x_2^{N-1} & x_3^{N-1} & .\,. & x_N^{N-1} \end{vmatrix} . \tag{16.22}$$

The determinant of an odd-dimensional skew-symmetric matrix vanishes. Given a skew-symmetric $(2N) \times (2N)$ matrix $\mathbf{A}$, its determinant has the form of $D(\mathbf{x})$. In this context, $p(\mathbf{x})$ is called the *Pfaffian* of $\mathbf{A}$, $p(\mathbf{x}) = Pf(\mathbf{A})$, $\det \mathbf{A} = [Pf(\mathbf{A})]^2$.

16.3 Polynomials

The usual lore on polynomial coefficients and their relations with the roots is translated into our language.

It is clear from (16.8) that the roots of $\Sigma[\mathbf{x}]$ are $r_j = -1/x_j$, which suggests that here, as in many cases, it is convenient to consider, simultaneously with an alphabet $\mathbf{x}$, its *reciprocal* alphabet $\mathbf{x}^* = \{x_1^*, x_2^*, x_3^*, \ldots x_N^*\}$, where each $x_j^* = -1/x_j$. Thus, the root-alphabet of $\Sigma[\mathbf{x}]$ is $\mathbf{r} = \mathbf{x}^*$. Notice that the operation of taking the reciprocal is an involution, $\mathbf{x}^{**} = \mathbf{x}$. The elementary symmetric functions of these alphabets are related by

$$(-)^j \, e_{N-j}\,[\mathbf{x}] = e_N\,[\mathbf{x}]\; e_j\,[\mathbf{x}^*]\;. \tag{16.23}$$

For $\mathbf{x}$ the eigenvalue-alphabet of an invertible matrix $\mathbf{A}$, this expression generalizes the identity $\det(\mathbf{A}^{-1}) = (\det \mathbf{A})^{-1}$, which is the case $j = N$. An immediate consequence is

$$\Sigma[\mathbf{x}; t] = e_N[\mathbf{x}]\; t^N\; \Sigma[\mathbf{x}^*; t^{-1}]\;. \tag{16.24}$$

Simple comparison shows that

$$\Sigma[\mathbf{x}; t] = \frac{1}{e_N(\mathbf{r})} \prod_{j=1}^{N} (r_j - t)\;. \tag{16.25}$$

A general polynomial of the form

$$P(z) = a_N z^N + a_{N-1} z^{N-1} + \cdots + a_1 z^1 + a_0 \tag{16.26}$$

can be written in many equivalent ways in terms of its zeros:

$$P(z) = a_N \prod_{j=1}^{N}(z - r_j) = (-)^N a_N\ e_N[\mathbf{r}]\ \Sigma[\mathbf{r}^*, z] \tag{16.27}$$

$$= (-)^N a_N e_N[\mathbf{r}] \sum_{j=0}^{N} e_j[\mathbf{r}^*] z^j = a_N \sum_{j=0}^{N} (-)^{N-j} e_{N-j}[\mathbf{r}] z^j \tag{16.28}$$

$$= (-)^N a_N\ e_N[\mathbf{r}] \prod_{j=1}^{N}(1 - z/r_j) = a_0 \prod_{j=1}^{N}(1 - z/r_j) \tag{16.29}$$

$$= P(0) \prod_{j=1}^{N}(1 - z/r_j)\ . \tag{16.30}$$

The last is Weierstrass' formula. We have thus

$$a_j = (-)^{N-j} a_N\ e_{N-j}\ [\mathbf{r}]\ ; \tag{16.31}$$

$$a_0 = P(0) = (-)^N a_N\ e_N[\mathbf{r}]\ . \tag{16.32}$$

Viète's formulas relating the coefficients of a polynomial $P(z)$ to the powers of its roots are, consequently,

$$a_j = a_N \frac{(-)^{N-j}}{(N-j)!} \sum_{r=0}^{N-j} \mathbb{B}_{N-j,r}\{(-)^{k-1}(k-1)!\ p_k[\mathbf{r}]\}\ . \tag{16.33}$$

Commentary 269.1. ► This means that the above list of expressions for a polynomial can be joined by still another,

$$P(z) = a_N \sum_{l=0}^{N} \frac{z^{N-l}}{l!} \sum_{r=0}^{l} \mathbb{B}_{l,r}\{-(k-1)!\ p_k[\mathbf{r}]\}\ . \tag{16.34}$$

◄

Commentary 269.2. ► Using the inversion formula (15.81), we have (for $j \geq 1$)

$$p_k[\mathbf{r}] = \frac{(-)^{k-1}}{(k-1)!} \sum_{j=1}^{k} (-)^{j-1}(j-1)!\ \mathbb{B}_{kj} \left\{(-)^j j!\ \frac{a_{N-j}}{a_N}\right\}\ . \tag{16.35}$$

◄

Commentary 269.3. ► Notice that $a_0 = P(0) = 1$ implies $a_N = \frac{(-)^N}{e_N[\mathbf{r}]} = e_N[\mathbf{r}^*]$. ◄

Commentary 269.4. ► The representation

$$\delta_{ab} = \tfrac{1}{N} \sum_{k=1}^{N} e^{i\frac{2\pi}{N}k(a-b)}$$

of the Kronecker delta comes from the simple result that the sum of the elementary roots $\{e^{i\frac{2\pi}{N}k}\}$ of the unit vanishes, as does the sum of a fixed power of them. Indeed, such zeros are those of $P(z) = z^N - 1$, whose only nonvanishing coefficients are $a_N = 1$ and $a_0 = -1$. Then, as $a_j = (-)^{N-j}\, e_{N-j}\,[\mathbf{r}]$, only $e_0\,[\mathbf{r}]$ and $e_N\,[\mathbf{r}]$ are $\neq 0$. In particular, $e_1\,[\mathbf{r}] = -a_{N-1} = 0$ is just that sum. Then all $p_j\,[\mathbf{r}] = \sum_{k=1}^{N} e^{i\frac{2\pi}{N}kj}$ vanish accordingly by (16.15), except $p_N[\mathbf{x}] = N$. ◄

The discriminant $\Delta(P)$ of a polynomial equation is the simplest alternating polynomial (16.22) in the roots, the Vandermonde determinant

$$\Delta(P) = \det(r_i^{N-j}) = (-)^N p[\mathbf{r}] = \begin{vmatrix} r_1^{N-1} & r_1^{N-2} & r_1^{N-3} & \dots & r_1^0 \\ r_2^{N-1} & r_2^{N-2} & r_2^{N-3} & \dots & r_2^0 \\ . & . & . & \dots & 1 \\ . & . & . & \dots & 1 \\ r_N^{N-1} & r_N^{N-2} & r_N^{N-3} & \dots & r_N^0 \end{vmatrix} . \quad (16.36)$$

Commentary 270.1. ► **Padé approximants** Take two polynomials $P_N(z) = \sum_{j=0}^{N} a_j z^j$ and $Q_R(z) = \sum_{j=0}^{R} b_j z^j$. We shall suppose that at least some of the b_j's are $\neq 0$. In this case they define the rational function $F_{N,R}(z) = \frac{P_N(z)}{Q_R(z)}$. If $b_0 \neq 0$, we can always choose the coefficients so as to have $b_0 = 1$. In this case,

$$\begin{aligned} \frac{1}{Q_R(z)} &= \frac{1}{1+\sum_{j=1}^{R} b_j z^j} = 1 + \sum_{k=1}^{\infty}\left(\sum_{j=1}^{R} b_j\ z^j\right)^k \\ &= 1 + \sum_{k=1}^{\infty} k! \sum_{n=k}^{\infty} \frac{z^n}{n!}\ \mathbb{B}_{nk}\{j!\ b_j\ \theta(R-j)\} \\ &= 1 + \sum_{n=1}^{\infty} \frac{z^n}{n!} \sum_{k=1}^{n} k!\ \mathbb{B}_{nk}\{j!\ b_j\ \theta(R-j)\} \\ &= \sum_{n=0}^{\infty} \frac{z^n}{n!} \sum_{k=0}^{n} k!\ \mathbb{B}_{nk}\{j!\ b_j\ \theta(R-j)\}\ . \end{aligned}$$

We have used the Heaviside step function:

$$\begin{cases} \theta(x) = 1 \ \text{ when } \ x \geq 0 \\ \theta(x) = 0 \ \text{ when } \ x < 0\,. \end{cases}$$

It follows from the expression above that

$$F_{N,R}(z) = (\sum_{j=0}^{N} a_j z^j) \sum_{n=0}^{\infty} \frac{z^n}{n!} \sum_{k=0}^{n} k!\ \mathbb{B}_{nk}\{j! b_j\ \theta(R-j)\}\ ,$$

or

$$F_{N,R}(z) = \sum_{r=0}^{\infty} \sum_{n=r-N}^{r} \frac{z^r}{n!}\ a_{r-n} \sum_{k=0}^{n} k!\ \mathbb{B}_{nk}(1!\ b_1, 2!\ b_2, \cdots, R!\ b_R, 0, 0, \cdots)\ , \quad (16.37)$$

an expression for the [N,R] Padé approximant [114]. ◄

16.4 Characteristic polynomials

The coefficients of the characteristic polynomial of **A** constitute a compact of all its invariants under similarity transformations.

If **A** is an $N \times N$ matrix, its characteristic polynomial has the form

$$\Delta_{\mathbf{A}}(\lambda) = \det[\lambda\mathbf{I} - \mathbf{A}] = \textstyle\sum_{k=0}^{N} a_k \lambda^k \,. \tag{16.38}$$

If λ_1, λ_2, λ_3, ..., λ_N are the eigenvalues of **A**, then

$$\Delta_{\mathbf{A}}(\lambda) = (\lambda - \lambda_1)(\lambda - \lambda_2)(\lambda - \lambda_3)...(\lambda - \lambda_N) = \textstyle\sum_{j=0}^{N} (-)^j \lambda^{N-j} e_j[\boldsymbol{\lambda}] \,. \tag{16.39}$$

Comparison with the various expressions (16.27)–(16.29) for a polynomial shows that $\Delta_{\mathbf{A}}(\lambda) = P(\lambda)$ with $a_N = 1$. Given the eigenvalue-alphabet $\boldsymbol{\lambda} = \text{Sp } \mathbf{A} = \{\lambda_1, \lambda_2, \lambda_3, \ldots, \lambda_N\}$, its reciprocal will be $\boldsymbol{\lambda}^* = \{-\lambda_1^{-1}, -\lambda_2^{-1}, -\lambda_3^{-1}, \ldots, -\lambda_N^{-1}\}$. We have trivially that $tr(\mathbf{A}^k) = p_k[\boldsymbol{\lambda}] = \sum_{j=0}^{N} \lambda_j^k$ and $\det \mathbf{A} = e_N[\boldsymbol{\lambda}]$.

Also the characteristic polynomial will have many faces:

$$\begin{aligned}
\Delta_{\mathbf{A}}(\lambda) = \det[\lambda\mathbf{I} - \mathbf{A}] &= \prod_{j=1}^{N} (\lambda - \lambda_j) \\
&= \sum_{j=0}^{N} \lambda^{N-j} \frac{(-)^j}{j!} \sum_{m=0}^{j} \mathbb{B}_{jm}\{(-)^{k-1}(k-1)!\; tr(\mathbf{A}^k)\} \\
&= \sum_{j=0}^{N} \lambda^{N-j} \frac{(-)^j}{j!} \sum_{m=0}^{j} \mathbb{B}_{jm}\{(-)^{k-1}(k-1)!\; p_k[\boldsymbol{\lambda}]\} \\
&= \sum_{j=0}^{N} \lambda^{N-j} (-)^j \; e_j[\boldsymbol{\lambda}] \,.
\end{aligned} \tag{16.40}$$

Commentary 271.1. ► Notice that, by taking the trace of the secular equation $\Delta_{\mathbf{A}}(\mathbf{A}) = 0$, we obtain (16.17). ◄

Commentary 271.2. ► Some other determinants are related to other generating functions. For example, the case

$$\det[\mathbf{I} + z\mathbf{A}] = \sum_{j=0}^{N} \frac{z^j}{j!} \sum_{m=0}^{j} \mathbb{B}_{jm} \; \{(-)^{k-1}(k-1)!\; p_k[\boldsymbol{\lambda}]\} = \sum_{j=0}^{N} z^j \; e_j[\boldsymbol{\lambda}] \tag{16.41}$$

will be of interest in Statistical Mechanics. The grand canonical partition function, and its related Fredholm determinant, will be of this form [see Eq.(18.166) and the discussion around it]. Notice $\det[\mathbf{I} + z\mathbf{A}] = \sum_{j=0}^{N} z^j \; e_j[\boldsymbol{\lambda}] = \Sigma[\boldsymbol{\lambda}; z] = \Delta_{\mathbf{A}}(1, z)$. The sum of all products of eigenvalues will be $\sum_{j=1}^{N} e_j[\boldsymbol{\lambda}] = \det[\mathbf{I} + \mathbf{A}] - 1 = \Sigma[\boldsymbol{\lambda}; 1] - 1 = \Delta_{\mathbf{A}}(1, 1) - 1$. ◄

When $\mathbf{A}^{-1}$ exists,

$$\det[\lambda\mathbf{I} + \mathbf{A}^{-1}] = \sum_{j=0}^{N} \lambda^{N-j} \frac{(-)^j}{j!} \sum_{m=0}^{j} \mathbb{B}_{jm}\{(-)^{k-1}(k-1)!\ tr([-\mathbf{A}^{-1}]^k)\}$$

$$= \sum_{j=0}^{N} \lambda^{N-j} \frac{(-)^j}{j!} \sum_{m=0}^{j} \mathbb{B}_{jm}\{(-)^{k-1}(k-1)!\ p_k[\boldsymbol{\lambda}^*]\} = \sum_{j=0}^{N} \lambda^{N-j}(-)^j e_j[\boldsymbol{\lambda}^*]\ . \tag{16.42}$$

Commentary 272.1. ► When $\lambda = 0$, $\Delta_{\mathbf{A}}(0) = \det[-\mathbf{A}]$, so that det $[-\mathbf{A}] = (-)^N$ det $\mathbf{A}$ is the coefficient of λ^0 in the characteristic polynomial: $a_0 = (-)^N$ det $\mathbf{A}$. In the present context, Eq.(16.4) states simply the well-known fact that the independent coefficient of the characteristic polynomial of $\mathbf{A}$ is, up to a sign, the determinant of $\mathbf{A}$. ◄

Let us apply the Cayley–Hamilton theorem, according to which a matrix satisfies formally its own characteristic equation. For an $N \times N$ matrix $\mathbf{A}$,

$$\Delta_{\mathbf{A}}(\mathbf{A}) = \sum_{j=0}^{N} \mathbf{A}^{N-j}\ (-)^j\ e_j[\boldsymbol{\lambda}] = 0\ . \tag{16.43}$$

This has important consequences:

(1) As $e_0 = 1$, we can isolate the highest power to obtain it in terms of the lower ones:

$$\mathbf{A}^N = \sum_{j=1}^{N} (-)^{j-1}\ \mathbf{A}^{N-j}\ e_j[\boldsymbol{\lambda}]\ . \tag{16.44}$$

All the larger-than-N powers can be obtained from the lower-than-N powers.

(2) Isolating the $\mathbf{A}^0 = \mathbf{I}$ term, we get

$$\mathbf{I}\ \det\mathbf{A} = \sum_{j=0}^{N-1} (-)^{N-j-1}\ \mathbf{A}^{N-j}\ e_j[\boldsymbol{\lambda}]\ . \tag{16.45}$$

(3) If $e_N[\boldsymbol{\lambda}] = \det\mathbf{A}$ is $\neq 0$, $\mathbf{A}^{-1}$ exists. Multiplying by $\mathbf{A}^{-1}$ the last expression above, we find the inversion formula

$$\mathbf{A}^{-1} = \frac{1}{\det\mathbf{A}} \sum_{j=0}^{N-1} (-\,\mathbf{A})^{N-j-1}\ e_j[\boldsymbol{\lambda}]\ . \tag{16.46}$$

This is a detailed version of Eq.(1.4): the inverse to an $N \times N$ matrix $\mathbf{A}$ is written in terms of the powers of $\mathbf{A}$, as an $(N-1)$-th order polynomial.

(4) Comparing with

$$(A^{-1})_{rs} = \frac{1}{\det \mathbf{A}} \left[\text{cof } A_{rs}\right]^T = \frac{1}{\det \mathbf{A}} \left[(-)^{r+s} \text{min } A_{rs}\right]^T , \tag{16.47}$$

we find expressions for the (rs)-cofactor and minor in terms of the (rs) entries of powers of $\mathbf{A}$:

$$(\text{cof } A)_{rs} = (-)^{r+s} (\text{min } A)_{rs} = \textstyle\sum_{j=0}^{N-1} (-)^{N+j-1} \left[A^{N-j-1}\right]_{sr} e_j[\boldsymbol{\lambda}] . \tag{16.48}$$

(5) Applying the inversion formula to a Bell matrix,

$$\begin{aligned}
\mathbb{B}^{-1}[g] &= \mathbb{B}[g^{<-1>}] \\
&= g_1^{-N(N+1)/2} \sum_{j=0}^{N-1} (-)^{N-j-1} \, \mathbb{B}^{N-j-1}[g] \; e_j[\mathbf{g} = \{g_1, g_1^2, ..., g_1^N\}] \\
&= g_1^{-N(N+1)/2} \sum_{j=0}^{N-1} (-)^j \, \mathbb{B}^j[g] \; e_{N-1-j}[\mathbf{g}] \\
&= \sum_{j=0}^{N-1} e_{j+1}[\mathbf{g}^*] \; \mathbb{B}^j[g] .
\end{aligned} \tag{16.49}$$

These expressions can be seen as alternatives to those given in Section 15.4.3.

Note. Taking traces in both sides of Eq.(16.45) gives

$$\det \mathbf{A} = \tfrac{1}{N} \sum_{j=0}^{N-1} (-)^{N+j-1} \, tr(\mathbf{A}^{N-j}) \; e_j[\boldsymbol{\lambda}] ,$$

which is an example of the property (16.17), here for an eigenvalue-alphabet. By Laplace's development formula, the determinant is the sum of the entries of a chosen row (or column) multiplied by the respective cofactors. Taking the first column, for example,

$$\det \mathbf{A} = \sum_{s=1}^{N} A_{1s} \, (\text{cof } A)_{1s} = \sum_{s=1}^{N} (-)^{s+1} A_{1s} \, (\text{min } A)_{rs} .$$

Multiplying Eq.(16.48) by A_{rs} and summing on both indices we get N times that value, so that

$$\det \mathbf{A} = \tfrac{1}{N} \sum_{r,s=1}^{N} A_{rs} \, (\text{cof } A)_{rs} = \tfrac{1}{N} \sum_{r,s=1}^{N} \sum_{j=0}^{N-1} (-)^{N+j-1} A_{rs} \left[A^{N-j-1}\right]_{sr} e_j[\boldsymbol{\lambda}],$$

just the first formula above.

Commentary 274.1. ► The following applications of the Cayley–Hamilton theorem (16.43) to a stochastic matrix **M** can be seen as exercises:

- take the entries of (16.43), and sum them up:

$$\Delta_{\mathbf{M}}(\mathbf{M}) = \sum_{j=0}^{N} \mathbf{M}^{N-j}\ (-)^j\ e_j[\boldsymbol{\lambda}] = 0 \Longrightarrow \sum_{j=0}^{N} \mathbf{M}_{rs}^{N-j}\ (-)^j\ e_j[\boldsymbol{\lambda}] = 0 \qquad (16.50)$$

$$\sum_{r=1}^{N}\sum_{j=0}^{N} \mathbf{M}_{rs}^{N-j}\ (-)^j\ e_j[\boldsymbol{\lambda}] = \sum_{j=0}^{N}(-)^j\ e_j[\boldsymbol{\lambda}] = 0\ . \qquad (16.51)$$

  Comparison with the characteristic equation $\sum_{j=0}^{N} \lambda^{N-j}(-)^j\ e_j[\boldsymbol{\lambda}] = 0$ shows that $\lambda = 1$ is always a root. Of course, the Perron-Frobenius theorems (page 42) say much more than that.

- past from future if $\det \mathbf{M} \neq 0$: apply both sides of (16.45), which is

$$\mathbf{I}\ \det \mathbf{M} = \sum_{j=0}^{N-1} (-)^{N+j-1}\ e_j[\boldsymbol{\lambda}]\ \mathbf{M}^{N-j}\ , \qquad (16.52)$$

  to an initial distribution $\mathbf{p}^{(0)}$:

$$\det \mathbf{M}\ \mathbf{p}^{(0)} = \sum_{j=0}^{N-1} (-)^{N+j-1}\ e_j[\boldsymbol{\lambda}]\ \mathbf{p}^{(N-j)}\ . \qquad (16.53)$$

  We see that, if $\det \mathbf{M} \neq 0$, the initial distribution is fixed by the knowledge of the next N steps. In that case, of course, $\mathbf{M}^{-1}$ exists and past distributions can be recovered from later ones step-by-step.

- from (16.44), the k-th component of $\mathbf{p}^{(N)}$ is a combination of the k-th components of the previous distributions:

$$\mathbf{M}^N \mathbf{p}^{(0)} = \mathbf{p}^{(N)} = \sum_{j=1}^{N}(-)^{j-1}\ e_j[\boldsymbol{\lambda}]\ \mathbf{p}^{(N-j)} \qquad (16.54)$$

$$\therefore\ \mathbf{p}_k^{(N)} = \sum_{j=1}^{N}(-)^{j-1}\ e_j[\boldsymbol{\lambda}]\ \mathbf{p}_k^{(N-j)}\ . \qquad (16.55)$$

◄

Let us say a few words on the invariants of a matrix, contenting ourselves for the time being with the rather vague requirement of "invariance under similarities". The case of invariance under a group will be examined in the next section. In that case **A** belongs to the group Lie algebra, and a group acts on its own Lie algebra by similarities, so that adaptation is immediate.

Clearly the trace of **A** is an invariant under the similarity transformation $\mathbf{A}' = \mathbf{g}^{-1}\mathbf{A}\mathbf{g}$, as $tr(\mathbf{g}^{-1}\mathbf{A}\mathbf{g}) = tr(\mathbf{g}\mathbf{g}^{-1}\mathbf{A}) = tr\mathbf{A}$. Actually, the trace of any power of **A** will be invariant, as $tr\mathbf{A}'^k = tr(\mathbf{g}^{-1}\mathbf{A}\mathbf{g}\mathbf{g}^{-1}\mathbf{A}\mathbf{g}\mathbf{g}^{-1}\mathbf{A}\mathbf{g} \cdots \mathbf{g}^{-1}\mathbf{A}\mathbf{g}) = tr(\mathbf{g}^{-1}\mathbf{A}^k\mathbf{g}) = tr\mathbf{A}^k$. The characteristic polynomial is an invariant: once written in the canonical form

$$\Delta_{\mathbf{A}}(\lambda) = \sum_{j=0}^{N} \lambda^{N-j} \phi_j(\mathbf{A}) \ , \tag{16.56}$$

each coefficient $\phi_j(\mathbf{A})$ is invariant because it can be written in terms of powers of the above traces, as seen in (16.40):

$$\phi_j(\mathbf{A}) = \frac{(-)^j}{j!} \sum_{m=0}^{j} \mathbb{B}_{jm}\{(-)^{k-1}(k-1)!\ tr(\mathbf{A}^k)\} \ . \tag{16.57}$$

These complete Bell polynomials contain only powers of traces of powers of $\mathbf{A}$ and are clearly invariant. Thus, the basic invariant of a matrix $\mathbf{A}$ are just the coefficients of its characteristic polynomial. Ultimately, they are symmetric functions on the eigenvalue-alphabet of $\mathbf{A}$. We call them "basic" because, of course, any function of invariant objects is also an invariant. Although the determinant is far from fixing the matrix $\mathbf{A}$, the whole set of invariants does determine its similarity class, as in principle they fix the spectrum and determine the matrix up to similarities $\mathbf{A}' = \mathbf{g}\mathbf{A}\mathbf{g}^{-1}$. Conversely, a set of matrices related by similarities have the same invariants, the same characteristic polynomial.

Commentary 275.1. ▶ With the notation of Section 16.1, we have

$$\Delta_{\mathbf{A}}(\lambda, z) = \sum_{j=0}^{N} (-z)^j \lambda^{N-j} \phi_j(\mathbf{A}) \ .$$

◀

16.5 Lie algebras invariants

When the matrix $\mathbf{A}$ belongs to a Lie algebra, it is possible to obtain a general treatment of the invariants from the characteristic polynomial of $\mathbf{A}$.

Let us apply what has been said in the previous section to matrices $\mathbf{A}$ belonging to a Lie algebra of a Lie group of transformations. Take then for the matrix a generic member

$$\mathbf{A} = \omega^a \mathbf{J}_a \tag{16.58}$$

of such a Lie algebra with generators $\mathbf{J}_a$, $a = 1, 2, \ldots, N$. These generators respect a commutation table $[\mathbf{J}_a, \mathbf{J}_b] = C^c{}_{ab}\mathbf{J}_c$, where the $C^c{}_{ab}$'s are the structure constants. The ω^a's are the transformation parameters. A group element will have the form $\mathbf{g} = \exp[\mathbf{A}]$. The group acts on $\mathbf{A}$ by the adjoint $N \times N$ representation, that is, by similarity transformations of the type $\mathbf{A} \to \mathbf{A}' = \mathbf{g}^{-1}\mathbf{A}\mathbf{g}$. As the characteristic polynomial does not change under these transformations, it is an example of an *invariant polynomial* on a Lie algebra. In the language of the previous section, $\Delta_{\mathbf{A}}(\lambda)$ is a polynomial in the variable λ whose coefficients are functions $\phi_j(\boldsymbol{\omega}) = \phi_j[\mathbf{A}]$ of the parameters, which are themselves Bell polynomials in the traces of powers of $\mathbf{A}$. These invariant functions are not necessarily independent. The rank of a matrix is the largest order found among all its nonvanishing minors. The rank of the algebra, a number of fundamental importance, is the number of coefficients $\phi_j(\boldsymbol{\omega})$ which are functionally independent, that is, the rank of the matrix $\left(\frac{\partial \phi_j}{\partial \omega^i}\right)$.

Some examples of these invariant polynomials are:

$$\begin{aligned}
\phi_0(\boldsymbol{\omega}) &= 1; \\
\phi_1(\boldsymbol{\omega}) &= -tr\mathbf{A}; \\
\phi_2(\boldsymbol{\omega}) &= \tfrac{1}{2}[-tr\mathbf{A}^2 + (tr\mathbf{A})^2]; \\
\phi_3(\boldsymbol{\omega}) &= -\tfrac{1}{3}tr\mathbf{A}^3 + \tfrac{1}{2}(tr\mathbf{A})(tr\mathbf{A}^2) - \tfrac{1}{3!}(tr\mathbf{A})^3; \\
\phi_4(\boldsymbol{\omega}) &= -\tfrac{1}{4}tr\mathbf{A}^4 + \tfrac{1}{3}(tr\mathbf{A})(tr\mathbf{A}^3) \\
&+ \tfrac{1}{8}(tr\mathbf{A}^2)^2 - \tfrac{1}{4}(tr\mathbf{A})^2(tr\mathbf{A}^2) + \tfrac{1}{4!}(tr\mathbf{A})^4; \\
&\cdots\cdots\cdots \\
\phi_N(\boldsymbol{\omega}) &= (-)^N \det\mathbf{A}\,.
\end{aligned} \tag{16.59}$$

The invariants ϕ_j are (up to eventual signs) nothing more than the elementary symmetric functions of the eigenvalue-alphabet of the generic Lie algebra member:

$$\phi_j[\mathbf{A}] = (-)^j e_j[\boldsymbol{\lambda}]\ . \tag{16.60}$$

These relations lead to the introduction of multilinear objects which are invariant symmetric tensors, such as

$$\eta^{(0)} = 1;\ \eta^{(1)}_a = -tr\ \mathbf{J}_a;\ \eta^{(2)}_{ab} = -\tfrac{1}{2}[tr(\mathbf{J}_a\mathbf{J}_b) - tr(\mathbf{J}_a)tr(\mathbf{J}_b)]\ ;$$

$$\eta^{(3)}_{abc} = -\tfrac{1}{3}[tr(\mathbf{J}_a\mathbf{J}_b\mathbf{J}_c) - \tfrac{3}{2}(tr\mathbf{J}_a)(tr\mathbf{J}_b\mathbf{J}_c) + \tfrac{1}{2}tr(\mathbf{J}_a)tr(\mathbf{J}_b)tr(\mathbf{J}_c)]\ .$$

The general expression is

$$\eta^{(n)}_{a_1a_2\cdots a_n} = \frac{(-)^n}{n!}\sum_{m=1}^{n} \mathbb{B}_{nm}\{(-)^{k-1}(k-1)!\ tr(\mathbf{J}_{a_1}\mathbf{J}_{a_2}\cdots\mathbf{J}_{a_k})\}\ . \tag{16.61}$$

These tensors produce invariant objects by contraction,

$$T = T^{a_1 a_2 \cdots a_n} \, \eta^{(n)}_{a_1 a_2 \cdots a_n} \, . \tag{16.62}$$

Most important are the invariant Casimir operators in the enveloping algebra:

$$C^{(n)} = \eta^{(n)}_{a_1 a_2 \cdots a_n} \mathbf{J}^{a_1} \mathbf{J}^{a_2} \cdots \mathbf{J}^{a_n} \, . \tag{16.63}$$

The best known of these tensors is the Killing form γ_{ab}. It is usually defined as $\eta^{(2)}_{ab}$ in the particular case in which $tr\ \mathbf{J}_a = 0$ for all the generators. Thus, it is $\gamma_{ab} = -\frac{1}{2} tr\ \mathbf{J}_a \mathbf{J}_b$ and leads to the Casimir operator $\gamma_{ab}\ \mathbf{J}^a \mathbf{J}^b$. The well-known operator $\mathbf{J}^2$, which classifies the angular momenta, is the Casimir operator for the Lie algebra of the rotation group in Euclidean 3-dimensional space (see Comment 10.2). In the adjoint representation, the generators are matrices $(J_a)^c{}_b = C^c{}_{ab}$ with the structure constants as entries, and $\gamma_{ab} = C^c{}_{ad} C^d{}_{bc}$. The Killing form plays a fundamental role in the classification of Lie groups [115]. For semisimple groups, γ_{ab} is a metric (that is, a symmetric nondegenerate bilinear form), the Killing–Cartan metric. This form will be of definite sign for compact groups, non-definite for non-compact groups and degenerate for non-semisimple groups.

The higher the rank (number of independent invariants) of the Lie algebra, the more multilinear forms there will be. The rotation group in Euclidean 3-dimensional space has rank 1, and the above mentioned Casimir operator of bilinear nature is its sole basic invariant. The Lorentz and the Poincaré groups are of rank 2 and have bilinear and quadrilinear forms. Another rank 2 case is the flavor and color group $SU(3)$ of special unitary 3×3 matrices, which has bilinear and trilinear invariants [144].

Of course, a function of invariants is also an invariant, and it is sometimes convenient to choose another set. Those given above are, however, because of their simple origin, the basic ones. Their properties may seem rather trivial when a diagonalized matrix is considered, but formulas (16.59) are independent of matrix basis. Some deep consequences can be found, related to the detailed algebraic structure of the Lie algebra. For example, a semisimple algebra is the direct product of simple sub-algebras and its characteristic polynomial is always of the form $\Delta(\lambda) = \prod_k \Delta_k(\lambda)$, where the Δ_k's are the characteristic polynomials of the simple subalgebras.

Orthogonal and pseudo-orthogonal groups are groups of transformations which preserve a real bilinear non-singular form η. If $\mathbf{g}$ is a group element, this means that $\mathbf{g}^T \eta \mathbf{g} = \eta$, where $\mathbf{g}^T$ is the transposed matrix. If $\mathbf{g} = \exp[\mathbf{A}]$, with $\mathbf{A}$ a member of the group Lie algebra, then this implies $\mathbf{A}^T = -\eta \mathbf{A} \eta^{-1}$ and the traces of the odd powers of $\mathbf{A}$ vanish. Actually, the odd coefficients in the characteristic polynomial (that is, the odd-order invariants) vanish. The argument is the following. Take the characteristic polynomial

$$\det[\mathbf{I}\,\lambda - \mathbf{A}] = \sum_{j=0}^{N} \frac{(-)^j\,\lambda^{N-j}}{j!} \sum_{m=0}^{j} \mathbb{B}_{jm}\{(-)^{k-1}(k-1)!\,tr(\mathbf{A}^k)\}\,. \quad (16.64)$$

The determinant for the transposed matrix must be the same: $\det[\lambda\mathbf{I} - \mathbf{A}] = \det[\lambda\mathbf{I} - \mathbf{A}^T]$. As $\mathbf{A}^T = -\eta\mathbf{A}\eta^{-1}$, a sign factor $(-)^k$ appears in the arguments of the Bell polynomials: $\mathbb{B}_{jm}\{(-)^{k-1}(k-1)!\;tr[(\mathbf{A}^T)^k]\} = \mathbb{B}_{jm}\{(-)^k(-)^{k-1}(k-1)!\;tr(\mathbf{A}^k)\} = (-)^j\;\mathbb{B}_{jm}\{(-)^{k-1}(k-1)!\;tr[(\mathbf{A}^T)^k]\}$ by (15.22) or (15.27). Consequently, we also have

$$\det[\mathbf{I}\,\lambda - \mathbf{A}] = \sum_{j=0}^{N} \frac{\lambda^{N-j}}{j!} \sum_{m=0}^{j} \mathbb{B}_{jm}\{(-)^{k-1}(k-1)!\;tr(\mathbf{A}^k)\}\,. \quad (16.65)$$

Comparison of the last two expressions shows that the coefficients must vanish for odd values of j: $\phi_{2n+1}(\boldsymbol{\omega}) = 0$. The invariants of (16.59) reduce to simpler forms, like

$$\phi_0(\boldsymbol{\omega}) = 1\;;\;\phi_1(\boldsymbol{\omega}) = 0\;;\;\phi_2(\boldsymbol{\omega}) = -\tfrac{1}{2}\;tr\mathbf{A}^2\;;\;\phi_3(\boldsymbol{\omega}) = 0\;;$$
$$\phi_4(\boldsymbol{\omega}) = -\tfrac{1}{4}\;tr\mathbf{A}^4 + \tfrac{1}{8}\;(tr\mathbf{A}^2)^2\,. \quad (16.66)$$

For unitary groups written in terms of hermitian generators, $\mathbf{A}^T = \mathbf{A}^*$, the only conclusion is that the invariants are real, $\phi_j(\boldsymbol{\omega})^* = \phi_j(\boldsymbol{\omega})$. The extra condition $\det[\mathbf{g} = \exp[i\mathbf{A}]] = 1$ for special unitary groups leads, of course, to $tr\mathbf{A} = 0$.

Besides their intrinsic interest, the above invariants are helpful as a controlling tool in complicated manipulations involving Lie algebras, such as those leading to the explicit construction of Lie group elements [13].

16.5.1 *Characteristic classes*

Matrices whose entries are differential forms are essential to modern Geometry. Even-order forms are commutative, but those of odd-degrees are not. The algebra of odd-order matrix forms is a bit more involved than that of usual matrices, but that of even-orders is just as usual. Connections are 1-forms, and their curvatures are 2-forms. Some of the most important invariants of Differential Geometry are just those given above, with curvatures in the role of the Lie algebra matrices.

Perhaps the most beautiful examples of the invariants described above are the differential forms describing the characteristic classes of fiber bundles on even-dimensional differentiable manifolds [116]. These eminently mathematical objects are of great interest in gauge theories and have received a lot of attention from physicists [117], [118]. We shall only consider briefly their formal aspects, and confine ourselves to Chern classes.

There are actually two interrelated kinds of classes going under Chern's name: the Chern classes proper and the Chern characters [119]. We shall find easily the closed, complete expressions of their interrelations. Consider the complex linear group $GL(N, \mathbb{C})$ of all invertible $N \times N$ matrices with complex entries and let $\mathbf{A}$ be an element of its Lie algebra $G'L(N, \mathbb{C})$. To conform to current mathematical notation it is enough to take formula (16.2) with $z = \frac{i}{2\pi}$, so that now the invariant polynomial functions are

$$\phi_k[\mathbf{A}] = \left(\frac{i}{2\pi}\right)^k \frac{1}{k!} \sum_{m=0}^{k} \mathbb{B}_{km}\{(-)^{r-1}(r-1)!\ tr(\mathbf{A}^r)\} . \tag{16.67}$$

Given a complex vector bundle E over a manifold M with typical fiber $\mathbb{C}^N$ and associated to a principal bundle P, the k-th Chern class $c_k(E)$ will be characterized by a closed differential form c_k of degree $2k$ on M. This form is that (unique) form whose pull-back is $\phi_k(\mathbf{F})$, where $\mathbf{F} = d\Gamma + \Gamma \wedge \Gamma$ is the curvature (a matrix 2-form) of a connection Γ (a matrix 1-form) on P. After they are found, the cohomology classes represented by the c_k's are shown to be independent of Γ and, consequently, related to the more basic, topological features of the bundle. If $\pi : P \to M$ is the bundle projection and π^* the corresponding pull-back, then the k-th Chern class is

$$\pi^*(c_k) = \phi_k(\mathbf{F}) = \frac{i^k}{(2\pi)^k k!} \sum_{m=0}^{k} \mathbb{B}_{km}\{(-)^{r-1}(r-1)!\ tr(\mathbf{F}^r)\} = e_k[i\mathbf{F}/2\pi] , \tag{16.68}$$

the elementary symmetric function of the alphabet whose letters are the eigenvalues of the matrix $(i\mathbf{F}/2\pi)$. Some examples:

$$\begin{aligned} c_0(E) &= 1\ ;\ c_1(E) = \frac{i}{2\pi}\ tr\mathbf{F}\ ;\ c_2(E) = \frac{1}{8\pi^2}\ \left[tr\mathbf{F}^2 - (tr\mathbf{F})^2\right]\ ; \\ c_3(E) &= -\ \frac{i}{8\pi^3}\ \left[\tfrac{1}{3}\ tr\mathbf{F}^3 + \tfrac{1}{2}\ (tr\mathbf{F})(tr\mathbf{F}^2) - \tfrac{1}{6}\ (tr\mathbf{F})^3\right]\ ; \\ &\cdots\cdots\cdots\cdots \\ c_N(E) &= (\frac{i}{2\pi})^N\ \det\mathbf{F}\ . \end{aligned} \tag{16.69}$$

Notice that the fiber dimension is $2N$. These relations are direct adaptations of (16.59). For orthogonal groups, the first invariants reduce to the simpler forms of (16.66), as in

$$c_0(E) = 1\ ; c_1(E) = 0\ ; c_2(E) = \frac{1}{8\pi^2}\ tr\mathbf{F}^2\ ; c_3(E) = 0\ ;$$

$$c_4(E) = -\ \frac{1}{64\pi^4}\ tr\mathbf{F}^4 + \frac{1}{128\pi^4}\ (tr\mathbf{F}^2)^2\ . \tag{16.70}$$

Handling matrices of forms in general requires some attention, because the non-commutativity of forms is added to that of matrices. Here only 2-forms appear, which are commutative. Of course, the higher-order classes are of limited physical interest, as products of forms vanish when the resulting degree surpasses the space dimension. On 4-dimensional spacetime, only invariant polynomials up to the second order in $\mathbf{F}$ are non-vanishing.

The Chern total character is defined as

$$ch(E) = \mathrm{tr}(\exp\ [\frac{i}{2\pi}\mathbf{F}]) = \sum_{k=0}^{\infty} \frac{i^k}{(2\pi)^k k!}\ tr\mathbf{F}^k = \sum_{k=0}^{N} \frac{1}{k!}\ p_k[i\mathbf{F}/2\pi]\ , \tag{16.71}$$

where the p_k's are the power symmetric functions, with the adopted conventions $p_{k>N}[i\mathbf{F}/2\pi] = 0$ and $p_0[i\mathbf{F}/2\pi] = N$. The summand

$$ch_k(E) = \frac{i^k}{(2\pi)^k k!}\ tr\mathbf{F}^k = \frac{1}{k!}\ p_k[i\mathbf{F}/2\pi] \tag{16.72}$$

is the k-th Chern character. The relations between classes and characters are given by Eqs.(15.80) and (15.81): for $j, k \geq 0$, they are

$$c_j(E) = \frac{1}{j!} \sum_{m=0}^{j} \mathbb{B}_{jm}\{(-)^{k-1}(k-1)!\ k!\ ch_k(E)\}\ ; \tag{16.73}$$

$$(-)^{k-1}(k-1)!\ ch_k(E) = \frac{1}{k!} \sum_{j=0}^{k}(-)^{j-1}(j-1)!\ \mathbb{B}_{kj}\{r!\ c_r(E)\}\ . \tag{16.74}$$

We have, consequently, $c_k = e_k[i\mathbf{F}/2\pi]$ from (16.68); and from (16.72), we get back $k! ch_k(E) = p_k[i\mathbf{F}/2\pi]$. We can therefore use (16.21) to get a determinant relation:

$$n! c_n(E) = \begin{vmatrix} 1!ch_1(E) & 1 & 0 & 0\,.. & 0 \\ 2!ch_2(E) & 1!ch_1(E) & 2 & 0\,.. & 0 \\ 3!ch_3(E) & 2!ch_2(E) & 1!ch_1(E) & 3\,.. & 0 \\ . & . & . & ... & . \\ . & . & . & ... & n-1 \\ n!ch_n(E) & (n-1)!ch_{n-1}(E) & . & ... & 1!ch_1(E) \end{vmatrix}\ . \tag{16.75}$$

Our concern here is simply to exhibit these closed expressions for the relationships. We have only considered the complex linear groups. The interested reader will find in the quoted references details on Pontryagin classes $g_{2k}(\mathbf{F}) = (-)^k f_{2k}(\mathbf{F})$, Euler classes, etc.

Relations (16.69) are, as said, simple adaptations of (16.59) and lead to the introduction of multilinear objects like (16.61), which are invariant tensors and allow to obtain more general invariants. The usual Lagrangian density for gauge fields, for example, is the invariant $tr\mathbf{F}\widetilde{\mathbf{F}} = \gamma_{ab}\mathbf{F}^a\widetilde{\mathbf{F}}^b$, where γ_{ab} is the Killing form and $\widetilde{\mathbf{F}}$ is the dual of $\mathbf{F}$ in Minkowski space, whose components are given by $\widetilde{\mathbf{F}}_{\mu\nu} = \frac{1}{2}\epsilon_{\mu\nu\rho\sigma}F^{\rho\sigma}$. The corresponding Chern class, $c_2(E)$, would be a "topological Lagrangian density". Current gauge theories use groups for which $tr\ \mathbf{F} = 0$, so that only the term in $tr\ \mathbf{F}^2$ appear in the Lagrangian. A gauge theory for the linear group $GL(4, R)$, for example, could have [120] an extra term in $(tr\ \mathbf{F})(tr\ \tilde{\mathbf{F}})$. And a gauge theory for the group $SU(3)$, like chromodynamics, could in principle have an extra trilinear term in the Lagrangian.

The approach throws a bridge towards the results on the relationship between invariant polynomials and von Neumann algebras [121]. We have taken for granted that functions of matrices, as long as they can be defined, are completely determined by their spectra. This is justified in a much more general context. Matrix algebras are very particular kinds of von Neumann algebras and it is a very strong result of the theory of Banach algebras [12], [82] that functions on such spaces, as long as defined, are indeed fixed by the spectra. Another point worth mentioning is that the infinite Bell matrices which constitute the true, complete representation of the group of invertible series will belong, as limits for $N \to \infty$ of $N \times N$ matrices, to a hyperfinite von Neumann algebra [12]. Our considerations here are purely formal from the mathematical point of view, as we are only discussing formal series. We are not concerned with the topological intricacies involved in convergence questions, though they surely deserve a detailed study. By the way, the infinite algebra generated by Bell matrices would provide a good guide in the study of function algebras with the composition operation.

Chapter 17

Projectors and iterates

The applications of some matrix functions, like the exponential $e^{\mathbf{A}}$, are well-known. Those of other functions are less popular. The real-power function $\mathbf{A}^t$, applied to a Bell matrix $\mathbb{B}[g]$, yields the continuous iterate $g^{<t>}$ of g and thereby helps bridging the gap between the two current approaches to dynamical systems: the continuous one, based on differential equations, and the discrete one, founded on the iteration of maps. A theory for the eigenvectors of $\mathbb{B}[g]$ turns up as a bonus.

17.1 Projectors, revisited

We give here closed expressions for the projectors of a matrix in terms of the symmetric functions of its eigenvalues.

Our aim here will be the study of functions of Bell matrices, specializing the treatment of Section 1.3. Given the spectrum $\{x_k = g_1^k\}$ of $\mathbb{B}[g]$, it will be convenient to consider both the alphabet $\mathbf{x} = \{x_1, x_2, x_3, \ldots, x_N\}$ and its reciprocal alphabet $\mathbf{x}^* = \{x_1^*, x_2^*, x_3^*, \ldots, x_N^*\}$ where each $x_j^* = -1/x_j$. Bell matrices are generically nondegenerate. The only exception turns up for the so-called unit series, which have $g_1 = 1$. The components are, consequently, projectors. General expressions for the projectors will be obtained for Bell matrices, which hold nevertheless for general nondegenerate matrices.

Only one kind of symmetric functions, the j-th elementary symmetric functions, will be needed in our present quest. We recall that $e_0[\mathbf{x}] = 1$ by convention, and that the symmetric functions of $\mathbf{x}$ and $\mathbf{x}^*$ are related by

$$(-)^j \; e_{N-j}[\mathbf{x}] = e_N[\mathbf{x}] \; e_j[\mathbf{x}^*] \; . \tag{17.1}$$

We recall also the general expression involving the generating function:

$$\prod_{j=1}^{N}(x_j - t) = e_N[\mathbf{x}] \sum_{j=0}^{N} e_j[\mathbf{x}^*] \, t^j = e_N[\mathbf{x}] \sum_{j=0}^{N} (-)^j \, e_{N-j} \, [\mathbf{x}] t^j \, . \tag{17.2}$$

The j-th eigenvalue is absent in the numerator of expression (1.18) for $\mathbf{Z}_j$. We shall consequently need some results involving an alphabet with one missing letter. Let $e_{ji}[\mathbf{x}]$ be the sum of all j-products of the alphabet $\mathbf{x}$, but excluding x_i — for example, $e_{Ni}[\mathbf{x}] = \prod_{k \neq i}^{N} x_k$. Simply excluding the terms involving x_i, we find that

$$e_{ki}[\mathbf{x}] = \sum_{p=0}^{k} (-)^p x_i^p e_{k-p}[\mathbf{x}] = \sum_{j=0}^{k} (-x_i)^{k-j} e_j[\mathbf{x}] \, . \tag{17.3}$$

In the absence of the i-th letter, (17.2) becomes

$$\prod_{k=1, k \neq i}^{N} (x_k - t) = e_{Ni}[\mathbf{x}] \sum_{k=0}^{N} e_{ki}[\mathbf{x}^*] t^k \, . \tag{17.4}$$

The projectors (1.18) can then be written

$$\mathbf{Z}_i[\mathbb{B}] = \prod_{k=1, k \neq i}^{N} \frac{x_k \mathbf{I} - \mathbb{B}}{x_k - x_i} = \frac{\sum_{k=0}^{N} e_{ki}[\mathbf{x}^*] \, \mathbb{B}^k}{\sum_{r=0}^{N} e_{ri}[\mathbf{x}^*] x_i^r} \, . \tag{17.5}$$

This is clearly written in the power basis $\{\mathbf{I}, \mathbb{B}, \mathbb{B}^2, \ldots, \mathbb{B}^{N-1}\}$. In our application to Bell matrices, it will be convenient to use instead the basis $\{\mathbb{B}, \mathbb{B}^2, \ldots, \mathbb{B}^N\}$. This is due to the fact that we shall prefer to start from the matrix $\mathbb{B}[g]$, corresponding to g, and not from the matrix $\mathbf{I}$, corresponding to the identity function. The mappings of interest, like the logistic map for example, have a general "$\cap$" aspect in the interval of import $[0, 1]$, and the identity map (whose aspect is "/") has not the same end-points. More pedantly, the identity map has Brouwer [3] degree 1, while the dynamical maps of interest can be characterized by their degree 0. The Cayley–Hamilton theorem implies $\sum_{j=0}^{N} e_j[\mathbf{x}^*] \, \mathbb{B}^j = 0$, from which we obtain the identity $\mathbf{I} = \mathbb{B}^0$ as

$$\mathbf{I} = -\sum_{j=1}^{N} e_j[\mathbf{x}^*] \, \mathbb{B}^j \, . \tag{17.6}$$

Each root x_i will satisfy the same equation, so that $1 = -\sum_{j=1}^{N} e_j[\mathbf{x}^*] \, x_i^j$. Replacing these identities in (17.5), the projectors are recast into another form,

$$\mathbf{Z}_i[\mathbb{B}] = \frac{\mathbf{I} + \sum_{k=1}^{N} e_{ki}[\mathbf{x}^*] \, \mathbb{B}^k}{1 + \sum_{r=1}^{N} e_{ri}[\mathbf{x}^*] x_i^r} = \frac{\sum_{k=1}^{N} \{e_{ki}[\mathbf{x}^*] - e_k[\mathbf{x}^*]\} \, \mathbb{B}^k}{\sum_{r=1}^{N} \{e_{ri}[\mathbf{x}^*] - e_r[\mathbf{x}^*]\} x_i^r} \, . \tag{17.7}$$

Using (17.3) and (17.1) we get

$$\mathbf{Z}_i[\mathbb{B}] = \frac{\sum_{k=1}^{N}\{\sum_{j=0}^{k-1} x_i^{j-k} e_j[\mathbf{x}^*]\}\ \mathbb{B}^k}{\sum_{r=0}^{N-1}(N-r)x_i^r e_r[\mathbf{x}^*]} = \frac{\sum_{k=1}^{N}\{\sum_{j=0}^{k-1} x_i^{j-k}(-)^j e_{N-j}[\mathbf{x}]\}\ \mathbb{B}^k}{\sum_{r=0}^{N-1}(N-r)x_i^r(-)^r e_{N-r}[\mathbf{x}]}\ , \tag{17.8}$$

where we have also used

$$\sum_{r=1}^{N}\left\{\sum_{j=0}^{r-1}(x_i)^j e_j[\mathbf{x}^*]\right\} = \sum_{j=0}^{N-1}(N-j)(x_i)^j e_j[\mathbf{x}^*]\ .$$

There is a good immediate check: replacing $\mathbb{B}$ by the eigenvalues we find, as expected,

$$\mathbf{Z}_i[x_k] = \delta_{ik}\mathbf{I}\ . \tag{17.9}$$

The projectors are now clearly in the basis $\{\mathbb{B}, \mathbb{B}^2, \ldots, \mathbb{B}^N\}$. Actually, each $\mathbf{Z}_i$ is now just that given in the basis $\{\mathbf{I}, \mathbb{B}, \mathbb{B}^2, \ldots, \mathbb{B}^{N-1}\}$ multiplied by $(\mathbb{B}/x_i)$: instead of that given in (1.18), it is

$$\mathbf{Z}_i[\mathbb{B}] = \frac{(\mathbb{B}-x_1)(\mathbb{B}-x_2)...(\mathbb{B}-x_{i-1})\ \mathbb{B}\ (\mathbb{B}-x_{i+1})...(\mathbb{B}-x_{N-1})(\mathbb{B}-x_N)}{(x_i-x_1)(x_i-x_2)...(x_i-x_{i-1})\ x_i\ (x_i-x_{i+1})...(x_i-x_{N-1})(x_i-x_N)}\ . \tag{17.10}$$

We can think about these results in the following way. Look at the $\mathbf{Z}_i$'s and the powers $\mathbb{B}^k$ as components of two formal column "vectors". The linear conditions $\mathbb{B}^n = \sum_{j=1}^{N} x_j^n \mathbf{Z}_j$ are then represented by a matrix $\Lambda = [x_j^n]$,

$$\begin{pmatrix} \mathbb{B}^1 \\ \mathbb{B}^2 \\ \mathbb{B}^3 \\ \cdot \\ \cdot \\ \mathbb{B}^N \end{pmatrix} = \begin{pmatrix} x_1 & x_2 & x_3 & \cdot\ \cdot & x_N \\ x_1^2 & x_2^2 & x_3^2 & \cdot\ \cdot & x_N^2 \\ x_1^3 & x_2^3 & x_3^3 & \cdot\ \cdot & x_{N^3} \\ \cdot & \cdot & \cdot & \cdot\ \cdot & \cdot \\ \cdot & \cdot & \cdot & \cdot\ \cdot & \cdot \\ x_1^N & x_2^N & x_3^N & \cdot\ \cdot & x_N^N \end{pmatrix} \begin{pmatrix} \mathbf{Z}_1 \\ \mathbf{Z}_2 \\ \mathbf{Z}_3 \\ \cdot \\ \cdot \\ \mathbf{Z}_N \end{pmatrix} \tag{17.11}$$

and what we have done has been to obtain its inverse:

$$\mathbf{Z}_i[\mathbb{B}] = \sum_{k=1}^{N}[\Lambda^{-1}]_{ik}\ \mathbb{B}^k\ , \tag{17.12}$$

$$[\Lambda^{-1}]_{ik} = \frac{\sum_{j=0}^{k-1} x_i^{j-k}(-)^j e_{N-j}[\mathbf{x}]}{\sum_{r=0}^{N-1}(N-r)\,x_i^r(-)^r e_{N-r}[\mathbf{x}]} = \frac{\sum_{j=0}^{k-1} x_i^{j-k} e_j[\mathbf{x}^*]}{\sum_{s=1}^{N} s\,x_i^{N-s} e_{N-s}[\mathbf{x}^*]}\ . \tag{17.13}$$

It seems a difficult task to improve the above expressions, as it would mean knowing a closed analytical expression for the recurrent summation of the form $\sum_{j=0}^{k-1} u^j\ e_j[\mathbf{x}^*]$. A closed expression for $e_j[\mathbf{x}^*]$ would be necessary and, even for the simple alphabet consisting of powers of a fixed letter a, which we shall find in the application to Bell matrices, this would be equivalent to solving a seemingly as yet unsolved problem in Combinatorics. In effect, in terms of such an alphabet $\{a^j\}$ with N letters, the symmetric function is given by

$$e_k = \sum_{j=1}^{N(N+1)/2} q_{j,k,N}\ a^j\ , \tag{17.14}$$

where $q_{j,k,N}$ = number of partitions of j into k unequal summands, each one $\leq N$. These conditional partition numbers have the generating function

$$\prod_{r=1}^{N}[1 + ua^r] = 1 + \sum_{j,k\geq 1} q_{j,k,N}\ a^j u^k\ , \tag{17.15}$$

which is a special case of (16.8). They have no known closed expression and are actually calculated by using this formula.[1]

Commentary 286.1. ► The reader may have qualms about the case where some root $x_0 = 0$. The good rule is to perform first the computation of the projectors with the symbols x_k, without putting in the values, and then multiply numerator and denominator by det $\mathbb{B} = \prod_k x_k$. This procedure will dispose of any problem with eventual vanishing eigenvalues. The closed expressions are more useful for larger matrices. For small ones they can, however, be used to check the results obtained by direct computation. We have found them very useful while checking calculations with the *Mathematica*^{TM} software system. ◄

As repeatedly said, Bell matrices are not normal, that is, they do not commute with their transposes. This means that Bell matrices cannot be put into diagonal form by similarity transformations. As it happens, this will not be a difficulty because the eigenvalues of an $N \times N$ matrix $\mathbb{B}[g]$ are known from the start: they are just g_1^k, $k = 1, 2, 3, ..., N$. Nevertheless, an important property of normal matrices will be missing: the column-eigenvectors of $\mathbb{B}[g]$ will not constitute a basis for the underlying vector space they belong to.

---

[1] See [96], Chapter II.3.

17.2 Continuous iterates

Because they transform function compositions into matrix products, Bell matrices provide a fair gateway into the notion of continuous iterate of a function. An application to a very idealized model of turbulence is shown. Continuous iteration has a strong link to the Schröder equations. This relationship leads to an unexpected result: solving a certain Schröder equation is equivalent to finding the eigenvectors of a Bell matrix.

As mentioned in Chapter 2, there are two main approaches to classical dynamical systems. In the first, time evolution is a continuous motion and leads to a curve. For example the trajectory, on phase space, of a point representative of the state of the system. In the second, evolution is described by the successive iterations of a mapping. The state is known at each step, as if the "time" parameter of the system were defined only at discrete values. We can go from the first approach to the second by taking the intersections leading to a Poincaré map. We present in what follows a converse procedure, going from a discrete to a continuous description. We shall "interpolate" between the discrete values in such a way that the notion of iteration keeps its meaning in the intervals.

In principle, any matrix describing a step-by-step evolution can be interpolated by its power. We have done so with the matrix governing a Markov process in Section 6.4. A Bell matrix provides a linear representation of a mapping, and its interpolation defines a time-interpolation of this map. The question, in the case, is whether or not such interpolation preserves the sense of iteration.

Iteration[2] is a particular case of function composition: given a basic map $f(x) = f^{<1>}(x)$, its first iterate is $f^{<2>}(x) = [f \circ f](x) = f[f(x)]$, its n-th iterate is $f^{<n>}(x) = f[f^{<n-1>}(x)] = f^{<n-1>}[f(x)]$, etc. The question is whether or not, given the set of functions $f^{<n>}(x)$, an interpolation $f^{<t>}(x)$ with real values of t can be found which represents the one-parameter continuous group (or semigroup) describing the dynamical flow of the system. In order to do it, $f^{<t>}(x)$ should satisfy the conditions

$$f^{<t>}[f^{<t'>}(x)] = f^{<t'>}[f^{<t>}(x)] = f^{<t+t'>}(x) \ ; \tag{17.16}$$

$$f^{<0>}(x) = Id(x) = x \ . \tag{17.17}$$

[2] For a general text on iterative functional equations, see [98].

We shall find in what follows a map interpolation $f^{<t>}(x)$ with these properties. It is a well-known fact, repeatedly used in previous sections, that Taylor series $f(x)$, $g(x)$ satisfying the conditions $f(0) = 0$, $g(0) = 0$ and with nonvanishing coefficients of x are invertible and constitute a group by composition [102]. Bell polynomials provide a clue to the question, as they attribute to every such a function f a matrix $\mathbb{B}[f]$, whose inverse represents the inverse function and such that the matrix product represents the composition operation. In other words, these matrices provide a linear representation of the group formed by the functions with the operation of composition. Composition is thus represented by matrix product and, consequently, iterations are represented by matrix powers. Furthermore, the representation is faithful, and the function f is completely determined by $\mathbb{B}[f]$. Now, in the matrix group there does exist a clear interpolation of discrete powers by real powers and the inverse way, going from matrices to functions, yields a map interpolation with the desired properties.

Commentary 288.1. ► An extremely simple case in which the iterates can be found in closed form is the function $f(x) = \frac{x}{1-x}$. It is related to the Lah numbers (Comments 236.2 and 244.1), as $\mathbb{B}_{jk}\left[\frac{x}{1-x}\right] = \mathbb{B}_{jk}\{r!\} = \binom{j-1}{k-1}\frac{j!}{k!}$. We can verify directly that $f^{<n>}(x) = \frac{x}{1-nx} = \frac{1}{n}f(nx)$. Consequently, the powers of the Bell matrix have entries $\mathbb{B}^n{}_{jk}\left[\frac{x}{1-x}\right] = \mathbb{B}_{jk}\left[\frac{x}{1-n\,x}\right] = n^{j-k}\mathbb{B}_{jk}\left[\frac{x}{1-x}\right] = n^{j-k}\binom{j-1}{k-1}\frac{j!}{k!}$. The (j,k) entry of the n-th power of a matrix $\mathbf{M}$ formed by Lah numbers is just the original entry M_{jk} multiplied by n^{j-k}. The particular case $n = -1$ gives exactly the inverse $f^{<-1>}(u) = \frac{u}{1+u}$. It is also directly found that $f^{<1/2>}(x) = \frac{x}{1-\frac{x}{2}}$ is the iterative square-root (see Section 17.2.1) of $f(x) = \frac{x}{1-x}$. It is a good guess that the continuous iteration is given by $f^{<t>}(x) = \frac{x}{1-t\,x}$, which indeed satisfies conditions (17.16, 17.17). ◄

Commentary 288.2. ► The inverse $g(x) = f^{<-1>}(x) = \frac{x}{1+x}$ in the above Comment is of fundamental interest in the theory of colloids. With x the initial number of particles, its continuous iterate

$$g^{<t>}(x) = \frac{x}{1 + x\tau} \tag{17.18}$$

is just Smoluchowski's simplified formula giving the total number of particles at time $\tau = 4\pi RDt$ (D is the diffusion coefficient; R is the radius of the "sphere of influence" of each agglomerate, inside which any other agglomerate fuses with it; both parameters are supposed the same for all sizes of agglomerates in this simplified version) in a coagulation process [103]. Its germ mapping,

$$g^{<1>}(x) = \frac{x}{1 + x}\,, \tag{17.19}$$

has another curious property: its reciprocal

$$h^{<1>}(x) = \frac{1 + x}{x}\,, \tag{17.20}$$

generates the Fibonacci numbers of Comment 237.2: putting $F_0 = 0$,

$$h^{<n>}(x) = \frac{x\,F_{n+1} + F_n}{x\,F_n + F_{n-1}} \; . \tag{17.21}$$

Matrix $M = \left(\begin{smallmatrix}1 & 1\\ 1 & 0\end{smallmatrix}\right)$ does the same job: $M^n = \left(\begin{smallmatrix}F_{n+1} & F_n\\ F_n & F_{n-1}\end{smallmatrix}\right)$. By the way, this matrix satisfies the same second-order equation of the Golden Ratio α: $M^2 = I + M$, and its eigenvalues are α and $1 - \alpha$. ◄

We have learned in Section 1.3 how to find the t-th power of any nondegenerate matrix. Compact versions of the necessary projectors have been given in the previous section. We shall now show that, given a function g, the function corresponding to the t-th power $\mathbb{B}^t[g]$ of $\mathbb{B}[g]$ is the continuous iterate $g^{<t>}(x)$. Many things — the spectrum of $\mathbb{B}[g]$ for example — will depend only on the first-order Taylor coefficient, so that we shall write $g_1 = a$ from now on. By (15.14), the letters in the eigenvalue-alphabet of $\mathbb{B}[g]$ will be simple powers of a and the alphabet $\mathbf{a} = (a^1, a^2, \cdots, a^N)$ will have the reciprocal $\mathbf{a}^* = (-a^{-1}, -a^{-2}, \cdots, -a^{-N})$. The matrix Λ has entries $(\Lambda)_{ik} = a^{ik}$ and the projectors (17.12) have now the coefficients

$$[\Lambda^{-1}]_{ik} = \frac{\sum_{j=0}^{k-1} a^{i(j-k)}(-)^j e_{N-j}[\mathbf{a}]}{\sum_{r=1}^{N}\sum_{s=0}^{r-1} a^{is}(-)^s e_{N-s}[\mathbf{a}]} \; . \tag{17.22}$$

Commentary 289.1. ► This is the same as

$$[\Lambda^{-1}]_{ik} = a^{-ik}\, \frac{\sum_{j=0}^{k-1} a^{ij} e_j[\mathbf{a}^*]}{\sum_{s=0}^{N-1}(N-s)a^{is}e_s[\mathbf{a}^*]} \; . \tag{17.23}$$

The characteristic equation for the Bell matrix,

$$\sum_{j=0}^{N} a^{ij} e_j[\mathbf{a}^*] = \prod_{r=1}^{N}\left[1 - \frac{a^i}{a^r}\right] = 0 \; , \tag{17.24}$$

will then say that (i) $\sum_{s=0}^{N-1}(N-s)a^{is}e_s[\mathbf{a}^*] = \sum_{s=0}^{N}(N-s)a^{is}e_s[\mathbf{a}^*] = -\sum_{s=0}^{N} s a^{is} e_s[\mathbf{a}^*]$ and (ii) $\sum_{j=0}^{k-1} a^{ij} e_j[\mathbf{a}^*] = -\sum_{j=k}^{N} a^{ij} e_j[\mathbf{a}^*]$. The expression above can consequently be put into the still more compact form

$$[\Lambda^{-1}]_{ik} = a^{-ik}\, \frac{\sum_{j=k}^{N} a^{ij} e_j[\mathbf{a}^*]}{\sum_{s=0}^{N} s a^{is} e_s[\mathbf{a}^*]} \; . \tag{17.25}$$

◄

Commentary 289.2. ► We can check directly $\sum_{k=1}^{N}[\Lambda^{-1}]_{ik} a^{km} = \delta_i^m$, again using (17.24). If we take expression (17.23) for Λ^{-1}, the numerator turning up is

$$\sum_{k=1}^{N} a^{-ik} a^{km} \sum_{j=0}^{k-1} a^{ij} e_j[\mathbf{a}^*] = \sum_{k=1}^{N} a^{k(m-i)} \sum_{j=0}^{k-1} a^{ij} e_j[\mathbf{a}^*] = \sum_{j=0}^{N-1} a^{ij} e_j[\mathbf{a}^*] \sum_{k=j+1}^{N} a^{k(m-i)}$$

$$= \sum_{j=0}^{N-1} a^{ij} e_j[\mathbf{a}^*] \begin{cases} a^{(m-i)} \frac{a^{(m-i)j} - a^{(m-i)N}}{1-a^{(m-i)}} & \text{if} \quad a^{(m-i)} \neq 1 \\ (N-j) & \text{if} \quad a^{(m-i)} = 1 \end{cases}$$

$$= \begin{cases} a^{(m-i)} \frac{\sum_{j=0}^{N-1} a^{mj} e_j[\mathbf{a}^*] - a^{(m-i)N} \sum_{j=0}^{N-1} a^{ij} e_j[\mathbf{a}^*]}{1-a^{(m-i)}} & \text{if} \quad a^{(m-i)} \neq 1 \\ \sum_{j=0}^{N-1} a^{ij} e_j[\mathbf{a}^*](N-j) & \text{if} \quad a^{(m-i)} = 1 \end{cases}$$

As we are supposing $a \neq 1$, case $a^{(m-i)} = 1$ in the last line means $m = i$. The numerator equals the denominator and the result is verified for this case. For case $m \neq i$, $a^{(m-i)} \neq 1$, we must use the characteristc polynomial to get $\sum_{j=0}^{N-1} a^{mj} e_j[\mathbf{a}^*] = -a^{mN} \frac{(-)^N}{a^{N(N+1)/2}}$. Then,

$$\frac{a^{(m-i)}}{1-a^{(m-i)}} \left\{ \sum_{j=0}^{N-1} a^{mj} e_j[\mathbf{a}^*] - a^{(m-i)N} \sum_{j=0}^{N-1} a^{ij} e_j[\mathbf{a}^*] \right\}$$

$$= \frac{a^{(m-i)}}{1-a^{(m-i)}} \left\{ -a^{mN} \frac{(-)^N}{a^{N(N+1)/2}} - a^{(m-i)N} \left[-a^{iN} \frac{(-)^N}{a^{N(N+1)/2}} \right] \right\} = 0 \ .$$

◄

We can easily verify that tr $\mathbf{Z}_i\,[\mathbb{B}] = 1$ and $\mathbb{B}\ \mathbf{Z}_i = a^i\ \mathbf{Z}_i$. A consequence of the latter is

$$f(\mathbb{B})\ \mathbf{Z}_i = f(a^i)\ \mathbf{Z}_i \ , \tag{17.26}$$

whose particular case

$$\mathbb{B}^t\ \mathbf{Z}_i = a^{it}\ \mathbf{Z}_i \tag{17.27}$$

will be helpful later on.

Commentary 290.1. ► We shall, in a later section (see 17.2.3), find another expression for the projectors of a Bell matrix [Eq.(17.72)], as well as for its eigenvectors (see page 303). ◄

To give an idea of their aspect, the projector matrices $\mathbf{Z}_i^{(N)}[\mathbb{B}[g]]$ for $N = 3$ are:

$$\mathbf{Z}_1^{(3)} = \begin{pmatrix} 1 & 0\,0 \\ \frac{g_2}{g_1(1-g_1)} & 0\,0 \\ \frac{3g_2^2+g_3(1-g_1)}{g_1(1-g_1)^2(1+g_1)} & 0\,0 \end{pmatrix} ; \ \mathbf{Z}_2^{(3)} = \begin{pmatrix} 0 & 0 & 0 \\ \frac{-g_2}{g_1(1-g_1)} & 1 & 0 \\ \frac{-3g_2^2}{g_1^2(1-g_1)^2} & \frac{3g_2}{g_1(1-g_1)} & 0 \end{pmatrix} ; \tag{17.28}$$

$$\mathbf{Z}_3^{(3)} = \begin{pmatrix} 0 & 0 & 0 \\ 0 & 0 & 0 \\ \frac{-3g_2^2-g_1g_3(1-g_1)}{g_1^2(1-g_1)^2(1+g_1)} & \frac{-3g_2}{g_1(1-g_1)} & 1 \end{pmatrix} . \tag{17.29}$$

Clearly, as they have not the form (15.75), they cannot be the Bell matrices of any function. They inherit, however, a good property of the Bell matrices: for each N, the projectors $\mathbf{Z}_j^{(N)}$ contain, in their higher rows, the projectors $\mathbf{Z}_j^{(k)}$ for all $k < N$. The upper-left 2×2 submatrices of the $\mathbf{Z}_k^{(3)}$ above are just the second-order projectors $\mathbf{Z}_k^{(2)}$. Another point comes from the fact that the projectors are constituted by columns which are their own eigenvectors with unit eigenvalues. We can read such eigenvectors in the examples above. We see that $\mathbf{Z}_1^{(3)}$ has only one nonvanishing eigenvector, while $\mathbf{Z}_2^{(3)}$ and $\mathbf{Z}_3^{(3)}$ have respectively two and three.

A certain family of functions will play an important role in what follows. We recall that the first column of a Bell matrix contains just the series coefficients. We shall take the first-column entries of the matrix (17.12) as Taylor coefficients, thereby defining the functions

$$R_i^{(N)}(x) = \sum_{r=1}^{\infty} \frac{x^r}{r!} \left[\mathbf{Z}_i[\mathbb{B}]\right]_{r1} \tag{17.30}$$

$$= \sum_{k=1}^{N} \Lambda_{ik}^{-1} g^{<k>}(x) \ . \tag{17.31}$$

To each projector $\mathbf{Z}_i$ corresponds such an "elementary function" $R_i^{(N)}(x)$, a relationship reflecting in part that between the series and their Bell matrices. Taking the summation of all $R_i^{(N)}(x)$ and using $\mathbf{I} = \sum_{i=1}^{N} \mathbf{Z}_i$, we find

$$\sum_{i=1}^{N} R_i^{(N)}(x) = \sum_{r=1}^{\infty} \frac{x^r}{r!} \left[\sum_{i=1}^{N} \mathbf{Z}_i[\mathbb{B}]\right]_{r1} = \sum_{r=1}^{\infty} \frac{x^r}{r!} \delta_{r1} = x \ . \tag{17.32}$$

Thus, just as the projectors give a decomposition of the identity matrix, the elementary functions provide a "resolution of the identity", a decomposition of the identity function $Id(x) = e(x) = x$:

$$Id(x) = \sum_{i=1}^{N} R_i^{(N)}(x) \ . \tag{17.33}$$

Applying to $\mathbb{B}[g]$ the general formula (1.9) for the function of a matrix, we have

$$\mathbb{B}^t = \sum_{i=1}^{N} a^{it} \mathbf{Z}_i[\mathbb{B}] = \sum_{k=1}^{N} C_k^{(N)}(t) \ \mathbb{B}^k \ , \tag{17.34}$$

where

$$C_k^{(N)}(t) = \sum_{i=1}^{N} a^{it} \Lambda_{ik}^{-1} \ . \tag{17.35}$$

The Taylor coefficients of the continuum iterate function

$$g^{<t>}(x) = \sum_{r=1}^{\infty} g_r^{<t>} \frac{x^r}{r!} \tag{17.36}$$

will be

$$g_r^{<t>}(x) = [\mathbb{B}^t[g]]_{r1} = \textstyle\sum_{i=1}^N a^{it}[\mathbf{Z}_i[\mathbb{B}]_{r1}$$
$$= \sum_{k=1}^{N} C_k^{(N)}(t) \mathbb{B}^k[g]_{r1} = \sum_{k=1}^{N} C_k^{(N)}(t) g_r^{<k>} \ . \tag{17.37}$$

Therefore,

$$g^{<t>}(x) = \sum_{k=1}^{N} C_k^{(N)}(t) g^{<k>}(x) = \sum_{k=1}^{N} \left[\sum_{j=1}^{N} a^{jt} \Lambda_{jk}^{-1} \right] g^{<k>}(x) \ . \tag{17.38}$$

Time dependence is factorized in the alternative expression

$$g^{<t>}(x) = \sum_{k=1}^{N} a^{kt} R_k^{(N)}(x) \ . \tag{17.39}$$

One of the required properties for an iterate, Eq.(17.17), follows immediately: $g^{<0>}(x) = Id(x)$. For $t = 1$, a suggestive decomposition of the function comes up:

$$g(x) = \sum_{k=1}^{N} a^k R_k^{(N)}(x) \ . \tag{17.40}$$

It remains to show that (17.39) does satisfy the other property, Eq.(17.16). Notice first that, from the definition (17.30) of $R_i^{(N)}$ and the multinomial theorem (15.4),

$$R_i^{(N)}[g^{<t>}(x)] = \sum_{r=1}^{\infty} [\mathbf{Z}_i[\mathbb{B}]]_{r1} \frac{[g^{<t>}(x)]^r}{r!} = \sum_{r=1}^{\infty} \sum_{j \geq r} \frac{x^j}{j!} \mathbb{B}_{jr}[g^{<t>}][\mathbf{Z}_i[\mathbb{B}]]_{r1}$$

$$= \sum_{j=1}^{\infty} \frac{x^j}{j!} \sum_{r=1}^{j} \mathbb{B}_{jr}^t[g] \ [\mathbf{Z}_i[\mathbb{B}]]_{r1} = \sum_{j=1}^{\infty} \frac{x^j}{j!} \left[\mathbb{B}^t \ \mathbf{Z}_i\right]_{j1} \ . \tag{17.41}$$

Therefore,

$$g^{<t'>}(g^{<t>}(x)) = \sum_{i=1}^{N} a^{it'} R_i^{(N)}[g^{<t>}(x)] = \sum_{r=1}^{\infty} \frac{x^r}{r!} \sum_{i=1}^{N} a^{it'} \left[\mathbb{B}^t \mathbf{Z}_i\right]_{r1}$$

$$= \sum_{r=1}^{\infty} \frac{x^r}{r!} \left[\mathbb{B}^t \sum_{i=1}^{N} a^{it'} \mathbf{Z}_i\right]_{r1} = \sum_{r=1}^{\infty} \frac{x^r}{r!} \left[\mathbb{B}^t \mathbb{B}^{t'}\right]_{r1} = \sum_{r=1}^{\infty} \frac{x^r}{r!} g_r^{<t+t'>} ,$$

which is $g^{<t+t'>}(x)$, just the result looked for. Notice also that, using (17.27) in (17.41), we obtain

$$R_k^{(N)}[g^{<t>}(x)] = a^{kt} R_k^{(N)}(x) \tag{17.42}$$

and, consequently,

$$R_k^{(N)}[g^{<t+t'>}(x)] = a^{k(t+t')} R_k^{(N)}(x) = a^{kt'} R_k^{(N)}[g^{<t>}(x)] . \tag{17.43}$$

The function decomposition is preserved in time, as we can write

$$g(x) = \sum_{k=1}^{N} R_k^{(N)}[g(x)] \tag{17.44}$$

and

$$g^{<t>}(x) = \sum_{k=1}^{N} R_k^{(N)}[g^{<t>}(x)] . \tag{17.45}$$

The expressions above hold for each value of N and give approximations to that order. Of course, exact results would only really come out when $N \to \infty$. The exact expressions would, as long as they are defined, take forms like

$$R_j(x) = \sum_{k \geq 1} [\Lambda^{-1}_{jk} g^{<k>}(x)] ; \tag{17.46}$$

$$g(x) = \sum_{k \geq 1} a^k R_k(x) ; \tag{17.47}$$

$$g^{<t>}(x) = \sum_{k \geq 1} a^{kt} R_k(x) . \tag{17.48}$$

Commentary 293.1. ▶ Notice *en passant* that this formula provides an alternative expression for the inverse series, which corresponds to the matrix inverse (1.15). It is enough to put $t = -1$:

$$g^{<-1>}(x) = \sum_{k \geq 1} a^{-k} R_k(x) . \tag{17.49}$$

We can, if we so wish, introduce the auxiliary two-variable function

$$F(z,x) = \sum_{k\geq 1} z^k R_k(x) \ .$$

It is such that

$$F(1,x) = x = Id(x); \ F(g_1,x) = g(x); \ F(g_1^n,x) = g^{<n>}(x); \ F(g_1^t,x) = g^{<t>}(x) \ .$$

◄

It is tempting to conjecture that the elementary functions R_i play, in function space and with the operation of composition, the role of projectors. In fact, this is not so. Their properties mirror only partially those of the projectors $\mathbf{Z}_i$. If we calculate $R_i[R_j(x)]$ using (17.30) and the multinomial theorem, we find $R_i[R_j(x)] = R_i(x^j/j!)$. The composition gives rise to a change in the variable. Thus, the functions R_i fail even to satisfy the defining idempotent property: in general, $R_i \circ R_i \neq R_i$. An explicit example of $R_j(x)$, for which all these properties can be tried out, is given below [see (17.61)].

We can introduce $\epsilon = \ln a$ and rewrite (17.39) as

$$x(t) = g^{<t>}(x_0) = \sum_{k\geq 1} e^{k\epsilon t} R_k(x_0) \ , \tag{17.50}$$

a decomposition of $g^{<t>}(x)$ into a sum of modes, each one evolving independently according to

$$x_k(t) = e^{k\epsilon t} x_k(0) = e^{k\epsilon t} R_k(x_0) \ . \tag{17.51}$$

The imaginary "frequency" $k\epsilon$ plays the role of a "modular Lyapunov exponent" of the k-th mode. We have thus a "multi-Hamiltonian" flow $x(t)$ with one Hamiltonian for each projector component: Eqs.(17.42) and (17.43) show $R_k(N)$ as a representation of the one-dimensional group engendered by the k-th dynamical flow

$$R_k[g^{<t>}(x)] = e^{k\epsilon t} R_k[g^{<0>}(x)] \ . \tag{17.52}$$

The function $g^{<t>}(x)$ is actually a Lagrange interpolation in the convenient variable $(g_1)^t = a^t$, which coincides with the discrete iterates at each integer value of t and *keeps the meaning of an iterate* in between. It is a homotopy of all the usual discrete iterations. As announced previously, we have used the basis $\{\mathbb{B}, \mathbb{B}^2, \mathbb{B}^3, \ldots, \mathbb{B}^N\}$, instead of $\{\mathbf{I}, \mathbb{B}, \mathbb{B}^2, \ldots, \mathbb{B}^{N-1}\}$, because in the latter the matrix $\mathbf{I}$ would correspond to the identity function $Id(x) = x$, which has not the same end-points of the maps of interest.

The variable t has been given the sense of a "time". If it is really time, $g^{<n>}(x)$ will give the n-th point of a Poincaré map. Now, there are in principle (infinitely) many dynamical flows corresponding to a given Poincaré map. The function $g^{<t>}(x)$, as given above, would correspond to a class of them, that of the flows with equal intervals of time between successive points on the Poincaré section. To consider other cases, we recall that any strictly monotonous function of a first-given "time" is another "time".

Finally, we cannot resist exhibiting a "thing of beauty". By the very manner it has been obtained, the expression (17.38) for the continuum iterate is equivalent to the highly mnemonic determinant equation

$$\begin{vmatrix} g^{<1>}(x) & a & a^2 & \dots\dots & a^N \\ g^{<2>}(x) & a^2 & a^4 & \dots\dots & a^{2N} \\ g^{<3>}(x) & a^3 & a^6 & \dots\dots & a^{3N} \\ \dots & \dots & \dots & \dots\dots & \dots \\ g^{<N>}(x) & a^N & a^{2N} & \dots\dots & a^{N^2} \\ g^{<t>}(x) & a^t & a^{2t} & \dots\dots & a^{Nt} \end{vmatrix} = 0 \,. \tag{17.53}$$

Commentary 295.1. ▶ Provided the discrete iterates are known, this formula (with $t = -1$) gives still another expression for the inverse series, basically the same as (17.49). ◀

Laplace expansions along the first column or the last row and comparison with (17.37) and (17.39) will give determinant expressions for $C_k^{(N)}(t)$ and $R_k^{(N)}(x)$.

Commentary 295.2. ▶ All the above approach fails, of course, at the value $a = 1$: the matrix Λ has no inverse in that case, all the eigenvalues of $\mathbb{B}[g]$ are degenerate, etc. Series g with $g_1 = 1$, sometimes called *unit series*, require a separate treatment (see, for example, reference [122]; we follow here reference [96]). They have, fortunately, a special property that renders such a treatment possible. Consider the $N \times N$ Bell matrix $\mathbb{B}$. Clearly,

$$\mathbb{B}^t = [\mathbb{I} + (\mathbb{B} - \mathbb{I})]^t = \sum_{j=0}^{\infty} \binom{t}{j} (\mathbb{B} - \mathbb{I})^j \,.$$

The mentioned property comes to light when we take the matrix elements. Matrix $(\mathbb{B} - \mathbb{I})$ is strictly lower-triangular and, consequently, nilpotent: $(\mathbb{B} - \mathbb{I})^M = 0$ for M high enough (and finite N). Actually, a simple direct check shows that $(\mathbb{B} - \mathbb{I})^j_{rs}$ is nonvanishing only for $r - s \geq j$. Consequently,

$$\mathbb{B}^t_{rs} = [\mathbb{I} + (\mathbb{B} - \mathbb{I})]^t_{rs} = \sum_{j=1}^{r-s} \binom{t}{j} (\mathbb{B} - \mathbb{I})^j_{rs}$$

and

$$g_r^{<t>} = \sum_{j=1}^{r-1} \binom{t}{j} (\mathbb{B} - \mathbb{I})^j_{r1} \,,$$

where use has been made of the convention (15.31), $\binom{t}{j} = \Gamma(t+1)/(\Gamma(t-j+1)\, j!)$. ◀

17.2.1 *Iterative roots*

The iterative square-root of function f is a function ϕ such that

$$\phi[\phi(x)] = f(x) \ .$$

When amenable to Bell matrices (that is, in the subdomain of functions represented by invertible series with vanishing independent term), the problem reduces to $\mathbb{B}^2[\phi] = \mathbb{B}[f]$. The concept extends naturally to higher orders [98]: the iterative n-th root of f is a function ϕ such that

$$\phi^{<n>}(x) = f(x) \ .$$

The most elementary case is given by the Babbage equation

$$\phi(\phi(x)) = x \ ,$$

whose solutions can be looked at as "roots of identity". Function ϕ must, in that case, be an involution, that is, its own inverse. A trivial example is $\phi(x) = 1 - x$. A less trivial is

$$\phi(x) = \frac{1}{1+x} - 1 = -\frac{x}{1+x} \ .$$

The corresponding Bell matrix $\mathbb{B}[-\frac{x}{1+x}] = \mathbb{B}\{(-)^k k!\}$ is also its own inverse, in this sense also a "root of identity": $\mathbb{B}^2[\frac{1}{1+x} - 1] = \mathbb{I}$. Notice the turning up of the "partially signed Lah numbers" $\mathbb{B}_{rs}[-\frac{x}{1+x}] = (-)^r \mathbb{B}_{rs}\{k!\} = (-)^r \binom{r-1}{s-1} \frac{r!}{s!}$. In a phrase: a matrix whose entries are partially signed Lah numbers is inverse to itself (see Comment 288.1). Function $\phi(x)$ above is actually a particular case of a more general 3-parameter solution,

$$\phi(x) = Id^{<1/2>}(x) = \frac{\alpha - \beta\, x}{\beta + \gamma\, x} \ , \ \forall\ \alpha, \beta, \gamma \ ,$$

which is "non-amenable" when $\alpha \neq 0$. In the "amenable" subdomain of functions, other solutions can be eventually found by using the Lagrange inversion formula for the case $g^{<-1>}(x) = g(x)$. Equation (15.99) becomes

$$g_n = \frac{1}{g_1^n} \left[\left(\frac{g_1 x}{g(x)} \right)^n - 1 \right]_{n-1} \ .$$

17.2.2 *An application to turbulence*

Let us describe an application of our basic result (17.48), which shows that a relationship may exist between harmonic oscillators and turbulence. Consider the special case of the extreme logistic map (see Comment 239.1)

$$g(x) = 4x(1-x) \ , \tag{17.54}$$

which is helpful in modeling the velocity field in fully developed turbulence (the "poor man's Navier–Stokes equation" [123]). It takes the interval $[0, 1]$ into itself. The coordinate transformation $x' = 1 - 2x$, $y' = 1 - 2y$, takes it into a map $[-1, 1] \to [-1, 1]$. There is a twist in the domains, immaterial for what follows.[3] We shall write $v(x) = 1 - 2x$, in terms of which the logistic map takes the form

$$g(x) = \tfrac{1}{2}\left[1 - T_2(1 - 2x)\right] , \tag{17.55}$$

where $T_2(z) = 2z^2 - 1 = \cos[2 \arccos z]$ is the second-order Chebyshev polynomial [124] of the first kind (see Eq.(15.117) and the text below it). Using $v^{<-1>}(u) = \frac{1}{2}(1 - u)$, we notice that

$$T_2(z) = v \circ g \circ v^{<-1>}(z) .$$

Iterating once this expression, we have $[T_2 \circ T_2](z) = v \circ g \circ v^{<-1>}[v] \circ g \circ v^{<-1>}(z) = T_2^{<2>}(z) = v \circ g^{<2>} \circ v^{<-1>}(z)$. In the same way we obtain the n-th iterate

$$T_2^{<n>}(z) = v \circ g^{<n>} \circ v^{<-1>}(z) = 1 - 2g^{<n>}[\tfrac{1}{2}(1 - z)] . \tag{17.56}$$

Commentary 297.1. ▶ A simpler parameterization would be obtained by introducing an angle ϕ and putting $x = \sin^2 \frac{\phi}{2}$. It would then follow $v(x) = \cos\phi, \phi = v^{<-1>}(x)$ and $g(x) = \sin^2 \phi$, which is just

$$g(x) = \sin^2[2 \arcsin \sqrt{x}] . \tag{17.57}$$

This would lead directly to Eq.(17.59) below. We shall use a longer procedure, because we shall later make use of the differential equation for the Chebyshev polynomials. ◀

Now, as said in Section 15.5.2, these Chebyshev polynomials transform composition into index-product:

$$T_n[T_m(z)] = [T_n \circ T_m](z) = T_{nm}(z) . \tag{17.58}$$

Therefore, $T_2^{<n>}(z) = T_{2^n}(z) = \cos[2^n \arccos z]$, and we obtain the closed expression $g^{<n>} = v^{<-1>} \circ [T_2^{<n>}(z)] \circ v$, or

$$g^{<n>}(x) = \tfrac{1}{2}\left[1 - T_2^{<n>}(1 - 2x)\right] .$$

Consequently, the n-th iterate of the extreme logistic map is

$$\begin{aligned} g^{<n>}(x) = & \\ & \tfrac{1}{2}\{1 - \cos[2^n \ \arccos(1 - 2x)]\} = \sin^2[2^{n-1} \ \arccos(1 - 2x)] \\ & = \sin^2[2^n \arcsin \sqrt{x}] . \end{aligned} \tag{17.59}$$

[3] We are thankful to Dr. K. Fujii, of the Department of Mathematical Sciences, Yokohama City University, for suggesting the derivation which follows, simpler than that of reference [19].

We proceed to find a continuum version of this iterate. The result will be

$$g^{<\alpha>}(x) = \tfrac{1}{2}\{1 - \cos[2^\alpha \ \arccos(1-2x)]\} = \sin^2[2^\alpha \arcsin\sqrt{x}\,] \ , \quad (17.60)$$

and can be naïvely anticipated, but we shall prove it. In the present case $a = 4$, the eigenvalue alphabet is $(4^1, 4^2, 4^3, \ldots)$ and the matrix ${\Lambda^{-1}}_{jn}$ in (17.13) is such that $\sum_{n\geq 1} {\Lambda^{-1}}_{kn}\, 4^{np} = \delta^p_k$. We use the second expression in (17.59) in the form $g^{<n>}(x) = -\frac{1}{2}\sum_{p\geq 1} \frac{(-)^p}{(2p)!}\, 4^{np}\, [\arccos(1-2x)]^{2p}$ and calculate (17.46):

$$R_k(x) = \sum_{n\geq 1} {\Lambda^{-1}}_{kn} g^{<n>}(x) = -\tfrac{1}{2}\,\frac{(-)^k}{(2k)!}\,[\arccos(1-2x)]^{2k} \qquad (17.61)$$

(notice that (17.32) is trivially satisfied). Equation (17.48) gives then

$$g^{<\alpha>}(x) = \sum_{k\geq 1} 4^{k\alpha}\, R_k(x) = -\tfrac{1}{2}\sum_{k\geq 1} \frac{(-)^k}{(2k)!}\,[2^\alpha \ \arccos(1-2x)]^{2k}, \ (17.62)$$

which is the same as (17.60).

Let us now recall a few facts about the double-oscillator. Given $x = \cos\omega t$ and $y = \cos(n\omega t)$, we have of course $y = T_n[x]$, which solves the differential equation $(1-x^2)y'' - xy' + n^2 y = 0$. This leads to the usual, "isolated", finite Lissajous curves. When the two frequencies are not commensurate, $y = \cos(\nu\omega t)$, the differential equation becomes

$$(1-x^2)y'' - xy' + \nu^2 y = 0 \ , \qquad (17.63)$$

whose solution $y = \cos[\nu \ \arccos x]$ is not a polynomial in the general case. It is a well-known fact that, when ν is not a rational number, each one of these transcendental, infinite Lissajous curves is dense [125] in the (x, y) square, and that the topology of the space of such trajectories is not even first-separable [54]. Now, the related functions

$$T_{2^\alpha}(z) = \cos[2^\alpha \ \arccos(1-2x)] = 1 - 2g^{<\alpha>}[\tfrac{1}{2}(1-z)] \qquad (17.64)$$

are just the continuum versions of the Chebyshev polynomials obtained from (17.60). For each value of α, $T_{2^\alpha}(z)$ solves (17.63) with $\nu = 2^\alpha$, and constitutes one of the Lissajous curves. The continuum parameter α, which can be interpreted as "time" for the logistic map, is a label for Lissajous curves from the point of view of the double oscillator. It is precisely the set of its values which constitutes a non-separable space: all trajectories corresponding to non-rational values of α are ultimately indistinguishable.

However rough the "poor man's" model may be, the result suggests that the velocity field in turbulent flow can assume a large amount of non-separable values.

In the passage from functions to Bell matrices, composition is translated into matrix product and iteration into matrix powers. Continuous powers of matrices have a sound meaning, which is translated back into continuous iteration and makes of it also a sound concept, respecting properties (17.16, 17.17). The usual discrete iterations are thus extended into a continuous flow. The term-by-term factorization of $g^{<t>}(x)$ into a factor depending only on x and another depending only on t reveals itself as a decomposition into independently evolving modes, one for each projector of the corresponding Bell matrix. Once it is established that iteration can have a continuous meaning and is, furthermore, a homotopy, new possibilities are made open to study, such as the use of the conservation of degrees like Brouwer's and Morse's. It is our contention that, with the continuum version of iterations, it will be possible to get a better understanding of the detailed unfolding of bifurcations and of the general relationship between the differential and the mapping approaches to chaotic dynamics.

Commentary 299.1. ► The general logistic map, which we have met in Comment 239.1, takes the interval $[0, 1]$ into itself, and reads

$$g(x) = ax(1-x)\ ,\ \ 0 \leq a \leq 4\ .$$

Function (17.54) is the extreme case $a = 4$. The wealth of information hidden in this simple map comes mainly from the changes in its behavior when a varies. It would be interesting to generalize the above approach, so as to have a general expression of the form $g(x) = f^{<-1>}[C\, f(x)]$, like (17.57), for any value of a. In that case, direct iteration shows that $g^{<n>}(x) = f^{<-1>}[C^n f(x)]$, like (17.59). From what has been found above, we see that

$$g_a(x) = \frac{a}{8}\,[1-\cos(2\arccos(1-2x))] = \frac{a}{4}\,\sin^2[2\arcsin\sqrt{x}] = \frac{a}{8}\,[1-T_2(1-2x)]$$

is an identity. Besides the above case $a = 4$, there is a simple solution for $a = 2$: define $f(x) = \ln(1-2x)$. Then,

$$\begin{aligned} g_2(x) &= \tfrac{1}{2}\,[1-\exp(2\ln[1-2x])] = \tfrac{1}{2}\,\left[1-[1-2x]^2\right] \\ &= \frac{1}{4}\,[1-T_2(1-2x)] = f^{<-1>}[2\,f(x)]\ . \end{aligned}$$

The continuous iterate follows immediately:

$$g_2^{<t>}(x) = \tfrac{1}{2}\,[1-\exp(2^t\ln[1-2x])] = \tfrac{1}{2}\,\left[1-[1-2x]^{2^t}\right].$$

◄

17.2.3 *The Schröder functional equation*

The equation to which $T_2(z)$ submits,

$$\arccos\left[T_2(z)\right] = 2\ \arccos(z)\ ,$$

is a particular case of a functional equation, the Schröder equation.[4] The Schröder equation for a function F is

$$F[g(x)] = c\,F(x)\ , \tag{17.65}$$

where c is a constant and $g(x)$ is a given function. The equation clearly determines F only up to a constant (example: for $g(x) = x^2$ and $c = 2$, $F(x) = \ln_b x$ in any base b) and F does not necessarily exist for every c and $g(x)$.

We shall consider a very particular case, in which $g(x)$ is of the form $g(x) = \sum_{j=1}^{\infty} g_j \frac{x^j}{j!}$, with $g_1 \neq 0$ and $c = g_1$:

$$F[g(x)] = g_1\ F(x)\ . \tag{17.66}$$

In that case, the equation has the so-called principal solutions [30], [98]

- a. if $g_1 < 1$,

$$F(x) = \lim_{n\to\infty} [g_1^{-n}\ g^{<n>}(x)]\ ; \tag{17.67}$$

- b. if $g_1 > 1$,

$$F(x) = \lim_{n\to\infty} [g_1^{n}\ g^{<-n>}(x)]\ . \tag{17.68}$$

It is significant that the solutions require all the discrete iterates.

Commentary 300.1. ▶ The result is more general: it supposes a fixed point of g, around which uniform convergence can be proven for F. Here, $x = 0$ satisfies the conditions, and is the case of interests for us. Notice that $c \neq 1$ implies $F(0) = 0$, justifying our looking for the Bell matrix of F. If $F[g(x)] = c\ F(x)$, c a constant, the general solution is $F(x) = f(x)\ \varphi(x)$ where $f(x)$ is a particular solution and $\varphi(x)$ is a general solution with $c = 1$ which is no particular solution with $c \neq 1$. ◀

In terms of $N \times N$ Bell matrices, the Schröder equation (17.66) is written

$$\mathbb{B}[g]\ [F] = \mathbb{B}[g_1 F]\ , \tag{17.69}$$

or

$$\mathbb{B}[g] = \mathbb{B}[g_1 F]\ \mathbb{B}^{-1}[F] = \mathbb{B}[g_1 F]\ \mathbb{B}[F^{<-1>}] = \mathbb{B}[F^{<-1>} \circ (g_1 F)]\ .$$

[4] We are thankful to Dr Keith M. Briggs, of the Department of Plant Sciences, University of Cambridge, for calling our attention to the Schröder equation. He has sent us the expression of the continuum iterate (17.73), which we derive here by using Bell matrices.

It follows that the matrix elements of the Bell matrix of g are given by a crossed-product:

$$\mathbb{B}_{nr}[g] = \sum_k \mathbb{B}_{nk}[g_1 F]\, \mathbb{B}^{-1}_{kr}[F] = \sum_k g_1^k\, \mathbb{B}_{nk}[F]\, \mathbb{B}_{kr}[F^{<-1>}] \,. \qquad (17.70)$$

The connection of the general equation with the iterates can be immediately seen from

$$\sum_j \mathbb{B}_{rj}[g]\; F_j = c\; F_r \;,$$

which leads to

$$\sum_j \mathbb{B}_{rj}[g^{<n>}]\; F_j = \sum_j \mathbb{B}^n_{rj}[g^{<n>}]\; F_j = c^n\; F_r \;.$$

Consider the set $\{\mathbf{P}_k\}$, $k = 1, 2, 3, \ldots, N$ matrices with entries

$$(P_k)_{rs} = \mathbb{B}_{rk}[F]\mathbb{B}_{ks}[F^{<-1>}] \qquad (17.71)$$

(fixed k). Such matrices are idempotents and orthogonal, as we easily see that

$$\mathbf{P}_k\; \mathbf{P}_{k'} = \delta_{kk'}\; \mathbf{P}_k \;.$$

As $\mathbb{B}[g] = \sum_k^N g_1^k\; \mathbf{P}_k$, a function φ of the matrix $\mathbb{B}$ will be given by $\varphi(\mathbb{B}[g])$ $= \sum_k^N \varphi(g_1^k)\; \mathbf{P}_k$. In particular, $\mathbb{B}^t_{nr}[g] = \sum_k^N g_1^{kt}\; (P_k)_{nr} = \sum_k^N g_1^{kt}\mathbb{B}_{nk}[F]$ $\mathbb{B}_{kr}[F^{<-1>}]$. This, and the fact that the $\mathbf{P}_k$'s satisfy the other properties of the projectors $\mathbf{Z}_k$, leads to the identification $\mathbf{P}_k = \mathbf{Z}_k$. We have thus another closed expression for the eigenprojectors of a Bell matrix:

$$(\mathbf{Z}_k[\mathbb{B}])_{nr} = \mathbb{B}_{nk}[F]\; \mathbb{B}_{kr}[F^{<-1>}] \;. \qquad (17.72)$$

Notice that matrix $\mathbb{B}[F]$, for $F(x)$ solving (17.66), plays for $\mathbb{B}[g]$ the role played by the unitary diagonalizing matrix $\mathbf{U}$ for normal matrix $\mathbf{M}$ in Eq.(1.31). Now it follows that

$$g_n^{<t>} = \mathbb{B}^t_{n1}[g] = \sum_k^N g_1^{kt}\; (P_k)_{n1} = \sum_k^N g_1^{kt}\; \mathbb{B}_{nk}[F]\; \mathbb{B}_{k1}[F^{<-1>}]$$

$$= \sum_k^N \mathbb{B}_{nk}[F]\; \mathbb{B}_{k1}[F^{<-1>}(g_1^t u)] = \sum_k^N \mathbb{B}_{nk}[g_1^t F]\; \mathbb{B}_{k1}[F^{<-1>}(u)] \;.$$

Consequently,

$$g^{<t>}(x) = \sum_{n\geq 1} \frac{x^n}{n!} \sum_k g_1^{kt}\mathbb{B}_{nk}[F]F_k^{<-1>}$$

$$= \sum_{n\geq 1} \frac{x^n}{n!} \sum_k \mathbb{B}_{nk}[g_1^t F] F_k^{<-1>} = \sum_{n\geq 1} \frac{x^n}{n!} \sum_k \mathbb{B}_{nk} \left[F^{<-1>} \left(g_1^t F\right)\right].$$

The continuum t-iterate of g,

$$g^{<t>}(x) = F^{<-1>} \left[g_1^t F(x)\right], \tag{17.73}$$

appears in this way as a consequence of the Schröder equation. By the way, Schröder's equation provides a formal way to find iterative roots (Section 17.2.1): the p-th iterative root of $f(x) = F^{<-1>}[c\,F(x)]$ will be $\phi(x) = F^{<-1>}[c^{1/p}\,F(x)]$.

From $g^{<t>}(x) = \sum_k g_1^{kt}\, R_k(x)$, we get

$$R_k(x) = \frac{1}{k!} \left(F(x)\right)^k F_k^{<-1>} \tag{17.74}$$

and

$$R_{kn} = \mathbf{Z}_k[\mathbb{B}]_{n1} = \mathbb{B}_{nk}[F]\; F_k^{<-1>} \; .$$

Note. Matrices $\mathbb{B}[F]$ play for the present case a rôle quite analogous to that of the unitary matrices which provide the projectors for normal matrices [compare, in particular, Eq.(17.72) above with Eq.(1.31), page 14].

A first trivial check is given by $\sum_{k\geq 1} R_k(x) = F^{<-1>}(F(x)) = x$. Another simple check involves the Stirling numbers. Consider as given the function

$$g(x) = \ln[1 + a\;(e^x - 1)] \tag{17.75}$$

and consider the Schröder equation

$$e^{g(x)} - 1 = a\;(e^x - 1)\;.$$

The solution is

$$F(x) = e^x - 1\;.$$

Well, $F^{<-1>}(x) = \ln(1+x)$, so that $F_k^{<-1>} = s_k^{(1)}$ (see Eq.(15.39)). As

$$\frac{1}{k!}[F(u)]^k = \sum_{n=k}^{\infty} \frac{u^n}{n!}\; S_n^{(k)}\;,$$

we find

$$R_{kn} = s_k^{(1)}\; S_n^{(k)} = (-)^{k-1}(k-1)!\; \mathbb{B}_{nk}\{1\}\;;$$

$$R_k(x) = s_k^{(1)} \sum_{n=k}^{\infty} \frac{x^n}{n!}\; S_n^{(k)}\;.$$

The continuous iterate is then

$$g^{<t>}(x) = \ln[1 + a^t\ (e^x - 1)]\ . \tag{17.76}$$

Equations (17.67) and (17.68) can be formally checked. For $a < 1$, for example: in $a^{-n} g^{<n>}(x) = \sum_{j \geq 1} \frac{(-)^{j-1}}{j} a^{n(j-1)} (e^x - 1)^n$, only the $j = 1$ term remains when $n \to \infty$, and consequently the solution is $F(x) = e^x - 1$. The same happens, when $a > 1$, for the other expression, $a^n g^{<-n>}(x) = \sum_{j \geq 1} \frac{(-)^{j-1}}{j}\ a^{n(j-1)} (e^x - 1)^n$.

We see from these considerations that, given any mapping f for which the inverse $f^{<-1>}$ exists, a closed expression for the continuous iterate of $g(x) = f^{-1}[af(x)]$ is at hand.

The Schröder equation can also be used to identify the eigenvectors of a Bell matrix. Taking matrix elements in (17.69), we obtain an eigenvalue equation,

$$\sum_j \mathbb{B}_{nj}[g]\ \mathbb{B}_{jr}[F] = g_1^r\ \mathbb{B}_{nr}[F]\ , \tag{17.77}$$

from which we see that the eigenvector related to the r-th eigenvalue g_1^r has, in its j-th row, the element

$$(v_r[\mathbb{B}])_j = \mathbb{B}_{jr}[F]\ . \tag{17.78}$$

The eigenvector $v_{(k)}$ of $\mathbb{B}$, corresponding to the k-th eigenvalue, will be a vector of components $v_{(k)s} = \mathbb{B}_{sk}[F]$. Checking directly,

$$\begin{aligned} [\mathbb{B}[g]\ v_{(k)}]_r &= \sum_s \mathbb{B}_{rs}[g]\ v_{(k)s} = \sum_s \mathbb{B}_{rs}[g]\ \mathbb{B}_{sk}[F] \\ &= \mathbb{B}_{rk}[F \circ g] = \mathbb{B}_{rk}[g_1 F] = g_1^k\ \mathbb{B}_{rk}[F] = g_1^k\ v_{(k)r}\ . \end{aligned}$$

Notice that F is fixed by the first eigenvector: $F_s = \mathbb{B}_{s1}[F] = v_{(1)s}$. As $\mathbb{B}[g]$ is not normal, the vectors $v_{(k)}$ will not constitute a linearly independent set. Their very special character lends to Bell matrices some of the properties of normal matrices. The above $\mathbb{B}[F^{<-1>}]$ has some analogy to the unitary diagonalizing matrix of Section 1.4. This is, however, only an analogy. If we try to use $\mathbb{B}[F^{<-1>}]$ to diagonalize $\mathbb{B}[g]$, we do get a diagonal matrix, but with the first eigenvalue g_1 everywhere in the diagonal.

Chapter 18

Gases: real and ideal

Statistical Mechanics of real gases provides some of the best applications of Bell polynomials. The whole formalism of cluster decomposition can be recast, advantageously, in their terms. Though, as a rule, the approach illustrates our frequent assertive that Bell polynomials are information-organizers and not creators of information, the approach gives much supplementary insight. Perhaps the main point is the presentation of all the results as properties of a certain basic matrix **A** — we call it the "Lee–Yang matrix" — which lies somewhat behind the scene, but must exist. All physical quantities (cluster integrals, pressure, density, virial coefficients) of the theory of real gases appear as invariants of this matrix under similarity transformations, a fact which brings to light the existence of a large group of invariance, which remains also implicit in the usual presentation. Furthermore, the Bell formalism gives closed expressions for all the relations involved. And, as sometimes a better organization of known information does produce something worthy, some recursion relations turn up and a better understanding comes forth in diagrammatics. For instance, we obtain a trivial proof of the Matsubara formula, according to which the contribution of all graphs to the grand canonical partition function is equivalent to the exponential of the contributions of the sole connected graphs. It should be said that, on the other hand, these physical applications give a much more intuitive view of the polynomials themselves. In the grand canonical formalism, the set of canonical partition functions contains the same information as the set of cluster integrals. The relationships between them show in which sense the polynomials are organizers, not creators of information. Let us then retrace the well-known *trivium* of the subject: the microcanonical, the canonical and the grand canonical ensembles.

18.1 Microcanonical ensemble

The microcanonical ensemble is, in Statistical Mechanics, the ensemble describing an isolated system, for which the energy is constant (see, for example, [126]). Energy is supposed to fluctuate in so small a range that its distribution is taken as a very narrow Gaussian centered at a value E. We shall use below the extreme limit of a narrow Gaussian, a Dirac delta function. Each microphysical state, describing the detailed situation of all the constituents (their position, momentum, spin, etc.) is supposed to have the same probability. The collective, macroscopic behavior of a system of N particles enclosed in a volume V is described by the three variables N, V and E. An average is taken over all the microscopical states, weighted by the probability of each state. As these probabilities are supposed equal *a priori*, the average over the states become simply a summation of their number. Discrete variables are easily taken into account, as summations over the values they can attain. For the continuum variables, summations reduce essentially to an integration on configuration space, times an integration on momentum space. This means the volume of the total phase space, an integration constrained by the requirement of constant energy or, in the relativistic case, of constant energy-momentum.

18.1.1 *Phase space volumes*

High-energy physics makes essential use of the microcanonical ensemble. Phase space volumes play an essential role in the calculation of cross-sections. In the absence of dynamical effects, only inert phase space remains, so that its volume is also important in assessing the meaning of experimental data. We shall consequently adopt the jargon and procedures of particle physicists [127], as well as their generous and smart use of delta functions.

Calculations of relativistic cross-sections for processes given by some scattering matrix T lead to integrals of the type

$$I_N(P) = \int \left[\prod_{i=1}^{N} \frac{d^3 p_i}{2e_i} \right] \delta^4(P - \sum_{i=1}^{N} p_i)\ T\ , \tag{18.1}$$

where $P = (E, \mathbf{p})$ is the total 4-momentum and $p_i = (e_i, \mathbf{p}_i)$ are the individual 4-momenta of the N particles involved. What is usually called *phase space volume* for a system of N particles is, in high-energy physics jargon, the expression

$$R_N^{(B)}(P, V) = \tfrac{1}{N!} \int \left[\prod_{i=1}^{N} \frac{d^3 x_i d^3 p_i}{h^3} \right] \delta^4(P - \sum_{i=1}^{N} p_i)$$

$$= \tfrac{1}{N!}\left(\frac{V}{h^3}\right)^N \int \left[\prod_{i=1}^{N} d^3p_i\right] \delta^4(P - \sum_{i=1}^{N} p_i) \; . \tag{18.2}$$

These expressions are clearly classical, that is, they suppose a Boltzmann statistics for the particles, though indistinguishability finds a partial allowance in the Gibbs factor $(1/N!)$. This is indicated by the upper index (B). Statistics is of course implicit in the normalization of the scattering matrix. We can, however, bring the normalization down to the phase space calculations. The use of Bell polynomials will allow

(1) to write the quantum phase space volumes in terms of the above classical, Boltzmann expression;
(2) to find a recursion relation [128] for the N-particle phase space R_N volume in terms of integrals of phase spaces R_k, with $k = 1, 2, 3, \ldots, (N-1)$.

A much used trick in Statistical Mechanics comes from noticing that, when going from the classical to the quantum description, the whole integration

$$\frac{1}{N!}\int \left[\prod_{i=1}^{N} \frac{d^3x_i d^3p_i}{h^3}\right]$$

is replaced by taking the trace. In the present case,

$$\delta^4(P - \sum_{i=1}^{N} p_i) = \delta(E - \sum_{i=1}^{N} e_i)\delta^3(\mathbf{P} - \sum_{i=1}^{N} \mathbf{p}_i) = \delta^3(\mathbf{P} - \sum_{i=1}^{N} \mathbf{p}_i)\mathrm{tr}(E - H_0) \; ,$$

with H_0 the free Hamiltonian operator. Then,

$$R_N(E, \mathbf{P}, V) = \mathrm{tr}_N^{\mathbf{P}} \delta(E - H_0) \; , \tag{18.3}$$

where the tags $\mathbf{P}$ and N are reminders that the trace of the formal operator is to be taken for N particles at fixed total momentum $\mathbf{P} = \sum_{i=1}^{N} \mathbf{p}_i$. When the whole system is at rest,

$$R_N(E, \mathbf{0}, V) = \tfrac{1}{N!}\int \left[\prod_{i=1}^{N} \frac{d^3x_i d^3p_i}{h^3}\right] \delta(E - H_0)\; \delta^3(\sum_{i=1}^{N} \mathbf{p}_i) \; . \tag{18.4}$$

In the laboratory system, we retain only

$$R_N(E, V) = \tfrac{1}{N!}\int \left[\prod_{i=1}^{N} \frac{d^3x_i d^3p_i}{h^3}\right] \delta(E - H_0) \; . \tag{18.5}$$

Now, it is possible to obtain the phase space volume for different statistics if the trace is taken with conveniently normalized state-representing kets. This points to the fact that the role of the symmetric group S_N in Statistical Mechanics is actually summed up in the ket normalization. This is much simpler than taking into detailed account the intricate real nature of the configuration space of identical particles (summarized in Section 18.4). The state of a system with n free particles is represented by an eigenket $|p_1, p_2, p_3, \cdots, p_n>$ of the momenta. For Boltzmann statistics, the ket normalization is given by

$$< p'_1, p'_2, p'_3, \cdots, p'_n | p_1, p_2, p_3, \cdots, p_n > \\ = \delta^3(p'_1 - p_1)\delta^3(p'_2 - p_2)\delta^3(p'_3 - p_3) \ \ldots \ \delta^3(p'_n - p_n) \, . \quad (18.6)$$

This comes directly from the normalization $< p'_k | p_k > = \delta^3(\mathbf{p}'_k - \mathbf{p}_k)$ for one-particle kets and from the fact that the total ket for a system of n free Boltzmann particles is a simple direct product of individual kets,

$$|p_1, p_2, p_3, \cdots, p_n> = |p_1> |p_2> |p_3> \cdots |p_n> ,$$

reflecting the complete independence of each particle. For bosons (upper sign) or fermions (lower sign), the ket is a sum over all permutations,

$$|p_1, p_2, p_3, \cdots, p_n> = \tfrac{1}{n!} \sum_{\{a_k\}} (\pm)^{a_k} |p_{a_1}> |p_{a_2}> |p_{a_3}> \cdots |p_{a_n}> , \quad (18.7)$$

where the sign factor indicates the permutation parity. The first examples are

$$|p_1, p_2> \ = \tfrac{1}{2!} \left[|p_1> |p_2> \pm |p_2> |p_1> \right] ;$$

$$|p_1, p_2, p_3> \ = \tfrac{1}{3!} \left[|p_1> |p_2> |p_3> \pm \ |p_2> |p_1> |p_3> \right. \quad (18.8)$$
$$\pm \ |p_1> |p_3> |p_2> \pm \ |p_3> |p_2> |p_1> \quad (18.9)$$
$$\left. + \ |p_2> |p_3> |p_1> + \ |p_3> |p_1> |p_2> \right] . \quad (18.10)$$

For fermions, Pauli's principle is automatically enforced: the kets vanish if, for example, $p_1 = p_2$.

Let us see in detail the normalizations for the $n = 2$ and $n = 3$ cases, in which the essentials already come out. Normalization for them is

$$< p'_1, p'_2 | p_1, p_2 > = \tfrac{1}{2!} [\delta^3(p_1 - p'_1)\delta^3(p_2 - p'_2) \pm \delta^3(p_2 - p'_1)\delta^3(p_1 - p'_2)] ;$$

$$< p'_1, p'_2, p'_3 | p_1, p_2, p_3 > =$$
$$\tfrac{1}{3!} [\delta^3(p_1 - p'_1)\delta^3(p_2 - p'_2)\delta^3(p_3 - p'_3) \pm \delta^3(p_1 - p'_2)\delta^3(p_2 - p'_1)\delta^3(p_3 - p'_3)$$
$$\pm \ \delta^3(p_1 - p'_1)\delta^3(p_2 - p'_3)\delta^3(p_3 - p'_2) \pm \delta^3(p_1 - p'_3)\delta^3(p_2 - p'_2)\delta^3(p_3 - p'_1)$$
$$+ \ \delta^3(p_1 - p'_2)\delta^3(p_2 - p'_3)\delta^3(p_3 - p'_1) + \delta^3(p_1 - p'_3)\delta^3(p_2 - p'_1)\delta^3(p_3 - p'_2)] \, .$$

Now two (long) remarks are of interest. The first is that the Hamiltonian for free particles is diagonal in this momentum basis, so that we should consider only the diagonal cases. For $n = 2$, we will only have

$$< p_1, p_2|p_1, p_2 > = \frac{1}{2!}[\delta^3(p_1-p_1)\delta^3(p_2-p_2)\pm\delta^3(p_2-p_1)\delta^3(p_1-p_2)] \ . \quad (18.11)$$

For fermions, this is the determinant

$$< p_1, p_2|p_1, p_2 > = \frac{1}{2!}\begin{vmatrix} \delta^3(p_1-p_1) & \delta^3(p_1-p_2) \\ \delta^3(p_2-p_1) & \delta^3(p_2-p_2) \end{vmatrix} \ . \quad (18.12)$$

For $n = 3$,

$$\begin{aligned} &< p_1, p_2, p_3|p_1, p_2, p_3 > = \\ &\tfrac{1}{3!}[\delta^3(p_1-p_1)\delta^3(p_2-p_2)\delta^3(p_3-p_3) \pm \delta^3(p_1-p_2)\delta^3(p_2-p_1)\delta^3(p_3-p_3) \\ &\pm \delta^3(p_1-p_1)\delta^3(p_2-p_3)\delta^3(p_3-p_2) \pm \delta^3(p_1-p_3)\delta^3(p_2-p_2)\delta^3(p_3-p_1) \\ &+ \delta^3(p_1-p_2)\delta^3(p_2-p_3)\delta^3(p_3-p_1) + \delta^3(p_1-p_3)\delta^3(p_2-p_1)\delta^3(p_3-p_2)] \ . \end{aligned}$$

For fermions, this is again a determinant:

$$< p_1, p_2, p_3|p_1, p_2, p_3 > = \frac{1}{3!}\begin{vmatrix} \delta^3(p_1-p_1) & \delta^3(p_1-p_2) & \delta^3(p_1-p_3) \\ \delta^3(p_2-p_1) & \delta^3(p_2-p_2) & \delta^3(p_2-p_3) \\ \delta^3(p_3-p_1) & \delta^3(p_3-p_2) & \delta^3(p_3-p_3) \end{vmatrix} \ . \quad (18.13)$$

Notice that for bosons the expressions are the same, but always with the positive sign. If we take a determinant and write for it the Laplace expansion as above, but taking always the positive sign, what we have is another object, called a permanent. Thus, we have determinants for fermions and permanents for bosons.

A widely used trick reveals something about the real meaning of the deltas in the above purely formal manipulations. It goes as follows: we recall the definition of the Dirac delta,

$$\delta^3(\mathbf{p}_1 - \mathbf{p}_2) = \int d^3x \ e^{i(\mathbf{p}_1-\mathbf{p}_2)\cdot\mathbf{x}} \ .$$

The equality $\mathbf{p}_1 = \mathbf{p}_2$ would give, of course, an infinite value for a true delta. But we notice the $\int d^3x$ and say that, actually, we only integrate over the volume of the system, so that

$$\delta^3(\mathbf{p}_1 - \mathbf{p}_1) = \int d^3x = V \ .$$

And then we use this systematically. For example,

$$< p_1, p_2|p_1, p_2 > = \frac{1}{2!}\left[V^2 \pm V\delta^3(\mathbf{p}_1 - \mathbf{p}_2)\right]$$

and

$$R_2(E,\mathbf{P},V) = \mathrm{tr}_2^{\mathbf{P}}\delta(E-H_0) = \int d^3p_1 d^3p_2 < p_1,p_2|\delta(E-H_0)|p_1,p_2 >$$
$$= \int d^3p_1 d^3p_2 \delta(E-e_1-e_2) < p_1,p_2|p_1,p_2 >$$
$$= \tfrac{1}{2!}\int d^3p_1 d^3p_2 \delta(E-e_1-e_2)\left[V^2 \pm V\delta^3(\mathbf{p}_1-\mathbf{p}_2)\right]. \quad (18.14)$$

In consequence the deltas, used in this way, are actually symbols of certain objects which tend to real Dirac deltas at very large (mathematically, infinite) values of V. Experiments using such results must prepare conditions in which the total volume is very large with respect to the other ("microscopic") scales involved. Here ends the first remark.

The second remark: all these expressions will be integrated over all the momenta. They will appear only in the integrand of $\int\prod_{i=1}^N d^3p_i$ and will consequently be symmetrised over the p_i's. This means that, in the above products of δ's, it is not important which particle is which. What remains is the way in which the particles associate. There are terms in which each particle couples to itself; terms in which a first particle couples with a second and that second couples to the first, forming a 2-particle cycle; and, in the $N=3$ case above, there are also 3-particle cycles: a first particle couples to a second, that one couples to a third which then couples to the first. What we want to say is that *only the cycle structure imports*, only distinct cycle structures will give different results after integration. Under the integration sign, the last expression for $< p_1,p_2,p_3|p_1,p_2,p_3 >$ gives the same result as

$$< p_1,p_2,p_3|p_1,p_2,p_3 > = \tfrac{1}{3!}[\delta^3(p_1-p_1)\delta^3(p_2-p_2)\delta^3(p_3-p_3)$$
$$\pm 3\delta^3(p_1-p_1)\delta^3(p_2-p_3)\delta^3(p_3-p_2) + 2\delta^3(p_1-p_2)\delta^3(p_2-p_3)\delta^3(p_3-p_1)]\,.$$

With the same trick previously described concerning the δ with vanishing argument,

$$< p_1,p_2,p_3|p_1,p_2,p_3 >$$
$$= \tfrac{1}{3!}\left[V^3 \pm 3V^2\delta^3(p_3-p_2) + 2\delta^3(p_1-p_2)\delta^3(p_2-p_3)\delta^3(p_3-p_1)\right]. \quad (18.15)$$

We can represent the above results (always with the proviso that the expressions only hold under the multiple integration symbol) in the compact notations

$$< p_1,p_2|p_1,p_2 > = \tfrac{1}{2!}\left[\hat{\delta}_1^2 \pm \hat{\delta}_2\right]; \quad (18.16)$$

$$< p_1, p_2, p_3 | p_1, p_2, p_3 > = \frac{1}{3!} \left[\hat{\delta}_1^3 \pm 3 \hat{\delta}_1 \hat{\delta}_2 + 2 \hat{\delta}_3 \right], \tag{18.17}$$

etc. The symbol $\hat{\delta}_k$ represents the product of k deltas forming a cycle. It has the general form

$$\hat{\delta}_k = \delta^3(p_{i+1} - p_{i+2}) \delta^3(p_{i+2} - p_{i+3}) \ldots \delta^3(p_{i+k} - p_{i+1}) \ .$$

For example, $\hat{\delta}_3 = \delta^3(p_{i+1} - p_{i+2}) \ \delta^3(p_{i+2} - p_{i+3}) \ \delta^3(p_{i+3} - p_{i+1})$.

These expressions are *cycle indicator polynomials* which, as we shall see, are Bell polynomials of a certain type. This spurs us into doing a short parenthesis on permutations and cycles.

18.1.1.1 *Cycle indicator polynomials*

Let us recall some well-known facts about permutations [129]. We have seen their matrix representation in Section 3.1. A general permutation P of particles with labels $x_1, x_2, \ldots, x_{N-1}, x_N$ is indicated by

$$P = \begin{pmatrix} x_1 & x_2 & \ldots & x_{N-1} & x_N \\ x_{p_1} & x_{p_2} & \ldots & x_{p_{N-1}} & x_{p_N} \end{pmatrix}$$

and can always be decomposed into the product of disjoint cycles, those particular "closed" permutations of type

$$P = \begin{pmatrix} x_1 & x_2 & \ldots & x_{r-1} & x_r \\ x_2 & x_3 & \ldots & x_r & x_1 \end{pmatrix} .$$

This is a cycle of length r, or an r-cycle. We may attribute a variable t_r to each r-cycle and indicate the cycle structure of a permutation by the monomial $t_1^{n_1} t_2^{n_2} \ldots t_r^{n_r}$, meaning that there is a number n_1 of 1-cycles, a number n_2 of 2-cycles, etc. Permutations of the same cycle type, that is, with the same set $\nu = \{n_j\}$, go into each other under the action of any element of the symmetric group S_N: they constitute conjugate classes. To all permutations of a fixed class will be attributed the same monomial above. In this sense, such monomials are invariants of S_N. The total number of permutations with such a fixed cycle configuration (that is, the number of members of the corresponding conjugate class) is

$$C_\nu = \frac{N!}{\prod_{j=1}^N n_j! j^{n_j}} \ . \tag{18.18}$$

The N-variable generating function for these numbers is called the *cycle indicator polynomial* [96] for S_N, written

$$C_N(t_1, t_2, \ldots, t_N) = \sum_{m=1}^N \sum_{\{n_i\}}'' \frac{N!}{\prod_{j=1}^N n_j! j^{n_j}} \ t_1^{n_1} t_2^{n_2} \ldots t_r^{n_r} \ , \tag{18.19}$$

where the double-prime in the second summation is a memento of two conditions: the summation is over all values of the n_i's satisfying the conditions $\sum_{j=1}^{N} n_j = m$ and $\sum_{j=1}^{N} jn_j = N$. These are just the same conditions (15.7) and (15.8), and simple comparison leads immediately to

$$C_N(t_1, t_2, \ldots, t_N) = \sum_{m=1}^{N} \mathbb{B}_{Nm}\{(j-1)!\, t_j\} \,. \tag{18.20}$$

Commentary 312.1. ▶ We can, as a test, calculate the number $p(N, m; \{n_i\})$ of permutations of type $\{n_i\}$ and whose number of orbits is fixed, $\sum n_i = m$:

$$p(N, m; \{n_i\}) = \sum_{\{n_i\}}'' \frac{N!}{\prod_{j=1}^{N} n_j! j^{n_j}} \,.$$

It is obtained by putting all $t_j = 1$ in $\mathbb{B}_{Nm}$. Recalling the relationship (15.37) between Bell polynomials and Stirling numbers,

$$p(N, m; \{n_i\}) = \mathbb{B}_{Nm}\{(j-1)!\} = |s_N^{(m)}| \,. \tag{18.21}$$

Thus, the number of N-permutations with m orbits is the modulus of the first-kind Stirling number (Cf. Comment 230.1). ◀

In view of (18.20), the recursion relation (15.107) leads to

$$C_{N+1}(t_1, t_2, \ldots, t_{N+1}) = \sum_{k=0}^{N} \frac{N!}{(N-k)!}\, t_{k+1}\, C_{N-k}(t_1, t_2, \ldots, t_{N-k}) \,, \tag{18.22}$$

with $C_0 = 1$.

Commentary 312.2. ▶ **Presentation** There is another means of introducing the symmetric group. Introduce the elementary transpositions (alluded to in Comment 24.1)

$$s_1 = \begin{pmatrix} x_1 & x_2 & x_3 & x_4 & \ldots & x_{N-2} & x_{N-1} & x_N \\ x_2 & x_1 & x_3 & x_4 & \ldots & x_{N-2} & x_{N-1} & x_N \end{pmatrix} ;$$

$$s_2 = \begin{pmatrix} x_1 & x_2 & x_3 & x_4 & \ldots & x_{N-2} & x_{N-1} & x_N \\ x_1 & x_3 & x_2 & x_4 & \ldots & x_{N-2} & x_{N-1} & x_N \end{pmatrix} ;$$

$$\ldots \quad \ldots \quad \ldots \quad \ldots \quad \ldots$$

$$s_{N-1} = \begin{pmatrix} x_1 & x_2 & x_3 & x_4 & \ldots & x_{N-2} & x_{N-1} & x_N \\ x_1 & x_2 & x_3 & x_4 & \ldots & x_{N-2} & x_N & x_{N-1} \end{pmatrix} .$$

Notice that s_1 exchanges the first and the second particles, s_2 exchanges the second and the third, and so on up to s_{N-1}, which exchanges the two last particles. Every permutation can be obtained as a product of such elementary transpositions, which are for that reason called *generators* of the symmetric group. They satisfy the product rules

$$s_i s_{i+1} s_i = s_{i+1} s_i s_{i+1} \;; \tag{18.23}$$

$$s_i s_j = s_j s_i \quad \text{for} \quad |i-j| \geq 2 \;; \tag{18.24}$$

$$s_i s_i = I \,. \tag{18.25}$$

The last rule has a very clear meaning: exchanging twice two particles of a fixed pair is the same as doing nothing, that is, applying the identity transformation. Waiving this condition would lead to quite a distinct group, the braid group we have found in Quantum Phase Space (Chapter 10). The Statistics of particles whose exchanges are governed by such an infinite group is significantly different from the Statistics (bosonic and fermionic) of particles whose exchanges are ruled by the symmetric group (see Section 18.4).

The rules above characterize completely the N-th symmetric group. Any group whose generators satisfy them is isomorphic to S_N. Discrete groups can be introduced in such a way, by

- defining the generators, elementary members of the group such that every other member is a product of their powers in some order;
- imposing constraints on some of their products.

This is called a *presentation*, and the group is said to have been "introduced through a presentation". ◀

Going back to the normalization problem: in the general case, the normalization of the N-particle state is given by

$$< p_1, p_2, \cdots, p_N | p_1, p_2, \cdots, p_N >$$
$$= \tfrac{1}{N!}\, C_N\{(\pm)^{j-1}\hat{\delta}_j\} = \tfrac{1}{N!} \sum_{m=1}^{N} \mathbb{B}_{Nm}\{(\pm)^{j-1}(j-1)!\; \hat{\delta}_j\}\,. \quad (18.26)$$

Relation (18.22), with the convention $\langle 0 \text{ particles} | 0 \text{ particles}\rangle = 1$, leads then to a recursion relation for the normalization factors:

$$\langle p_1, p_2, \cdots, p_N | p_1, p_2, \cdots, p_N \rangle$$
$$= \frac{1}{N} \sum_{r=0}^{N-1} (\pm)^{N-1-r} \hat{\delta}_{N-r}\; \langle p_1, p_2, \cdots, p_r | p_1, p_2, \cdots, p_r \rangle\,. \quad (18.27)$$

We insist that these expressions are not to be taken at their face value: they only hold under the multiple integration sign. Notice also that, once it is understood that $< p_1 | p_1 > = \delta^3(\mathbf{p}_1 - \mathbf{p}_1) = \delta^3(\mathbf{0}) = V$, what we have here is a recursion relation: normalization of any n-particle state can be obtained from the 1-particle normalization, in particular (18.16) and (18.17). For example, (18.16) is

$$\langle p_1, p_2 | p_1, p_2 \rangle = \tfrac{1}{2}\left[\hat{\delta}_1 \langle p_1 | p_1 \rangle \pm \hat{\delta}_2\right] = \tfrac{1}{2}\left[V^2 \pm V \delta^3(\mathbf{p}_1 - \mathbf{p}_2)\right].$$

The phase space volume, or the microcanonical partition function, is then

$$R_N(E, \mathbf{P}, V) = \tfrac{1}{N!} \int \left[\prod_{i=1}^{N} d^3 p_i\right] \delta(E - \sum_{i=1}^{N} e_i)\; \delta(\mathbf{P} - \sum_{i=1}^{N} \mathbf{p}_i)$$
$$\times \sum_{m=1}^{N} (\pm)^{N-m}\; \mathbb{B}_{Nm}\{(j-1)!\; \hat{\delta}_j\}\,. \quad (18.28)$$

The integration variables appear in the arguments $g_j = (j-1)!\,\hat{\delta}_j$ of the Bell polynomials. A detailed consideration of the low-order cases will convince the reader that it is possible to introduce carefully the integrations "inside" the arguments. The result is the Prakash–Sudarshan formula, an expression for the phase space volume of a quantum system of N particles, bosons (upper sign) or fermions (lower sign), in terms of the Boltzmann phase space volumes. In our notation, it reads

$$\begin{aligned} R_N^{(b,f)}(P,V) &= \tfrac{1}{N!}\sum_{m=1}^{N}(\pm)^{N-m}\sum_{\nu}{}'' \frac{C_\nu}{\prod_{j=1}^{N} j^{3n_j}}\, R_m^{(B)}(P,V) \\ &= \tfrac{1}{N!}\sum_{m=1}^{N}(\pm)^{N-m}\sum_{\nu}{}'' \frac{N!}{\prod_{j=1}^{N} n_j!\, j^{4n_j}}\, R_m^{(B)}(P,V) \\ &= \tfrac{1}{N!}\sum_{m=1}^{N}(\pm)^{N-m}\, R_m^{(B)}(P,V)\, \mathbb{B}_{Nm}\{j^{-3}(j-1)!\}\,. \end{aligned} \tag{18.29}$$

Now, using $t_j = (\pm)^{j-1}\hat{\delta}_j$ in (18.22) leads to a recursion formula for the quantum phase space volumes. With the convention

$$R_0(P,V) = \delta(E)\delta^3(\mathbf{p})$$

the formula reads

$$R_N(P,V) = \tfrac{1}{N}\,\tfrac{V}{h^3}\sum_{k=1}^{N}(\pm)^{k-1}\int d^3q R_{N-k}(P-lq,V)\,. \tag{18.30}$$

18.1.2 *Towards the canonical formalism*

The canonical partition function of an ideal quantum gas of N particles, to be examined in the next section, can be obtained as a Laplace transform of $R_N(E,\mathbf{P}=0,V)$ in the energy:

$$Q_N(\beta,V) = \int dE\; e^{-\beta E} R_N(E,V)\,, \tag{18.31}$$

where $\beta = 1/kT$ is the inverse temperature in natural units, with k the Boltzmann constant. Expression (18.30) leads then to

$$Q_N(\beta,V) = \tfrac{1}{N}\sum_{k=0}^{N-1}(\pm)^k\, Q_1[(k+1)\beta,V]\, Q_{N-1-k}(\beta,V)\,. \tag{18.32}$$

This gives, in the non-relativistic limit, some expressions we shall meet below, such as

$$Q_N(\beta,V) = \tfrac{1}{N}\,\tfrac{V}{\lambda^3}\sum_{k=0}^{N-1}\frac{(\pm)^k}{(N-k)^{3/2}}\,Q_k(\beta,V)\,, \tag{18.33}$$

and

$$Q_N(\beta,V) = \tfrac{1}{N}\,\tfrac{V}{\lambda^3}\sum_{k=0}^{N-1}(N-k)\;b_{N-k}\;Q_k(\beta,V)\;, \tag{18.34}$$

where [see Eq.(18.85)]

$$b_j = \frac{(\pm)^{j-1}}{j^{5/2}} \tag{18.35}$$

for bosons (upper sign) and fermions (lower sign), and

$$\lambda = \sqrt{\frac{h^2}{2\pi mkT}} \tag{18.36}$$

is the non-relativistic mean thermal wavelength for a particle of mass m. These are particular cases of a much more general expression, to be seen in the next section. Let us here only say that the canonical partition function of a real non-relativistic gas of N particles contained in a d-dimensional volume V is a cycle indicator polynomial for the symmetric group S_N, with the (combinatorially meaningless) variables $t_j = j\;b_j\frac{V}{\lambda^d}$ giving the contribution of the j-th order cluster integral:

$$Q_N(\beta,V) = \tfrac{1}{N!}C_N\left\{jb_j\frac{V}{\lambda^d}\right\}, \tag{18.37}$$

where in the general case b_j is the j-th cluster integral, in principle computable from the interaction potential between the constituents [31]. For 3-dimensional ideal quantum gases, statistics is accounted for by an effective interaction for which the j-th cluster integral is (18.35). We shall meet later the canonical partition function for a Bose/Fermi gas in the 2-dimensional case [Eq.(18.125)].

Quantities describing a gas of N identical particles must be invariant under the transformations of the symmetric group S_N. We are quite used to this idea. We shall later be able to exhibit a certain matrix $\mathbf{A}$ (to be called the "Lee–Yang matrix") to whose characteristic polynomial the grand canonical partition function is related. The virial coefficients are then related to traces of powers of $\mathbf{A}$ and the partial canonical partition functions to the minors of $\mathbf{A}$. Ultimately, everything of physical meaning can be written in terms of traces, and is consequently invariant under transformations of the unitary group $U(N)$, which appears then as an underlying symmetry.

18.2 The canonical ensemble

The canonical ensemble represents the system as immersed in a thermal bath, which fixes the temperature. The temperature, consequently, replaces the energy as the preserved quantity. The macroscopic variables are N, V and T.

As said in the previous section, the canonical partition function for a gas without interactions can be seen as a Laplace transform of the microcanonical partition function:

$$Q_N(\beta, \mathbf{P}, V) = \mathrm{tr}_N e^{\beta H_0} = \int dE e^{-\beta E} \mathrm{tr}_N \delta(E - H_0) \ .$$

We can use the momentum eigenstates and take the trace at fixed total momentum $\mathbf{P}$. This would give

$$Q_N(\beta, \mathbf{P}, V) = \int dE e^{-\beta E} \ R_N(E, \mathbf{P}, V) \ .$$

In the standard treatment of Statistical Mechanics, the physical system as a whole is supposed at rest. In that case,

$$Q_N(\beta, V) = Q_N(\beta, \mathbf{0}, V) = \int dE e^{-\beta E} R_N(E, \mathbf{0}, V) \ .$$

In the presence of interactions, the free Hamiltonian is replaced by the total Hamiltonian H and, in the rest system,

$$R_N(E, V) = R_N(E, \mathbf{0}, V) = \tfrac{1}{N!} \int \left[\prod_{i=1}^{N} \tfrac{d^3x_i d^3p_i}{h^3}\right] \delta(E - H) \ . \quad (18.38)$$

Thus,

$$Q_N(\beta, V) =$$
$$\int dE e^{-\beta E} R_N(E, V) = \tfrac{1}{N!} \int dE e^{-\beta E} \int \left[\prod_{i=1}^{N} \tfrac{d^3x_i d^3p_i}{h^3}\right] \delta(E - H) \ ,$$

or

$$Q_N(\beta, V) = \tfrac{1}{N!} \int \left[\prod_{i=1}^{N} \tfrac{d^3x_i d^3p_i}{h^3}\right] e^{-\beta H} \ . \quad (18.39)$$

This is the semiclassical expression of the canonical partition function. For the general quantum case with interactions between the constituent particles, $Q_N(\beta, V) = \mathrm{tr}\, e^{-\beta H}$. In the usual non-relativistic, semiclassical approach, the Hamiltonian is

$$H = \sum_{i=1}^{N} \frac{\mathbf{p_i}^2}{2m} + \sum_{i=1}^{N-1} \sum_{j=i+1}^{N} U(\mathbf{r}_i - \mathbf{r}_j) \ ,$$

where $U(\mathbf{r}_{ij}) = U(\mathbf{r}_i - \mathbf{r}_j)$ is the potential between particles i and j. Abrupt and/or divergent potential behavior are smoothed down through the introduction of the Mayer functions $f_{ij} = f(\mathbf{r}_{ij})$, which are such that

$$e^{-\beta U(\mathbf{r}_{ij})} = 1 + f(\mathbf{r}_{ij}) .$$

The only effect of the independent integrations on the momenta is to engender factors in the (cubes of) the thermal wavelength, so that

$$Q_N(\beta, V) = \frac{1}{N!\lambda^{3N}} \int \left[\prod_{k=1}^{N} d^3 r_k\right] \prod_{i=1}^{N-1} \prod_{j=i+1}^{N} [1 + f(\mathbf{r}_{ij})] . \quad (18.40)$$

Expanding the product, a succession of terms of the form $\sum\sum\sum f_{ij} f_{ik} f_{rs} \ldots$ appears, on which the integrations are to be performed. In this way $Q_N(\beta, V)$ is written as a sum of contributions of subsystems. There will be one term taking isolated particles into account, there will be terms taking into account all subsystems with 2 particles, terms taking into account all subsystems with 3 particles, and so on up to a term in which the N particles are considered. There is a diagram corresponding to each kind of linkage between particles. Each particle is represented by a point, and a line is drawn linking r and s if a factor f_{rs} is present. We shall see a simple example below [see Figures 18.1 and 18.2]. A diagram is connected when we can go from a particle to any other by following a succession of links, and disconnected otherwise.

We call a *k-cluster* any subset of k particles, which can be united to each other by some succession of common indices or not (a cluster can be connected or not). The procedure is systematized with the introduction of the *cluster integrals*, defined as

$$b_n = \frac{1}{n!\lambda^{3(n-1)}V} \int \prod_{k=1}^{n} d^3 r_k \sum \{\text{products of } f_{ij}\text{'s involving n particles}\}. \quad (18.41)$$

b_n accounts for contributions of the subsystems of n particles. We have consequently to separate the N particles into subsets: ν_1 isolated particles, ν_2 particles which constitute pairs, ν_k particles forming a k-cluster, etc. Each cluster of k particles will contribute a factor $k!\lambda^{3(k-1)}Vb_k$ to the expression of Q_N in (18.40). As there are ν_k of such sets, the set of all k-clusters will contribute a factor $[k!V\lambda^{3(k-1)}b_k]^{\nu_k}$.

How many ways are there to separate N indistinguishable particles into the subsets above? If the particles were distinguishable, there would be $N!$ permutations between them. The particles inside each k-cluster are k in number. They are indistinguishable, so that we have to divide $N!$ by $k!$ As there are ν_k of such clusters, we actually have to divide by $(k!)^{\nu_k}$. But particles can also be exchanged in block, by exchanging entire k-clusters, which are by themselves identical. Thus, we must still divide by ν_k! The desired number is consequently

$$\frac{N!}{\prod_{k=1}^{N}(k!)^{\nu_k}\nu_k!} . \tag{18.42}$$

Commentary 318.1. ▶ In Quantum Field Theory, this number is well known as the number of Feynman diagrams "topologically equivalent to each other". ◀

We have then

$$Q_N(\beta, V) = \frac{1}{N!\lambda^{3N}}\sum_{\{\nu_k\}}{}' \frac{N!}{\prod_{k=1}^{N}(k!)^{\nu_k}\nu_k!}\prod_{k=1}^{N}(k!V\lambda^{3(k-1)}b_k)^{\nu_k} .$$

The sum $\sum'_{\{\nu_k\}}$ is meant to take place over all the possible sets of numbers ν_k, that is, on all the solutions of the equation $N = \sum_{k=1}^{N} k\nu_k$. The number of clusters will be $m = \sum_{k=1}^{N} \nu_k$. We recognize the expressions and conditions defining Bell polynomials. We can, as seen in Comment 224.1, separate the summation into two parts, so as to have $\sum'_{\{\nu_k\}} = \sum_{m=1}^{N} \sum''_{\{\nu_k\}}$.

Commentary 318.2. ▶ Braving the danger of boring repetition (of what has just been said and, for example, of Comment 224.2), we can now reinterpret the coefficients in the detailed expression of the Bell polynomial in (15.6). First, the factor $N!/(\prod_j [j!]^{\nu_j})$ counts the number of ways to have N identical particles distributed in m clusters, with ν_j clusters of j particles. Division by $j!$ means that the j particles in each of the ν_j clusters are identical, so that their permutations must not be counted. The total number of distinct configurations of this kind is $(j!)^{\nu_j}$, for each j. Configurations differing by permutations of entire clusters of "size" j must also be excluded. There are $\nu_j!$ of such for each j, wherefrom the further division by the factor $\prod_j \nu_j!$ ◀

Finally, the general expression for the canonical partition function in terms of the cluster integrals is (with $v = \frac{V}{\lambda^3}$) a complete Bell polynomial:

$$Q_N(\beta, V) = \tfrac{1}{N!}\sum_{m=1}^{N}\sum_{\{\nu_i\}}{}'' \frac{N!}{\prod_{j=1}^{N}\nu_j!}(vb_1)^{\nu_1}(vb_2)^{\nu_2}\dots(vb_N)^{\nu_N} \tag{18.43}$$

$$= \tfrac{1}{N!}\sum_{m=1}^{N}\mathbb{B}_{Nm}\{j!vb_j\} = \tfrac{1}{N!}\sum_{m=1}^{N}\mathbb{B}_{Nm}\left[v\sum_{k=1}^{\infty}b_k t^k\right] \tag{18.44}$$

$$= \tfrac{1}{N!}\mathbb{Y}_N\left[\frac{V}{\lambda^3};\ \sum_{k=1}^{\infty} b_k z^k\right] . \tag{18.45}$$

As the cluster integrals appear always multiplied by v, we shall sometimes use the simplifying notation

$$b_j^* = vb_j = \frac{V}{\lambda^3}\, b_j .$$

And, once Bell polynomials have been identified, we can use all their machinery to get some results. For example, now we have not only

$$Q_N(\beta, V) = \tfrac{1}{N!}\sum_{m=1}^{N} \mathbb{B}_{Nm}\{j!\ b_j^*\} = \tfrac{1}{N!}\sum_{m=1}^{N} \mathbb{B}_{Nm}\left[\sum_{k=1}^{\infty} b_k^*\ t^k\right] \quad (18.46)$$

but also its inverse [from Eq.(15.81)],

$$b_n^* = \tfrac{1}{n!}\sum_{k=1}^{n} (-)^{k-1}(k-1)!\ \mathbb{B}_{nk}\{j!\,Q_j\} = \tfrac{1}{n!}L_n\{j!\,Q_j\} \quad (18.47)$$

in terms of the logarithmic polynomials of Comment 238.1. This is the same as

$$b_n^* = \tfrac{1}{n!\lambda^{3n}} \int \left[\prod_{i=1}^{n} d^3 r_i\right] \sum_{k=1}^{n} (-)^{k-1}(k-1)!\ \mathbb{B}_{nk}\left\{\prod_{i=1}^{j-1}\prod_{s=i+1}^{j} e^{-\beta U(\mathbf{r}_{is})}\right\}. \quad (18.48)$$

Commentary 319.1. ▶ For relativistic systems, potentials lose their meaning. Interactions are, for them, described by the S-matrix. The relationship between the b_k's and the S-matrix elements have been extensively studied by Dashen, Bernstein, Ma and Rajaraman (see [130]). ◀

There is a remarkable kinship between expressions (18.46), (18.47) and the pair of equations (16.4), (16.5). This fact suggests the presence of a matrix behind the scene. It will be seen in Section 18.5.2 that an $N \times N$ matrix $\mathbf{A}$ actually exists such that

$$tr(\mathbf{A}^k) = (-)^{k-1} k\ b_k^*\ , \quad \text{and}$$

$$Q_N(\beta, V) = \det \mathbf{A}\ .$$

The analogy also helps to understand the meaning of "sub-partition functions" Q_j in a gas with $N > j$ particles, which will be particularly important in the grand canonical formalism. From the discussion following Eq.(16.6), Q_j is the sum of all the partition functions of the subsets with j particles. The Fredholm formalism (Section 18.6) will show them as the minors of a Fredholm determinant which coincides with the grand partition function.

And here comes one of the best illustrations of the meaning of Bell polynomials which Physics can provide. Imagine a gas with N particles. The partition function Q_N counts all the ways by which the N particles can be exchanged, or are interacting with each other, or still the manners in which they can be symmetrized or antisymmetrized. The particles are grouped into subsets ("clusters") of ν_k particles, with the ν_k satisfying (15.7) and (15.8). At a given moment, there will be ν_1 isolated, non-interacting particles; ν_2 sets of particles which are interacting two-by-two; ν_3 are interacting three-by-three, ν_j are interacting j-by-j. Such a situation, characterized by the set of partitions (see Comment 224.1) $\{\nu_k\}$, is called a "configuration". The number $m = \sum_{j=1}^{N} \nu_j$ is the number of clusters in that configuration. The possible cases for $N = 3$,

$$Q_3 = \tfrac{1}{3!}\{v^3 + 3b_1\ 2b_2 v^2 + 3!b_3 v\} \ ,$$

are shown in the diagrams of Figures 18.1 and 18.2. Points joined by a line represent particles in binary interaction. A set of points directly or indirectly linked to each other constitutes a cluster. The number m appears clearly as the number of clusters in a configuration. The upper-left configuration of Figure 18.1 has 3 "clusters" of 1 particle each. This is what $\nu_1 = 3$ says. In the upper-right configurations the particles interact in pairs. There are thus 2 clusters in each case, one with 1 particle (so that $\nu_1 = 1$) and another with 2 particles (so that $\nu_2 = 1$). Finally, in the configurations of Figure 18.2, the 3 particles are joined by interactions: the diagrams are connected. They form only one cluster ($m = 1$) of 3 particles ($\nu_3 = 1$). Putting what has been said above in other words, for an arbitrary total number N of particles each configuration $\{\nu_k\}$ will be a solution of conditions (15.7) and (15.8).

Figure 18.2 shows also that a closed (doubly-connected) 3-cycle is counted in the same configuration as the three connected but open diagrams with 3 particles. The latter are not really cycles, but b_3 counts all of them together. Thus, though we speak of cycle decomposition, in the virial

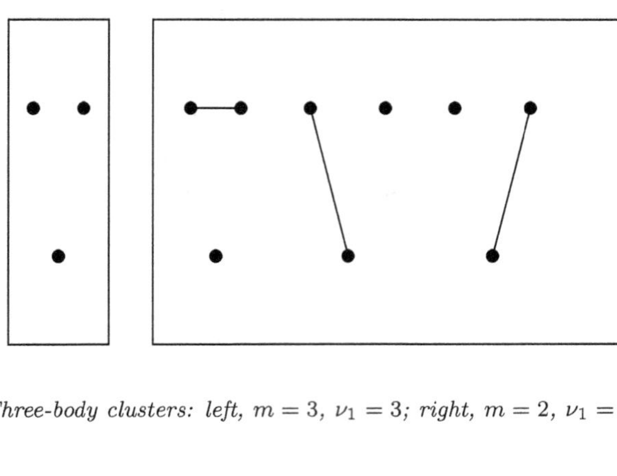

Fig. 18.1 *Three-body clusters: left, $m = 3$, $\nu_1 = 3$; right, $m = 2$, $\nu_1 = 1$ and $\nu_2 = 1$.*

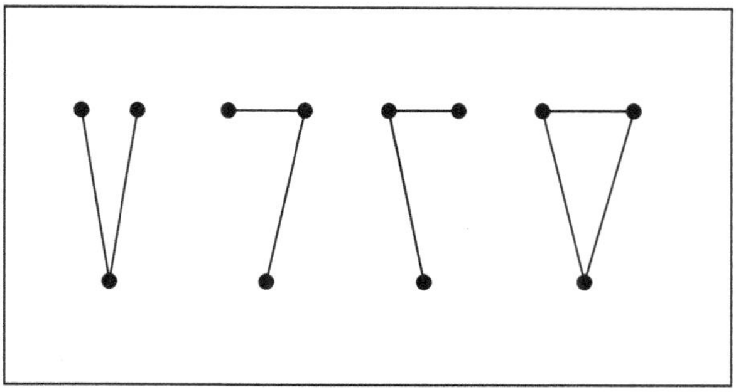

Fig. 18.2 *Three-body clusters: $m = 1$, $\nu_3 = 1$.*

approach the cycles are actually lost in the integrations. Complete cycle decompositions are recovered in the Fredholm formalism (Section 18.6).

The graphs with $m = 1$ correspond to the case in which each particle interacts with every other, that is, to the irreducible contribution of the N-body interaction. All the particles are interacting and the graph is completely connected. The contribution of the completely connected graphs to Q_N will consequently be the $m = 1$ term in (18.46):

$$Q_N^{(c)} = v\, b_N \ . \tag{18.49}$$

When two particles are exchanged, it is as if they were interacting for everything concerning graphs. There is an effective interaction ("exchange interaction") which simulates the effect of the quantum correlations induced by symmetrization or antisymmetrization. For an ideal non-relativistic gas of bosons or fermions, with the b_j's given in (18.35),

$$Q_N(\beta, V) = \tfrac{1}{N!} \sum_{m=1}^{N} \mathbb{B}_{Nm} \left[g \sum_{k=1}^{\infty} \frac{(\pm)^{k-1}}{k}\, Q_1(\beta k, V) t^k \right] \tag{18.50}$$

$$= \tfrac{1}{N!} \sum_{m=1}^{N} \mathbb{B}_{Nm} \left[g \frac{V}{\lambda^3} \sum_{k=1}^{\infty} \frac{(\pm)^{k-1}}{k^{5/2}} t^k \right] \tag{18.51}$$

$$= \tfrac{1}{N!} \sum_{m=1}^{N} (\pm)^{N-m} \left(\frac{gV}{\lambda^3} \right)^m \mathbb{B}_{Nm} \left\{ \frac{k!}{k^{5/2}} \right\} . \tag{18.52}$$

The parameter g is the number of values taken by some extra, "internal", degree of freedom (such as the spin). The relativistic case will be examined in Section 18.2.1.1. The case $m = N$ has only one configuration, that is, only one solution of the conditions (15.7) and (15.8): $\nu_1 = N$, $\nu_{k \neq 1} =$

0. Each particle is a cluster by itself, there is no exchange at all. The corresponding contribution is

$$\tfrac{1}{N!}\,\mathbb{B}_{NN}\left[g\textstyle\sum_{k=1}^{\infty}(\pm)^{k-1}(k-1)!\;Q_1(k\beta)t^k\right] = \tfrac{1}{N!}\,\left[g\;Q_1(\beta)\right]^N ,$$

that is, just what would come up for a Boltzmann gas. In the opposite extreme case, $m = 1$, the only solution is $\nu_N = 1$, $\nu_{k\neq N} = 0$. This means that all particles together form a single cluster.

It is actually much simpler to get the cluster integrals from the generating function of the Q_N's, which is the grand canonical partition function (see Comment 330.1). From the "recursion" formula (15.107) for the Bell polynomials with $Q_0 = 1$, we get

$$Q_N(\beta, V) = \frac{v}{N}\sum_{p=0}^{N-1}(N-p)\;b_{N-p}\;Q_p\;, \tag{18.53}$$

an expression which we have found previously [see Eq.(18.34)] for the particular case of ideal bosons and fermions. Some of the first examples are:

$$Q_1 = b_1 v\;\;;\;\;Q_2 = \tfrac{1}{2!}\left[b_1^2v^2 + 2b_2v\right]$$

$$Q_3 = \tfrac{1}{3!}\{b_1^3v^3 + 3!b_1b_2v^2 + 3!b_3v\}\;;$$

$$Q_4 = \tfrac{1}{4!}\{b_1^4v^4 + 12b_1^2b_2v^3 + 12b_2^2v^2 + 24b_1b_3v^2 + 4!b_4v\}\;.$$

Notice that $b_1 = 1$ in all cases. We can easily check that

$$vb_2 = \tfrac{1}{2!}\{2!Q_2 - Q_1^2\}\;,\quad vb_3 = \tfrac{1}{3!}\{3!Q_3 - 3!Q_1Q_2 + 2Q_1^3\}\;,$$

etc., which illustrates Eqs.(18.46) and (18.47). Also the fact that these relations are only an information rearrangement is clear: the sets $\{Q_p\}$ and $\{b_j\}$ have the same information content.

Comparing (18.44) with the identity (16.19), we find Widom's determinant form[1] of the canonical partition function,

$$Q_N(\beta,V) = \frac{1}{N!}\begin{vmatrix} b_1v & 1 & 0 & 0 & \cdots & 0 \\ -2b_2v & b_1v & 2 & 0 & \cdots & 0 \\ 3b_3v & -2b_2v & b_1v & 3 & \cdots & 0 \\ \cdots & \cdots & \cdots & \cdots & \cdots & . \\ \cdots & \cdots & \cdots & \cdots & \cdots & N-1 \\ (-)^{N-1}Nb_Nv & (-)^N(N-1)b_{N-1}v & \cdots & \cdots & \cdots & b_1v \end{vmatrix}. \tag{18.54}$$

[1] An expression of this type was first found by B. Widom for the ideal Bose gas case [131].

This can be seen as a solution of the recursion equation (18.53).

Commentary 323.1. ▶ Typically, determinant $Q_N(\beta, V) = \frac{1}{N!}\sum_{m=1}^{N} \mathbb{B}_{Nm}\{j!\, v\, b_j\}$ solves the convolution equation $Q_N(\beta, V) = \frac{v}{N}\sum_{p=0}^{N-1}(N-p)\; b_{N-p}Q_p$ (in the sense given in Comment 252.1. ◀

Commentary 323.2. ▶ The role of the symmetric group S_N, already discussed in the context of phase spaces, is also exhibited in the above cluster decomposition of the canonical partition function Q_N, which happens to be an invariant polynomial of S_N [105]. The decomposition has been so built as to be invariant under the exchange of particles. Thus, $Q_N(\beta, V)$ is a multi-variate polynomial (in the variables b_n) which is invariant under the action of S_N. ◀

Once the partition function is given, it is of interest to introduce a thermodynamic potential, a function of $Q_N(\beta, V)$ from which physical quantities (internal energy, specific heats, etc.) are obtained as derivatives. For the canonical ensemble, the potential usually taken is the Helmholtz free energy F, which is given by $Q_N(\beta, V) = \exp[-\beta F]$, or

$$F = E - TS = -\,\frac{1}{\beta}\,\ln Q_N(\beta, V)\ .$$

The internal energy is then given by $E = -\frac{\partial}{\partial\beta}\ln Q_N(\beta, V)$, the pressure by $p = kT\frac{\partial}{\partial V}\ln Q_N(\beta, V)$, etc.

18.2.1 *Distribution functions*

In the process of calculating cluster integrals it is sometimes necessary to take integrations *into the arguments* of Bell polynomials, or to extract them. As anyhow no higher order integrations have ever been achieved for the simplest interaction potentials (the record is Boltzmann's calculation of the 4-th virial coefficient for a gas of hard spheres), we shall use the solvable, simple case of the exchange interaction to illustrate the procedure. We shall actually consider the distribution functions, whose non-triviality measures quantum correlations between particles in a gas. Relativistic gases are more willing to exhibit such effects. We shall therefore consider their case [106].

18.2.1.1 *Relativistic gases, starting*

Some preliminaries are in order. When considering a gas of relativistic particles, it is extremely convenient to change to adequate units. In particular, there are two lengths of major interest:

- Given a particle of mass m, its static Compton length

$$\lambda_C = \frac{\hbar c}{mc^2} \tag{18.55}$$

will be a natural unit of length.

- At inverse temperature $\beta = 1/kT$, (the cube of) its relativistic thermal wavelength is given by

$$\Lambda^3(\beta) = 2\,\pi^2\,\beta mc^2\,\frac{e^{-\beta mc^2}}{K_2(\beta mc^2)}\left(\frac{\hbar c}{mc^2}\right)^3 . \tag{18.56}$$

Here $K_2(x)$ is the modified Bessel function of second order. To ease some calculations, and to obtain the non-relativistic and the ultra-relativistic limits, it is useful to recall the asymptotic behaviors of both $K_2(x)$ and $K_1(x)$,

- for $x >> 1$,

$$K_2(x) \approx \left(\frac{\pi}{2x}\right)^{1/2} e^{-x}\left(1 + \frac{15}{8x} + \ldots\right)$$

$$K_1(x) \approx \left(\frac{\pi}{2x}\right)^{1/2} e^{-x}\left(1 + \frac{3}{8x} + \ldots\right)$$

- for $x << 1$,

$$K_2(x) \approx \frac{2}{x^2}$$

$$K_1(x) \approx \frac{1}{x}$$

- as well as the differential relationship between them:

$$x\,\frac{dK_2(x)}{dx} = -2K_2(x) - xK_1(x) . \tag{18.57}$$

If we are interested in high-temperature behavior, it is also natural to use the variable $\tau = (kT/mc^2)$ as the temperature variable. For example,

$$\Lambda^3(\tau) = \frac{2\,\pi^2}{\tau}\,\frac{e^{-1/\tau}}{K_2(1/\tau)}\,\lambda_C^3 . \tag{18.58}$$

- The non-relativistic limit gives the usual expression:

$$\Lambda_{NR}(\beta) = \lambda = \hbar\sqrt{\frac{2\pi\beta}{m}} = \sqrt{\frac{2\pi}{\tau}}\,\lambda_C . \tag{18.59}$$

- The ultra-relativistic limit turns out to be

$$\Lambda_{UR}(\beta) = \pi^{2/3}\ \beta\hbar c = \pi^{2/3}\ \frac{\lambda_C}{\tau}\ . \tag{18.60}$$

Commentary 324.1. ▶ Eq.(18.57) always turn out to be necessary in the calculation of the energy. For the time being, let us only notice that

$$\frac{1}{\Lambda^3}\frac{\partial}{\partial\beta}\ \Lambda^3 = \frac{3}{\beta} + mc^2\left[\frac{K_1(\beta mc^2)}{K_2(\beta mc^2)} - 1\right] = \begin{cases} \frac{3}{2}kT \text{ (NR limit)} \\ 3kT \text{ (UR limit)} \end{cases} \ . \tag{18.61}$$

◀

In the absence of interactions, the Hamiltonian operator depends only on the momenta and consequently its exponential can be written, in the momentum basis,

$$e^{-\beta H} = \int d^3p_1 d^3p_2 \dots d^3p_N |p_1, p_2, \cdots, p_N > e^{-\beta\sum_{i=1}^N \epsilon_i(p)} < p_1, p_2, \cdots, p_N|\ .$$

It is only natural to take the trace in the same diagonal momentum basis $\{|p_1, p_2, \cdots, p_N >\}$:

$$\begin{aligned} Q_N(\beta, V) &= tre^{-\beta H} \\ &= \int d^3p_1 d^3p_2 \dots d^3p_N < p_1, p_2, \cdots, p_N|e^{-\beta H}|p_1, p_2, \cdots, p_N > \\ &= \int d^3p_1 d^3p_2 \dots d^3p_N e^{-\beta\sum_{i=1}^N \epsilon_i(p)} < p_1, p_2, \cdots, p_N|p_1, p_2, \cdots, p_N >\ . \end{aligned}$$

We can then use (18.26) and introduce carefully the integrations in the arguments of the Bell polynomials. A personal contact with the first lower cases is of great help to get the gist of how the cyclic δ's work to simplify the calculations. It is, however, necessary to have beforehand some knowledge of the kinds of integrals turning up. The relativistic energy of a particle of 3-momentum $\mathbf{p}$ is $\epsilon(\mathbf{p}) = \sqrt{\mathbf{p}^2c^2 + m^2c^4}$, and the integrals are of the type

$$\frac{1}{h^3}\int d^3p\ e^{-n\beta[(p^2c^2+m^2c^4)^{1/2}-mc^2]}\ .$$

It is better to consider the integral

$$f(\mathbf{r}, \beta) = \frac{1}{h^3}\int d^3p\ e^{-\frac{i}{\hbar}\mathbf{p}\cdot\mathbf{r}}\ e^{-\beta[(p^2c^2+m^2c^4)^{1/2}-mc^2]}\ , \tag{18.62}$$

which will be important in the ensuing discussion of distribution functions. Its value is

$$f(\mathbf{r},\beta) = \frac{1}{\Lambda^3(\beta)}\,\frac{1}{1+(\frac{r}{\hbar\beta c})^2}\,\frac{K_2\left[\beta mc^2\sqrt{1+(\frac{r}{\hbar\beta c})^2}\right]}{K_2(\beta mc^2)}\ . \qquad (18.63)$$

Λ is the relativistic thermal wavelength (18.56) and $r = |\mathbf{r}|$. Notice that $f(\mathbf{r},0) = \delta^3(\mathbf{r})$. Of course, what really appears in the partition function is simply

$$f(\mathbf{0},n\beta) = \frac{1}{\Lambda^3(n\beta)}\ .$$

Once in possession of the result above, the integrations inside the Bell polynomials can be performed, leading to the canonical partition function

$$Q_N(V,\beta) = \frac{1}{N!}\sum_{m=1}^{N}\mathbb{B}_{Nm}\left\{(\pm)^{k-1}(k-1)!\frac{V}{\Lambda^3(k\beta)}\right\}\ . \qquad (18.64)$$

A recursion formula follows then from Eq.(18.53):

$$Q_N(V,\beta) = \frac{V}{N}\sum_{m=1}^{N}\frac{(\pm)^{m-1}}{\Lambda^3(m\beta)}\,Q_{N-m}(V,\beta)\ . \qquad (18.65)$$

We shall have more to say on relativistic ideal gases in the context of the grand canonical ensemble (see Section 18.3.1.1).

18.2.1.2 *Quantum correlations*

It is sometimes interesting to use some other basis to take the trace of the exponential operator $e^{-\beta H}$. The most usual alternative is the configuration space basis $\{|r_1, r_2, \cdots, r_N>\}$. In that case,

$$Q_N(\beta,V) = \mathrm{tr} e^{-\beta H} = \frac{1}{N!h^{3N}}\int d^3r_1d^3r_2\ldots d^3r_N\ \times$$

$$\int d^3p_1d^3p_2\ldots d^3p_N e^{-\beta\sum_{i=1}^{N}\epsilon_i(p)}\left|\langle r_1,r_2,\cdots,r_N|p_1,p_2,\cdots,p_N\rangle\right|^2. \qquad (18.66)$$

This is what must be done when we want to calculate the N-point distribution function $F_N(r_1, r_2, \cdots, r_N)$, which is such that

$$F_N(r_1,r_2,\cdots,r_N)d^3r_1d^3r_2\ldots d^3r_N$$

is the probability of finding a particle in a region of volume d^3r_1 around r_1, another particle in a region of volume d^3r_2 around r_2, etc. In that case, it is enough to refrain from integrating on space in (18.66), and to normalize to 1:

$$F_N(r_1, r_2, \cdots, r_N) = \frac{1}{N! h^{3N} Q_N} \times$$

$$\int d^3p_1 d^3p_2 \ldots d^3p_N e^{-\beta \sum_{i=1}^{N} \epsilon_i(p)} \left| \langle r_1, r_2, \cdots, r_N | p_1, p_2, \cdots, p_N \rangle \right|^2 . \tag{18.67}$$

Use of the δ function definition allows to write the cyclic δ's in the form

$$\hat{\delta}_j = \int d^3r_1 d^3r_2 \ldots d^3r_j e^{-\frac{i}{\hbar}[\mathbf{p}_1 \cdot (\mathbf{r}_1 - \mathbf{r}_j) + \mathbf{p}_2 \cdot (\mathbf{r}_2 - \mathbf{r}_1) + \ldots + \mathbf{p}_j \cdot (\mathbf{r}_j - \mathbf{r}_{j-1})]} .$$

Once this is introduced in Eq.(18.26), the integrations $\int d^3r_k$ can be extracted from the Bell polynomials to put the partition function in the form

$$Q_N(V, \beta) = \tfrac{1}{N!} \int d^3r_1 d^3r_2 \ldots d^3r_N \sum_{m=1}^{N} \mathbb{B}_{Nm} \left\{ (\pm)^{j-1} (j-1)! f_{12} f_{23} \ldots f_{j1} \right\} , \tag{18.68}$$

where each f_{ij} stands here for $f(\mathbf{r}_i - \mathbf{r}_j, \beta)$, given by (18.62). Comparison with (18.64) suggests that

$$\int d^3r_1 d^3r_2 \ldots d^3r_j f_{12} f_{23} \ldots f_{j1} = \frac{V}{\Lambda^3(j\beta)} , \tag{18.69}$$

a result which has indeed been found, in another guise, a long time ago [106].

The distribution function can then be immediately read:

$$F_N(r_1, r_2, \cdots, r_N) = \frac{1}{N! Q_N} \sum_{m=1}^{N} \mathbb{B}_{Nm} \left\{ (\pm)^{j-1} (j-1)! f_{12} f_{23} \ldots f_{j1} \right\} . \tag{18.70}$$

The Fredholm formalism provides a determinant form for this expression (see Comment 349.1). Again a recursion formula follows from Eq.(18.53), with the convention $F_0 = 1$:

$$F_N(r_1, r_2, \cdots, r_N)$$
$$= \frac{V}{N} \sum_{m=1}^{N} (\pm)^{m-1} f_{12} f_{23} \ldots f_{m1} \, Q_{N-m} F_N(r_{m+1}, r_{m+2}, \cdots, r_N) . \tag{18.71}$$

From this N-point distribution, correlations can be calculated. The most usual, the pair correlation function, is given by $\nu(\mathbf{r}) = g(\mathbf{r}) - 1$, where $g(\mathbf{r})$ is defined by the $N \to \infty$ limit of

$$N^2 g(\mathbf{r}_1 - \mathbf{r}_2) = N(N-1) \int d^3r_3 d^3r_4 \ldots d^3r_N F_N(r_1, r_2, \cdots, r_N) .$$

It is here that recursion (18.71) shows its usefulness. Not very informative by itself, it greatly eases the calculations of the integrals. To compute the above expression, we separate the terms including only $\mathbf{r}_1$ and $\mathbf{r}_2$ and use the trivial results

$$\int d^3r_3 d^3r_4 \dots d^3r_N F_{N-1}(r_2, r_3, \cdots, r_N) = \tfrac{1}{V}$$
$$\int d^3r_3 d^3r_4 \dots d^3r_N F_{N-2}(r_3, r_4, \cdots, r_N) = 1$$

and the non-trivial

$$\int d^3r_3 d^3r_4 \dots d^3r_k f_{23}(\beta) f_{34}(\beta) \dots f_{k1}(\beta) = f_{12}[(k-1)\beta] \; .$$

The result is

$$g(\mathbf{r}) = \frac{1}{NQ_N}\left\{\frac{Q_{N-1}}{V\Lambda^3(\beta)} + f(\mathbf{r},\beta)\sum_{m=2}^{N}(\pm)^{m-1}Q_{N-m}f(\mathbf{r},(m-1)\beta)\right\}. \tag{18.72}$$

18.3 The grand canonical ensemble

The grand canonical ensemble represents a system S immersed in a far larger system which, besides playing the role of a thermal bath, is also able to exchange particles with S. There are, consequently, three macroscopic variables: the volume V, the temperature T and the chemical potential μ.

The grand canonical partition function for a gas of identical non-interacting quantum particles with chemical potential μ is given by the trace

$$\Xi(\beta, V, \mu) = \mathrm{tr}\left[e^{-\beta\sum_i(\epsilon_i-\mu)\hat{n}_i}\right], \tag{18.73}$$

where each $\hat{n}_i$ is the occupation number operator corresponding to the energy level ϵ_i. The thermodynamic potential for the grand canonical ensemble is the Gibbs potential $J = -pV$, such that $\Xi(\beta, V, \mu) = \exp[-\beta J]$.

We shall use the occupation number representation. Calculating the trace above means, in that case, simply summing over all the possible sets $\{n_j\}$ of occupation numbers at each fixed level. Thus,

$$\Xi(\beta, V, \mu) = \sum_{\{n_j\}} \langle n_0 n_1 n_2 \dots | e^{-\beta\sum_i(\epsilon_i-\mu)\hat{n}_i} | n_0 n_1 n_2 \dots \rangle$$
$$= \sum_{n_0}\sum_{n_1}\sum_{n_2} \dots e^{-\beta(\epsilon_0-\mu)n_0}\, e^{-\beta(\epsilon_1-\mu)n_1}\, e^{-\beta(\epsilon_2-\mu)n_2} \dots$$

$$= \prod_i \sum_n e^{-\beta(\epsilon_i - \mu)n} = \prod_\epsilon \sum_n e^{-\beta(\epsilon-\mu)n} . \tag{18.74}$$

We see that, in this non-interacting case, each level contributes an independent factor. The system can also have internal degrees of freedom, which will likewise contribute separately. Suppose there is a single degree of freedom taking g possible values. The partition function will, in that case, be

$$\Xi(\beta, V, \mu) = \prod_\epsilon \left[\sum_n e^{-\beta(\epsilon-\mu)n} \right]^g . \tag{18.75}$$

The type of statistics appears in the summation, which is over the possible values of the occupation number n, from $n = 0$ up to the maximum number of particles allowed in each state: 1 for fermions, ∞ for bosons. The product can be transformed into a summation by using the formal identity

$$\prod_\epsilon \{\ldots\} = \prod_\epsilon [\exp(\ln\{\ldots\})] = \exp \sum_\epsilon \ln\{\ldots\} .$$

To treat bosons and fermions at the same time, we adopt as always the convention upper signs for bosons, lower signs for fermions. It is also eventually convenient to use the fugacity variable $z = e^{\beta\mu}$. The above formulas lead to

$$\Xi(\beta, V, \mu) = e^{\left\{ g \sum_\epsilon \ln\left[1 \mp e^{-\beta(\epsilon-\mu)}\right]^{\mp 1} \right\}} = e^{\left\{ \mp g \sum_\epsilon \ln\left[1 \mp z e^{-\beta\epsilon}\right] \right\}} . \tag{18.76}$$

Expanding the logarithm and collecting like terms, we find

$$\Xi^{(\pm)}(V, \beta, \mu) = \exp \left\{ g \sum_\epsilon \sum_{k=1}^{\infty} \frac{(\pm 1)^{k-1}}{k} \, e^{-\beta(\epsilon-\mu)k} \right\} . \tag{18.77}$$

We might wish now to change $\sum_\epsilon$ into an integral over the momenta. In principle, the prescription is $\sum_\epsilon \to \frac{\int d^3x d^3p}{h^3}$, but a preliminary step would be necessary, as this rule gives zero weight to the zero-energy states. These states must be treated separately first. If this is done, they are given by the factor in curly braces of the expression

$$\Xi^{(\pm)}(V, \beta, \mu) = \left\{ \begin{matrix} \frac{1}{1-z} \\ 1+z \end{matrix} \right\}^g \exp \left[g \sum_{k=1}^{\infty} \frac{(\pm 1)^{k-1}}{k} z^k \sum_\epsilon e^{-k\beta\epsilon} \right] .$$

The prescription can now be applied. The sum $\sum_\epsilon e^{-k\beta\epsilon}$ is just the particular function $Q_1(k\beta, V)$, so that we recognize the canonical partition function (18.50). We arrive in this way at

$$\Xi^{(\pm)}(V, \beta, \mu) = (1 \mp z)^{\mp g} \sum_{N=0}^{\infty} z^N Q_N . \tag{18.78}$$

As long as the zero-energy terms can be neglected, the grand canonical partition function appears as a generating function for the canonical partition functions in terms of the fugacity:

$$\Xi(\beta, V, \mu) = 1 + \sum_{N=1}^{\infty} z^N Q_N(\beta, V) \,. \tag{18.79}$$

The same procedure can be applied to a gas with interacting constituents, with the same formal results. In the non-relativistic case, the partition function $\Xi(\beta, V, \mu)$ can be written, in terms of the cluster integrals b_j, as

$$\Xi(\beta, V, \mu) = e^{v \sum_{j=1}^{\infty} b_j z^j} = 1 + \sum_{N=1}^{\infty} \frac{z^N}{N!} \sum_{m=1}^{N} \mathbb{B}_{Nm} \left[v \sum_{j=1}^{\infty} b_j z^j \right] . \tag{18.80}$$

Notice that we have, on one hand,

$$\ln \Xi(\beta, V, \mu) = \sum_{n=1}^{\infty} z^n \, b_n^* \, ;$$

on the other,

$$\ln \Xi = \ln \left[1 + \sum_{N=1}^{\infty} z^N \, Q_N \right] = \sum_{k=1}^{\infty} (-)^{k-1} (k-1)! \frac{1}{k!} \left[\sum_{N=1}^{\infty} z^N \, Q_N \right]^k$$

$$= \sum_{k=1}^{\infty} (-)^{k-1} (k-1)! \sum_{n=k}^{\infty} \frac{z^n}{n!} \, \mathbb{B}_{nk} \{ j! \, Q_j \}$$

$$= \sum_{n=1}^{\infty} \frac{z^n}{n!} \sum_{k=1}^{n} (-)^{k-1} (k-1)! \, \mathbb{B}_{nk} \{ j! \, Q_j \} \, .$$

Comparing the two expressions for $\ln \Xi$, we get just (18.47),

$$b_n^* = \frac{1}{n!} \sum_{k=1}^{n} (-)^{k-1} (k-1)! \, \mathbb{B}_{nk} \{ j! \, Q_j \} \, .$$

18.3.1 *The Mayer formula*

In the cluster approach to Statistical Mechanics [31] of gases and liquids, the pressure and the density are given by

$$\frac{pV}{kT} = \ln \, \Xi = v \sum_{j=1}^{\infty} b_j z^j \, ; \tag{18.81}$$

$$N(z) = nV = z\frac{\partial}{\partial z}\ln \Xi = v\sum_{j=1}^{\infty} jb_j z^j \ . \tag{18.82}$$

Clearly, the concentration is given by $n(z) = z\frac{d}{dz}\frac{p(z)}{kT}$, an equation which can be integrated by homotopy [3] to give $\frac{p(z)}{kT} = \int_0^1 dt\ \frac{n(tz)}{t}$.

Commentary 330.1. ► As an example, let us give these functions for the case of ideal quantum gases. These come up in terms of Bose and Fermi functions, defined as

$$\left.\begin{matrix} g_\alpha(z) \\ f_\alpha(z) \end{matrix}\right\} = \frac{1}{\Gamma(\alpha)}\int_0^\infty \frac{x^{\alpha-1}dx}{z^{-1}e^x \mp 1} = \sum_{k=1}^{\infty}(\pm)^{k-1}\frac{z^k}{k^\alpha} \ . \tag{18.83}$$

Then,

$$\ln \Xi(z) = \frac{V}{\lambda^3}\left\{\begin{matrix} g_{5/2}(z) \\ f_{5/2}(z) \end{matrix}\right. ; \quad N(z) = \frac{V}{\lambda^3}\left\{\begin{matrix} g_{3/2}(z) \\ f_{3/2}(z) \end{matrix}\right. . \tag{18.84}$$

It is immediate to recognize the formal cluster integrals:

$$b_n = \frac{(\pm)^{n-1}}{n^{5/2}} \ . \tag{18.85}$$

◄

The "virial equations" (18.81), (18.82) can be thought of as a parametric form of the equation of state, which can in principle be found by inverting (18.82) to get $z = z(n\lambda^3)$, which is then introduced into the pressure expression. The resulting "virial equation" is a series[2]

$$\frac{pV}{NkT} = 1 + \sum_{k\geq 1} a_{k+1}\left(n\lambda^3\right)^k \ , \tag{18.86}$$

where the a_k's are the "virial coefficients", related to the b_k's by the Mayer formula [105]

$$a_{k+1} = \sum_{j=1}^{k}(-)^j \frac{(k+j-1)!}{(k+1)!(k-1)!}\ \mathbb{B}_{kj}(2!b_2, 3!b_3, \ldots) \ , \tag{18.87}$$

which we shall prove in the following. The closed expression for the equation of state in terms of the computable b_k's is, therefore

$$\frac{pV}{NkT} = 1 + \sum_{k\geq 1}\left(n\lambda^3\right)^k \sum_{j=1}^{k}(-)^j \frac{(k+j-1)!}{(k+1)!(k-1)!}\ \mathbb{B}_{kj}(2!b_2, 3!b_3, \ldots) \ . \tag{18.88}$$

It is simpler to prove (18.87) by analytical methods (see, for example, [104] and references therein). We shall give here a more combinatorial

[2] For a strong criticism of this approach, see [32].

proof, which is rather lengthy. We need beforehand two combinatorial identities [34], [95]:

$$\sum_{r=j}^{n} \frac{(r-1)!}{(r-j)!} = (j-1)!\binom{n}{j} ; \tag{18.89}$$

$$\sum_{j=1}^{n} (-)^{j-1} \binom{n}{j} \binom{n+j}{i} = \binom{n}{i} \qquad (\text{for } \; i < n) . \tag{18.90}$$

Commentary 332.1. ► Only to get a feeling of how such beautiful and esoteric results are proved, let us examine (18.89): it is the same as $\sum_{r=j}^{n} \binom{r-1}{j-1} = \binom{n}{j}$. Multiply the left-hand side by x^{j-1} and sum over j: $\sum_{j=1}^{n} \sum_{r=j}^{n} \binom{r-1}{j-1} x^{j-1} = \sum_{r=1}^{n} \sum_{j=1}^{r} \binom{r-1}{j-1} x^{j-1}$ $= \sum_{r=1}^{n} (1+x)^{r-1} = \sum_{k=0}^{n-1} (1+x)^k = \frac{1-(1+x)^n}{1-(1+x)} = \frac{1}{x}[(1+x)^n - 1]$. On the right-hand side, we have $\sum_{j=1}^{n} \binom{n}{j} x^{j-1} = \frac{1}{x} \sum_{j=1}^{n} \binom{n}{j} x^j$, and the same expression obtains. As it holds for an arbitrary $x < 1$, the coefficients must coincide. ◄

It is convenient to introduce the two formal series

$$f(z) = \frac{p\lambda^3}{kT} = \sum_{r=1}^{\infty} b_r z^r = \sum_{r=1}^{\infty} \frac{f_r}{r!} z^r ; \tag{18.91}$$

$$g(z) = \frac{N\lambda^3}{V} = z \frac{\partial}{\partial z} f(z) = \sum_{r=1}^{\infty} r b_r z^r = \sum_{r=1}^{\infty} \frac{g_r}{r!} z^r . \tag{18.92}$$

The equation of state (18.86) is

$$A(g) = \frac{pV}{NkT} = 1 + \sum_{n=1}^{\infty} a_{n+1} g^n , \tag{18.93}$$

and is in principle obtained by elimination of the fugacity z in the parametric form (18.91), (18.92). The direct procedure would consist in inverting (18.92) to get the function $z(g)$ and then insert the result in (18.91). As $\sum_{n=1}^{\infty} a_n g^n = gA(g) = f[z(g)]$, the composition would lead to

$$a_n = \frac{1}{n!} \sum_{r=1}^{n} r! \; b_r \; \mathbb{B}_{nr}(z_1, z_2, \ldots, z_{n-r+1}) . \tag{18.94}$$

The coefficients z_r of the series inverse to (18.92) are given by Lagrange's formula (15.97):

$$z_1 = 1 ; \quad z_{n\geq 2} = \sum_{j=1}^{n-1} (-)^j n^j \; \mathbb{B}_{n-1,j} \left[\ln \frac{g(t)}{t} \right] . \tag{18.95}$$

This approach does not lead to the closed form (18.87).

Commentary 332.2. ▶ By using what has been said in Comment 232.1,

$$A = \frac{f[z(g)]}{g} = \sum_{k=1}^{\infty} \frac{g^{k-1}}{k!} \sum_{j=1}^{k} \mathbb{B}_{kj}[f \circ g^{<-1>}] = 1 + \sum_{n=1}^{\infty} \frac{g^n}{(n+1)!} \sum_{j=1}^{n+1} \mathbb{B}_{n+1,j}[f \circ g^{<-1>}] \ ,$$

$$\therefore \ a_{n+1} = \frac{1}{(n+1)!} \sum_{j=1}^{n+1} \mathbb{B}_{n+1,j}[f \circ g^{<-1>}] \ .$$

◀

Let us begin with a method going back to Kahn [132]. It gives the inverse to what we want, that is, it gives the b_k's in terms of the a_n's. The Lagrange formula gives, now for series $g(z)$ in terms of its inverse,

$$g_1 = 1 \ ; \quad g_{n \geq 2} = \sum_{j=1}^{n-1} (-)^j n^j \ \mathbb{B}_{n-1,j} \left[\ln \frac{z(t)}{t} \right] . \tag{18.96}$$

We see that what matters is the function $\ln \frac{z(t)}{t}$. As

$$g(z) = z \frac{d}{dz} f(z) = z \frac{dg}{dz} \left[A(g) + g \frac{dA}{dg} \right] \text{ and consequently } \frac{1}{z} \frac{dz}{dg} = \frac{1}{g} \frac{d}{dg}[gA] \ ,$$

it comes out from integrating $\frac{d}{dg} \ln \frac{z}{g} = \frac{1}{z} \frac{dz}{dg} - \frac{1}{g} = \frac{1}{g} [\frac{d}{dg}[gA] - 1]$ that

$$\ln \frac{z(t)}{t} = \sum_{j=1}^{\infty} \frac{j+1}{j} \ a_{j+1} t^j \ . \tag{18.97}$$

The expression

$$\beta_{j+1} = - \ \frac{j+1}{j} \ a_{j+1} \tag{18.98}$$

has an interest by itself, and is called the "virial coefficient for the connected graphs". As $g_n = n \ n! \ b_n$, Eq.(18.96) becomes

$$b_n = \frac{1}{n \, n!} \sum_{j=1}^{n-1} (-)^j n^j \ \mathbb{B}_{n-1,j} \left[\ln \frac{z(t)}{t} \right] = \frac{1}{n \, n!} \sum_{j=1}^{n-1} (-)^j n^j \ \mathbb{B}_{n-1,j} \left\{ \frac{(k+1)! \, a_{k+1}}{k} \right\}$$
$$= \frac{1}{n \, n!} \sum_{j=1}^{n-1} (-)^j n^j \ \mathbb{B}_{n-1,j} \left(2! \ a_2, \frac{3!}{2} \ a_3, \frac{4!}{3} \ a_4, \ldots \right)$$
$$= \frac{1}{n \, n!} \sum_{j=1}^{n-1} (-)^j n^j \ \mathbb{B}_{n-1,j} \left\{ - \ j! \ \beta_{j+1} \right\} . \tag{18.99}$$

What is necessary now is to obtain the inverse to these expressions. We notice, to begin with, that (18.97) can be put into the form

$$\frac{(n+1)!}{n}\, a_{n+1}$$

$$= \mathbb{B}_{n1}\left[\ln\left(1+\frac{z(t)}{t}-1\right)\right] = \sum_{j=1}^{n} \mathbb{B}_{nj}\left[\ln\left(\frac{z(t)}{t}-1\right)\right] \mathbb{B}_{j1}\left[\ln\left(1+u\right)\right]$$

$$= \sum_{j=1}^{n} (-)^{j-1}\,(j-1)!\;\mathbb{B}_{n,j}\left(\frac{z_2}{2},\frac{z_3}{3},\ldots\right). \tag{18.100}$$

Use of the multinomial theorem gives

$$\mathbb{B}_{nj}\left[\ln\left(\frac{z(t)}{t}-1\right)\right]$$

$$= \mathbb{B}_{n,j}\left(\frac{z_2}{2},\frac{z_3}{3},\ldots\right) = n!\sum_{j=1}^{k}\frac{(-)^k}{(k-j)!(n+j)!}\,\mathbb{B}_{n+j,j}\left[z(t)\right]. \tag{18.101}$$

It is now that the combinatorial identity (18.89) comes in handy. We use it, after taking the last result into (18.100), to arrive at

$$\frac{(n+1)!}{n}\,a_{n+1} = n!\sum_{j=1}^{n}\frac{(-)^{j-1}\,(j-1)!}{(n+j)!}\binom{n}{j}\,\mathbb{B}_{n+j,j}\left[z(t)\right]. \tag{18.102}$$

We use now the Lagrange formula (15.97), which in the present case reads

$$\mathbb{B}_{n+j,j}\left[z(t)\right] = \binom{n+j-1}{j-1}\sum_{k=1}^{n}(-)^k\,(n+j)^k\,\mathbb{B}_{n,k}\left[\ln\frac{g(t)}{t}\right],$$

to get, after an inversion of the summation order,

$$\frac{(n+1)!}{n}\,a_{n+1} = \sum_{k=1}^{n}(-)^k\mathbb{B}_{n,k}\left[\ln\frac{g(t)}{t}\right]\sum_{j=1}^{n}(-)^{j-1}\binom{n}{j}(n+j)^{k-1}. \tag{18.103}$$

The form of the last summation can be modified, by using the generating function (15.38) for the second-kind Stirling numbers, into

$$\sum_{i=0}^{k-1} S_{k-1}^{(i)}\; i!\;\sum_{j=1}^{n}(-)^{j-1}\binom{n}{j}\binom{n+j}{i} = \sum_{i=0}^{k-1} S_{k-1}^{(i)}\; i!\,\binom{n}{i},$$

where use has been made of the second combinatorial identity (18.90), which applies because $k-1<n$. It is enough to look again to (15.38) to recognize the above expression as n^{k-1}. Consequently,

$$a_{n+1} = \tfrac{1}{(n+1)!}\textstyle\sum_{k=1}^{n}(-)^k\mathbb{B}_{nk}\left[\ln\frac{g(t)}{t}\right]\; n^k$$

$$= \frac{1}{(n+1)!}\sum_{k=1}^{n}(-n)^k\sum_{j=k}^{n}\mathbb{B}_{nj}\left[\frac{g(t)}{t}-1\right]\;\mathbb{B}_{jk}\left[\ln(1+u)\right].$$

Using (15.56), we find finally

$$a_{n+1} = \frac{1}{(n+1)!} \sum_{j=1}^{n} (-)^j \, \frac{(n+j-1)!}{(n-1)!} \, \mathbb{B}_{nj} \left\{(k+1)! b_{k+1}\right\}, \qquad (18.104)$$

which is (18.87).

Commentary 335.1. ▶ Take the grand canonical partition function for bosons as

$$\Xi^B(z) = \prod_{k=1}^{\infty} \frac{1}{1 - z e^{-\beta \epsilon_k}}.$$

Suppose a very special system, for which the energy levels are $\epsilon_k = k\epsilon$, with ϵ some constant. Referring to Comment 263.1, we see that $\Xi^B(1) = \prod_{k=1}^{\infty} \frac{1}{1-x^k} = \sum_{n=0}^{\infty} p(n) x^n$, leading to the interpretation $Q_n = p(n) e^{-k\beta\epsilon}$. The canonical partition function appears as the number of partitions of n (see Comment 224.1) weighted by the Boltzmann factor. ◀

Commentary 335.2. ▶ The energy has not been considered because its expression is much more involved:

$$\bar{E} = -\left[\frac{\partial}{\partial \beta} \ln \Xi^{(\pm)}(V, \beta, \mu)\right]_{V,z} = \frac{pV}{kT} \frac{1}{\Lambda^3} \frac{\partial \Lambda^3}{\partial \beta} - \frac{V}{\Lambda^3} \sum_{k=1}^{\infty} z^k \frac{\partial b_k}{\partial \beta} . \qquad (18.105)$$

Use can be made of Eq.(18.61) to obtain

$$\bar{E} = pV \left\{3 + \beta mc^2 \left[\frac{K_1(\beta mc^2)}{K_2(\beta mc^2)} - 1\right]\right\} - \frac{V}{\Lambda^3} \sum_{k=1}^{\infty} z^k \frac{\partial b_k}{\partial \beta} . \qquad (18.106)$$

◀

18.3.1.1 *Relativistic gases, continued*

In the relativistic case, interaction potentials lose their meaning. Unless we are willing to use the involved S-matrix formalism mentioned in Comment 319.1, all we can do is to study the non-interacting gas. The integrals appearing are those already seen, and the result is

$$\Xi^{(\pm)}(V, \beta, \mu) = \left\{ \begin{matrix} \frac{1}{1-z} \\ 1+z \end{matrix} \right\}^g \exp\left\{ gV \sum_{j=1}^{\infty} \frac{(\pm 1)^{j-1}}{j} z^j \frac{1}{\Lambda^3(j\beta)} \right\}. \qquad (18.107)$$

The pressure comes immediately,

$$p = \frac{kT}{V} \ln \Xi^{(\pm)}(V, \beta, \mu) = gkT \sum_{n=1}^{\infty} \frac{(\pm)^{n-1}}{n} \frac{1}{\Lambda^3(n\beta)} z^n , \qquad (18.108)$$

where we have dropped the zero-energy terms, which vanish in the thermodynamic limit. It is also easy to obtain the number density,

$$n = \frac{1}{V} \left[z \frac{\partial}{\partial z} \ln \Xi^{(\pm)}(V, \beta, \mu)\right]_{V,\beta} = \frac{g}{V} \left\{ \begin{matrix} \frac{z}{1-z} \\ \frac{z}{1+z} \end{matrix} \right\} + g \sum_{j=1}^{\infty} \frac{(\pm)^{j-1}}{\Lambda^3(j\beta)} z^j , \qquad (18.109)$$

where the zero-energy contribution may resist the thermodynamic limit in the bosonic case. To obtain the average energy, however, use must be made of the relation (18.57). Its expression is

$$\begin{aligned}\bar{E} &= -\left[\frac{\partial}{\partial\beta}\ln\Xi^{(\pm)}(V,\beta,\mu)\right]_{V,z} \\ &= 3pV + g\frac{4\pi V}{h^3c^3}\frac{(mc^2)^3}{\beta}\sum_{k=1}^{\infty}\frac{(\pm)^{k-1}}{k}e^{k\beta mc^2}z^kK_1(k\beta mc^2)\ . \end{aligned} \tag{18.110}$$

It is comforting that the virial equations can be given the same aspect of the usual, non-relativistic expressions, provided $\lambda(k\beta)$ is everywhere replaced by $\Lambda^3(k\beta)$ and the effective cluster integral

$$B_k = \frac{(\pm 1)^{k-1}}{k}\ \frac{\Lambda^3(\beta)}{\Lambda^3(k\beta)} \tag{18.111}$$

is used instead of b_k. We find indeed

$$\frac{p}{kT} = \frac{g}{\Lambda^3}\sum_{k=1}^{\infty}B_kz^k\ ; \tag{18.112}$$

$$n = n_{(0)} + \frac{g}{\Lambda^3}\sum_{k=1}^{\infty}kB_kz^k\ . \tag{18.113}$$

The virial equation has then the form

$$\frac{p}{nkT} = \sum_{k=1}^{\infty}A_k\left(\Lambda^3 n\right)^{k-1}\ , \tag{18.114}$$

the Mayer formula being now

$$A_{k+1} = \sum_{j=1}^{k}(-)^j\frac{(k+j-1)!}{(k+1)!(k-1)!}\ \mathbb{B}_{kj}(2!B_2,3!B_3,\ldots)\ . \tag{18.115}$$

All the above expressions tend, of course, to the previous non-relativistic formulas in the due limit. On the other hand, the ultrarelativistic limits are

$$p_{UR} = g\ \frac{8\pi}{c^3h^3\beta^4}\sum_{j=1}^{\infty}\frac{(\pm)^{j-1}}{j^4}z^j\ ;$$

$$n_{UR} = n_{(0)UR} + g\ \frac{8\pi}{c^3h^3\beta^3}\sum_{j=1}^{\infty}\frac{(\pm)^{j-1}}{j^3}z^j\ ;$$

$$\bar{E}_{UR} = 3p_{UR}V\ .$$

They lead to the well-known formulas for the photon gas ($g=2, z=1$) in terms of Riemann zeta functions,

$$p_\gamma = \frac{2}{\pi^2 c^3 \hbar^3 \beta^4}\,\zeta(4)\,, n_\gamma = n_{(0)\gamma} + \frac{2}{\pi^2 c^3 \hbar^3 \beta^3}\,\zeta(3)\,,$$

and for the massless neutrino gas ($g=1, z=1$),

$$n_\nu = \frac{3}{4\pi^2 c^3 \hbar^3 \beta^3}\,\zeta(3)\,.$$

We have mentioned in the previous section that relativistic gases more easily exhibit quantum effects. Indeed, quantum degeneracy is present when the number of particles inside a cube with sides equal to the thermal wavelength, the "degeneracy coefficient"

$$d = n\Lambda^3$$

is of the order of 1. In the ultrarelativistic case, $d = g\sum_{n=1}^{\infty}\frac{(\pm 1)^{n-1}}{n^3}z^n$. For photons and massless neutrinos this gives $2\zeta(3)$ and $\frac{3}{4}\zeta(3)$. As $\zeta(3)\approx 1.2$, such gases are always extremely quantal.

18.3.2 *Matsubara: only connected graphs matter*

Recall that the contribution of the completely connected graphs to the canonical partition function Q_N will be given by Eq.(18.49), $Q_N^{(c)} = v b_N$. Let us call $\Xi^{(c)}$ the contribution of such connected graphs to Ξ. As there is no sense in speaking on connectivity for zero particles, we define

$$\Xi^{(c)}(\beta, V, \mu) = \sum_{N=1}^{\infty} z^N\, Q_N^{(c)} = \sum_{N=1}^{\infty} z^N\, b_N^*\,. \tag{18.116}$$

This is just $\ln \Xi(\beta, V, \mu)$. We arrive thus at a version of an important result, the Matsubara formula [133]:

$$\Xi(\beta, V, \mu) = e^{\Xi^{(c)}(\beta, V, \mu)}\,. \tag{18.117}$$

There is a graphic treatment of the problem, in which of course $Q_N^{(c)}$ is given by a totally connected diagram. The theorem is then stated as:

> The contribution of all the graphs is the exponential of the contributions of the connected graphs.

This is an important property of the several diagram techniques used in Physics. There is an analogue for the Feynman diagrams of Quantum Field

Theory [134]. Here, a remarkable consequence is that

$$\frac{pV}{kT} = \Xi^{(c)}(\beta, V, \mu) \; . \tag{18.118}$$

For an ideal non-relativistic Boltzmann gas, $\Xi^{(c)}(\beta, V, \mu) = vz = \frac{V}{\lambda^3} z$. As in that case $z = n\lambda^3$, it follows that $\Xi^{(c)} = N$. For a gas without interactions, the contributions of the "connected" graphs to the grand canonical partition functions is just the number of particles.

Introducing in a similar way the contribution of the connected graphs to the expectation value of N, we have

$$N^{(c)} = z \frac{\partial}{\partial z} \ln \Xi^{(c)} = z \frac{\partial}{\partial z} \ln \ln \Xi = \frac{\sum_n n \; b^*_n \; z^n}{\sum_n b^*_n \; z^n} = \frac{N}{\Xi^{(c)}} \; .$$

Thus, for the ideal gas, $N^{(c)} = 1$. We see in this way how the connected graphs come to provide corrections to the ideal gas law: the equation of state takes the form

$$\frac{PV}{kT} = \frac{N}{N^{(c)}} \; . \tag{18.119}$$

18.4 Braid statistics

We have repeatedly said that particle exchange, for a gas of N particles in ambient space, is ruled by the symmetric group S_N. This is true if by "ambient space" we understand our usual home-space, the 3-dimensional Euclidean space $\mathbf{E}^3$. Things are different if the particles are somehow confined to a 2-dimensional space: the group responsible for the exchange of particles is, in that case, the braid group B_N. As mentioned in 14.3.1, this is an infinite group, which contains S_N as a subgroup and leads to a different statistics. This kind of statistics is believed to be of interest in the study of superconductivity, as electrons in a superconductor are forced to stay at its surface.

Take two strings. It is possible to interlace them so as to get back the initial position of the extremities. The difference with respect to the simple permutation of the extremities is that there are infinite ways of getting back to the initial positions. What would be the identity transformation in the permutation group can now be any transformation in which one of the strings winds around the other any complete number of times [93]. Each way of coming back to the original situation is different. As with permutations, a phase appears in the wave function as a result of an exchange. A transposition of two particles is given by $|p_2p_1> = \sigma_1|p_1p_2>$ $= e^{i\phi}|p_1p_2>$. However, while performing twice the same transposition of S_N leads back to the initial wave function [$s_1^2 = I$, see Eq.(18.25)], doing twice the same braid elementary transformation only doubles the phase: $\sigma_1^2|p_1p_2> = e^{i2\phi}|p_1p_2>$. Each transposition of the symmetric group is equal to its own inverse, which is not true for a braid transposition.

Commentary 339.1. ▶ The braid group can be introduced by a presentation (see Comment 312.2). Its generators satisfy, as mentioned in Section 14.3.1,

$$\sigma_i\sigma_{i+1}\sigma_i = \sigma_{i+1}\sigma_i\sigma_{i+1} \; ; \tag{18.120}$$

$$\sigma_i\sigma_j = \sigma_j\sigma_i \text{ for } |i-j| \geq 2 \; , \tag{18.121}$$

which shows a lot in common with the symmetric group: these are exactly the same conditions (18.24). The difference comes from the absence of condition (18.25). ◀

The presence of the braid groups in 2-dimensional spaces is a topological effect. Quantum Mechanics of a given system is controlled by the group of classes of closed loops (the fundamental group) of its configuration space [136]. Quantum Statistical Mechanics will, of course, take that in consideration [76]. The configuration space of a system of N identical particles will have a non-trivial fundamental group [137], that is, will be multiply connected.

Let M be the configuration space of each particle in a gas. The configuration space of a gas of N distinguishable particles contained in M will be the set M^N, the cartesian product of manifold M by itself N times. M is usually a box in $\mathbf{E}^3$, whose volume V is taken to infinity at the thermodynamic limit. When the particles are indistinguishable, two points x and x' of M^N are equivalent if the sets $\{\mathbf{x}_1, \mathbf{x}_2, \ldots, \mathbf{x}_N\}$ and $\{\mathbf{x'}_1, \mathbf{x'}_2, \ldots, \mathbf{x'}_N\}$ differ only by a permutation, a transformation belonging to the symmetric group S_N, and the configuration space becomes the quotient M^N/S_N. If M $= \mathbf{E}^3$, and is in consequence topologically trivial, the fundamental group, indicated $\pi_1[M^N/S_N]$, is just S_N. M^N will be the universal covering of the real configuration space and the N-particle wave function will belong in the carrier space of some representation of S_N. Things are more involved when the particles are also impenetrable. In that case, coincident positions must be excluded. This is done by introducing the set

$$D_N = \{\mathbf{x}_1, \mathbf{x}_2, \ldots, \mathbf{x}_N \quad \text{such} \quad \text{that} \quad \mathbf{x}_i = \mathbf{x}_j \quad \text{for} \quad \text{some} \quad i, j\}$$

and taking its complement in M^N,

$$F_N M = M^N \backslash D_N \; .$$

This would be the configuration space for N distinct impenetrable particles, as a classical hard-sphere gas. The fundamental group of this space, $P_N = \pi_1[F_N M]$, is the pure braid group of N strings. If the particles are impenetrable and indistinguishable, both the quotients above are to be put together. Let $B_N M$ be the space obtained by identifying all equivalent points of $F_N M$, its quotient by S_N:

$$B_N M = F_N M / S_N = [M^N \backslash D_N]/S_N \ .$$

The fundamental group of this configuration space, $B_N = \pi_1[B_N M]$, is the full braid group, or simply braid group.

Now, what is important for us here is that the combinatorics for the braid statistics is the same as that of the symmetric group. This means that a recursion relation similar to (18.26) will hold and be instrumental provided we are able to find the normalization for some initial state. In the case, a detailed examination [93] shows that, for the 2-particle case, Eq.(18.16) is replaced by

$$< p_1, p_2 | p_1, p_2 > = \tfrac{1}{2!} \left[\hat{\delta}_1^2 + \cos\phi \ \hat{\delta}_2 \right] . \tag{18.122}$$

The angle ϕ is, in superconductors, somehow fixed by the medium. We see that bosons and fermions are recovered when $\phi = 0$ and π, respectively. A rule of thumb turns out: the factor $(\pm)$, found for bosons and fermions in S_N statistics, is to be replaced by $\cos\phi$ in braid statistics. Thus, for example, the effective configuration integral (18.35) would become

$$b_j = \frac{(\cos\phi)^{j-1}}{j^{5/2}} \ .$$

Actually, this is not quite so. Recall that the fundamental group is a group formed by (classes of) loops. Every loop which can, in a punctured space, be reduced to a point in a continuous way is related to the identity. In a 3-dimensional space, point-holes are of no great importance for these loops, as they can be displaced around them. It turns out that, for a space of dimension ≥ 3, the braid group reduces to the permutation group: $B_N M = S_N M$. Consequently, braid statistics is of interest only for 2-dimensional configuration spaces. The exponent $5/2$ in the formula above comes from the kinematics of a 3-dimensional space. In a 2-dimensional kinematics it reduces to 2, so that the effective configuration integral is actually

$$b_j = \frac{(\cos\phi)^{j-1}}{j^2} \ . \tag{18.123}$$

The recursion relation (18.26) becomes then

$$< p_1, p_2, \cdots, p_N | p_1, p_2, \cdots, p_N > = \tfrac{1}{N!} \sum_{m=1}^{N} (\cos\phi)^{N-m} \mathbf{B}_{Nm} \left\{ (j-1)! \hat{\delta}_j \right\}.$$

Higher order normalization can be calculated starting from (18.122). A 3-particle state, for example, will have

$$< p_1, p_2, p_3 | p_1, p_2, p_3 > = \tfrac{1}{3!} \left[\hat{\delta}_1^3 + 3\cos\phi\ \hat{\delta}_1 \hat{\delta}_2 + 2\cos^2\phi\ \hat{\delta}_3 \right]. \quad (18.124)$$

Usual quantities change from 3-dimensional to 2-dimensional kinematics. Volume V is replaced by surface area S, the cube λ^3 of the thermal wavelength becomes the square λ^2, etc. The canonical partition function, whose expression for a Bose/Fermi 2-dimensional gas would be

$$Q_N(\beta, S) = \tfrac{1}{N!} C_N \left\{ \frac{(\pm)^{j-1}}{j} \frac{S}{\lambda^2} \right\} = \tfrac{1}{N!} \sum_{m=1}^{N} \mathbb{B}_{Nm} \left\{ \frac{(\pm)^{j-1}}{j} (j-1)! \frac{S}{\lambda^2} \right\}, \quad (18.125)$$

becomes

$$Q_N(\beta, S) = \frac{1}{N!} \sum_{m=1}^{N} \mathbb{B}_{Nm} \left\{ \frac{(\cos\phi)^{j-1}}{j} (j-1)! \frac{S}{\lambda^2} \right\} \quad (18.126)$$

for a braid gas.

18.5 Condensation theories

As more detailed knowledge is obtained on a subject, it tends to earn a proper name. Second order phase transitions are nowadays commonly called "critical phenomena". Condensation is the paradigmatic example of first order phase transition, and sometimes lends its name to the whole class. Here, we shall use "condensation" in its old, less well-defined, sense. We shall actually consider only two specific theories, because they provide an intuitive, "physical" vision of the different elements of a Bell polynomial. As an appetizer, let us give a short account of Mayer's model for condensation, which has a lot of heuristic interest.

18.5.1 *Mayer's model for condensation*

Let us derive the canonical partition function (18.44),

$$Q_N(\beta, V) = \frac{1}{N!} \sum_{m=1}^{N} \left(\frac{V}{\lambda^3} \right)^m \mathbb{B}_{Nm} \{k! b_k\},$$

with respect to the volume, to obtain

$$V\left[\frac{\partial}{\partial V}\ln Q_N\right]_{N,T}=\frac{1}{Q_N}\frac{1}{N!}\sum_{m=1}^{N} m\ \mathbb{B}_{Nm}\{j!\ b_j\frac{V}{\lambda^3}\}=\langle m\rangle\ .$$

As m is the number of clusters for each configuration, this expression gives a kind of "average value" for this number. This relates, by Eq.(15.127), volume variations to the homogeneity in the virial coefficients:

$$V\left[\frac{\partial}{\partial V}\ \ln Q_N\right]_{N,T}=\frac{1}{N!}\sum_{m=1}^{N} D\ \mathbb{B}_{Nm}\{j!\ b_j\frac{V}{\lambda^3}\}\ . \tag{18.127}$$

We now take a bold and risky step: we *suppose* that fluctuations are negligible, so that the results of different ensembles can be used simultaneously. That assumed, we find

$$\frac{pV}{kT}=V\left[\frac{\partial}{\partial V}\ \ln Q_N\right]_{N,T}=\langle m\rangle$$

and arrive at the curious general equation of state

$$pV=\langle m\rangle\ kT\ . \tag{18.128}$$

This is the Boyle–Mariotte equation with $\langle m\rangle$ instead of N. It is as if we had always an ideal gas of clusters. Of course, as the signs of the b_k are not necessarily positive, the "weights" in the above "average" are not positive and $\langle m\rangle$ is no real average after all. But the expression is too suggestive to be simply put aside, and Mayer used this line of thought [138] to obtain a heuristic theory for condensation. With the above "equation of state", $\langle m\rangle$ relates to the connected graphs:

$$\Xi^{(c)}=\frac{V}{\lambda^3}\sum_{N=0}^{\infty} z^N b_N=\frac{pV}{kT}=\langle m\rangle\ . \tag{18.129}$$

Using the detailed form of the Bell polynomials (15.6), we find in the same way, and with analogous interpretation, that

$$\langle\nu_j\rangle=b_j\ \frac{\partial}{\partial b_j}\ \ln Q_N\ .$$

We get then

$$\langle m\rangle=\sum_{j=1}^{N}\langle\nu_j\rangle\ ,$$

and arrive at $\frac{\partial}{\partial b_j}\ \ln Q_N=\frac{V}{\lambda^3}z^j$. Consequently,

$$\langle\nu_j\rangle=\frac{V}{\lambda^3}\ z^j\ b_j\ . \tag{18.130}$$

Pressure, the Lagrange multiplier associated to volume preservation, is obtained by taking the homogeneous derivative $V\frac{\partial}{\partial V}$. This leads to an analogue interpretation of the $\langle \nu_k \rangle$: they can be seen as variables conjugate to the b_k in the same sense that pressure is conjugate to the volume.

Knowledge on condensation, involving large clusters, requires knowledge of virial coefficients of very high orders. The result above is usually obtained by supposing that, for $N >> 1$, only one term dominates the summation (18.44) for Q_N. That is, a single configuration, or set of numbers ν_j, all of which very large, is such that

$$Q_N = \frac{1}{N!}\,\mathbb{B}_{Nm}\left\{k!\; b_k\;\frac{V}{\lambda^3}\right\}.$$

The set $\{\nu_j\}$ which maximizes $\ln Q_N$ is looked for, with the condition $\sum_{j=1}^{N} j\;\nu_j = N$ being imposed via a Lagrange multiplier. Of course, the "average" is also here unreliable: the $\langle \nu_j \rangle$ of (18.130) might be negative.

For example, if we calculate the average energy from (18.44) for an ideal gas of bosons or fermions, we get

$$\langle E \rangle = U = \tfrac{3}{2}\; kT\; \langle m \rangle \tag{18.131}$$

(actually, this expression holds whenever the b_k's are temperature-independent). Thus, from $u = \frac{3}{2}\; kT\; \frac{\langle m \rangle}{N}$ we see that $\frac{N - \langle m \rangle}{N}$ measures the breaking of energy equipartition (which holds true when $\langle m \rangle = N$). We can also relate the specific heat with the fluctuations of $\langle m \rangle$:

$$C_V = \tfrac{3}{2}\; k\langle m \rangle + \tfrac{9}{4}\; k\; [\langle m^2 \rangle - \langle m \rangle^2] = \frac{U}{\langle m \rangle} + \tfrac{9}{4}\; k\; [\langle m^2 \rangle - \langle m \rangle^2]\;. \tag{18.132}$$

When $\langle m \rangle = N$ in the canonical ensemble, we have the law of Dulong–Petit. The deviation from $C_V = U/N$ comes from the fact that particles behave in a collective way, in clusters, and we have less agglomerates than particles. What is the behavior of b_j, and consequently of $\langle \nu_j \rangle$, for large values of j? Well, let us suppose that $b_j \approx c_j\; b^j$ for large j's, with b a general constant and c_j some coefficient. This can be reasonably justified [31], [139]. Under this assumption, $\langle \nu_j \rangle = \frac{V}{\lambda^3}\; z^j\; b_j \approx \frac{V}{\lambda^3}\; c_j\; (bz)^j$. Then, for $z < b^{-1}$, other numbers of large clusters will be small; but for $z > b^{-1}$ this number will tend to become larger and larger. There is thus a change of regime at some critical value $z_c = b^{-1}$. The critical volume is given by

$$\frac{1}{v_c} = \frac{1}{\lambda^3}\sum_{j=1}^{\infty} j b_j z_c^j = \frac{1}{\lambda^3}\sum_{j=1}^{\infty} j\; b_j b^{-j}\;. \tag{18.133}$$

As soon as $z \geq z_c$, the series $\frac{1}{v} = \frac{1}{\lambda^3}\sum_{j=1}^{\infty} j b_j z^j$ grows quickly and the specific volume falls down.

Let us here only retain some qualitative aspects. The picture emerging from Mayer's approach is that of a free gas of clusters, of a mixture of gases: one gas with ν_1 really free particles, another with ν_2 pairs, still another with ν_3 triples, etc. In the canonical partition function (18.44), we sum over m in $\mathbb{B}_{Nm}\{k!\, b_k\}$. We have seen that pressure gives a measure, however faulty, of the average value of m. With an analogous interpretation, $\langle \nu_j \rangle$ measures the partial pressure related to clusters of order j.

18.5.2 *The Lee–Yang theory*

The Lee–Yang approach [140] establishes clearly the relationship between phase transitions and the thermodynamic limit. It involves a detailed account of the order in which the limits are taken, and stresses the great generality of the problem. The details of the interaction are relegated to a secondary role.

The physical values of the fugacity $z = e^{\beta\mu}$, of the canonical partition function $Q_N(\beta, V) = \sum_m e^{-\beta E_m}$ and of the grand canonical partition function

$$\Xi(\beta, V, \mu) = 1 + \sum_{N=1}^{\infty} z^N Q_N(\beta, V)$$

are real positive numbers. Stability conditions require the two-particle interaction potential to include a hard core of some radius a. For a finite volume V, the summation above is actually restricted by the number $M = N_{max}$ corresponding to maximal packing, $M \approx V/v_0$, with $v_0 \propto a^3$ the hard-core volume of each particle; for $N > M$, the total potential is necessarily infinite positive and $Q_{N>M}(\beta, V) = 0$. As a consequence, the grand canonical partition function is a positive polynomial of order M, monotonously increasing with z:

$$\Xi(\beta, V, z) = \sum_{N=0}^{M} z^N Q_N(\beta, V) \,. \tag{18.134}$$

The meaning of the partition functions Q_N appearing here in a gas with $M > N$ particles has been discussed in Section 18.2, below Eq.(18.47).

As the coefficients Q_N are all positive and $Q_0 = 1$, it follows that $\Xi \geq 1$. Furthermore,

- there are no zeros on the positive real line of the variable z;
- $\ln \Xi(\beta, V, z)$, being the logarithm of a polynomial, is a well behaved function of z;

- it has furthermore well behaved derivatives in the three variables and, for finite V, will present no singularity;
- the relationship between phase transitions and the thermodynamic limit comes handy: only when $V \to \infty$ and $M \to \infty$ can a singular behavior show up.

We shall not be interested here in the limiting behavior. We shall only recall that the theory examines

$$P(z,T) = kT \lim_{V\to\infty} (\frac{1}{V} \ln \Xi) \tag{18.135}$$

and

$$\frac{1}{v(z,T)} = \lim_{V\to\infty} (\frac{1}{V} \frac{\partial \ln \Xi}{\partial \ln z}) = \lim_{V\to\infty} \left[\frac{\partial}{\partial \ln z} (\frac{1}{V} \ln \Xi) \right] \tag{18.136}$$

in the *complex plane* of the variable z.

Consider then the series to be finite, and take for the grand canonical partition function the polynomial $\Xi^{(M)}(z) = \sum_{N=0}^{M} z^N Q_N$. Its roots will be indicated by r_k:

$$\Xi^{(M)}(r_k) = \sum_{N=0}^{M} r_k^N \; Q_N(\beta, V) = 0 \; . \tag{18.137}$$

Let us recall the Weierstrass representation (16.30) for a polynomial,

$$\Xi^{(M)}(z) = \sum_{N=0}^{M} z^N Q_N(\beta, V) = \prod_{k=1}^{M} \left(1 - \frac{z}{r_k}\right) . \tag{18.138}$$

The roots will depend on V and β, $r_k = r_k(\beta, V)$. By the above considerations on the good behavior of $\ln \Xi$, there are no roots on the real positive axis. They will be either real negative or complex. In the last case they will come by pairs with opposite imaginary parts. The particle number M will tend to infinite with V, so that the number of roots will go to infinite in the thermodynamic limit. When we approach the thermodynamic limit, it may happen that some root (simultaneously with its complex conjugate) touches a point in the physical real positive line. Their imaginary parts go to zero. The signal for a phase transition lies in this "pinching" of the real axis by a zero or more. The "pinched" point z_c on the real line is said to be a critical value. $P(z,T)$ will have a singularity at z_c.

For the lattice gas, Lee and Yang have shown their "third" theorem, the *circle theorem*: in the thermodynamic limit, all the roots of $\Xi = 0$ display themselves on the unit circle of the complex fugacity plane. We may speculate on the generality of this result [141]. It is known that, for a 1-dimensional hard "sphere" gas (the Potts gas), the roots crowd on the negative real axis, but no phase transitions are expected in this case (van Hove theorem[3]).

Let us go back to Eq.(18.138). Expanding the logarithm in

$$\ln \Xi^{(M)}(z) = \sum_{k=1}^{M} \ln\left(1 - \frac{z}{r_k}\right) = \frac{V}{\lambda^3} \sum_{i=1}^{M} b_i z^i ,$$

we find

$$\frac{V}{\lambda^3}\, j\, b_j = - \sum_{k=1}^{M} r_k^{-j} . \qquad (18.139)$$

Consider the alphabet $\mathbf{r} = (r_1, r_2, \ldots, r_M)$, whose letters are the roots of $\Xi(\beta, V, z)$. As usual, also its reciprocal will play an important role. Recalling the power-sum symmetric functions $p_j[\mathbf{r}^*] = \sum_{k=1}^{M} (-)^{-j} r_k^{-j}$, it is clear from the above that

$$v\, j\, b_j = (-)^{j-1} p_j[\mathbf{r}^*] . \qquad (18.140)$$

Also,

$$\begin{aligned} N(z) &= v \textstyle\sum_{j=1}^{M}\, j\, b_j z^j \\ &= - \sum_{j=1}^{M} \sum_{k=1}^{M} r_k^{-j} z^j = \sum_{j=1}^{M} (-)^{j+1} p_j[\mathbf{r}^*]\, z^j = z\, P[\mathbf{r}^*, -z] , \end{aligned} \qquad (18.141)$$

where use has been made of (16.11) in the last step. On the other hand, from (18.134),

$$Q_M = \frac{1}{M!} \sum_{m=1}^{M} \mathbb{B}_{Mm} \left\{ -(j-1)! \sum_{k=1}^{M} r_k^{-j} \right\} . \qquad (18.142)$$

Consequently,

$$Q_M = (-)^M \prod_{k=1}^{M} r_k^{-1} . \qquad (18.143)$$

If the circle theorem holds, $Q_M = 1$. Consider now the diagonal matrix

$$\mathbf{A}_M = \text{diag}\left[-r_1^{-1}, -r_2^{-1}, -r_3^{-1}, \ldots, -r_M^{-1}\right] \qquad (18.144)$$

and its powers. We have clearly

$$Q_M = \det \mathbf{A}_M \qquad (18.145)$$

[3] The van Hove theorem, as well as other classics on the subject, can be found in Frisch and Lebowitz [142].

and

$$v\, j\, b_j = (-)^{j-1}\, tr\, \mathbf{A}_M^j \ . \tag{18.146}$$

This suggests an interpretation of the exponent k in $\mathbf{A}_M^k$: when we take the k-th power of $\mathbf{A}_M$, we are taking into account interactions between up to k particles: two-by-two interactions given by $2vb_2 = -\ \mathrm{tr}\ \mathbf{A}_M^2$, three-by-three interactions given by $3vb_3 = \mathrm{tr}\ \mathbf{A}_M^3$, etc. Thus, at least as long as M is finite, there is a "Lee–Yang matrix" $\mathbf{A}_M$ whose characteristic polynomial is related to the grand canonical partition function. The characteristic polynomial determines a (normal) matrix only up to the (unitary) transformations leading to its diagonalization. Thus, the whole description, based on the partition function, will have an underlying invariance under such transformations, which constitute a group $U(M)$.

Commentary 347.0. ► Notice that, account taken of (18.146), the formal series of Eqs.(18.91) and (18.92) become

$$f(z) = \tfrac{1}{v}\,\mathrm{tr}\,\ln(\mathbf{I} + z\,\mathbf{A}_M) = \tfrac{1}{v}\,\ln\det(\mathbf{I} + z\,\mathbf{A}_M)\ ;\ g(z) = \tfrac{1}{v}\,\mathrm{tr}\,\frac{z\,\mathbf{A}_M}{\mathbf{I} + z\,\mathbf{A}_M}\ .$$

◄

Summing up: a matrix $\mathbf{A}$ with the characteristics given above should exist for each real gas. However, the description of the system is not really dependent on $\mathbf{A}$. It depends only on its similarity class. If the circle theorem holds for the system, $\mathbf{A}$ must be unitary in the thermodynamic limit. In the next section [Eq.(18.166) and at the very end], we shall see how the grand partition function can be assimilated to a Fredholm determinant.

18.6 The Fredholm formalism

The formalism of the Fredholm integral equations, and in particular its finite, matrix version, provides a most detailed view of the decomposition into cycles. It helps to see, for example, which particulars are lost by the virial decomposition and by Bell polynomials in general. The grand canonical partition function can be assimilated to a Fredholm determinant, and the partial canonical partition functions to the Fredholm minors.

We recall now the main lines of Fredholm theory.[4] The theory is concerned with a current type of integral equations and, as such, finds its place

[4] The author is grateful to M.A. Ozório de Almeida for calling his attention to reference [143] and, more generally, to the possible interest of the Bell formalism to Fredholm theory.

in many chapters of Physics. Applications to semiclassical quantization of chaotic systems have deserved special attention recently (see, for example, [143]). As usual, this compact will have a physicists's bias, with little concern on the deeper mathematical questions like convergence, or the spaces the involved objects belong to. The presentation given is purely descriptive and utterly simplified. It mixes the continuous and the discrete points of view, but has in mind, ultimately, finite $N \times N$ matrices.

Suppose an equation of type

$$\mathbf{X} = \mathbf{J} + \lambda \, \mathbf{K} \, \mathbf{X} \, . \tag{18.147}$$

We shall in what follows take this as a matrix equation. Nevertheless, if taken as a symbolic expression, it might as well represent a Fredholm integral equation of the second kind,

$$\mathbf{X}(x) = \mathbf{J}(x) + \lambda \int dx' K(x, x') \, \mathbf{X}(x') \, . \tag{18.148}$$

$\mathbf{X}$ and $\mathbf{J}$ are members of some Hilbert or Banach space, on which $\mathbf{K}$ is a well-behaved kernel. Objects $\mathbf{J}$ and $\mathbf{K}$ are supposed to be known and defined on some finite domain. In case (18.148), if $\mathbf{J}$ and $\mathbf{K}$ are well-behaved enough (say, continuous, or square-integrable), then the Fredholm alternative holds:

- either there exists a unique, equally nice solution $\mathbf{X}$(x), or
- the homogeneous equation ($\mathbf{J} = 0$) has a solution.

In the last case, (18.148) becomes an eigenvalue equation. The set of eigenvalues for which the solution is not unique is discrete. The main result of Fredholm theory is that the results we shall report, talking on matrices, are valid also for a large class of integral-functional cases.

Given (18.147) we can write, at least formally,

$$\mathbf{X} = \frac{\mathbf{I}}{\mathbf{I} - \lambda \mathbf{K}} \, \mathbf{J} \, . \tag{18.149}$$

The resolvent operator or matrix $\mathbf{R}(\lambda)$ (the name has been previously applied in another sense, see Sections 1.1 and 1.3) is defined by

$$\mathbf{X} = (\mathbf{I} + \mathbf{R}) \, \mathbf{J} \, , \tag{18.150}$$

so that

$$\mathbf{R}(\lambda) = \sum_{n=1}^{\infty} \lambda^n \, \mathbf{K}^n = \frac{\lambda \, \mathbf{K}}{\mathbf{I} - \lambda \, \mathbf{K}} \, . \tag{18.151}$$

Everything works as if, in the equation, $\mathbf{X}$ was repeatedly replaced by its own expression, in an infinite recurrence. The above comment on eigenvalues can now be restated as:

The inverse $(\mathbf{I}-\lambda\mathbf{K})^{-1}$ exists for all values of λ except a discrete set $\{\lambda_k\}$.

This means that the spectrum of $\mathbf{K}$ is discrete. The inverse of a matrix is always some matrix divided by its determinant [see Eqs.(16.46), (16.47)]. The expression

$$(\mathbf{I}-\lambda\mathbf{K})^{-1} = \frac{1}{D(\lambda)}\,\hat{\mathbf{N}} \tag{18.152}$$

introduces the Fredholm determinant

$$D(\lambda) = \det(\mathbf{I}-\lambda\mathbf{K}) \ . \tag{18.153}$$

This determinant can be written as

$$D(\lambda) = 1 + \sum_{n=1}^{\infty,N} \lambda^n D_n \ , \tag{18.154}$$

with

$$D_n = \frac{(-)^n}{n!}\int ds_1 \int ds_2 \cdots \int ds_n \begin{vmatrix} K_{s_1s_1} & K_{s_1s_2} & \cdots & K_{s_1s_n} \\ K_{s_2s_1} & K_{s_2s_2} & \cdots & K_{s_2s_n} \\ \cdot & \cdot & \cdots & \cdot \\ \cdot & \cdot & \cdots & \cdot \\ K_{s_ns_1} & K_{s_ns_2} & \cdots & K_{s_ns_n} \end{vmatrix} . \tag{18.155}$$

In the discrete version with $N\times N$ matrices [see Eq.(16.7)], it can be directly verified that

$$\sum_{s_1=1}^{N}\sum_{s_2=1}^{N}\cdots\sum_{s_n=1}^{N} \begin{vmatrix} K_{s_1s_1} & K_{s_1s_2} & \cdots & K_{s_1s_n} \\ K_{s_2s_1} & K_{s_2s_2} & \cdots & K_{s_2s_n} \\ \cdot & \cdot & \cdots & \cdot \\ \cdot & \cdot & \cdots & \cdot \\ K_{s_ns_1} & K_{s_ns_2} & \cdots & K_{s_ns_n} \end{vmatrix}$$

$$= \sum_{m=1}^{n} \mathbb{B}_{nm}\{(-)^{j-1}(j-1)!\ \mathrm{tr}\mathbf{K}^j\} \ . \tag{18.156}$$

Consequently,

$$D_n = \frac{(-)^n}{n!}\sum_{m=0}^{n} \mathbb{B}_{nm}\{(-)^{j-1}(j-1)!\ \mathrm{tr}\mathbf{K}^j\} \ . \tag{18.157}$$

Commentary 349.1. ► There is an obvious analogy with the canonical quantum ideal partition function. We see from (18.68) that $Q_N = (\pm)^N D_N$ if we identify above $K_{r_ir_j}$ with the function f_{r_i,r_j} there defined. The distribution function, which is the integrand of Q_N normalized to one, will be

$$F_N(r_1,r_2,\cdots,r_N) = \frac{(\mp)^N}{N!Q_N} \begin{vmatrix} f_{r_1r_1} & f_{r_1r_2} & \cdots & f_{r_1r_N} \\ f_{r_2r_1} & f_{r_2r_2} & \cdots & f_{r_2r_N} \\ \cdot & \cdot & \cdots & \cdot \\ \cdot & \cdot & \cdots & \cdot \\ f_{r_Nr_1} & f_{r_Nr_2} & \cdots & f_{r_Nr_N} \end{vmatrix} .$$

◄

All this follows from the definition (18.153) using (16.2). We have extended the summation down to $m = 0$, which is consistent with the convention $D_0 = 1$ and is irrelevant for $n \neq 0$. A few lower-order examples:

$$D_1 = \frac{(-)^1}{1!} \operatorname{tr} \mathbf{K} \; ;$$

$$D_2 = \frac{(-)^2}{2!} \left[(\operatorname{tr} \mathbf{K})^2 - \operatorname{tr} \mathbf{K}^2\right] = \frac{(-)^2}{2!} \begin{vmatrix} \operatorname{tr} \mathbf{K} & 1 \\ \operatorname{tr} \mathbf{K}^2 & \operatorname{tr} \mathbf{K} \end{vmatrix} = \det{}_{(2)} \mathbf{K} \; ;$$

$$D_3 = \frac{(-)^3}{3!} \left[(\operatorname{tr} \mathbf{K})^3 + 2\operatorname{tr} \mathbf{K}^3 - 3(\operatorname{tr} \mathbf{K}^2)(\operatorname{tr} \mathbf{K})\right]$$

$$= \frac{(-)^3}{3!} \begin{vmatrix} \operatorname{tr} \mathbf{K} & 1 & 0 \\ \operatorname{tr} \mathbf{K}^2 & \operatorname{tr} \mathbf{K} & 2 \\ \operatorname{tr} \mathbf{K}^3 & \operatorname{tr} \mathbf{K}^2 & \operatorname{tr} \mathbf{K} \end{vmatrix} = -\det{}_{(3)} \mathbf{K} \; .$$

These expressions are identical to those of the Lie algebra invariants (16.59), and also reminiscent of several other formulas previously found. It should be borne in mind however that, in the trace computations of those formulas, say Eq.(16.20), the indices run from 1 to N. D_3, for example, is not a determinant unless $N = 3$. Notice

$$\frac{1}{D(\lambda)} = \frac{1}{1 + \sum_{j \geq 1} \lambda^j D_j} = \sum_{n=0}^{\infty} \frac{\lambda^n}{n!} \sum_{m=0}^{n} \mathbb{B}_{nm}\{-D_j\} \; . \tag{18.158}$$

The relations between minors, determinants, traces of powers and symmetric functions have been discussed in Section 16.2, but the expressions (18.155) and (18.157) help clarifying the minors appearing in many expressions of Chapter 14. We shall see that, in the same line, they will shed light on the meaning of the partial canonical partition functions turning up in the grand canonical formalism.

It comes not as a surprise, in view of the "recursion" relation (15.107), or more specifically of Eq.(16.16),

$$e_n = \frac{1}{n} \sum_{k=1}^{n} (-)^{k-1} p_k e_{n-k} \; ,$$

that one of the main ingredients of the Fredholm formalism is the property

$$D_n = -\frac{1}{n} \sum_{j=1}^{n} p_j \, D_{n-j} \; , \tag{18.159}$$

where $p_j = \operatorname{tr} \mathbf{K}^j$.

The Fredholm first minor is of fundamental importance. It is defined as

$$D(x, y; \lambda) = 1 + \sum_{n=1}^{N} \lambda^n D_n(x, y) \ , \tag{18.160}$$

where

$$D_n(x, y) = \frac{(-)^n}{n!} \int ds_1 \int ds_2 \cdots \int ds_n \begin{vmatrix} K_{xy} & K_{xs_1} & K_{xs_2} & \cdots & K_{xs_n} \\ K_{s_1y} & K_{s_1s_1} & K_{s_1s_2} & \cdots & K_{s_1s_n} \\ K_{s_2y} & K_{s_2s_1} & K_{s_2s_2} & \cdots & K_{s_2s_n} \\ \cdot & \cdot & \cdot & \cdots & \cdot \\ \cdot & \cdot & \cdot & \cdots & \cdot \\ K_{s_ny} & K_{s_ns_1} & K_{s_ns_2} & \cdots & K_{s_ns_n} \end{vmatrix} . \tag{18.161}$$

$D(x, y; \lambda)$ is important because the theory shows that

$$R(x, y; \lambda) = R_{xy}(\lambda) = \frac{1}{D(\lambda)} \ D(x, y; \lambda) \ . \tag{18.162}$$

The first cases of $D_n(x, y)$ are:

$$D_1(x, y) = \frac{(-)^1}{1!}[(\text{tr } \mathbf{K})K_{xy} - K_{xy}^2] \ ;$$

$$D_2(x, y) = \frac{(-)^2}{2!} \left(K_{xy}[(\text{tr } \mathbf{K})^2 - \text{tr } \mathbf{K}^2] + 2K_{xy}^3 - 2(\text{tr } \mathbf{K})K_{xy}^2\right) .$$

If we move the first row and column to the last positions in (18.161), it becomes

$$D_n(x, y) = \frac{(-)^n}{n!} \sum_{s_1=1}^{N} \sum_{s_2=1}^{N} \cdots \sum_{s_n=1}^{N} \begin{vmatrix} K_{s_1s_1} & K_{s_1s_2} & \cdots & K_{s_1s_n} & K_{s_1y} \\ K_{s_2s_1} & K_{s_2s_2} & \cdots & K_{s_2s_n} & K_{s_2y} \\ \cdot & \cdot & \cdots & \cdot & \cdot \\ \cdot & \cdot & \cdots & \cdot & \cdot \\ K_{s_ns_1} & K_{s_ns_2} & \cdots & K_{s_ns_n} & K_{s_ny} \\ K_{xs_1} & K_{xs_2} & \cdots & K_{xs_n} & K_{xy} \end{vmatrix} .$$

And if we now put $x = y = s_{n+1}$, we find

$$D_n(s_{n+1}, s_{n+1}) = \frac{(-)^n}{n!} \sum_{s_1=1}^{N} \sum_{s_2=1}^{N} \cdots \sum_{s_n=1}^{N} \begin{vmatrix} K_{s_1s_1} & K_{s_1s_2} & \cdots & K_{s_1s_n} & K_{s_1s_{n+1}} \\ K_{s_2s_1} & K_{s_2s_2} & \cdots & K_{s_2s_n} & K_{s_2s_{n+1}} \\ \cdot & \cdot & \cdots & \cdot & \cdot \\ \cdot & \cdot & \cdots & \cdot & \cdot \\ K_{s_ns_1} & K_{s_ns_2} & \cdots & K_{s_ns_n} & K_{s_ns_{n+1}} \\ K_{s_{n+1}s_1} & K_{s_{n+1}s_2} & \cdots & K_{s_{n+1}s_n} & K_{s_{n+1}s_{n+1}} \end{vmatrix} . \tag{18.163}$$

To get D_{n+1} it would be enough to add a factor $\left(-\frac{1}{n+1}\right)$ and sum over s_{n+1}: $-\frac{1}{n+1}\sum_{s_{n+1}} D_n(s_{n+1}, s_{n+1}) = D_{n+1}$, which can be rewritten as

$$D_n = -\frac{1}{n}\sum_x D_{n-1}(x,x) \ . \tag{18.164}$$

As long as $\mathbf{K}$ is well-behaved, $D(\lambda)$ is an entire function (an absolutely convergent series) of λ. In the $N \times N$ matrix case, $D(\lambda)$ will be an N-th order polynomial.

The similarity of the above expressions to those we have found in Statistical Mechanics of real gases will not have escaped the reader. Indeed, recall the expression of the grand canonical partition function for the M-particle real gas:

$$\Xi^{(M)}(z) = \det\left[\mathbf{I} + z\mathbf{A}_M\right] = \sum_{N=0}^{M} z^N Q_N \ . \tag{18.165}$$

$\mathbf{A}_\mathbf{M}$ is the matrix (18.144) which shows up in the Lee–Yang theory, whose eigenvalues $\{-1/r_i\}$ are related to the zeros r_i of $\Xi^{(M)}$ on the complex fugacity plane. We obtain a complete analogy with the Fredholm case by putting $\mathbf{K} = \mathbf{A}$ and $z = -\lambda$:

$$D(\lambda) = \Xi(-\lambda) \ . \tag{18.166}$$

The analogy between D_n and the canonical partition function is also evident. Indeed, Eqs.(18.140) and (18.53), which are

$$v\, j\, b_j = (-)^{j-1} \operatorname{tr} \mathbf{A}_M^j \ ,$$

$$Q_N(\beta, V) = \frac{v}{N}\sum_{j=1}^{N} j\, b_j Q_{N-j} \ ,$$

clearly relate to (18.159): it is enough to put $Q_N = (-)^N D_N$ and $p_j = \operatorname{tr} \mathbf{A}_M^j$. It would then be possible to trace the contribution of each term to the pressure (18.81) and the density (18.82)

$$\frac{pV}{kT} = \ln\, \Xi = v\sum_{j=1}^{\infty} b_j z^j \ ;$$

$$N(z) = nV = z\,\frac{\partial}{\partial z}\, \ln \Xi = v\sum_{j=1}^{\infty}\, j\, b_j z^j \ .$$

It can be easily seen that the summands in (18.155) provide a very detailed cycle decomposition, including eventually asymmetric interactions. Most details are lost after summations are applied and, consequently, erased in the cluster integrals. This is an opportunity to see in detail what is retained and what is lost. Because it is the lowest order exhibiting all the main aspects, let us examine in detail the case $N = 4$:

$$D_4 = \frac{1}{4!}\sum_{s_1}\sum_{s_2}\sum_{s_3}\sum_{s_4}\begin{vmatrix} K_{s_1s_1} & K_{s_1s_2} & K_{s_1s_3} & K_{s_1s_4} \\ K_{s_2s_1} & K_{s_2s_2} & K_{s_2s_3} & K_{s_2s_4} \\ K_{s_3s_1} & K_{s_3s_2} & K_{s_3s_3} & K_{s_3s_4} \\ K_{s_4s_1} & K_{s_4s_2} & K_{s_4s_3} & K_{s_4s_4} \end{vmatrix}.$$

This is to be assimilated to the contribution Q_4 in a gas with N particles. Let us drop $\sum_{s_1}\sum_{s_2}\sum_{s_3}\sum_{s_4}$ and look at the summands. A diagram indicating the particle associations will be drawn below each term in the expansion of the above determinant. We shall consider the terms three by three.

A convention must be established. We shall suppose that K_{ij} means, as in Field Theory scattering matrices or Green's functions (and as our stochastic matrices), a transition $j \to i$. Let us also recall that, in $\mathbb{B}_{Nm}$, m is the number of clusters and ν_k is the number of clusters with k particles, so that $m = \sum_k \nu_k$. Now, let us examine term by term:

$$K_{s_1s_1}K_{s_2s_2}K_{s_3s_3}K_{s_4s_4} + K_{s_1s_1}K_{s_2s_3}K_{s_3s_4}K_{s_4s_2} + K_{s_1s_1}K_{s_2s_4}K_{s_4s_3}K_{s_3s_2}$$

$$\begin{array}{ccc} 1\bullet & & \bullet\,2 \\ & & \\ 4\bullet & & \bullet\,3 \end{array}\qquad\qquad \begin{array}{ccc} 1\bullet & & \bullet\,2 \\ & \swarrow & \uparrow \\ 4\bullet & \rightarrow & \bullet\,3 \end{array}\qquad\qquad \begin{array}{ccc} 1\bullet & & \bullet\,2 \\ & \nearrow & \downarrow \\ 4\bullet & \leftarrow & \bullet\,3 \end{array}$$

The first term accounts for the non-interacting particles ($\nu_1 = 4$, $m = 4$). After summation, it will give $(\mathrm{tr}\ \mathbf{K})^4$. The next two will both give $\mathrm{tr}\ \mathbf{K}(\mathrm{tr}\ \mathbf{K})^3$ after summation, but they are differently arrowed. This detail, which would account for possible differences between K_{ij} and K_{ji}, is blurred out by the summations. They correspond to $\nu_1 = 1$, $\nu_3 = 1$, $m = 2$.

$$-K_{s_1s_1}K_{s_2s_4}K_{s_3s_3}K_{s_4s_2} - K_{s_1s_1}K_{s_3s_4}K_{s_4s_3}K_{s_2s_2} - K_{s_1s_1}K_{s_4s_4}K_{s_3s_2}K_{s_2s_3}$$

$$\begin{array}{ccc} 1\bullet & & \bullet\,2 \\ & \swarrow\!\!\!\nearrow & \\ 4\bullet & & \bullet\,3 \end{array}\qquad\qquad \begin{array}{ccc} 1\bullet & & \bullet\,2 \\ & & \\ 4\bullet & \leftrightarrow & \bullet\,3 \end{array}\qquad\qquad \begin{array}{ccc} 1\bullet & & \bullet\,2 \\ & & \downarrow\uparrow \\ 4\bullet & & \bullet\,3 \end{array}$$

Three cases with binary clusters: $\nu_1 = 2$, $\nu_2 = 1$, $m = 3$ in each term. Summations will lead to $(\mathrm{tr}\ \mathbf{K})^2\mathrm{tr}\ \mathbf{K}^2$ in each case.

$$-K_{s_1s_2}K_{s_2s_1}K_{s_3s_3}K_{s_4s_4} - K_{s_1s_2}K_{s_2s_3}K_{s_3s_4}K_{s_4s_1} - K_{s_1s_2}K_{s_2s_4}K_{s_4s_3}K_{s_3s_1}$$

$$\begin{array}{ccc} 1\bullet & \leftrightarrow & \bullet\,2 \\ & & \\ 4\bullet & & \bullet\,3 \end{array}\qquad\qquad \begin{array}{ccc} 1\bullet & \leftarrow & \bullet\,2 \\ \downarrow & & \uparrow \\ 4\bullet & \rightarrow & \bullet\,3 \end{array}\qquad\qquad \begin{array}{ccc} 1\bullet & \leftarrow & \bullet\,2 \\ & \nearrow\!\!\!\!\searrow & \\ 4\bullet & \leftarrow & \bullet\,3 \end{array}$$

The first has again a binary, $\nu_1 = 2$, $\nu_2 = 1$, $m = 3$ terms, leading to $(\mathrm{tr}\ \mathbf{K})^2\mathrm{tr}\ \mathbf{K}^2$. The two other are completely connected diagrams, $\nu_4 = 1$, $m = 1$, with the summations producing $\mathrm{tr}\ \mathbf{K}^4$.

$$+K_{s_1s_2}K_{s_2s_4}K_{s_3s_3}K_{s_4s_1} + K_{s_1s_2}K_{s_3s_4}K_{s_4s_3}K_{s_2s_1} + K_{s_1s_2}K_{s_4s_4}K_{s_3s_1}K_{s_2s_3}$$

The first and the third are distinct contributions with one triad and one isolated particle. They are thus of type $\nu_3 = 1$, $m = 2$, tr $\mathbf{K}$tr $\mathbf{K}^3$. The middle term has two binaries, so that $\nu_2 = 2$, $m = 2$, (tr $\mathbf{K}^2)^2$.

$$+K_{s_1s_3}K_{s_2s_1}K_{s_3s_2}K_{s_4s_4} + K_{s_1s_3}K_{s_2s_2}K_{s_3s_4}K_{s_4s_1} + K_{s_1s_3}K_{s_2s_4}K_{s_4s_2}K_{s_3s_1}$$

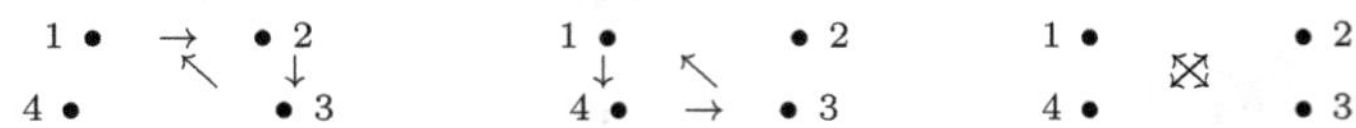

Two cases with triads, and the last containing two binaries.

$$-K_{s_1s_3}K_{s_2s_4}K_{s_3s_2}K_{s_4s_1} - K_{s_1s_3}K_{s_3s_4}K_{s_4s_2}K_{s_2s_1} - K_{s_1s_3}K_{s_4s_4}K_{s_3s_1}K_{s_2s_2}$$

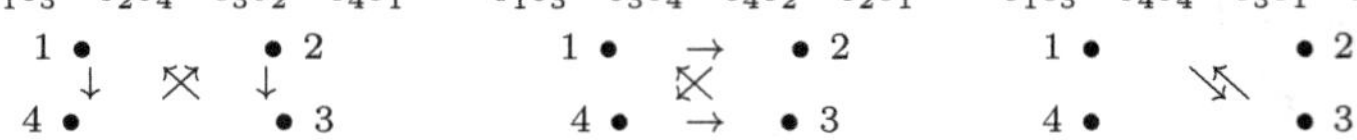

The last case is a binary, the first two are connected diagrams, with $m = 1$.

$$-K_{s_1s_4}K_{s_2s_1}K_{s_3s_2}K_{s_4s_3} \; -K_{s_1s_4}K_{s_2s_2}K_{s_3s_3}K_{s_4s_1} \; -K_{s_1s_4}K_{s_2s_3}K_{s_4s_2}K_{s_3s_1}$$

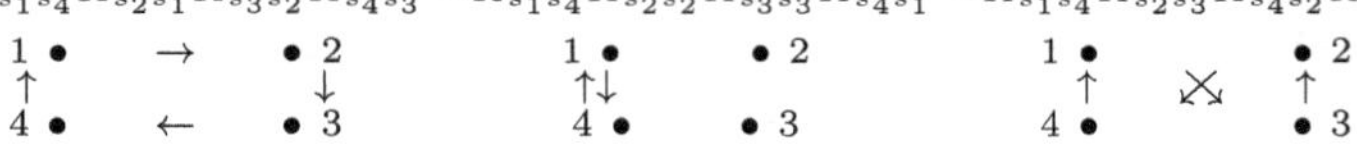

The clockwise cycle is, of course an $m = 1$ case, as is the third term.

$$+K_{s_1s_4}K_{s_2s_3}K_{s_3s_2}K_{s_4s_1} \; +K_{s_1s_4}K_{s_3s_3}K_{s_4s_2}K_{s_2s_1} \; +K_{s_1s_4}K_{s_4s_3}K_{s_3s_1}K_{s_2s_2}$$

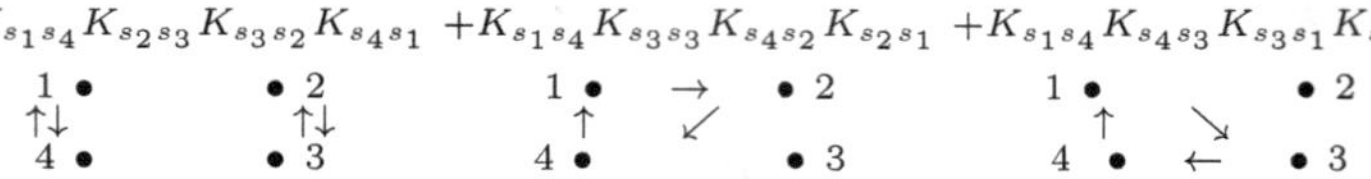

It should be noticed that the summands show only simple cycles. Each particle has an incoming line and one outgoing line. No terms of the type

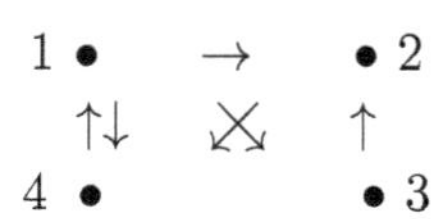

turns up. Such terms, with one particle coupled to 3 or more others, do appear in the usual virial graphs. The above formalism provides a real cycle decomposition, precisely that division of the symmetric group into cyclic factors which, as we have said, should exist. The summation of all the contributions above gives

$$\frac{1}{4!}\left\{-6\mathrm{tr}\,\mathbf{K}^4 + 8(\mathrm{tr}\,\mathbf{K})\mathrm{tr}\,\mathbf{K}^3 - 6(\mathrm{tr}\,\mathbf{K})^2\mathrm{tr}\,\mathbf{K}^2 + 3(\mathrm{tr}\,\mathbf{K}^2)^2 + (\mathrm{tr}\,\mathbf{K})^4\right\} ,$$

as it would be expected from (18.156). This decomposition: (i) is really cyclic; (ii) is arrowed: would work even if $K_{ij} \neq K_{ji}$.

All this suggests that, besides the existence of a general invariance under a unitary group, the formalism of real gases indicates the presence of a Fredholm equation with the Lee–Yang matrix in the role of kernel. That equation would have a unique solution for all values of the fugacity which are not zeros of the grand canonical partition function. The meaning (and interest, if any) of these solutions is as yet unknown.

It is tempting to examine the circle theorem from this point of view. In the thermodynamic limit $M \to \infty$, the matrix $\mathbf{K} = \mathbf{A}$ would become unitary. There are two interesting points. One is a general result on the eigenvalues of a Fredholm operator: *the eigenvalues have no finite accumulation point.* Another seems more physical: for a phase transition to occur, it is enough that one single pair of roots come close to pinch the real positive axis in the limit [32].

Chapter 19

Carleman matrices

19.1 Introduction

A series with an independent constant term can also be attributed a matrix, by generalizing the Bell matrix definition in a way which enlarges the *power polynomials* of Eq.(15.71): calling $G(x) = g_0 + g(x) = g_0 + \sum_{j=1}^{\infty} \frac{g_j}{j!} x^j$,

$$\frac{1}{r!}\left[G(x)\right]^r = \frac{1}{r!}\left(g_0 + \sum_{j=1}^{\infty} \frac{g_j}{j!}\, x^j\right)^r = \frac{1}{r!}\sum_{m=0}^{r} \binom{r}{m} g_0^{r-m} [g(x)]^m$$
$$= \sum_{m=0}^{r} \frac{g_0^{r-m}}{(r-m)!} \sum_{n=m}^{\infty} \frac{x^n}{n!} \mathbb{B}_{n,m}[g] \ , \tag{19.1}$$

where the Bell matrix of $g(x) = G(x) - g_0$ turns up in a natural way. We can then define the matrix elements

$$\mathbb{C}_{n,r}[G] = \sum_{m=0}^{\min[r,n]} \frac{g_0^{r-m}}{(r-m)!}\, \mathbb{B}_{n,m}[g] \ . \tag{19.2}$$

They are such that

$$\frac{1}{r!}\left[G(x)\right]^r = \sum_{n=0}^{\infty} \frac{x^n}{n!}\, \mathbb{C}_{n,r}[G] \ . \tag{19.3}$$

We shall call $\mathbb{C}$ a “Carleman matrix”, though this name is frequently used for its transpose. And we shall reserve the name “Bell matrices” for the series considered previously, with no independent term. The following procedure gives what seems to be the simplest way to work with Carleman matrices:

(1) enlarge the Bell matrix $\mathbb{B}[g]$ to a matrix $\hat{\mathbb{B}}[g]$, by adding an extra "zero-th" row and an extra "zero-th" column, with entries $\hat{\mathbb{B}}_{0k}[g] = \hat{\mathbb{B}}_{k0}[g] = 0$ and $\hat{\mathbb{B}}_{00}[g] = 1$. $\hat{\mathbb{B}}[g]$ will have the form

$$\hat{\mathbb{B}}[g] = \begin{pmatrix} 1 & 0 \\ 0 & \mathbb{B} \end{pmatrix} = \begin{pmatrix} 1 & 0 & 0 & \cdots & 0 \\ 0 & \mathbb{B}_{11}[g] & 0 & \cdots & 0 \\ 0 & \mathbb{B}_{21}[g] & \mathbb{B}_{22}[g] & \cdots & 0 \\ \cdots & \mathbb{B}_{31}[g] & \mathbb{B}_{32}[g] & \cdots & 0 \\ 0 & \mathbb{B}_{n1}[g] & \mathbb{B}_{n2}[g] & \cdots & \mathbb{B}_{nn}[g] \end{pmatrix} ; \tag{19.4}$$

(2) define the upper-triangular matrix $\mathbb{M}[g_0]$ with entries

$$\mathbb{M}_{nm} = \frac{g_0^{m-n}}{(m-n)!} , \tag{19.5}$$

whose $N = 2$ and $N = 3$ examples are

$$\begin{pmatrix} 1 & g_0 & \frac{g_0^2}{2} \\ 0 & 1 & g_0 \\ 0 & 0 & 1 \end{pmatrix} \quad \textit{and} \quad \begin{pmatrix} 1 & g_0 & \frac{g_0^2}{2} & \frac{g_0^3}{6} \\ 0 & 1 & g_0 & \frac{g_0^2}{2} \\ 0 & 0 & 1 & g_0 \\ 0 & 0 & 0 & 1 \end{pmatrix} . \tag{19.6}$$

$\mathbb{M}[g_0]$ has properties $\det \mathbb{M}[g_0] = 1$ and $\mathbb{M}^n[g_0] = \mathbb{M}[ng_0]$.

(3) then,

$$\mathbb{C}[G] = \hat{\mathbb{B}}[g] \ \mathbb{M}[g_0] , \tag{19.7}$$

which, by the way, shows that $\det \mathbb{C}[G] = \det \mathbb{B}[g]$ - and that $\mathbb{C}[G]$ is invertible when and only when $\mathbb{B}$ is invertible. Notice that the matrix product above includes the contributions of the zero-th rows and columns: $\mathbb{C}_{nm}[G] = \sum_{j=0}^{N} \hat{\mathbb{B}}_{nj}[g]\mathbb{M}_{jm}[g_0]$ are the entries of an $(N+1) \times (N+1)$ matrix.

Two lower-order examples of Carleman matrices are

$$\begin{pmatrix} 1 & g_0 & \frac{g_0^2}{2} \\ 0 & g_1 & g_0 g_1 \\ 0 & g_2 & g_1^2 + g_0 g_2 \end{pmatrix} = \begin{pmatrix} 1 & 0 & 0 \\ 0 & g_1 & 0 \\ 0 & g_2 & g_1^2 \end{pmatrix} + g_0 \begin{pmatrix} 0 & 1 & 0 \\ 0 & 0 & g_1 \\ 0 & 0 & g_2 \end{pmatrix} + \frac{g_0^2}{2} \begin{pmatrix} 0 & 0 & 1 \\ 0 & 0 & 0 \\ 0 & 0 & 0 \end{pmatrix} \tag{19.8}$$

and

$$\begin{pmatrix} 1 & g_0 & \frac{g_0^2}{2} & \frac{g_0^3}{3!} \\ 0 & g_1 & g_0 g_1 & \frac{g_0^2 g_1}{2} \\ 0 & g_2 & g_1^2 + g_0 g_2 & g_0 g_1^2 + \frac{g_0^2 g_2}{2} \\ 0 & g_3 & 3g_1 g_2 + g_0 g_3 & g_1^3 + 3g_0 g_1 g_2 + \frac{g_0^2 g_3}{2} \end{pmatrix} = \begin{pmatrix} 1 & 0 & 0 & 0 \\ 0 & g_1 & 0 & 0 \\ 0 & g_2 & g_1^2 & 0 \\ 0 & g_3 & 3g_1 g_2 & g_1^3 \end{pmatrix} +$$

$$g_0 \begin{pmatrix} 0 & 1 & 0 & 0 \\ 0 & 0 & g_1 & 0 \\ 0 & 0 & g_2 & g_1^2 \\ 0 & 0 & g_3 & 3g_1 g_2 \end{pmatrix} + \frac{g_0^2}{2} \begin{pmatrix} 0 & 0 & 1 & 0 \\ 0 & 0 & 0 & g_1 \\ 0 & 0 & 0 & g_2 \\ 0 & 0 & 0 & g_3 \end{pmatrix} + \frac{g_0^3}{3!} \begin{pmatrix} 0 & 0 & 0 & 1 \\ 0 & 0 & 0 & 0 \\ 0 & 0 & 0 & 0 \\ 0 & 0 & 0 & 0 \end{pmatrix} . \tag{19.9}$$

The second column of $\mathbb{C}[G]$ can be read into the series $G(x) = g_0 + g(x)$. Notice that, when isolating progressive powers of g_0, matrix $\hat{\mathbb{B}}[g]$ appears: first itself, and then, at each order, dislocated one more step to the right. An operational version is obtained if we introduce $(N+1) \times (N+1)$ matrices $\mathbb{W}$ whose entries are all zero, except those in the first-right diagonal, which are $= 1$. Thus, for $N = 3$, $\mathbb{W} = \begin{pmatrix} 0 & 1 & 0 & 0 \\ 0 & 0 & 1 & 0 \\ 0 & 0 & 0 & 1 \\ 0 & 0 & 0 & 0 \end{pmatrix}$. Their right - action on another matrix causes an overall shift to the right, destroying the last column and zeroing the first. Then, $\mathbb{M}[g_0] = \sum_{j=0}^{N} \frac{\mathbb{W}^j g_0^j}{j!}$ and

$$\mathbb{C}[G] = \hat{\mathbb{B}}[g] \exp_N[\mathbb{W} g_0] \tag{19.10}$$

with $\exp_N$ the truncated exponential introduced in Comment 225.3.

Carleman matrices maintain the relationship between composition and matrix right-product: it is indeed enough to put $t = F(x) = f_0 + f(x)$ in (A.7.1) to show that, for $(N+1) \times (N+1)$ matrices,

$$\mathbb{C}_{n,m}[G \circ F] = \sum_{r=0}^{N} \mathbb{C}_{n,r}[F] \; \mathbb{C}_{r,m}[G] \; .$$

This means that we can apply here the procedure of Section 1.3 to get functions of matrices. We shall now have

$$\mathbb{C} = \sum_{j=0}^{N} \lambda_j \; \mathbb{Z}_{(j)}[\mathbb{C}] \; ; \tag{19.11}$$

$$F(\mathbb{C}) = \sum_{j=0}^{N} F(\lambda_j) \; \mathbb{Z}_{(j)}[\mathbb{C}] \; ; \tag{19.12}$$

$$\mathbb{C}^{\alpha} = \sum_{j=0}^{N} \lambda_j^{\alpha} \, \mathbb{Z}_{(j)}[\mathbb{C}] \; ; \tag{19.13}$$

$$G^{<\alpha>}(x) = \sum_{k=0}^{\infty} \mathbb{C}^{\alpha}_{k1} \frac{x^k}{k!} = \sum_{k=0}^{\infty} \left[\sum_{j=1}^{N} \lambda_j^{\alpha} \, \mathbb{Z}_{(j)k1}[\mathbb{C}] \right] \frac{x^k}{k!} \; ; \tag{19.14}$$

$$\therefore \quad G_k^{<\alpha>} = \sum_{j=0}^{N} \lambda_j^{\alpha} \, \mathbb{Z}_{(j)k1}[\mathbb{C}] \; . \tag{19.15}$$

A difficulty comes now from the eigenvalues λ_j of $\mathbb{C}$, whose general expressions are, unlike those of $\mathbb{B}$, highly non-trivial. Already for the $N = 2$ matrix, they are $\lambda_0 = 1$ and $\lambda_{1,2} = \frac{g_1 + g_1^2 + g_0 g_2 \mp \sqrt{(g_1 + {g_1}^2 + g_0 g_2)^2 - 4{g_1}^3}}{2}$.

We can alternatively define the projector–related functions

$$R_j(x) = \sum_{k=0}^{\infty} \mathbb{Z}_{(j)k1}[\mathbb{C}] \, \frac{x^k}{k!} \; , \tag{19.16}$$

so that the α– and the x–dependences are, in a sense, factorized:

$$G^{<\alpha>}(x) = \sum_{j=1}^{N} \lambda_j^{\alpha} \, R_j(x) \; . \tag{19.17}$$

More generally, Eq.(19.12) defines a functional $\mathcal{F}[G]$ by

$$\mathcal{F}[G](x) = \sum_{k=0}^{\infty} \left(\sum_{j=1}^{N} F(\lambda_j) \, \mathbb{Z}_{(j)k1}[\mathbb{C}] \right) \frac{x^k}{k!} = \sum_{j=1}^{N} F(\lambda_j) \, R_j(x) \; , \tag{19.18}$$

or by the Taylor coefficients

$$\mathcal{F}[G]_k = \sum_{j=1}^{N} F(\lambda_j) \, \mathbb{Z}_{(j)k1} \left[\mathbb{C}[G] \right] \; . \tag{19.19}$$

19.2 A trivial example

Let us begin with a rather trivial example: consider the two linear functions $G(x) = g_0 + g_1 x$, $F(y) = f_0 + f_1 y$, and to them make correspond matrices as follows:

$$G(x) = g_0 + g_1 x \quad \longleftrightarrow \quad \mathbb{C}(G) = \begin{pmatrix} 1 & g_0 \\ 0 & g_1 \end{pmatrix} \; , \tag{19.20}$$

$$F(x) = f_0 + f_1 x \quad \longleftrightarrow \quad \mathbb{C}(F) = \begin{pmatrix} 1 & f_0 \\ 0 & f_1 \end{pmatrix} \; . \tag{19.21}$$

Notice that the function coefficients can in each case be read in the second columns of the matrices. Taking now the function composition $[F \circ G](x) = F[G(x)] = f_0 + f_1 g_0 + f_1 g_1 x$, and comparing with the matrix product in inverse order (right-product), we find

$$\mathbb{C}(G)\mathbb{C}(F) = \begin{pmatrix} 1 & f_0+f_1 g_0 \\ 0 & f_1 g_1 \end{pmatrix} = \mathbb{C}(F \circ G) \ . \tag{19.22}$$

The composition coefficients are again read in the second column of the resulting matrix. The relationship function $\longleftrightarrow$ matrix is consequently maintained, with function composition corresponding to matrix right-product. It is trivially verified that this result holds on for higher orders of composition: $\mathbb{C}(H)\mathbb{C}(G)\mathbb{C}(F) = \mathbb{C}(F \circ G \circ H)$, etc. In particular, function iterations

$$G \circ G, G \circ G \circ G, \ldots, \underbrace{G \circ G \circ \ldots \circ G \circ G}_{\text{k times}},$$

for which we shall use notations $G^{<2>}$, $G^{<3>}$, $\ldots$, $G^{<k>}$, are represented by matrix powers:

$$\mathbb{C}(G^{<2>}) = \mathbb{C}^2(G) \ , \quad \mathbb{C}(G^{<k>}) = \mathbb{C}^k(G) \ . \tag{19.23}$$

Matrix $\mathbb{C}(G)$ has eigenvalues $\lambda_0 = 1, \lambda_1 = g_1$ and (provided $\lambda_0 \neq \lambda_1$) can be decomposed as

$$\mathbb{C}(G) = \lambda_0 \times \mathbb{Z}_{(0)}(\mathbb{C}) + \lambda_1 \times \mathbb{Z}_{(1)}(\mathbb{C}) = 1 \times \begin{pmatrix} 1 & \frac{g_0}{1-g_1} \\ 0 & 0 \end{pmatrix} + g_1 \begin{pmatrix} 0 & \frac{-g_0}{1-g_1} \\ 0 & 1 \end{pmatrix} . \tag{19.24}$$

The two "components" $\mathbb{Z}_{(0)}$ and $\mathbb{Z}_{(1)}$, which appear multiplied by the corresponding eigenvalues, are idempotent eigen-projectors "orthogonal" to each other: they satisfy $\mathbb{C}\mathbb{Z}_{(i)} = \lambda_i \mathbb{Z}_{(i)}$, $\mathbb{Z}_{(i)}\mathbb{Z}_{(i)} = \mathbb{Z}_{(i)}$ and, for $i \neq j$, $\mathbb{Z}_{(i)}\mathbb{Z}_{(j)} = \mathbb{O}$ (the zero matrix). It is then immediate to verify that

$$\mathbb{C}^2(G) = \lambda_0^2 \times \mathbb{Z}_{(0)}(\mathbb{C}) + \lambda_1^2 \times \mathbb{Z}_{(1)}(\mathbb{C}) \ ; \quad \text{and} \tag{19.25}$$

$$\mathbb{C}^k(G) = \lambda_0^k \times \mathbb{Z}_{(0)}(\mathbb{C}) + \lambda_1^k \times \mathbb{Z}_{(1)}(\mathbb{C}) \ . \tag{19.26}$$

This is actually a very particular example of a very general property: an arbitrary function f of matrix $\mathbb{C}(G)$ will be written

$$f[\mathbb{C}(G)] = f[\lambda_0] \times \mathbb{Z}_{(0)}(\mathbb{C}) + f[\lambda_1] \times \mathbb{Z}_{(1)}(\mathbb{C}) \ . \tag{19.27}$$

This is easily proven for f defined via a power series and, for non-analytic functions, is taken as the definition of $f[\mathbb{C}(G)]$. Notice that (19.27) holds in this simple form only for non-degenerate $\mathbb{C}(G)$, whose eigenvalues are all distinct of each other ($1 \neq g_1$ in the case).[1]

Continuous function iteration (say, $G^{<\alpha>}(x)$ of order α) can then be obtained [always for the trivial function $G(x) = g_0 + g_1 x$] without any ado: (i) obtain the matrix

$$\mathbb{C}^{\alpha}(G) = \lambda_0^{\alpha} \times \mathbb{Z}_{(0)}(\mathbb{C}) + \lambda_1^{\alpha} \times \mathbb{Z}_{(1)}(\mathbb{C}) = \begin{pmatrix} 1 & \frac{g_0(1-g_1^{\alpha})}{1-g_1} \\ 0 & g_1^{\alpha} \end{pmatrix} \tag{19.28}$$

and (ii) read the corresponding function in the second column:

$$G^{<\alpha>}(x) = \frac{g_0(1-g_1^{\alpha})}{1-g_1} + g_1^{\alpha} x \; . \tag{19.29}$$

All this holds in this simple finite way because function G is of first order in the variable x — and so are its iterations. Higher order functions require a far more involved procedure, which is our aim to describe below. Let us by now only retain the idea that a function can be represented by a matrix, and that in such a way that function composition is translated into matrix right–multiplication.

Only functions that can be represented by Taylor series will be considered here. We shall actually use a rather loose language, interchanging expressions "functions" and "formal series". And our approach will be "umbral", in the sense of formal series: we shall not be concerned with convergence problems. Just as functions are in general represented by infinite series, they can also be represented by matrices. This matrix representation will actually require — a drawback — infinite matrices. Nevertheless, finite sections of these infinite matrices will retain the character of approximations, quite analogous to that of approximating an analytic function, equivalent to an infinite series, by a finite approximation to a certain order.

Carleman matrices actually "over-represent" the corresponding series at each order, as they contain a lot of redundant information. For the above linear example a 2×2 matrix is enough because iteration of a linear function is a linear function. Instead of Eq.(19.20), for example, the complete matrix would have the aspect

[1] The more involved approach needed for degenerate matrices can be found in references [3] or [10].

$$\begin{pmatrix} 1 & g_0 & \frac{g_0^2}{2} & \frac{g_0^3}{3!} & \dots \\ 0 & g_1 & g_0 g_1 & \frac{g_0^2 g_1}{2} & \dots \\ 0 & 0 & g_1^2 & g_0 g_1^2 & \dots \\ 0 & 0 & 0 & g_1^3 & \dots \\ \dots & \dots & \dots & \dots & \dots \\ \dots & \dots & \dots & \dots & \dots \end{pmatrix} , \tag{19.30}$$

but all rows and columns not present in Eq.(19.20) are irrelevant.

19.3 Tetration

Tetration will be the fourth operation of arithmetic, after addition, multiplication and exponentiation. Its explicit form would come from the solution of a Schröder equation, which becomes, once written in terms of Carleman matrices, an eigenvalue equation.

The simplest of arithmetic operations is addition: starting with integers, $a + b$ is defined for any real or complex numbers a and b by an implicit interpolation process: integer $n + m \to$ real $\to$ complex (and to more general fields). Multiplication starts by na, summing n times the value a, and then interpolating the integers $n \to b$ to define ba. Next in this hierarchy comes exponentiation: a is multiplied by itself n times to give a^n, and then n is interpolated to continuum and complex values to give a^b. Tetration [153] would be next: $a \to a^a \to a^{a^a} \to \dots a^{a^{\cdot^{\cdot^{\cdot a}}}}$, a exponentiated to a n times; and integer n is then interpolated to real/complex values.

How to proceed to this last interpolation? As $a^x = e^{(\ln a)x}$, it will be enough to consider the exponential function with basis e, $e^x = \exp(x)$. The formal solution would be given by solving the Schröder equation [98] for that exponential function, that is, by finding an invertible function F such that

$$F[\exp(x)] = KF(x) \tag{19.31}$$

with $K \neq 1$ some real constant. In that case, the n-th iterate would be $\exp^{<n>}(x) = F^{<-1>}[K^n F(x)]$ and its continuous interpolation,

$$\exp^{<t>}(x) = F^{<-1>}[K^t F(x)] \ , \tag{19.32}$$

would give the t-th order tetration of variable x. The problem is that a complete solution would only come from an inspired guess of function F. In absence of such, we are condemned to resort to an algorithmic approach. Before going into that, let us notice that continuous iteration defines one–parameter "families" of functions. For example, the logarithm $\ln(x) =$

$\ln^{<1>}(x) = \exp^{<-1>}(x)$ belongs to the same family as the exponential, as do $\exp^{<t>}(x)$ for any value of parameter t and the inevitable member of every family, the identity function $\mathrm{Id}(x) = \exp^{<0>}(x) = \ln^{<0>}(x) = x$. An easy check shows that no member on this "exponential family" can be the solution F of Eq.(19.31).

A systematic procedure to obtain continuous interpolations of order-by-order iterations of certain functions – while keeping the sense of iteration – have been given many years ago [19]. It uses Bell matrices, and is restricted to functions $g(x)$ whose formal Taylor series starts at order x: $g(x) \approx g_1 x + \frac{g_2}{2} x^2 + ...$, that is to say, to series $g(x)$ such that $g(0) = 0$. Its main interest turns up when considering dynamical systems – interpolations between successive Poincaré sections of evolving structures. For the particular "extreme" logistic map $g^{<1>}(x) = 4x(1-x)$, for example, it gives a well-defined result. It is enough to introduce a ϕ such that $x = \sin^2 \phi$ to find that $g^{<1>}(x) = \sin^2 [2 \arcsin \sqrt{x}]$ and that $g^{<k>}(x) = \sin^2 \left[2^k \arcsin \sqrt{x}\,\right]$; it follows that $g^{<t>}(x) = \sin^2 \left[2^t \arcsin \sqrt{x}\,\right]$.

The exponential function being not of type $g(0) = 0$, it will require the use of Carleman matrices. Actually, the procedure leads to notions much more general than tetration.

The relationship between functions and matrices have been shown in Section 19.2. We in the following recall the method to obtain the continuous iterate of a function with vanishing independent Taylor coefficient. Carleman matrices, unlike Bell matrices, have complicated eigenvalues. A generalization of the exponential is then examined through lower-order examples. The main problem turning up is that of convergence: in a way analogous to the series for the exponential function, good numerical approximations require matrices of higher and higher orders for higher and higher values of the arguments.

For functions with vanishing independent term, an equation like (19.31) can be transcribed into the matrix approach as

$$\sum_{r=s}^{k} \mathbb{B}_{kr}[g]\, \mathbb{B}_{rs}[f] = \mathbb{B}_{ks}[Kf] = K^s\, \mathbb{B}_{ks}[f]\ ; \quad \text{in particular,} \tag{19.33}$$

$$\sum_{r=1}^{k} \mathbb{B}_{kr}[g]\, \mathbb{B}_{r1}[f] = \sum_{r=1}^{k} \mathbb{B}_{kr}[g]\, f_r = K\, \mathbb{B}_{k1}[f] = K\, f_k\ . \tag{19.34}$$

This eigenvalue problem allows a step-by-step calculation, but seldom leads to a closed result.

The expressions for the $\mathbb{C}[G]$ entries become rapidly very long with increasing N. It will be instructive to examine the exponential function, for which all $g_k = 1$. For case $N = 2$, $\mathbb{C}[e^x = 1 + x + \frac{x^2}{2} + O(x^3)]$ is written in terms of its projectors and respective eigenvalues as

$$\begin{pmatrix} 1 & 1 & \frac{1}{2} \\ 0 & 1 & 1 \\ 0 & 1 & 2 \end{pmatrix} = 1 \times \begin{pmatrix} 1 & \frac{1}{2} & -1 \\ 0 & 0 & 0 \\ 0 & 0 & 0 \end{pmatrix} + \\ + \frac{3-\sqrt{5}}{2} \times \begin{pmatrix} 0 & -\frac{1+\sqrt{5}}{4} & \frac{1}{2} \\ 0 & \frac{5+\sqrt{5}}{10} & -\frac{1}{\sqrt{5}} \\ 0 & -\frac{1}{\sqrt{5}} & \frac{5-\sqrt{5}}{10} \end{pmatrix} + \frac{3+\sqrt{5}}{2} \times \begin{pmatrix} 0 & -\frac{1-\sqrt{5}}{4} & \frac{1}{2} \\ 0 & \frac{5-\sqrt{5}}{10} & \frac{1}{\sqrt{5}} \\ 0 & \frac{1}{\sqrt{5}} & \frac{5+\sqrt{5}}{10} \end{pmatrix}. \tag{19.35}$$

A function F of this non-degenerate matrix will be

$$F\left[\begin{pmatrix} 1 & 1 & \frac{1}{2} \\ 0 & 1 & 1 \\ 0 & 1 & 2 \end{pmatrix}\right] = F[1] \times \begin{pmatrix} 1 & \frac{1}{2} & -1 \\ 0 & 0 & 0 \\ 0 & 0 & 0 \end{pmatrix} + \\ + F\left[\frac{3-\sqrt{5}}{2}\right] \times \begin{pmatrix} 0 & -\frac{1+\sqrt{5}}{4} & \frac{1}{2} \\ 0 & \frac{5+\sqrt{5}}{10} & -\frac{1}{\sqrt{5}} \\ 0 & -\frac{1}{\sqrt{5}} & \frac{5-\sqrt{5}}{10} \end{pmatrix} + F\left[\frac{3+\sqrt{5}}{2}\right] \times \begin{pmatrix} 0 & -\frac{1-\sqrt{5}}{4} & \frac{1}{2} \\ 0 & \frac{5-\sqrt{5}}{10} & \frac{1}{\sqrt{5}} \\ 0 & \frac{1}{\sqrt{5}} & \frac{5+\sqrt{5}}{10} \end{pmatrix}. \tag{19.36}$$

Examples are

(i) the identity matrix, which comes for $F(u) = u^0 = 1$, and whose 2nd column gives just $G(x) = x$;

(ii) the identity function $F(u) = u^1 = u$, leading to $F\left[\begin{pmatrix} 1 & 1 & \frac{1}{2} \\ 0 & 1 & 1 \\ 0 & 1 & 2 \end{pmatrix}\right] = \begin{pmatrix} 1 & 1 & \frac{1}{2} \\ 0 & 1 & 1 \\ 0 & 1 & 2 \end{pmatrix}$, whose 2nd column gives $G(x) = 1 + x + \frac{x^2}{2}$;

(iii) and the arbitrary power

$$\begin{pmatrix} 1 & 1 & \frac{1}{2} \\ 0 & 1 & 1 \\ 0 & 1 & 2 \end{pmatrix}^t = \begin{pmatrix} 1 & \frac{1}{2} & -1 \\ 0 & 0 & 0 \\ 0 & 0 & 0 \end{pmatrix} + \\ + \left[\frac{3-\sqrt{5}}{2}\right]^t \times \begin{pmatrix} 0 & -\frac{1+\sqrt{5}}{4} & \frac{1}{2} \\ 0 & \frac{5+\sqrt{5}}{10} & -\frac{1}{\sqrt{5}} \\ 0 & -\frac{1}{\sqrt{5}} & \frac{5-\sqrt{5}}{10} \end{pmatrix} + \left[\frac{3+\sqrt{5}}{2}\right]^t \times \begin{pmatrix} 0 & -\frac{1-\sqrt{5}}{4} & \frac{1}{2} \\ 0 & \frac{5-\sqrt{5}}{10} & \frac{1}{\sqrt{5}} \\ 0 & \frac{1}{\sqrt{5}} & \frac{5+\sqrt{5}}{10} \end{pmatrix}. \tag{19.37}$$

It is easily verified that $t = -1$ does provide the inverse to matrix (19.35). By the way, this illustrates also a statement made above: the procedure is self-consistent at each order. At the N-th level, the inverse matrix found is just the inverse to the Carleman $(N+1) \times (N+1)$ matrix. Reading along the 2nd columns in (19.37),

$$G^{<t>}(x) = \tfrac{1}{2} - \tfrac{1-\sqrt{5}}{4}\left(\tfrac{3+\sqrt{5}}{2}\right)^t - \tfrac{1+\sqrt{5}}{4}\left(\tfrac{3-\sqrt{5}}{2}\right)^t + \left[\tfrac{5-\sqrt{5}}{10}\left(\tfrac{3+\sqrt{5}}{2}\right)^t + \tfrac{5+\sqrt{5}}{10}\left(\tfrac{3-\sqrt{5}}{2}\right)^t\right]x + \tfrac{1}{\sqrt{5}}\left[\left(\tfrac{3+\sqrt{5}}{2}\right)^t - \left(\tfrac{3-\sqrt{5}}{2}\right)^t\right]\frac{x^2}{2}, \quad (19.38)$$

or

$$G^{<t>}(x) = \tfrac{1}{2} + \left(\tfrac{3+\sqrt{5}}{2}\right)^t\left[\tfrac{5-\sqrt{5}}{10}\,x - \tfrac{1-\sqrt{5}}{4} + \tfrac{1}{\sqrt{5}}\tfrac{x^2}{2}\right] + \left(\tfrac{3-\sqrt{5}}{2}\right)^t\left[\tfrac{5+\sqrt{5}}{10}x - \tfrac{1+\sqrt{5}}{4} - \tfrac{1}{\sqrt{5}}\tfrac{x^2}{2}\right]. \quad (19.39)$$

For integer values of t, this gives the approximate expressions: $G^{<0>}(x) = x$, $G^{<1>}(x) = 1 + x + \frac{x^2}{2}$, $G^{<2>}(x) = \frac{1}{2}\left(5 + 4x + 3x^2\right) = 1 + G^{<1>}(x) + \frac{1}{2}(G^{<1>}(x))^2$.

Equations (19.38, 19.39) give the tetration values for function e^x to order $N = 2$. If we want e exponentiated by e itself t times, it is enough to put $x = e$. And to have e exponentiated by e itself "e times", it is enough to put also $t = e$. Approximations to order $N = 2$ are, in this case, very poor: we find $G^{<1>}(e) = 7.41281$, to be compared to $e^e \approx 15.1543$, Thus, the found value $G^{<e>}(e) = 37.5795$ is not trustworthy. But this is consistent with the exponential function itself: to evaluate e^e with less than 1% error, order $N = 7$ would be necessary.

It is fascinating that expression (19.39) can be rewritten in terms of the Golden Ratio $\alpha = \frac{1+\sqrt{5}}{2}$:

$$G^{<t>}(x) = \tfrac{1}{2} + (1+\alpha)^t\left[\tfrac{3-\alpha}{5}\,x - \tfrac{1-\alpha}{2} + \tfrac{x^2}{4\alpha-2}\right] + (2-\alpha)^t\left[\tfrac{2+\alpha}{5}\,x - \tfrac{\alpha}{2} - \tfrac{x^2}{4\alpha-2}\right]. \quad (19.40)$$

Of course, in terms of Carleman matrices, solving the Schröder equation $\mathbb{C}[g]\mathbb{C}[F] = \mathbb{C}[KF]$ becomes finding eigenvectors: as $\mathbb{C}_{n1}[f] = f_n$, it becomes

$$\sum_{s=0}^{\infty} \mathbb{C}_{rs}[g]\mathbb{C}_{s1}[F] = \mathbb{C}_{r1}[KF]\ , \quad \text{which is} \quad (19.41)$$

$$\sum_{s=0}^{\infty} \mathbb{C}_{rs}[g]F_s = KF_r\ . \quad (19.42)$$

In particular, for tetration we should find the general eigenvectors of

$$\mathbb{C}_{nj}[e^x] = \frac{j^n}{j!} \ . \tag{19.43}$$

A curious point is that the sum of all such matrix elements in a line is related to Stirling numbers. Recall that, as seen in Commentary 231.1, the Bell number $\omega(n) = \sum_{k=1}^{n} S_n^{(k)}$ is the number of ways of partitioning a set of n members. Taking by convention $\omega(0) = 1$, the Dobiński formula 15.50 gives

$$\sum_{j=0}^{\infty} \mathbb{C}_{nj}[e^x] = e\ \omega(n) \ , \tag{19.44}$$

the sum of all the elements in the n-th row of $\mathbb{C}_{nj}[e^x]$. For the 0-th row it gives, of course, just exp(1). We have then that

$$G^{<2>}(x) = e^{e^x} = e \sum_{j=0}^{\infty} \omega(n)\ \frac{x^n}{n!} \ . \tag{19.45}$$

Continuous iteration is, of course, a much more general subject [149]. We have here restricted ourselves to its application to tetration [150], [151], [152]. It would be desirable to examine the differential equations which would come out of its use. Continuous iteration involves a "non-locality", a global feature in function space. Functions constitute families. The "exponential family" of functions $\exp^{<\alpha>}(x) = \ln^{<-\alpha>}(x)$ will include the exponential $\exp(x) = \exp^{<1>}(x)$, the logarithm $\ln(x) = \exp^{<-1>}(x)$ and the inevitable member of every family, the identity function $\mathrm{Id}(x) = \exp^{<0>}(x) = \ln^{<0>}(x) = x$. Let us remember that many physical phenomena, from the decay of subatomic particles to the cosmological expansion of the Universe, could not be described before the introduction of exponentiation in the XVIII-th century. It is to be hoped that this "fourth operation" come to be of interest to the description of some as yet untamed phenomena. The appearance of Stirling numbers — and the alluring presence of the Golden Ratio in the first approximations — is an intimation of relationship to complex evolving systems.

Appendix A

Formulary

The formulas below have been used in checking the text results and can be instrumental in completing the examples and comments. The assortment is by no means exhaustive on any topic. A few of them may even be new. Readers interested in more detail should consult the sources indicated at the end.

A.1 General formulas

(1) Useful finite and infinite sums

sums of powers

$$\sum_{k=j}^{N} x^k = \left(x^j - x^{N+1}\right) \sum_{k=0}^{\infty} x^k = \begin{cases} \frac{x^j - x^{N+1}}{1-x} & \text{if } x \neq 1 \\ N-j+1 & \text{if } x = 1 \end{cases}$$

(whenever the infinite series in the middle member makes sense)

in particular, for $x \neq 1$,

$$\sum_{k=0}^{N} x^k = \frac{1-x^{N+1}}{1-x}; \quad \sum_{k=0}^{n-1} x^k = \frac{1-x^n}{1-x}; \sum_{k=1}^{N} x^k = x\,\frac{1-x^N}{1-x}$$

$$1 - x^N = (1-x) \sum_{k=0}^{N-1} x^k$$

the N-th roots of the identity obey

$$\sum_{k=1}^{N} e^{i\frac{2\pi}{N}k} = \sum_{k=0}^{N-1} e^{i\frac{2\pi}{N}k} = 0$$

the Kronecker identity

$$\tfrac{1}{N}\sum_{k=1}^{N} e^{i\frac{2\pi}{N}k(r-s)} = \delta^{rs}$$

$$\frac{1}{2\pi}\int_{-\pi}^{\pi} d\phi e^{i\phi(r-s)} = \delta_{rs}$$

combinatorial identities

$$\textstyle\sum_{j=0}^{N}\binom{N}{j}\binom{M}{r-j} = \binom{N+M}{r}\ ;\ \sum_{j=1}^{N} j\,\binom{N}{j} = N\ 2^{N-1}$$

$$\textstyle\sum_{r=j}^{n}\frac{(r-1)!}{(r-j)!} = (j-1)!\binom{n}{j}$$

$$\textstyle\sum_{j=1}^{n}(-)^{j-1}\binom{n}{j}\binom{n+j}{i} = \binom{n}{i} \qquad (\text{for}\ \ i<n)$$

$$\textstyle\sum_{j=p}^{n}(-)^{j-p}\binom{n}{j}\binom{j}{p} = \delta_{np}\ .$$

binomial formulae

$$\textstyle(1+t)^{\alpha} = \sum_{r=0}^{\infty}\binom{\alpha}{r}t^{r}\ ;\ (1+t)^{m} = \sum_{r=0}^{m}\binom{m}{r}t^{r}$$

$$\textstyle(1-t)^{-m} = \sum_{r=0}^{\infty}\binom{m+r-1}{r}t^{r}\ ;\ \displaystyle\sum_{n=m}^{\infty}\binom{n}{m}t^{n} = \frac{t^{m}}{(1-t)^{m+1}}$$

$$\frac{1}{\sqrt{1-4x}} = \sum_{j=0}^{\infty}\frac{(2j)!}{(j!)^{2}}x^{j}\ .$$

(2) Gamma function

$$\Gamma(z) = \lim_{n\to\infty}\frac{1\cdot 2\cdot 3\ldots(n-1)n^{z}}{z(z+1)(z+2)\ldots(z+n-1)}$$

$$= \frac{1}{z}\prod_{n=1}^{\infty}(1+\frac{1}{n})(1+\frac{z}{n})^{-1} = \int_{0}^{\infty} dt\, e^{-t}t^{z-1}$$

$$\Gamma(z+1) = z\ \Gamma(z)\ ;\ \Gamma(z)\Gamma(1-z) = \frac{\pi}{\sin\pi z}$$

$$\Gamma(\tfrac{1}{2}+z)\Gamma(\tfrac{1}{2}-z) = \frac{\pi}{\cos\pi z}$$

$$\Gamma(n+1) = n!\ ;\ \Gamma(\tfrac{1}{2}-m) = (-)^{m}\ \frac{2^{m}\sqrt{\pi}}{(2m-1)!!}$$

$$\Gamma(1) = \Gamma(2) = 1;\ \Gamma(\tfrac{1}{2}) = \sqrt{\pi};\ \Gamma(\tfrac{3}{2}) = \tfrac{1}{2}\sqrt{\pi}\ ;\ \Gamma(\tfrac{5}{2}) = \tfrac{3}{4}\sqrt{\pi}$$

Stirling formulae:

$$\Gamma(z) \approx \sqrt{\frac{2\pi}{z}} e^{-z} z^z \left[1 + \frac{1}{12z} + ...\right]$$

$$\ln N! = N \ln N - N + o(1/N) \; .$$

general factorial

$$x! = \Gamma(x+1), \quad \binom{x}{k} = \frac{x!}{k!(x-k)!}$$

general binomials

$$\binom{r}{s} = \begin{cases} \frac{r!}{(r-s)!s!} & r = \text{non-negative integer} \\ (-)^s \frac{(|r|+s-1)!}{(|r|-1)!s!} & r = -\,|r| = \text{negative integer} \\ = \frac{\Gamma(r+1)}{\Gamma(r-s+1)\Gamma(s+1)} & r, s \text{ complex numbers} \end{cases}$$

(3) Exponential and related series

$$e^x = \lim_{m\to\infty} (1 + x/m)^m = \sum_{n=0}^{\infty} \frac{x^n}{n!}; \quad \ln(1+x) = \sum_{n=1}^{\infty} \frac{(-)^{n-1} x^n}{n}$$

$$\sin(z) = \sum_{k=1}^{\infty} \frac{(-)^k}{(2k-1)!} z^{2k-1} \quad ; \quad \cos(z) - 1 = \sum_{k=1}^{\infty} \frac{(-)^k}{(2k)!} z^{2k}$$

$$\arccos z = \frac{\pi}{2} - z - \frac{z^3}{3!} + \dots \; ; \quad \arccos(1-2x) = 2\sqrt{x} + \frac{x^{3/2}}{3} + \frac{x^{5/2}}{20} + \dots$$

$$\arcsin z = \pi/2 - \arccos z$$

$$= \sum_{k=0}^{\infty} \frac{(2k)!}{2^{2k}(k!)^2(2k+1)} z^{2k+1}$$

$$= zF(1/2, 1/2, 3/2, z^2)$$

ready-to-use inverse trigonometric relations

$$\arccos(1-2x) = 2 \arcsin \sqrt{x}; \quad \cos[\arcsin \sqrt{x}] = \sqrt{1-x}$$

(4) Hypergeometric function $F(a, b, c; x) = F(b, a, c; x)$: solves equation

$$x(1-x) \frac{d^2}{dx^2} F + [c - (a+b+1)x] \frac{d}{dx} F - ab\, F = 0 \, .$$

With the symbols $(f)_n = \frac{\Gamma(f+n)}{\Gamma(f)} = (f+n-1)(f+n-2)(f+n-3)\ldots(f+1)f$, F is the Gauss hypergeometric series $F(a,b,c;x) = \sum_{n=0}^{\infty} \frac{(a)_n(b)_n}{(c)_n}\frac{x^n}{n!} = \sum_{n=0}^{\infty} \frac{\Gamma(a+n)}{\Gamma(a)}\frac{\Gamma(b+n)}{\Gamma(b)}\frac{\Gamma(c)}{\Gamma(c+n)}\frac{x^n}{n!}$.

Example:

inverse trigonometric: $\arcsin z = zF(1/2, 1/2, 3/2, z^2)$

When either a or b is a negative integer $= -m$, $F(a,b,c;x)$ is a polynomial of order m:

$F(-m,b,c;x) = \sum_{n=0}^{m} \frac{(-m)_n(b)_n}{(c)_n}\frac{x^n}{n!}$

$F(a,b,1;x) = \sum_{n=0}^{\infty} \frac{\Gamma(a+n)}{\Gamma(a)}\frac{\Gamma(b+n)}{\Gamma(b)}\frac{1}{n!}\frac{x^n}{n!}$

Examples:

Chebyshev: $T_n(z) = F(-n, n, \frac{1}{2}; \frac{1-z}{2})$

Legendre: $P_n(z) = F(-n, n+1, \frac{1}{2}; \frac{1-z}{2})$

Gegenbauer: $C_n^{\alpha}(z) = \frac{(2\alpha)_n}{n!} F(-n, n+2\alpha, \frac{1}{2}+\alpha; \frac{1-z}{2})$

Jacobi: $P_n^{(\alpha,\beta)}(z) = \frac{(1+\alpha)_n}{n!} F(-n, n+1+\alpha+\beta, 1+\alpha; \frac{1-z}{2})$.

(5) Gaussian sums

$$\sum_{k=-N}^{N} e^{i\frac{2\pi}{N}k^2} = 1 + \sqrt{N}(1+i)[1+(-i)^N]$$

$$\sum_{k=0}^{N-1} e^{i\frac{2\pi}{N}k^2} = \sum_{k=1}^{N} e^{i\frac{2\pi}{N}k^2} = \sqrt{N}\,\frac{1+i^{-N}}{1+i^{-1}}$$

Jacobi Theta function

$$\Theta_3(\phi; q) = \sum_{r=-\infty}^{\infty} q^{r^2/2} e^{ir\phi}$$

(6) Gauss integrals and related

$$\int_0^{\infty} dx\, e^{-\alpha x^2} = \frac{1}{2}\sqrt{\frac{\pi}{\alpha}}\ ;\ \int_{-\infty}^{\infty} dx\, e^{-\alpha x^2+\beta x} = e^{\beta^2/4\alpha}\sqrt{\frac{\pi}{\alpha}}$$

$$\int_{-\infty}^{\infty} dy\ e^{-\alpha[(x-y)^2+(y-z)^2]} = \sqrt{\frac{\pi}{2\alpha}}\ e^{-\alpha(x-z)^2/2}\ .$$

Dirac delta representations

$$\delta(x) = \frac{1}{2\pi}\int_{-\infty}^{\infty} dp\, e^{ipx}$$

$$\delta(x) = \lim_{\epsilon\to 0} \frac{1}{\sqrt{4\epsilon\pi}}\, e^{-\frac{x^2}{4\epsilon}}\,;\;\; \delta(x) = \lim_{\epsilon\to 0} \frac{1}{\pi}\, \frac{\epsilon}{\epsilon^2 + x^2}\,;\;\; \delta(x) = \lim_{\epsilon\to 0} \frac{e^{-|\frac{x}{\epsilon}|}}{2\epsilon}$$

Heavyside step distribution

$$\theta(x) = \begin{cases} 1 & \text{if } x \geq 0\,; \\ 0 & \text{if } x < 0\,; \end{cases}$$

$$\theta(x) = \lim_{\epsilon\to 0} \frac{i}{2\pi} \int_{-\infty}^{\infty} \frac{dp\, e^{-ipx}}{p + i\epsilon},\;\; \delta(x) = \frac{d}{dx}\theta(x)$$

Error function:

$$\mathrm{erf}(x) = \frac{2}{\sqrt{\pi}} \int_0^x e^{-t^2}\, dt = \frac{2}{\sqrt{\pi}} \sum_{n=0}^{\infty} \frac{(-)^n x^{2n+1}}{n!(2n+1)}$$

$$\frac{d^n}{dx^n}\, \mathrm{erf}(x) = (-)^{n-1} \frac{2}{\sqrt{\pi}}\, e^{-x^2}\, H_{n-1}(x)$$

($H_n(x)$: Hermite polynomial).

(7) Laplace transform of a function $g(x)$

$$\hat{g}(\beta) = \int_0^{\infty} dt e^{-\beta t} g(t)$$

inverse (Bromwich integral): $\displaystyle \frac{1}{2\pi i} \int_{c-i\infty}^{c+i\infty} d\beta e^{\beta t} \hat{g}(\beta) = \begin{cases} g(t) & \text{for } t > 0\,; \\ 0 & \text{for } t < 0\,; \end{cases}$

Examples:

$g(t) = \frac{1}{\sqrt{\pi t}}\, e^{-\,x^2/4t} \Rightarrow \hat{g}(\beta) = \frac{1}{\sqrt{\beta}}\, e^{-\,x\sqrt{\beta}}$ (for $x > 0$);

$g(t) = \sin(x\sqrt{t}) \Rightarrow \hat{g}(\beta) = \sqrt{\frac{\pi}{2}}\, \frac{x}{\beta^{3/2}} e^{-\,x^2/4\beta}$

Borel transform $\mathcal{B}$ of function $\phi(x)$ is

$\mathcal{B}[\phi](z) = \int_0^{\infty} du e^{-u} \phi(uz) = \frac{1}{z} \int_0^{\infty} dt e^{-t/z} \phi(t) = \frac{1}{z}\, \hat{\phi}(\frac{1}{z})$

Examples: $\mathcal{B}[\phi(u) = u^n](z) = n! z^n$; $\mathcal{B}[\phi(u) = \exp u](z) = \frac{1}{1-z}$

(8) Bernoulli polynomials $B_n(x)$ and numbers B_n

$$\frac{te^{xt}}{e^t - 1} = \textstyle\sum_{n=0}^{\infty} B_n(x)\, \displaystyle\frac{t^n}{n!} \;\; (\text{for } \; |t| < 2\pi)\;;\;\; B_0(x) = 1;\;\; B_n = B_n(0)$$

$$B_n(x+1) - B_n(x) = nx^{n-1}; B_n(x) = \sum_{k=0}^{n} \binom{n}{k} B_k x^{n-k}$$

$$B_0 = 1; B_1 = -\frac{1}{2}; B_2 = \frac{1}{6}; B_4 = -\frac{1}{30}; B_{2n+1} = 0 \quad (\text{for } n = 1, 2, 3, \ldots)$$

In terms of Bell polynomials,

$$B_n = \sum_{m=1}^{n} (-)^m\, m!\, \mathbb{B}_{nm} \left\{ \tfrac{1}{j+1} \right\} = \sum_{m=1}^{n} (-)^m\, m!\, \mathbb{B}_{nm} \left\{ \tfrac{1}{2}, \tfrac{1}{3}, \tfrac{1}{4}, \ldots, \tfrac{1}{n-m+2} \right\}$$

See below for a relation to Stirling numbers.

Euler–MacLaurin formula

$$h \sum_{k=1}^{n} f_k = \int_0^n dx\, f(x) + \tfrac{h}{2}[f(n) - f(0)] + \sum_{k=1}^{m} \frac{B_{2k}}{(2k)!} h^{2k} \left[f_n^{(2k-1)} - f_0^{(2k-1)} \right]$$
$$- \theta\, n\, \frac{B_{2m+2}}{(2m+2)!}\, h^{2m+3} \max_{0<x<n} \left[f^{(2m+2)}(x) \right] \qquad (-1 \leq \theta \leq 1)$$

sum of k-powers of integers

$$S_k = \textstyle\sum_{j=1}^{N} j^k = \dfrac{B_{k+1}(N+1) - B_{k+1}(1)}{k+1}$$

$$S_0 = N \; ; \quad S_1 = N(N+1)/2 \; ;$$

$$S_2 = N(N+1)(2N+1)/6 \; ; \quad S_3 = N^2(N+1)^2/4 \; .$$

(9) Euler polynomials $E_n(x)$ and numbers E_n

$$\frac{2e^{xt}}{e^t + 1} = \textstyle\sum_{n=0}^{\infty} E_n(x)\, \dfrac{t^n}{n!} \quad (\text{for } \; |t| < \pi) \; ; \; E_n = 2^n E_n(1/2)$$

$$\frac{1}{\cosh t} = \textstyle\sum_{n=0}^{\infty} E_n\, \dfrac{t^n}{n!} \quad (\text{for } \; |t| < \pi/2)$$

$$E_0 = 1; E_2 = -1; E_4 = 5; E_6 = -61; E_{2n+1} = 0 \quad (\text{for } n = 0, 1, 2, 3, \ldots).$$

relation to Bernouilli: $2^n B_n(x) = \sum_{k=0}^{n} \binom{n}{k} B_{n-k} E_k(2x)$

(10) Riemann's zeta function and related

$$\zeta(z) = \sum_{j=1}^{\infty} \frac{1}{j^z} = \prod_p \frac{1}{1 - p^{-z}} \quad (p = \text{all prime numbers})$$
$$= \frac{1}{\Gamma(z)} \int_0^{\infty} dt \frac{t^{z-1}}{e^t - 1}$$

$$\eta(s) = \sum_{j=1}^{\infty} \frac{(-)^{j-1}}{j^s} = (1 - 2^{1-s})\,\zeta(s) \quad (\mathrm{Re}\, s > 0)$$

some values:

$$\zeta(0) = -1/2;\ \zeta(-2n) = 0;\ \zeta(2) = \pi^2/6;\ \zeta(4) = \pi^4/90;$$

$$\eta(2) = \pi^2/12;\ \eta(4) = 7\pi^4/720;$$

$$\zeta(1-2n) = -\frac{B_{2n}}{2n};\ \zeta(2n) = (-)^{n+1}\frac{(2\pi)^{2n}}{2(2n)!}B_{2n}.$$

Dirichlet series

$$D(s) = \textstyle\sum_{j=1}^{\infty} a_j j^{-s}$$

the Bose function is the case $g_n(z) = D(n)$, with $a_j = z^j$.

polylogarithmic functions, or Polylogs:[2]

$$Li_n(z) = \sum_{k=1}^{\infty} \frac{z^k}{k^n}$$

With this definition the Bose function $g_n(z)$ is an n-Log: $g_n(z) = Li_n(z)$. Furthermore, $Li_n(1) = \zeta(n)$, $Li_n(-1) = -\eta(n)$.

(11) Inverse and identity series

$$[g^{<-1>} \circ g](x) = [g \circ g^{<-1>}](x) = e(x)\ .$$

(12) Stirling and related

Stirling numbers of the first kind $s_n^{(k)}$

$$s_n^{(k)} = \mathbb{B}_{nk}\{(-)^{j-1}(j-1)!\} = \mathbb{B}_{nk}[\ln(1+x)]$$

$$x(x-1)(x-2)\ldots(x-k+1) = \binom{x}{k}k! = \sum_{j=0}^{k} s_k^{(j)} x^j$$

$$s_{n+1}^{(m)} = s_n^{(m-1)} - n\, s_n^{(m)} \qquad (s_j^{(0)} = \delta_j^0)$$

Stirling numbers of the second kind $S_n^{(k)}$

$$S_n^{(k)} = \mathbb{B}_{nk}(1,1,1,\ldots,1) = \mathbb{B}_{nk}\{1\} = \mathbb{B}_{nk}[e^x - 1]$$

[2] This is Eric Weinstein's [100] definition. Abramowitz & Stegun give instead, for the Dilogarithm, $Li_2(1-z) = \sum_{k=1}^{\infty} \frac{z^k}{k^2}$.

$S_0^{(0)} = 1$ by convention; for $n \geq 1$:

$$x^n = \sum_{j=1}^{n} S_n^{(j)} x(x-1)(x-2)\ldots(x-j+1) = \sum_{j=1}^{n} S_n^{(j)}\, j! \binom{x}{j}$$

$$\frac{1}{k!}\,(e^u - 1)^k = \sum_{n=k}^{\infty} \frac{u^n}{n!}\; S_n^{(k)}$$

$$S_n^{(k)} = \frac{1}{k!} \sum_{j=0}^{k} (-)^{k-j} \binom{k}{j} j^n$$

relation to the Poisson distribution

$$\sum_{m=1}^{N} x^m S_N^{(m)} = \sum_{k=0}^{\infty} k^N e^{-x} \frac{x^k}{k!}$$

relation to Bernoulli numbers: use $\frac{t}{e^t-1} = \frac{\ln[1+(e^t-1)]}{e^t-1}$

$$B_m = \sum_{k=0}^{m} \frac{(-)^k k!}{k+1}\, S_m^{(k)} = \sum_{k=0}^{m} \frac{(-)^k k!}{k+1}\, \mathbb{B}_{mk}\{1\}\,;\, \sum_{m=0}^{n} s_n^{(m)}\, B_m = \frac{(-)^n n!}{n+1}$$

(13) Bessel functions

differential equation: $\frac{d^2}{du^2} Z_\nu(u) + \frac{1}{u}\,\frac{d}{du} Z_\nu(u) + (1 - \frac{\nu^2}{u^2}) Z_\nu(u) = 0$

Bessel functions of neighboring orders satisfy

$$uZ_{\nu-1}(u) + uZ_{\nu+1}(u) = 2\,\nu Z_\nu(u)$$

$$Z_{\nu-1}(u) - Z_{\nu+1}(u) = 2\,\frac{d}{du} Z_\nu(u)$$

$$\therefore\; uZ_{\nu\pm1}(u) = \left[\nu \mp u\frac{d}{du}\right]\, Z_\nu(u)\;.$$

first kind Bessel function:

$$J_n(z) = \frac{1}{2\pi} \int_{-\pi}^{\pi} d\zeta\; e^{i\,n\zeta}\, e^{i\,z\sin\zeta};\; J_\nu(2z) = z^\nu \sum_{n=0}^{\infty} \frac{(-)^n}{n!\,\Gamma(\nu+n+1)}\,(z)^{2n}$$

$$e^{iz\sin\zeta} = \sum_{n=-\infty}^{\infty} J_n(z) e^{in\zeta}; \sin(z\sin\zeta) = 2\sum_{n=0}^{\infty} J_{2n+1}(z)\sin[(2n+1)\zeta]$$

second kind Bessel function (or Neumann functions):

$$N_\nu(z) = \frac{1}{\sin \nu\pi} \left[\cos \nu\pi J_\nu(z) - J_{-\nu}(z)\right]$$

third kind Bessel function (or Hankel functions):

$$H_\nu^{(1)}(z) = J_\nu(z) + iN_\nu(z); H_\nu^{(2)}(z) = J_\nu(z) - iN_\nu(z)$$

"imaginary" Bessel function $I_\nu(z) = e^{-i\nu\pi/2} J_\nu(iz)$:

$$I_n(z) = e^{-in\pi/2} J_n(iz) = \frac{(-i)^n}{2\pi} \int_{-\pi}^{\pi} d\zeta\, e^{i\,n\zeta}\, e^{-\,z\sin\zeta}$$

$$I_n(z) = \frac{i^n}{2\pi} \int_0^{2\pi} du\, e^{i\,nu}\, e^{z\sin u} = \tfrac{1}{2\pi} \int_0^{2\pi} dv\, e^{-inv}\, e^{z\cos v}; I_0'(z) = I_1(z)$$

$$I_n(2z) = \sum_{p=0}^{\infty} \frac{(z)^{2p+n}}{p!(p+n)!}$$

"modified" Bessel function

$$K_\nu(z) = \frac{i\pi}{2}\, e^{i\nu\pi/2} H_\nu^{(1)}(iz) = \frac{i\pi}{2}\, e^{-i\nu\pi/2} H_{-\nu}^{(2)}(iz)$$

extreme cases of $K_n(x)$:

for $x >> 1$

$$K_2(x) \approx \left(\frac{\pi}{2x}\right)^{1/2} e^{-x} \left(1 + \frac{15}{8x} + \dots\right)$$

$$K_1(x) \approx \left(\frac{\pi}{2x}\right)^{1/2} e^{-x} \left(1 + \frac{3}{8x} + \dots\right)$$

for $x << 1$

$$K_n(x) \approx 2^{n-1} \frac{(n-1)!}{x^n}\ ; \quad K_2(x) \approx \frac{2}{x^2}\ ; \quad K_1(x) \approx \frac{1}{x}\ .$$

A.2 General matrices

(1) characteristic equation of an $N \times N$ matrix $\mathbf{M}$

$$\Delta_{\mathbf{M}}(\lambda) = \det(\lambda\mathbf{I} - \mathbf{M}) = \prod_{k=1}^{N} (\lambda - \lambda_k) = 0$$

(2) Cayley–Hamilton theorem

$$\Delta_{\mathbf{M}}(\mathbf{M}) = \prod_{k=1}^{N}(\mathbf{M} - \lambda_k) = 0$$

(3) components of a matrix

$$\mathbf{M} = \sum_{j=1}^{N} \lambda_j \, \mathbf{Z}_j[\mathbf{M}]$$

(4) function of a matrix

$$\mathbf{F}(\mathbf{M}) = \sum_{j=1}^{N} F(\lambda_j) \, \mathbf{Z}_j[\mathbf{M}].$$

A.3 Stochastic matrices

(1) stochastic $N \times N$ matrix $\mathbf{M}$: defining requirements

(i) $0 \leq M_{ab} \leq 1 \;, \forall \; a, b = 1, 2, \ldots N$

(ii) $\sum_{a=1}^{N} M_{ab} = 1$

property: $\mathbf{M}$ has at least one eigenvalue $= 1$; if there is only one state with unit modulus, $\mathbf{M}$ is totally regular

the discrete finite Chapman–Kolmogorov equation is

$$\sum_{a=1}^{N} {M^2}_{ab} = \sum_{a=1}^{N} \sum_{c=1}^{N} M_{ac} M_{cb} = 1.$$

(2) Markov chain

a column vector $\mathbf{p}$ is a probability distribution if

$$\mathbf{p} \geq \mathbf{0} \quad \text{and} \quad \sum_{a=1}^{N} p_a = 1$$

evolution equation (Markov condition)

$$p^{(n+1)}{}_a = \sum_{b=1}^{N} M_{ab} \, p^{(n)}{}_b$$

average of f on a distribution $\mathbf{p}$

$$\bar{f}_{\mathbf{p}} = \sum_a f_a p_a$$

detailed balancing

$$M_{ab} \, p_b = M_{ba} \, p_a$$

a state a is positive-recurrent if it is recurrent and its expected return time $t_a = \inf\{n \mid M^n{}_{aa} > 0\}$ is finite; a chain is positive-recurrent if all its states are;

for a totally regular positive-recurrent chain holds the ergodic theorem:

$$\lim_{n\to\infty}\frac{1}{n}\sum_{k=0}^{n-1}\bar{f}_{\mathbf{p}^{(k)}} = \bar{f}_{\mathbf{p}^{(\infty)}} = \sum_a f_a\; p_a^{(\infty)}.$$

A.4 Circulant matrices

(1) general form: $A_{ij} = a_{i-j}$, or

$$\mathbf{A} = \begin{pmatrix} a_N & a_{N-1} & a_{N-2} & \dots & a_1 \\ a_1 & a_N & a_{N-1} & \dots & a_2 \\ a_2 & a_1 & a_N & \dots & a_3 \\ . & . & . & \dots & . \\ . & . & . & \dots & . \\ a_{N-1} & a_{N-2} & a_{N-3} & \dots & a_N \end{pmatrix}$$

(2) cyclic convention

$$a_N = a_0 \quad \text{and} \quad a_{N+r} = a_r = a_{r-N}$$

(3) relations between eigenvalues λ_k and entries a_j:

$$\lambda_k = \textstyle\sum_{j=1}^N a_j\; e^{i(2\pi/N)kj}$$

$$a_j = \tfrac{1}{N}\textstyle\sum_{k=1}^N e^{-i(2\pi/N)kj}\lambda_k$$

(4) with $\omega = e^{i(2\pi/N)}$, the eigenvectors are $u_{(k)s} = \frac{\omega^{-ks}}{\sqrt{N}}$

(5) projectors

$$\left[P_{(k)}\right]_{rs} = u_{(k)r}u^*_{(k)s} = M_{rk}\; M^{-1}_{ks} = \tfrac{1}{N}\;\omega^{-k(r-s)}$$

(6) convolution

$$(a^{*2})_r = (a*a)_r = \sum_{s=1}^N a_s\; a_{r-s} \;\;;\;\; (a^{*m})_r = \sum_{s=1}^N a_s^{*(m-1)}\; a_{r-s}$$

$$(a*b)_n = \textstyle\sum_{m=1}^N a_m b_{n-m}$$

(7) the powers of $\mathbf{A}$ have entries $(\mathbf{A}^m)_{rs} = (a^{*m})_{r-s}$

(8) relations between powers

$$\lambda_k^m = \sum_{j=1}^N \omega^{kj} a_j^{*m} \quad ; \quad a_j^{*m} = \tfrac{1}{N} \sum_{k=1}^N \omega^{-kj} \lambda_k^m$$

(9) shift operator **U**:

$$U_{rs} = \delta_{r,s+1} \quad ; \quad (\mathbf{UA})_{ij} = A_{i-1,j}$$

any circulant is of the form

$$\mathbf{A} = \sum_{n=1}^N a_n \mathbf{U}^n$$

(10) projectors and shifts

$$\mathbf{P}_{(k)} = \tfrac{1}{N} \sum_{n=1}^N \omega^{-kn} \mathbf{U}^n \quad ; \quad \mathbf{U}^n = \sum_{k=1}^N \omega^{kn} \mathbf{P}_{(k)} \ .$$

(11) product of two circulants **A** and **B**:

$$\mathbf{AB} = \sum_{n=1}^N (a * b)_n \mathbf{U}^n$$

(12) the **V** operator: $\mathbf{V} = \mathbf{U}_{diag}$, or

$$\mathbf{V} = \begin{pmatrix} \omega & 0 & 0 & 0 & \cdots & \cdots & \cdots & \cdots & 0 \\ 0 & \omega^2 & 0 & \cdots & \cdots & \cdots & \cdots & \cdots & 0 \\ 0 & 0 & \omega^3 & 0 & \cdots & \cdots & \cdots & \cdots & 0 \\ 0 & 0 & 0 & \omega^4 & 0 & \cdots & \cdots & \cdots & 0 \\ \cdots & \cdots & \cdots & \cdots & \cdots & \cdots & \cdots & \cdots & \cdots \\ 0 & 0 & 0 & 0 & \cdots & \cdots & \cdots & 0 & \omega^N \end{pmatrix}$$

(13) Weyl realizations of the Heisenberg group

$$\mathbf{V}^m \ \mathbf{U}^n = \omega^{nm} \ \mathbf{U}^n \ \mathbf{V}^m$$

(14) **U** and **V** operators

$$\mathbf{U}|v_k\rangle = |v_{k+1}\rangle \quad ; \quad \mathbf{V}|u_k\rangle = |u_{k-1}\rangle$$

(15) the Schwinger operators

$$\mathbf{S}_{(m,n)} = e^{i\frac{\pi}{N}mn} \ \mathbf{U}^m \mathbf{V}^n = \omega^{\frac{mn}{2}} \ \mathbf{U}^m \mathbf{V}^n$$

entries $\left(\mathbf{S}_{(m,n)}\right)_{rs} = \delta_{r-s,m}\omega^{n(r-m/2)}$

properties

$$\mathbf{S}_{(0,0)} = \mathbf{I}\ ,\ \mathbf{S}^{-1}_{(m,n)} = \mathbf{S}^{\dagger}_{(m,n)} = \mathbf{S}_{(-m,-n)}\ .$$

$$\left(\mathbf{S}_{(m,n)}\mathbf{S}_{(r,s)}\right)\ \mathbf{S}_{(k,l)} = \mathbf{S}_{(m,n)}\ \left(\mathbf{S}_{(r,s)}\mathbf{S}_{(k,l)}\right)$$

$$\mathbf{S}_{(N,p)} = (-)^p\mathbf{S}_{(0,p)}\ ;\quad \mathbf{S}(p,N) = (-)^p\mathbf{S}(p,0)\ ;\quad \mathbf{S}_{(N,N)} = (-)^N\mathbf{S}_{(0,0)}$$

$$\mathbf{S}^{p}_{(m,n)} = \mathbf{S}_{(pm,pn)}\quad ;\quad \text{tr}\ \mathbf{S}_{(m,n)} = N\delta_{n0}\delta_{m0}$$

general matrix $\mathbf{A}$ is written

$$\mathbf{A} = \tfrac{1}{N}\sum_{(m,n)} A_{(m,n)}\mathbf{S}_{(m,n)}$$

$$A_{(m,n)} = tr[\mathbf{S}^{\dagger}_{(m,n)}\mathbf{A}]$$

notation

$\mathbf{m} = (m_1, m_2)$, $\mathbf{n} = (n_1, n_2)$, $\mathbf{r} = (r_1, r_2)$, $\mathbf{0} = (0,0)$, $-\mathbf{m} = (-m_1, -m_2)$, $\mathbf{m} + \mathbf{r} = (m_1 + r_1, m_2 + r_2)$, $\mathbf{m} \times \mathbf{r} = (m_1r_2 - m_2r_1)$, $\mathbf{m} \cdot \mathbf{r} = (m_1r_1 + m_2r_2)$

$$\mathbf{S_m} = e^{i(\pi/N)m_1m_2}\mathbf{U}^{m_1}\mathbf{V}^{m_2}$$

$$\mathbf{A} = \tfrac{1}{N}\sum_{\mathbf{m}} A^{\mathbf{m}}\ \mathbf{S_m}$$

(16) in terms of the cochains

$$\mathbf{S_m}\ |v_k\rangle = e^{i\alpha_1(k;\mathbf{m})}\ |v_{k+m_1}\rangle\quad ;\quad \alpha_1(k;\mathbf{m}) = \frac{\pi}{N}\ (2k + m_1)\ m_2$$

$$\mathbf{S_r}\ \mathbf{S_m} = e^{2i\alpha_2(\mathbf{m},\mathbf{r})}\ \mathbf{S_m}\ \mathbf{S_r} = e^{i\alpha_2(\mathbf{m},\mathbf{r})}\ \mathbf{S_{m+r}}$$

$$\alpha_2(\mathbf{m},\mathbf{r}) = \frac{\pi}{N}\ (m_1r_2 - m_2r_1) = \frac{\pi}{N}\ (\mathbf{m} \times \mathbf{r}).$$

product of two operators is

$$\mathbf{AB} = \tfrac{1}{N^2}\ \textstyle\sum_{\mathbf{m}}\sum_{\mathbf{n}} A^{\mathbf{m}}B^{\mathbf{n}}e^{i\alpha_2(\mathbf{n},\mathbf{m})}\ \mathbf{S_{m+n}}$$

$$= \tfrac{1}{N}\sum_{\mathbf{p}}\left[\frac{1}{N}\sum_{\mathbf{m}} A^{\mathbf{m}}B^{\mathbf{p}-\mathbf{m}}e^{i\alpha_2(\mathbf{p},\mathbf{m})}\right]\ \mathbf{S_p}$$

(17) Weyl–Wigner transformations

$$A_W(q,p) = \int da\ db\ e^{i(aq+bp)}\ A(a,b)$$

$$\mathbf{A}(\mathbf{q},\mathbf{p}) = \int da\ db\ e^{i(a\mathbf{q}+b\mathbf{p})}\ A(a,b)$$

Weyl prescription for the commutator

$$[\mathbf{A},\mathbf{B}] = \frac{1}{N^2}\sum_{\mathbf{p}}\sum_{\mathbf{m}} A^{\mathbf{m}}B^{\mathbf{p}-\mathbf{m}}\ 2i\sin\left[\alpha_2(\mathbf{p},\mathbf{m})\right]\mathbf{S}_{\mathbf{p}}$$

continuum versions

$$[\mathbf{S}_{\mathbf{a}},\mathbf{S}_{\mathbf{b}}] = 2i\sin\left[-\hbar\,\frac{\mathbf{a}\times\mathbf{b}}{2}\right]\mathbf{S}_{\mathbf{a}+\mathbf{b}}$$

$$[\mathbf{A},\mathbf{B}]^{\mathbf{p}} = \frac{2i}{\hbar}\int d\mathbf{a}\int d\mathbf{b}\ A(\mathbf{b})\ B(\mathbf{a}-\mathbf{b})\sin\left[-\hbar\,\frac{\mathbf{a}\times\mathbf{b}}{2}\right]\mathbf{S}_{\mathbf{a}}\ .$$

(18) Matrix differential geometry: exterior derivatives defined by

$$d\mathbf{M}(\mathbf{X}) = \mathbf{X}(\mathbf{M}) = ad_{\mathbf{X}}\mathbf{M} = [\mathbf{X},\mathbf{M}]$$

$$(p+1)(d\Xi)(e_1,e_2,\dots,e_{p+1}) = e_1\Xi(e_2,e_3,\dots,e_{p+1}) -$$

$$e_2\Xi(e_1,e_3,\dots,e_{p+1}) + \cdots + (-)^p e_{p+1}\Xi(e_1,e_2,\dots,e_p)$$

$$-\Xi([e_1,e_2],e_3,\dots,e_{p+1}) + \Xi([e_1,e_3],e_2,\dots,e_{p+1})$$

$$-\Xi([e_1,e_4],e_2,\dots,e_{p+1}) + \dots + (-)^p\Xi([e_1,e_{p+1}],e_2,\dots,e_p)$$

$$+\Xi(e_1,[e_2,e_3],\dots,e_{p+1}) + \cdots + (-)^p\Xi(e_1,e_2,e_3,\dots,[e_p,e_{p+1}])$$

natural Poisson bracket

$$\{\mathbf{A},\mathbf{B}\}_M = i[\mathbf{A},\mathbf{B}]$$

(19) in basis $\mathbf{S}_{\mathbf{m}}$ the derivative algebra is generated by the operators $e_{\mathbf{m}} = \mathrm{ad}(i\mathbf{S}_{\mathbf{m}}) = \mathrm{ad}_{i\mathbf{S}_{\mathbf{m}}}$ (for all $\mathbf{S}_{\mathbf{m}} \neq \mathbf{S}_{\mathbf{0}}$)

the general commutators are

$$[e_{\mathbf{r}},e_{\mathbf{m}}] = 2\sin[\alpha_2(\mathbf{r},\mathbf{m})]\ e_{\mathbf{m}+\mathbf{r}}\ .$$

structure coefficients

$$C^{\mathbf{p}}{}_{\mathbf{rm}} = 2\sin[\alpha_2(\mathbf{r},\mathbf{m})]\ \delta^{\mathbf{p}}_{\mathbf{m}+\mathbf{r}}$$

(20) quantum symplectic form

$$\Omega = \frac{1}{2} \sum_{\mathbf{p},\mathbf{m},\mathbf{r}} \mathbf{S}_{\mathbf{p}} \, C^{\mathbf{p}}{}_{\mathbf{mr}} \, \theta^{\mathbf{m}} \, \theta^{\mathbf{r}} = \sum_{\mathbf{p},\mathbf{m}} \sin[\alpha_2(\mathbf{m},\mathbf{p})] \, \mathbf{S}_{\mathbf{p}} \, \theta^{\mathbf{m}} \, \theta^{\mathbf{p}-\mathbf{m}}$$

components

$$\Omega_{\mathbf{rm}} = \sum_{\mathbf{p}} C^{\mathbf{p}}{}_{\mathbf{rm}} \mathbf{S}_{\mathbf{p}} = i \, [\mathbf{S}_{\mathbf{r}}, \mathbf{S}_{\mathbf{m}}]$$

hypermatrix $\hat{\Omega} = (\Omega_{\mathbf{rm}})$ has an inverse

$$r^{\mathbf{ij}} = \frac{i}{N^2} \, \mathbf{S}^{\mathbf{j}} \, \mathbf{S}^{\mathbf{i}} = \frac{i}{N^2} \, e^{i\alpha_2(\mathbf{i},\mathbf{j})} \, \mathbf{S}^{\mathbf{i}+\mathbf{j}} = \frac{i}{N^2} \, e^{i(\pi/N)\mathbf{i}\times\mathbf{j}} \, \mathbf{S}^{\mathbf{i}+\mathbf{j}}$$

property:

$$r^{\mathbf{ij}} = e^{2i\alpha_2(\mathbf{i},\mathbf{j})} \, r^{\mathbf{ji}} \, .$$

(21) the $(N-1)$ generators $\{\sigma_j\}$ of the N-strand braid group B_N obey

$$\sigma_i \sigma_{i+1} \sigma_i = \sigma_{i+1} \sigma_i \sigma_{i+1} \quad \text{for } i = 1, 2, \ldots, N-1$$

$$\sigma_i \sigma_j = \sigma_j \sigma_i \quad \text{for } |i-j| \geq 2$$

for $N = 3$, this is equivalent to the Yang–Baxter equation

$$R^{jk}{}_{ab} R^{ib}{}_{cr} R^{ca}{}_{mn} = R^{ij}{}_{ca} R^{ck}{}_{mb} R^{ab}{}_{nr} \, .$$

with $R^{ij}{}_{mn} = B^{ji}{}_{mn}$, this is the same as

$$\mathbf{B}_{12} \mathbf{B}_{23} \mathbf{B}_{12} = \mathbf{B}_{23} \mathbf{B}_{12} \mathbf{B}_{23} \, .$$

A.5 Bell polynomials

(1) Faà di Bruno formula

$$f[g(x)] = \sum_{n=1}^{\infty} \frac{x^n}{n!} \sum_{k=1}^{n} \mathbb{B}_{nk}(g_1, g_2, \ldots, g_{n-k+1}) \, f_k$$

$$\left[\frac{d^n}{dx^n} f[g(x)]\right]_{x=0} = \sum_{k=1}^{n} \mathbb{B}_{nk}(g_1, g_2, \ldots, g_{n-k+1}) \, f_k$$

(2) Multinomial theorem

$$\frac{1}{k!} \left[\sum_{j=1}^{\infty} \frac{g_j}{j!} \, x^j\right]^k = \sum_{n=k}^{\infty} \frac{x^n}{n!} \, \mathbb{B}_{nk}(g_1, g_2, \ldots, g_{n-k+1})$$

(3) Complete Bell polynomial

$$\mathbb{Y}_n[u; g] = \sum_{j=1}^{n} u^j \, \mathbb{B}_{nj}(g_1, g_2, \ldots, g_{n-j+1})$$

(4) Generating function

$$e^{ug(x)} = 1+\sum_{n=1}^{\infty} \frac{x^n}{n!} \mathbb{Y}_n[u;g] = 1+\sum_{n=1}^{\infty} \frac{x^n}{n!} \sum_{k=1}^{n} u^k \, \mathbb{B}_{nk}(g_1, g_2, \dots, g_{n-k+1}) \,.$$

(5) Notation

$$\mathbb{B}_{nk}(g_1, g_2, \dots, g_{n-k+1}) = \mathbb{B}_{nk}[g(x)] = \mathbb{B}_{nk}[g] = \mathbb{B}_{nk}\{g_j\}$$

$$\mathbb{B}_{n0}[g] = \delta_{n0}$$

(6) Properties

$$\mathbb{B}_{Nm}[g] = \left(\frac{1}{m!}[g(x)]^m\right)_N = \frac{1}{m!}\, \mathbb{B}_{N1}[g^m]$$

$$\mathbb{B}_{n1}[g] = g_n \; ; \; \mathbb{B}_{n2}[g] = \tfrac{1}{2}\textstyle\sum_{j=1}^{n-1} \binom{n}{j} \, g_j \, g_{n-j}$$

$$\mathbb{B}_{nn}[g] = g_1^n \; ; \; \mathbb{B}_{n,n-1}[g] = \binom{n}{2} \, g_1^{n-2} g_2$$

$$\mathbb{B}_{nk}[af(x)] = a^k \, \mathbb{B}_{nk}[f(x)] \; ; \; \mathbb{B}_{nk}[f(bx)] = b^n \, \mathbb{B}_{nk}[f(x)]$$

$$\mathbb{B}_{nk}[f(x)+g(x)] = \sum_{j=1}^{n} \sum_{m=1}^{k} \binom{n}{j} \, \mathbb{B}_{jm}[f(x)] \, \mathbb{B}_{n-j,k-m}[g(x)]$$

$$\mathbb{B}_{Nm}[e(x)=x] = \mathbb{B}_{Nm}(1,0,\dots,0) = \delta_{Nm}$$

$$m \, \mathbb{B}_{nm}[g] = \textstyle\sum_{k=m-1}^{n-1} \binom{n}{k} g_{n-k} \, \mathbb{B}_{k,m-1}[g]$$

(7) Stirling numbers: generating functions

first kind

$$x(x-1)(x-2)...(x-k+1) = \tbinom{x}{k}k! = \textstyle\sum_{j=0}^{k} s_k^{(j)} \, x^j$$

$$\tfrac{1}{k!}\,[\ln(1+x)]^k = \textstyle\sum_{n=k}^{\infty} \frac{x^n}{n!} \, s_n^{(k)}$$

second kind

$$x^k = \textstyle\sum_{j=0}^{k} S_k{}^{(j)} x(x-1)(x-2)\dots(x-j+1)$$

$$\tfrac{1}{k!}\,(e^u-1)^k = \textstyle\sum_{n=k}^{\infty} \frac{u^n}{n!} \, S_n{}^{(k)}$$

closed form:

$$S_n{}^{(m)} = \frac{1}{m!}\sum_{k=0}^{m}(-)^{m-k}\binom{m}{k}k^n$$

relation to the Poisson distribution

$$\sum_{m=1}^{n} x^m S_n{}^{(m)} = \sum_{k=0}^{\infty} k^n e^{-x}\,\frac{x^k}{k!}$$

relations between the two kinds

$$\textstyle\sum_{k=m}^{n} s_n^{(k)} S_k{}^{(m)} = \sum_{k=m}^{n} S_n{}^{(k)} s_k^{(m)} = \delta_n^m$$

(8) Stirling numbers: relations with Bell polynomials

$$\mathbb{B}_{nk}[\ln(1+x)] = \mathbb{B}_{nk}(0!, -1!, 2!, -3!, \dots)$$
$$= \mathbb{B}_{nk}\{(-)^{j-1}(j-1)!\} = s_n^{(k)}$$

$$\mathbb{B}_{nk}(0!, 1!, 2!, 3!, \dots) = \mathbb{B}_{nk}\{(j-1)!\} = |s_n^{(k)}| = (-)^{n+k}\ s_n^{(k)}$$

$$\mathbb{B}_{nk}[e^u - 1] = \mathbb{B}_{nk}(1, 1, 1, \dots, 1) = \mathbb{B}_{nk}\{1\} = S_n{}^{(k)}$$

(9) Bell number

$$\omega(n) = \textstyle\sum_{k=1}^{n} S_n{}^{(k)}\ ;\ \ \omega(n+1) = \sum_{k=0}^{n}\binom{n}{k}\ \omega(n)$$

(10) Reciprocal series

$$(1+g(x))^{-n} - 1 = \textstyle\sum_{m=1}^{\infty}\frac{x^m}{m!}\sum_{j=1}^{m}(-)^j\frac{(n+j-1)!}{(n-1)!}\ \mathbb{B}_{mj}[g]\ .$$

(11) Logarithmic Bell polynomials

$$L_n[g] = \sum_{k=1}^{n}(-)^{k-1}(k-1)!\,\mathbb{B}_{nk}[g]\ ;\ \ln[1+g(x)] = \sum_{n=1}^{\infty}\frac{x^n}{n!}L_n[g]$$

(12) Power Bell polynomials

$$\left(1+\sum_{j=1}^{\infty}\frac{g_j}{j!}\,t^j\right)^k = 1+\sum_{n=1}^{\infty}\frac{t^n}{n!}\,P_n^{(k)}[g]\ ;\ P_n^{(k)}[g] = \sum_{j=1}^{n}\frac{k!}{(k-j)!}\,\mathbb{B}_{nj}[g]$$

$$P_n^{(k)}[g] = \binom{n+k}{k}^{-1} \mathbb{B}_{n+k,k}(1, 2g_1, 3g_2, \ldots)$$

$$k!\ \mathbb{B}_{nk}[g] = \sum_{r=0}^{k} \binom{k}{r} (-)^{k-r} P_n^{(r)}[g]\ ;\ L_n[g] = \sum_{k=1}^{n} \frac{(-)^{k-1}}{k}\ P_n^{(k)}[g]$$

$$\tfrac{1}{m!}\left(g_0 + \textstyle\sum_{j=1}^{\infty} \frac{g_j}{j!} t^j\right)^m = \textstyle\sum_{N=0}^{\infty} \frac{t^N}{N!} \sum_{k=0}^{N} \frac{g_0^k}{k!}\ \mathbb{B}_{N,m-k}[g]\ .$$

(13) Recursive relations

$$\textstyle\sum_{k=1}^{n} \mathbb{B}_{nk}[g] = \sum_{p=0}^{n-1} \binom{n-1}{p} g_{n-p} \sum_{m=0}^{p} \mathbb{B}_{pm}[g]$$

$$= \tfrac{1}{n} \textstyle\sum_{j=1}^{n} \binom{n}{j} j g_j \sum_{r=0}^{n-j} \mathbb{B}_{n-j,r}[g]$$

$$\mathbb{Y}_n[u; g] = \textstyle\sum_{j=1}^{n} \mathbb{B}_{nj}(ug_1, ug_2, \ldots, ug_{n-j+1})$$

$$= \textstyle\sum_{p=0}^{n-1} \binom{n-1}{p} g_{n-p} \sum_{m=0}^{p} \mathbb{B}_{pm}[ug]$$

$$= \sum_{p=0}^{n-1} \binom{n-1}{p} g_{n-p} \sum_{m=0}^{p} u^m\ \mathbb{B}_{pm}[g] = \sum_{p=0}^{n-1} \binom{n-1}{p} g_{n-p} \mathbb{Y}_p[u; g]$$

(14) Leibniz rule

$$\mathbb{B}_{nk}[F(t)\ G(t)] = \begin{cases} k! \sum_{j=k}^{n-k} \binom{n}{j} \mathbb{B}_{jk}[F]\ \mathbb{B}_{n-j,k}[G] & \text{if } n \geq 2k \\ 0 \text{ for } k \leq n < 2k\ . \end{cases}$$

$$\mathbb{B}_{n1}[F(t)G(t)] = \sum_{j=1}^{n-1} \binom{n}{j} \mathbb{B}_{j1}[F]\ \mathbb{B}_{n-j,1}[G] = \sum_{j=1}^{n-1} \binom{n}{j} F_j\ G_{n-j}$$

(15) Lagrange inversion formula

$$g_n^{<-1>} = \frac{1}{g_1^n} \sum_{k=1}^{n-1} (-)^k \frac{(n+k-1)!}{(n-1)!}\ \mathbb{B}_{n-1,k} \left(\frac{g_2}{2g_1}, \frac{g_3}{3g_1}, \frac{g_4}{4g_1}, \ldots \right)$$

$$g_n^{<-1>} = \tfrac{1}{g_1} \textstyle\sum_{k=1}^{n-1} (-)^k \frac{(n+k-1)!}{(n-1)!}\ \mathbb{B}_{n-1,k} \left\{ \frac{g_{k+1}}{(k+1) g_1^{k+1}} \right\}$$

$$g_n^{<-1>} = \tfrac{1}{g_1^n} \textstyle\sum_{k=1}^{n-1} (-)^k n^k\ \mathbb{B}_{n-1,k} \left[\ln \frac{g(t)}{g_1 t} \right]$$

$$\mathbb{B}_{nm}[g^{<-1>}] = \tfrac{1}{g_1^n} \binom{n-1}{m-1} \textstyle\sum_{k=0}^{n-m} \mathbb{B}_{n-m,k} \left[-n \ln \left(\frac{g(t)}{g_1 t} \right) \right]\ .$$

A.5.1 *Orthogonal polynomials*

(1) Hermite polynomials $H_n(x)$ (with the conventions of [34])

$$e^{-t^2+2tx} - 1 = \sum_{n=1}^{\infty} \frac{t^n}{n!}\, H_n(x)$$

$$H_n(x) = \mathbb{Y}_n[1; 2tx - t^2] = \sum_{j=1}^{n} \mathbb{B}_{nj}[2tx - t^2]$$

$$= \sum_{j=1}^{n} \mathbb{B}_{nj}(2x, -2, 0, 0, \dots, 0) = 2xH_{n-1}(x) - 2(n-1)H_{n-2}(x)$$

(2) Legendre polynomials $P_n(x)$

$$\frac{1}{\sqrt{1-2tx+t^2}} = \sum_{n=0}^{\infty} P_n(x)\, t^n$$

$$\exp\left[\frac{1}{\sqrt{1-2tx+t^2}}\right] - 1 = \sum_{N=1}^{\infty} \frac{t^N}{N!} \sum_{m=1}^{N} \mathbb{B}_{Nm}\{j!\; P_j(x)\}\,.$$

(3) Gegenbauer polynomials $C_N^{\alpha}(x)$

$$(1 - 2tx + t^2)^{-\alpha} = \sum_{N=0}^{\infty} \frac{t^n}{N!}\, C_N^{\alpha}(x)$$

$$C_N^{\alpha}(x) = \sum_{m=[N/2]}^{N} \frac{(-)^m\, \Gamma(1-\alpha)}{(1-\alpha-m)!}\, \mathbb{B}_{Nm}(2x, -2, 0, 0, \dots, 0)$$

$$P_N(x) = \frac{1}{N!} \sum_{m=[N/2]}^{N} \frac{1}{2^m\; (2m-1)!}\, \mathbb{B}_{Nm}(2x, -2, 0, 0, \dots, 0)$$

$$= \frac{1}{N!} \sum_{m=[N/2]}^{N} \frac{1}{2^m\; (2m-1)!}\, \mathbb{B}_{Nm}[2x\; t - t^2]$$

(4) Chebyshev polynomials of the first kind $T_n(x)$

$$\frac{xt - t^2}{1 - 2tx + t^2} = \sum_{n=1}^{\infty} T_n(x)\, t^n \;\; ; \;\; T_n(x) = \cos\,[n \arccos x]$$

$$T_n(x) = F(-n, n, 1/2, \frac{1-x}{2});\;\; F(-n, n, 1/2, x) = T_n(1-2x)$$

$$2\,T_m(x)\,T_n(x) = T_{n+m}(x) + T_{n-m}(x)$$

$$\text{differential equation: } (1-x^2)\frac{d^2}{dx^2}T_n(x) - x\frac{d}{dx}T_n(x) + n^2\,T_n(x) = 0$$

$$T_m[T_n(x)] = [T_m \circ T_n](x) = T_{mn}(x) = T_n[T_m(x)] \ ;$$

$$T_{n^m}(x) = [\underbrace{T_n \circ T_n \circ T_n \ldots \circ T_n}_{m \ times}](x) = T_n^{<m>}(x) \ .$$

When defined,

$$\mathbf{B}[T_n^{<m>}] = \mathbf{B}[\underbrace{T_n \circ T_n \circ T_n \ldots \circ T_n}_{m \ times}] = \mathbf{B}^m[T_n] \ .$$

A.5.2 *Differintegration, derivatives of Bell polynomials*

$$\frac{d^q}{dx^q} f[g(x)] = \frac{x^{-q}}{\Gamma(1-q)} f[g(x)] + \sum_{n=1}^{\infty} \binom{q}{n} \frac{x^{n-q}}{\Gamma(n+1-q)} \sum_{m=1}^{n} \mathbb{B}_{nm}[g] f_m \ .$$

$$\frac{\partial}{\partial g_i} \mathbb{B}_{nk}[g] = \binom{n}{i} \ \mathbb{B}_{n-i,k-1}[g]$$

$$m \ \mathbb{B}_{nm}[g] = \sum_{j=1}^{n-m+1} g_j \frac{\partial}{\partial g_j} \ \mathbb{B}_{n,m}[g] \ .$$

A.6 Determinants, minors and traces

$$\det \mathbf{X} = \exp\{\mathrm{tr}[\ln \mathbf{X}]\}$$

$$\det\left[\mathbf{I} \ \lambda + z \ \mathbf{A}\right] = \sum_{j=0}^{N} \frac{z^j \ \lambda^{N-j}}{j!} \sum_{m=0}^{j} \mathbb{B}_{jm}\{(-)^{k-1}(k-1)! \ \mathrm{tr}(\mathbf{A}^k)\}$$

$$\det \mathbf{A} = \tfrac{1}{N!} \ \textstyle\sum_{m=0}^{N} \mathbb{B}_{Nm}\{(-)^{k-1}(k-1)! \ \mathrm{tr}(\mathbf{A}^k)\}$$

$$tr(\mathbf{A}^n) = \tfrac{(-)^{n-1}}{(n-1)!} \ \textstyle\sum_{k=1}^{n} (-)^{k-1}(k-1)! \ \mathbb{B}_{nk}\{j! \ \det_j \mathbf{A}\}$$

$$\det_j \mathbf{A} = \tfrac{1}{j!} \ \textstyle\sum_{m=0}^{j} \mathbb{B}_{jm}\{(-)^{k-1}(k-1)! \ \mathrm{tr}(\mathbf{A}^k)\}$$

$$\det_j \mathbf{A} = \frac{1}{j!} \sum_{s_1=1}^{N} \sum_{s_2=1}^{N} \cdots \sum_{s_j=1}^{N} \begin{vmatrix} A_{s_1 s_1} & A_{s_1 s_2} & \cdots & A_{s_1 s_j} \\ A_{s_2 s_1} & A_{s_2 s_2} & \cdots & A_{s_2 s_j} \\ \cdot & \cdot & \cdots & \cdot \\ \cdot & \cdot & \cdots & \cdot \\ A_{s_j s_1} & A_{s_j s_2} & \cdots & A_{s_j s_j} \end{vmatrix} .$$

A.6.1 *Symmetric functions*

$$\Sigma[\mathbf{x};t] = \sum_{n=0}^{N} e_n[\mathbf{x}]t^n = \prod_{n=1}^{N}(1+x_n t)$$

$$P[\mathbf{x};z] = \sum_{j=1}^{N} p_j[\mathbf{x}]z^{j-1} = \sum_{j=1}^{N} x_j(1-x_j z)^{-1} = -\frac{d}{dz}\ln \Sigma[\mathbf{x};-z]$$

$$H[\mathbf{x};t] = \sum_{j=0}^{N} h_j[\mathbf{x}]t^j = \prod_{j=1}^{N}(1-x_j t)^{-1} = \frac{1}{\Sigma[\mathbf{x};-t]}$$

$$p[\mathbf{x},z] = \sum_{j=1}^{N} e^{zx_j} = \sum_{n=0}^{N}\frac{1}{n!}\sum_{j=1}^{N}(x_j\ z)^n = \sum_{n=0}^{N}\frac{1}{n!}\,p_n\,[\mathbf{x}]\,z^n$$

$$e_n[\mathbf{x}] = \frac{1}{n!}\sum_{j=1}^{n}\mathbb{B}_{nj}\{(-)^{k-1}(k-1)!\,p_k[\mathbf{x}]\}$$

$$p_k[\mathbf{x}] = \frac{(-)^{k-1}}{(k-1)!}\sum_{j=1}^{k}(-)^{j-1}(j-1)!\,\mathbb{B}_{kj}\{n!\,e_n[\mathbf{x}]\}\quad (\text{for } k\geq 1)$$

$$n!e_n = \begin{vmatrix} p_1 & 1 & 0 & 0 & .. & 0 \\ p_2 & p_1 & 2 & 0 & .. & 0 \\ p_3 & p_2 & p_1 & 3 & .. & 0 \\ . & . & . & . & .. & 0 \\ p_n & p_{n-1} & . & . & .. & p_1 \end{vmatrix}$$

$$\sum_{j=0}^{n}(-)^{n-j}p_{n-j}e_j = \sum_{i=1}^{n}\sum_{j=0}^{n}(-)^{n-j}x_i^{n-j}e_j = 0$$

$$\sum_{m=1}^{N}\mathbb{B}_{Nm}\{(-)^{j-1}(j-1)!\,a_j\} = \begin{vmatrix} a_1 & 1 & 0 & 0 & .. & 0 \\ a_2 & a_1 & 2 & 0 & .. & 0 \\ a_3 & a_2 & a_1 & 3 & .. & 0 \\ . & . & . & . & .. & 0 \\ a_N & a_{N-1} & . & . & .. & a_1 \end{vmatrix}$$

$$\sum_{m=1}^{N}\mathbb{B}_{Nm}\{g_j\} = \begin{vmatrix} g_1 & 1 & 0 & 0 & .. & 0 \\ -g_2 & g_1 & 2 & 0 & .. & 0 \\ g_3/2! & -g_2 & g_1 & 3 & .. & 0 \\ . & . & . & . & .. & 0 \\ (-)^{N-1}\frac{g_N}{(N-1)!} & (-)^{N}\frac{g_{N-1}}{(N-2)!} & . & . & .. & g_1 \end{vmatrix}$$

$$\det \mathbf{A} = \frac{1}{N!} \begin{vmatrix} \mathrm{tr}\,\mathbf{A} & 1 & 0 & 0 & . & . & 0 \\ \mathrm{tr}\,\mathbf{A}^2 & \mathrm{tr}\,\mathbf{A} & 2 & 0 & . & . & 0 \\ \mathrm{tr}\,\mathbf{A}^3 & \mathrm{tr}\,\mathbf{A}^2 & \mathrm{tr}\,\mathbf{A} & 3 & . & . & 0 \\ . & . & . & . & . & . & . \\ . & . & . & . & . & . & N-1 \\ \mathrm{tr}\,\mathbf{A}^N & \mathrm{tr}\,\mathbf{A}^{N-1} & . & . & . & \mathrm{tr}\,\mathbf{A}^2 & \mathrm{tr}\,\mathbf{A} \end{vmatrix} .$$

A.6.2 *Polynomials*

- expressions

$$P(z) = a_N z^N + a_{N-1} z^{N-1} + \cdots + a_1 z^1 + a_0$$

$$= a_N \prod_{j=1}^{N} (z - r_j) = (-)^N a_N e_N[\mathbf{r}]\ \Sigma[\mathbf{r}^*, z]$$

$$= (-)^N a_N e_N[\mathbf{r}] \sum_{j=0}^{N} e_j[\mathbf{r}^*] z^j = a_N \sum_{j=0}^{N} (-)^{N-j} e_{N-j}[\mathbf{r}] z^j$$

$$= P(0)\ \prod_{j=1}^{N} (1 - z/r_j) = (-)^N a_N\ e_N[\mathbf{r}]\ \prod_{j=1}^{N} (1 - z/r_j)$$

$$= a_0 \prod_{j=1}^{N} (1 - z/r_j) = a_N \sum_{l=0}^{N} \frac{z^{N-l}}{l!} \sum_{r=0}^{l} \mathbb{B}_{l,r}\{-(k-1)!\ p_k[\mathbf{r}]\}$$

- coefficients

$$a_j = (-)^{N-j} a_N\ e_{N-j}\ [\mathbf{r}]\ \ ;\ \ a_0 = P(0) = (-)^N a_N\ e_N[\mathbf{r}]$$

$$a_j = a_N\ \frac{(-)^{N-j}}{(N-j)!}\ \sum_{r=0}^{N-j} \mathbb{B}_{N-j,r}\{(-)^{k-1}(k-1)!\ p_k[\mathbf{r}]\}$$

- Viète's formula

$$a_j = a_N\ \frac{(-)^{N-j}}{(N-j)!}\ \sum_{r=0}^{N-j} \mathbb{B}_{N-j,r}\{(-)^{k-1}(k-1)!\ p_k[\mathbf{r}]\}$$

- discriminant

$$\Delta(P) = \det(r_i^{N-j}) = (-)^N p[\mathbf{r}] = \begin{vmatrix} r_1^{N-1} & r_1^{N-2} & r_1^{N-3} & \ldots & r_1^0 \\ r_2^{N-1} & r_2^{N-2} & r_2^{N-3} & \ldots & r_2^0 \\ . & . & . & \ldots & 1 \\ . & . & . & \ldots & 1 \\ r_N^{N-1} & r_N^{N-2} & r_N^{N-3} & \ldots & r_N^0 \end{vmatrix} .$$

A.6.3 *Characteristic polynomials and classes*

$$\Delta_{\mathbf{A}}(\lambda) = \det[\lambda\mathbf{I} - \mathbf{A}] = \textstyle\sum_{k=0}^{N} a_k\lambda^k$$

$$= \textstyle\prod_{j=1}^{N}(\lambda - \lambda_j) = \sum_{j=0}^{N}(-)^j\lambda^{N-j}e_j[\lambda]$$

$$= \textstyle\sum_{j=0}^{N} \lambda^{N-j}\frac{(-)^j}{j!}\sum_{m=0}^{j}\mathbb{B}_{jm}\{(-)^{k-1}(k-1)!\ \mathrm{tr}(\mathbf{A}^k)\}$$

$$= \sum_{j=0}^{N}\lambda^{N-j}\frac{(-)^j}{j!}\sum_{m=0}^{j}\mathbb{B}_{jm}\{(-)^{k-1}(k-1)!\ p_k[\boldsymbol{\lambda}]\} = \sum_{j=0}^{N}\lambda^{N-j}(-)^j\ e_j[\boldsymbol{\lambda}]$$

(1) Lie Algebras Invariants

multilinear invariant symmetric tensors

$$\eta^{(n)}_{a_1a_2\cdots a_n} = \frac{(-)^n}{n!}\sum_{m=1}^{n}\mathbb{B}_{nm}\{(-)^{k-1}(k-1)!\ tr(J_{a_1}J_{a_2}\cdots J_{a_k})\}$$

(2) Characteristic Classes

if $\mathbf{F}$ is a matrix curvature 2-form, the k-th Chern class is

$$\pi^*(c_k) = \phi_k(\mathbf{F}) = e_k[i\mathbf{F}/2\pi]$$

$$= \textstyle\frac{i^k}{(2\pi)^k k!}\sum_{m=0}^{k}\mathbb{B}_{km}\{(-)^{r-1}(r-1)!\ \mathrm{tr}(\mathbf{F}^r)\}$$

Chern total character

$$ch(E) = \mathrm{tr}(\exp\ [\frac{i}{2\pi}\mathbf{F}]) = \sum_{k=0}^{\infty}\frac{i^k}{(2\pi)^k k!}\ \mathrm{tr}\mathbf{F}^k = \sum_{k=0}^{N}\frac{1}{k!}\ p_k[i\mathbf{F}/2\pi]$$

k-th Chern character

$$\frac{i^k}{(2\pi)^k k!}\ \mathrm{tr}\mathbf{F}^k = \frac{1}{k!}\ p_k[i\mathbf{F}/2\pi]$$

relations between classes and characters (for $j, k \geq 0$)

$$c_j(E) = \frac{1}{j!}\sum_{m=0}^{j}\mathbb{B}_{jm}\{(-)^{k-1}(k-1)!\ k!\ ch_k(E)\}$$

$$(-)^{k-1}(k-1)!\ ch_k(E) = \frac{1}{k!}\ \sum_{j=0}^{k}(-)^{j-1}(j-1)!\ \mathbb{B}_{kj}\{r!\ c_r(E)\}.$$

A.7 Bell matrices

(1) Fundamental property

$$\mathbf{B}[g]\,\mathbf{B}[f] \;=\; \mathbf{B}[f \circ g]$$

(2) Eigenvectors of $\mathbf{B}[g]$

$$\mathbf{B}[g]\; v_{(k)} = g_1^k\; v_{(k)}\;;\; [\mathbf{B}[g]\; v_{(k)}]_r = g_1^k\; v_{(k)r}$$

$$v_{(k)s} = \mathbb{B}_{sk}[F]\; with\; F[g(x)] = g_1 F(x)]$$

$$\mathbf{B}^{-1}[g]\; v_{(k)} = g_1^{-k}\; v_{(k)}$$

(3) Projectors $(a = g_1)$ (see also §A.7.2)

$$(\mathbf{Z}_k)_{nr} = \mathbb{B}_{nk}[F]\;\mathbb{B}_{kr}[F^{<-1>}] = v_{(k)n}\;\mathbb{B}_{kr}[F^{<-1>}]$$

$$\mathbf{Z}_i[\mathbf{B}] = \sum_{k=1}^{N}[\Lambda^{-1}]_{ik}\;\mathbf{B}^k$$

$$[\Lambda^{-1}]_{ik} = \frac{\sum_{j=0}^{k-1} a^{i(j-k)}(-)^j e_{N-j}[\mathbf{a}]}{\sum_{r=1}^{N}\sum_{s=0}^{r-1} a^{is}(-)^s e_{N-s}[\mathbf{a}]} = a^{-ik}\,\frac{\sum_{j=k}^{N} a^{ij} e_j[\mathbf{a}^*]}{\sum_{s=0}^{N} s a^{is} e_s[\mathbf{a}^*]}$$

(4) arbitrary power of $\mathbf{B}$

$$\mathbf{B}^t = \sum_{i=1}^{N} a^{it}\;\mathbf{Z}_i[\mathbf{B}] = \sum_{k=1}^{N} C_k^{(N)}(t)\;\mathbf{B}^k$$

$$C_k^{(N)}(t) = \sum_{i=1}^{N} a^{it}\Lambda_{ik}^{-1}$$

(5) continuum iterate function

$$g_r^{<t>} = \sum_{k=1}^{N} C_k^{(N)}(t) g^{<k>}(x) = \sum_{k=1}^{N}\left[\sum_{i=1}^{N} a^{it}\Lambda_{ik}^{-1}\right] g^{<k>}(x)$$

$$= \sum_{i=1}^{N} a^{it} R_i^{(N)}(x)$$

$$R_i^{(N)}(x) = \sum_{r=1}^{\infty}\tfrac{x^r}{r!}\,[\mathbf{Z}_i[\mathbf{B}]]_{r1} = \sum_{k=1}^{N}\Lambda_{ik}^{-1} g^{<k>}(x)\;.$$

A.7.1 *Carleman matrices*

$$\frac{1}{r!}\,[G(x)]^r = \frac{1}{r!}\left(g_0 + \sum_{j=1}^{\infty}\frac{g_j}{j!}\,x^j\right)^r = \frac{1}{r!}\sum_{m=0}^{r}\binom{r}{m} g_0^{r-m}[g(x)]^m\;,$$

$$\frac{1}{r!}\,[G(x)]^r = \sum_{n=0}^{\infty} \frac{x^n}{n!}\; \mathbb{C}_{n,r}[G]\ ,$$

$$\mathbb{C}_{n,r}[G] = \sum_{m=0}^{\min[r,n]} \frac{g_0^{r-m}}{(r-m)!}\; \mathbb{B}_{n,m}[g]\ ,$$

$$\mathbb{C}_{n,m}[G \circ F] = \sum_{r=0}^{N} \mathbb{C}_{n,r}[F]\; \mathbb{C}_{r,m}[G]\,,$$

$$\sum_{j=0}^{\infty} \mathbb{C}_{nj}[e^x] = \sum_{j=0}^{\infty} \frac{j^n}{j!} = e\;\omega(n)\ , e^{e^x} = e \sum_{n=0}^{\infty} \omega(n)\, \frac{x^n}{n!}\ .$$

A.7.2 *Schröder equation*

(1) for a function F, given $g(x)$ and a constant c, the equation is $F[g(x)] = c\,F(x)$

(2) matrix version

$$\mathbf{B}[g]\;\mathbf{B}[F] = \mathbf{B}[g_1 F]$$

(3) relation to projectors

$$(\mathbf{Z}_k[\mathbf{B}])_{nr} = \mathbb{B}_{nk}[F]\;\mathbb{B}_{kr}[F^{<-1>}].$$

A.7.3 *Fredholm theory*

(1) equation

$$\mathbf{X} = \mathbf{J} + \lambda\;\mathbf{K}\;\mathbf{X} = \frac{\mathbf{I}}{\mathbf{I} - \lambda\mathbf{K}}\;\mathbf{J}$$

(2) Fredholm determinant

$$D(\lambda) = \det(\mathbf{I} - \lambda\mathbf{K}) = 1 + \sum_{n=1}^{\infty, N} \lambda^n D_n$$

$$D_n = -\;\frac{1}{n}\;\sum_{j=1}^{n} \mathrm{tr}\mathbf{K}^j\; D_{n-j} = \frac{(-)^n}{n!}\;\sum_{m=0}^{n} \mathbb{B}_{nm}\{(-)^{j-1}(j-1)!\mathrm{tr}\mathbf{K}^j\}.$$

(3) Resolvent and Green's function

$$\mathbf{R}(\lambda) = \sum_{n=1}^{\infty} \lambda^n \, \mathbf{K}^n = \frac{\lambda \, \mathbf{K}}{\mathbf{I} - \lambda \, \mathbf{K}} \quad ; \quad (\mathbf{I} - \lambda \mathbf{K})^{-1} = \frac{1}{D(\lambda)} \, \hat{\mathbf{N}}.$$

A.8 Statistical mechanics

A.8.1 *Microcanonical ensemble*

(1) double signs: upper, Bose; lower, Fermi

(2) ideal quantum gas: normalization of the N-particle state

$$< p_1, p_2, \cdots, p_N | p_1, p_2, \cdots, p_N > = \tfrac{1}{N!} \sum_{m=1}^{N} \mathbb{B}_{Nm} (\pm)^{j-1} \{ (j-1)! \; \hat{\delta}_j \}$$

$$= \frac{1}{N} \sum_{r=0}^{N-1} (\pm)^{N-1-r} \hat{\delta}_{N-r} \; \langle p_1, p_2, \cdots, p_r | p_1, p_2, \cdots, p_r \rangle$$

(3) phase space volume

$$R_N(E, \mathbf{P}, V) = \tfrac{1}{N!} \int \left[\prod_{i=1}^{N} d^3 p_i \right] \delta(E - \sum_{i=1}^{N} e_i) \; \delta(P - \sum_{i=1}^{N} \mathbf{p}_i)$$

$$\times \sum_{m=1}^{N} (\pm)^{N-m} \; \mathbb{B}_{Nm} \{ (j-1)! \; \hat{\delta}_j \} \, .$$

Prakash–Sudarshan formula (b, f, B: bosons, fermions and Boltzmann particles)

$$R_N^{(b,f)}(P, V) = \tfrac{1}{N!} \sum_{m=1}^{N} (\pm)^{N-m} \sum_{m=1}^{N} R_m^{(B)}(P, V) \; \mathbb{B}_{Nm} \{ \, j^3 (j-1)! \}$$

recursion relation

$$R_N(P, V) = \tfrac{1}{N} \, \tfrac{V}{h^3} \, \sum_{k=1}^{N} (\pm)^{k-1} \int d^3 q R_{N-k}(P - lq, V) \, .$$

A.8.2 *Canonical ensemble*

(1) ideal quantum gas: partition function

$$Q_N(\beta, V) = \int dE\; e^{-\beta E} R_N(E, V)$$

formal cluster integral for bosons and fermions: $b_j = \frac{(\pm)^{j-1}}{j^{5/2}}$

(2) real gas: semiclassical partition function

$$Q_N(\beta, V) = \tfrac{1}{N!} \int \left[\prod_{i=1}^{N} \frac{d^3x_i d^3p_i}{h^3} \right] e^{-\beta H}$$

$$= \frac{1}{N!\lambda^{3N}} \int \left[\prod_{i=1}^{N} d^3r_i \right] \prod_{i=1}^{N-1} \prod_{j=i+1}^{N} [1 + f(\mathbf{r}_{ij})]$$

above, H and the Mayer functions $f_{ij} = f(\mathbf{r}_{ij})$ are

$$H = \sum_{i=1}^{N} \frac{\mathbf{p}^2}{2m} + \sum_{i=1}^{N-1} \sum_{j=i+1}^{N} U(\mathbf{r}_i - \mathbf{r}_j) \quad ; \quad f(\mathbf{r}_{ij}) = e^{-\beta V(\mathbf{r}_{ij})} - 1$$

cluster integral (thermal wavelength: $\lambda = \sqrt{\frac{\hbar^2}{2\pi mkT}}$)

$$b_n = \frac{1}{n!\lambda^{3(n-1)}V} \int d^3r_1 d^3r_2 \ldots d^3r_n \prod_{i=1}^{n-1} \prod_{j=i+1}^{n} f_{ij}$$

notation: $v = \frac{V}{\lambda^3}$, $b_j^* = vb_j = \frac{V}{\lambda^3} b_j$

(3) relations between partition function and cluster integrals

$$Q_N(\beta, V) = \frac{1}{N!} \mathbb{Y}_N \left[\frac{V}{\lambda^3} \sum_{k=1}^{\infty} b_k z^k \right] = \frac{v}{N} \sum_{p=0}^{N-1} (N-p)\; b_{N-p} Q_p$$

$$Q_N(\beta, V) = \tfrac{1}{N!} \sum_{m=1}^{N} \mathbb{B}_{Nm}\{j!\; b_j^*\} = \tfrac{1}{N!} \sum_{m=1}^{N} \mathbb{B}_{Nm} \left[\sum_{k=1}^{\infty} b_k^*\; t^k \right]$$

$$b_n^* = \tfrac{1}{n!} \textstyle\sum_{k=1}^{n} (-)^{k-1} (k-1)!\; \mathbb{B}_{nk}\{j!\; Q_j\}.$$

A.8.3 *Grand canonical ensemble*

(1) expressions for the partition function (fugacity: $z = e^{\beta\mu}$)

$$\Xi(\beta, V, \mu) = \mathrm{tr}\left[e^{-\beta\sum_i(\epsilon_i-\mu)\hat{n}_i}\right] = \prod_\epsilon \left[\sum_n e^{-\beta(\epsilon-\mu)n}\right]^g$$

$$= \exp\left\{g \sum_\epsilon \sum_{j=1}^\infty \frac{(\pm 1)^{j-1}}{j} e^{-\beta(\epsilon-\mu)j}\right\}$$

$$= \left\{\begin{matrix}\frac{1}{1-z}\\ 1+z\end{matrix}\right\} \exp\left\{g \sum_{j=1}^\infty \frac{(\pm 1)^{j-1}}{j} z^j \sum_\epsilon e^{-j\beta\epsilon}\right\}$$

neglecting zero-energy terms

$$\Xi(\beta, V, \mu) = 1 + \sum_{N=1}^\infty z^N Q_N(\beta, V)$$

Matsubara's formula

$$\Xi(\beta, V, \mu) = e^{\Xi^{(c)}(\beta, V, \mu)}$$

(2) finite approach (M particles)

$$\Xi^{(M)}(r_k) = \sum_{N=0}^{M} r_k^N\, Q_N = 0$$

$$\mathbf{A}_M = diag\ \left[-r_1^{-1}, -r_2^{-1}, -r_3^{-1}, \ldots, -r_M^{-1}\right]$$

$$Q_M = \det \mathbf{A}_M = (-)^M \prod_{i=1}^{M} \frac{1}{r_i}$$

$$v\ j\ b_j = (-)^{j-1}\ \mathrm{tr}\ \mathbf{A}_M^j = -\sum_{i=1}^{M} \frac{1}{r_i^j}$$

with D = Fredholm determinant with $\mathbf{K} = \mathbf{A}$,

$$\Xi^{(M)}(z) = \sum_{N=0}^{M} z^N Q_N = \det[\mathbf{I} + z\mathbf{A}_M] = \prod_{k=1}^{M}\left(1 - \frac{z}{r_k}\right) = D(-z)$$

(3) virial approach

density and pressure related by

$$\frac{pV}{kT} = \ln\,\Xi = v \sum_{j=1}^{\infty} b_j z^j \quad ; \quad N(z) = nV = z\frac{\partial}{\partial z}\ln\Xi = v\sum_{j=1}^{\infty} j b_j z^j$$

virial equation

$$\tfrac{pV}{NkT} = 1 + \textstyle\sum_{k\geq 1} a_{k+1}\left(n\lambda^3\right)^k$$

Mayer's formula

$$a_{k+1} = \textstyle\sum_{j=1}^{k} (-)^j \tfrac{(k+j-1)!}{(k+1)!(k-1)!}\, \mathbb{B}_{kj}(2!b_2, 3!b_3, \ldots).$$

A.8.4 *Ideal relativistic quantum gases*

(1) parameters

Compton length $\lambda_C = \frac{\hbar c}{mc^2}$

cube of relativistic thermal wavelength

$$\Lambda^3(\beta) = 2\,\pi^2\,\beta mc^2\,\frac{e^{-\beta mc^2}}{K_2(\beta mc^2)}\left(\frac{\hbar c}{mc^2}\right)^3$$

an integral:

$$f(\mathbf{r},\beta) = \frac{1}{h^3}\int d^3p\, e^{-\frac{i}{\hbar}\mathbf{p}\cdot\mathbf{r}}\; e^{-\beta[(p^2c^2+m^2c^4)^{1/2}-mc^2]}$$

$$= \frac{1}{\Lambda^3(\beta)}\,\frac{1}{1+(\frac{r}{\hbar\beta c})^2}\frac{K_2\left[\beta mc^2\sqrt{1+(\frac{r}{\hbar\beta c})^2}\right]}{K_2(\beta mc^2)}$$

(2) canonical partition function

$$Q_N(V,\beta) = \tfrac{V}{N}\textstyle\sum_{m=1}^{N}\tfrac{(\pm)^{m-1}}{\Lambda^3(m\beta)}\, Q_{N-m}(V,\beta)$$

$$= \tfrac{1}{N!}\textstyle\sum_{m=1}^{N}\mathbb{B}_{Nm}\left\{(\pm)^{k-1}(k-1)!\tfrac{V}{\Lambda^3(k\beta)}\right\}$$

(3) distribution function (with $f_{ij} = f(\mathbf{r}_i - \mathbf{r}_j, \beta)$)

$$F_N(r_1, r_2, \cdots, r_N) = \frac{1}{N!Q_N} \sum_{m=1}^{N} \mathbb{B}_{Nm} \left\{(\pm)^{j-1}(j-1)! f_{12} f_{23} \ldots f_{j1}\right\}$$

$$= \frac{V}{N} \sum_{m=1}^{N} (\pm)^{m-1} f_{12} f_{23} \ldots f_{m1}\ Q_{N-m} F_N(r_{m+1}, r_{m+2}, \cdots, r_N)$$

(4) pair correlation function

$$g(\mathbf{r}) = \frac{1}{NQ_N} \left\{ \frac{Q_{N-1}}{V\Lambda^3(\beta)} + f(\mathbf{r}, \beta) \sum_{m=2}^{N} (\pm)^{m-1} Q_{N-m} f(\mathbf{r}, (m-1)\beta) \right\}$$

(5) grand canonical partition function

$$\Xi^{(\pm)}(V, \beta, \mu) = \left\{ \begin{array}{c} \frac{1}{1-z} \\ 1+z \end{array} \right\} \exp\left\{ gV \sum_{j=1}^{\infty} \frac{(\pm 1)^{j-1}}{j} z^j \frac{1}{\Lambda^3(j\beta)} \right\}$$

(6) two-dimensional gas

Bose and Fermi statistics

$$Q_N(\beta, S) = \tfrac{1}{N!} C_N \left\{ \frac{(\pm)^{j-1}}{j} \frac{S}{\lambda^2} \right\} = \tfrac{1}{N!} \sum_{m=1}^{N} \mathbb{B}_{Nm} \left\{ \frac{(\pm)^{j-1}}{j} (j-1)!\, \frac{S}{\lambda^2} \right\}$$

braid statistics

$$Q_N(\beta, S) = \frac{1}{N!} \sum_{m=1}^{N} \mathbb{B}_{Nm} \left\{ \frac{(\cos\phi)^{j-1}}{j} (j-1)!\, \frac{S}{\lambda^2} \right\}.$$

Sources

Besides Gradshteyn & Ryzhik [34], Riordan [95], and the site [100], we have extensively used Abramowitz & Stegun [146], Prudnikov–Brychkov–Marichev [147], Terras [39] and Karatsuba [148].

Bibliography

[1] F. Caruso, J.G. Pereira and A. Santoro (Editors) (2024), "Tribute to Ruben Aldrovandi", *Livraria da Fìsica*, São Paulo, Brazil. ISBN: 987-65-5563-418-1

[2] R. Kerner, A.L. Barbosa and D.V. Gal'tsov. (2001) "Topics in Born-infeld electrodynamics", *AIP Conference Proceedings*, Vol. **589**, No. 1, 377-389). *American Institute of Physics.*

[3] R. Aldrovandi and J.G. Pereira, *An Introduction to Geometrical Physics*, World Scientific, Singapore, 1995.

[4] R. Aldrovandi and J.G. Pereira, (2012). Teleparallel gravity: an introduction (Vol. **173**), Springer Science and Business Media.

[5] R. Aldrovandi, (2001) Special matrices of mathematical physics: stochastic, circulant, and Bell matrices. World Scientific.

[6] R. Aldrovandi and R. Kerner, (2000) in *Proceedings of XIVth Max Born Symposium*, Eds. Frydryszak et al., World Scientific, 153-169.

[7] Kerner, R. (2007) Models of Agglomeration and Glass Transition, (book), *Imperial College Press.*

[8] R. Aldrovandi and A.L. Barbosa (2003) "Spacetime algebraic skeleton", Proceedings of the 24th ICGTMP Colloquium (Paris 2002), IOP Conference Series, Number 173, pp. 229-232.

[9] R. Kerner and R. Aldrovandi, *Stochastic Matrices as a Tool for Biological Evolution Models*, International Symposium on Mathematical and Computational Biology, Brasilia, D.F., Brazil, August, 2009; in "BIOMAT-2009", Ed. Rubem P. Mondaini, ps. 87-109, World Scientific, Singapore, 2010.

[10] F.R. Gantmacher, *The Theory of Matrices*, Chelsea, New York, 1990 (2 vols.).

[11] D.K. Faddeev and V.N. Faddeeva, *Computational Methods of Linear Algebra*, W.H. Freeman and Co., San Francisco, 1963.

[12] O. Bratelli and D.W. Robinson, *Operator Algebras and Quantum Statistical Mechanics I*, Springer, New York, 1979.

[13] R. Aldrovandi, A.L. Barbosa and L.P. Freitas, *Int. J. Theor. Phys.* **36** (1997) 3021.

[14] J.L. McCauley, *Classical Mechanics*, Cambridge University Press, Cambridge, 1997.
[15] V.I. Arnold, *Mathematical Methods of Classical Mechanics*, Springer-Verlag, New York, 1968.
[16] J.L. McCauley, *Chaos, Dynamics and Fractals*, Cambridge University Press, Cambridge, 1993.
[17] E. Ott, *Chaos in Dynamical Systems*, Cambridge University Press, Cambridge, 1994.
[18] S.N. Rasband, *Chaotic Dynamics of Nonlinear Systems*, J. Wiley, New York, 1989.
[19] R. Aldrovandi and L.P. Freitas, *J. Math. Phys.* **39** (1998) 5324.
[20] Jean Zinn-Justin, *Quantum Field Theory and Critical Phenomena*, Clarendon Press, Oxford, 1997.
[21] R. Kerner, *Models of Agglomeration and Glass Transition*, Imperial College Press, London, 2007.
[22] J.R. Norris, *Markov Chains*, Cambridge University Press, Cambridge, 1997.
[23] K. Binder and W. Kob, *Grassy Materials and Disordered Solids*, World Scientific, Singapore, 2005.
[24] J.H. Gibbs and E.A. Di Marzio, *J. Chem. Phys.* **28** (1958) 373; E.A. Di Marzio and J.H. Gibbs, *J. Polym. Sci.* **60** (1959) 121; *J. Res. Nat. Bur. Stand.* **68A** (1964) 611; E.A. Di Marzio, *J. Polym. Sci.* **A1** (1963) 1417.
[25] J.C. Phillips, *J. Non-Cryst. Solids* **34** (1979) 153.
[26] M.F. Thorpe, *J. Non-Cryst. Solids* **57** (1983) 355; H. He and M.F. Thorpe, *Phys. Rev. Lett.* **54** (1985) 2107.
[27] Xingwei Feng, W.J. Bresser, and P. Boolchand, *Phys. Rev. Lett.* **78** (1997) 4422.
[28] A.N. Sreeram, D.R. Swiler, and A.K. Varshneya, *J. Non-Cryst. Solids* **127** (1991) 287.
[29] R. Kerner, *J. Non-Cryst. Solids* **135** (1991) 155; R. Kerner and M. Micoulaut, *J. Non-Cryst. Solids* **176** (1994) 271; *J. Non-Cryst. Solids* **210** (1997) 298; R. Kerner, *Physica* **B215** (1995) 267; G.G. Naumis and R. Kerner, *J. Non-Cryst. Solids* **231** (1998) 111; R.A. Barrio, R. Kerner, M. Micoulaut, and G.G. Naumis, *Condens. Matter* **9** (1997) 9219; R. Aldrovandi & Richard Kerner, *Stochastic matrix algebras applied to models of agglomeration and growth*, in "New Symmetries and Integrable Models", Eds. Andrzej Frydryszak, Jerzy Lukierski and Ziemowit Popowicz; Proceedings of the 14th Max Born Symposium, Karpac, Poland, 1999; World Scientific, Singapore, 2000, ps. 153 - 169.
[30] Kyiosi Itô (ed.), *Encyclopedic Dictionary of Mathematics*, MIT Press, 1996.
[31] R.K. Pathria, *Statistical Mechanics*, Pergamon, Oxford, 1972.
[32] K. Huang, *Statistical Mechanics*, J. Wiley, New York, 1987 (2nd ed.).
[33] D.A. McQuarrie, *Statistical Mechanics*, Harper & Row, New York, 1976.
[34] I.S. Gradshteyn and I.M. Ryzhik, *Table of Integrals, Series and Products*, Academic Press, New York, 1980 (edition by Alan Jeffrey).

[35] L. Mirsky, *An Introduction to Linear Algebra*, Oxford University Press, Oxford, 1963.

[36] A. Papoulis, *Signal Analysis*, McGraw-Hill, New York, 1977.

[37] P.J. Davis, *Circulant Matrices*, J. Wiley, New York, 1979.

[38] A. Ekert and R. Jozsa, *Rev. Mod. Phys.* **68** (1996) 733.

[39] A. Terras, *Fourier Analysis on Finite Groups and Applications*, Cambridge University Press, 1999.

[40] A.T. Bharucha-Reid, *Elements of the Theory of Markov Processes and their Applications*, Dover, New York, 1997.

[41] R. Balescu, *Statistical Dynamics*, Imperial College Press, London, 1997.

[42] S. Karlin and J.L. McGregor, *Linear growth, birth and death processes*, *J. Math. Mech.* **7** (1957) 489-546.

[43] S. Karlin and J.L. McGregor, *The differential equations of birth-and-death processes, and the Stieljes moment problem*, *Trans. Amer. Math. Soc.* **85** (1958) 643-662.

[44] S. Karlin and J.L. McGregor, *Random walks*, *Ill. J. Math.* **3** (1959) 66-81.

[45] C.W. Gardiner, *Handbook of Stochastic Methods*, Springer, Berlin, 1997.

[46] Y. Pomeau P. Bergé and C. Vidal. *L'ordre dans le chaos*. Hermann, Paris, 1988.

[47] See, for example, L. D. Landau and E. M. Lifshitz, *The Classical Theory of Fields* (Butterworth-Heinemann, Oxford, 1999), Fourth Revised Edition.

[48] Gregor Mendel, *Experiments in Plant-Hybridization*, in James Newman (ed.), *The World of Mathematics*, Dover, Mineola, New York, 2000, vol.2, ps. 937-49. Also in the web address www.mendelweb.org/Mendel.html.

[49] L.L. Cavalli-Sforza and W.F. Bodmer, *The genetics of human populations*, Dover, Mineola, New York, 1999.

[50] L.L. Cavalli-Sforza, *Genes, Peoples and Languages*, Penguin Books, London, 2001.

[51] D. Galetti and A.F.R. Toledo Piza, *Physica* **A 149** (1988) 267; **A 214** (1995) 207.

[52] P.A.M. Dirac, *Proc. Roy. Soc.* **A109** (1926) 642.

[53] R. Abraham and J. Marsden, *Foundations of Mechanics*, Benjamin–Cummings, Reading, Mass., 2nd edition, 1978.

[54] A. Connes, *Non-Commutative Geometry*, Academic Press, New York, 1994.

[55] Yu. I. Manin, *Comm. Math. Phys.* **123** (1989) 163.

[56] M. Dubois-Violette, in C. Bartocci, U. Bruzzo U, and R. Cianci, editors, *Differential Geometric Methods in Theoretical Physics*, Berlin, 1991. 1990 Rapallo Conference, Springer, Springer Lecture Notes in Physics 375.

[57] R. Coquereaux, *J. Geom. Phys.* **6** (1989) 425.

[58] F. Bayen, M. Flato, C. Fronsdal, A. Lichnerowicz, and D. Sternheimer, *Ann. Phys.* **111** (1978) 61 and 111.

[59] Bo Gao, *Phys. Rev. Lett.* **83** (1999) 4225.

[60] R. Aldrovandi and D. Galetti, *J. Math. Phys.* **31** (1990) 2987.

[61] J.E. Moyal, *Proc. Cambridge Phys. Soc.* **45** (1949) 99.

[62] T. Hakioglu, *J. Phys.* **A31** (1998) 6975.
[63] H. Weyl, *Theory of Groups and Quantum Mechanics*, E. P. Dutton, New York, 1932.
[64] J. Schwinger, *Proc. Nat. Acad. Sci.* **46** (1960) 570 and **47** (1961) 1075; both papers, and much more, can be found in *Quantum Kinematics and Dynamics*, Benjamin, New York, 1970.
[65] D. Galetti and A.F.R. Toledo Piza, *Physica* **A186** (1992) 513. Dynamical aspects have been developed by these authors in *Physica* **A264** (1999) 473; *J. Phys.* **A33** (2000) 1065 and 2799.
[66] D. Galetti and M. Ruzzi, *Physica* **A 264** (1999) 473; *J. Phys.* **A 33** (2000) 2799.
[67] J. Patera and H. Zassenhaus, *J. Math. Phys.* **29** (1988) 665.
[68] L.D. Faddeev, *Lett. Math. Phys.* **34** (1995) 249.
[69] R. Aldrovandi and L.A. Saeger, *Int. J. Theor. Phys.* **36** (1997) 345 and **36** (1997) 573.
[70] G.A. Baker, *Phys. Rev.* **109** (1958) 2198.
[71] G.S. Agarwal and E. Wolf, *Phys. Rev.* **D2** (1970) 2161.
[72] K.C. Liu, *J. Math. Phys.* **17** (1976) 859.
[73] G.W. Mackey, *Unitary Group Representations in Physics, Probability and Number Theory*, Benjamin/Cummings, Reading, Mass., 1978.
[74] J.S. Birman, *Math. Intell.* **13** (1991) 52.
[75] V.F.R. Jones, *Intern. J. Mod. Phys.* **B4** (1990) 701.
[76] C.N. Yang and M.L. Ge, *Braid Group, Knot Theory and Statistical Mechanics*, World Scientific, Singapore, 1989.
[77] R.J. Baxter, *Exactly Solved Models in Statistical Mechanics*, Academic Press, London, 1982.
[78] V.G. Drinfeld, *Sov. Math. Dokl.* **27** (1983) 222.
[79] S.L. Woronowicz, *Commun. Math. Phys.* **111** (1987) 613.
[80] M. Dubois-Violette and G. Launer, *Phys. Lett.* **B245** (1990) 175.
[81] R.H. Crowell and R.H. Fox, *Introduction to Knot Theory*, Springer, Berlin, 1963.
[82] A. Kirillov, *Éléments de la Théorie des Représentations*, MIR, Moscow, 1974.
[83] P. Fletcher, *Phys. Lett.* **B248** (1990) 323.
[84] E.G. Floratos, *Phys. Lett.* **B228** (1989) 335; *Phys. Lett.* **B233** (1989) 395; *Phys. Lett.* **B252** (1990) 97.
[85] D.B. Fairlie, P. Fletcher, and C.Z. Zachos, *Phys. Lett.* **B218** (1989) 203.
[86] V.I. Arnold, *Ann. Inst. Fourier* **16** (1966) 319.
[87] C.J. Isham, In B. S. deWitt and R. Stora, editors, *Relativity, Groups and Topology II*, Amsterdam, 1984. Les Houches Summer School, North-Holland.
[88] R. Kerner, M. Dubois-Violette and J. Madore, *J. Math. Phys.* **31** (1990) 316.
[89] R. Aldrovandi and J.G. Pereira, in V. C. Aguilera-Navarro, D. Galetti,

B. M. Pimentel, and L. Tomio, editors, *Topics in Theoretical Physics*, pp. 277–291, São Paulo, 1995. I.F.T.

[90] M. Dubois-Violette, *C. R. Acad. Sci. Paris* **307** (1988) 403.

[91] N. Jacobson, *Lie Algebras*, Dover, New York, 1962.

[92] R. Aldrovandi, *J. Math. Phys.* **32** (1991) 2503.

[93] R. Aldrovandi, *Fortschr. Physik* **40** (1992) 631.

[94] J. Riordan, *An Introduction to Combinatorial Analysis*, J. Wiley, New York, 1968.

[95] J. Riordan, *Combinatorial Identities*, J. Wiley, New York, 1968.

[96] L. Comtet, *Advanced Combinatorics*, Reidel, Dordrecht, 1974.

[97] G.E. Andrews, *The Theory of Partitions*, Cambridge University Press, Cambridge, 1998.

[98] M. Kuczma, B. Choczewski and R. Ger, *Iterative Functional Equations*, Encyclopedia of Mathematics and its Applications, Cambridge University Press, Cambridge, 1990.

[99] L. Kindermann and P. Georgiev, *Modelling iterative roots of mappings in multidimensional spaces*; L. Kindermann and A. Lewandowski, *Natural Interpolation of Time Series*; both reports of the Riken Brain Science Institute, 2-1, Hirosawa, Wako-shi, Saitama, 351-0198, Japan. See sites http://www.brain.riken.jp/labs/bsis and http://www.awi-bremerhaven.de/People/show?lkinderm.

[100] E. Weisstein, *Eric Weisstein's World of Mathematics*, whose web address is http://mathworld.wolfram.com.

[101] E.T. Whittaker and G.N. Watson, *A Course in Modern Analysis*, Cambridge University Press, Cambridge, 1969.

[102] P. Henrici, *Applied and Computational Complex Analysis*, J. Wiley, New York, 1974.

[103] S. Chandrasekhar, *Stochastic problems in physics and astronomy*, *Rev. Modern Phys.* **15** (1943) 1–89.

[104] John D. Ramshaw, *J. Chem. Phys.* **57** (1972) 2070.

[105] R. Aldrovandi and I. Monte Lima, *J. Phys.* **A13** (1980) 3685.

[106] R. Aldrovandi and I. Monte Lima, *Astrophys. Space Sci.* **90** (1983) 179.

[107] K.B. Oldham and J. Spanier, *The Fractional Calculus*, Academic Press, New York, 1974.

[108] B.B. Mandelbrot, *The Fractal Geometry of Nature*, W. H. Freeman, New York, 1983.

[109] P. Blasiak, K.A. Penson and A. Solomon, *J.Phys. A: Math.Gen.* **36** (2003) L273 and references therein.

[110] D.E. Littlewood, *The Theory of Group Characters*, Clarendon Press, Oxford, 1950.

[111] I.G. MacDonald, *Symmetric Functions and the Hall Polynomials*, Oxford Math. Monographs. Oxford University Press, Oxford, 1979.

[112] I.M. Gelfand, D. Kroband, A. Lascoux, B. Leclerc, V.S. Retakh, and J.-Y. Thibon, *Noncommutative symmetric functions*, HEP-TH/9407124,

preprint LITP 94.39, 1994.

[113] I.M. Gelfand and V.S. Retakh, *Noncommutative Vieta theorem and symmetric functions*, preprint Rutgers University, 1995, contribution to the Gelfand Seminar Volume.

[114] J.E. Baker Jr, *The Padé Approximation*, volume 13 and 14 of *Encyclopedia of Mathematics and its Applications*, Addison-Wesley, New York, 1981.

[115] S.I. Goldberg, *Curvature and Homology*, Dover, New York, 1962.

[116] S. Kobayashi and K. Nomizu, *Foundations of Differential Geometry*, Interscience, New York, 1963, (2 vols.).

[117] T. Eguchi, P.B. Gilkey, and A.J. Hanson, *Phys. Rep.* **66** (1980) 213.

[118] B. Zumino, in B.S. DeWitt and R. Stora, editors, *Relativity, Groups and Topology II*, Amsterdam, 1985. Les Houches Summer School, North-Holland, vol. 3.

[119] M. Nakahara, *Geometry, Topology and Physics*, Institute of Physics Pub., Bristol, 1990.

[120] R. Aldrovandi and E. Stedile, *Int. J. Theor. Phys.* **23** (1984) 301.

[121] V.F.R. Jones, *Subfactors and Knots*, AMS, Providence, Rhode Island, 1991.

[122] G. Labelle, *Europ. J. Combinatorics* **1** (1980) 113.

[123] U. Frisch, *Turbulence*, Cambridge University Press, Cambridge, 1996.

[124] T.J. Rivlin, *The Chebyshev Polynomials*, J. Wiley, New York, 1974.

[125] R. Balescu, *Equilibrium and Nonequilibrium Statistical Mechanics*, J. Wiley, New York, 1975.

[126] R. Kubo M. Toda and N. Saitô, *Statistical Physics I*, Springer, Berlin, 2nd edition, 1998.

[127] R. Hagedorn, *Relativistic Kinematics*, Benjamin, New York, 1963; E. Byckling and K. Kajantie, *Particle Kinematics*, J. Wiley, New York, 1973.

[128] R. Aldrovandi and N.R. Teixeira Filho, *Lett. Nuovo Cimento* **15** (1976) 591 and **17** (1976) 9.

[129] M. Hamermesh, *Group Theory and its Applications to Physical Problems*, Addison-Wesley, Reading, Mass., 1962.

[130] R.F. Dashen, S.K. Ma, and H.J. Bernstein, *Phys. Rev.* **187** (1969) 345 and *Phys. Rev.* **A 6** (1972) 851; R.F. Dashen and R. Rajaraman, *Phys. Rev.* **D10** (1974) 694.

[131] B. Widom, *Phys. Rev.* **96** (1954) 16.

[132] B. Kahn, Thesis, In J. de Boer and G. E. Uhlenbeck, editors, *Studies in Statistical Mechanics 3*, Amsterdam, 1965. North-Holland, Translation of 1938 Utrecht Thesis.

[133] T. Matsubara, *Prog. Theor. Phys.* **14** (1955) 351.

[134] P. Ramond, *Field Theory A Modern Primer*, Benjamin-Cummings, Reading, Mass., 1981.

[135] J.S. Birman, *Braids, Links, and Mapping Class Groups*, Princeton University Press, Princeton, 1975.

[136] L. Schulman, *Phys. Rev.* **176** (1968) 1558; M.G.G. Laidlaw and C. deWitt Morette, *Phys. Rev.* **D3** (1971) 1375; C. deWitt Morette, *Comm. Math.*

Phys. **28** (1972) 47; C. deWitt Morette, A. Masheshwari, and B. Nelson, *Phys. Rep.* **50** (1979) 255.

[137] J.M. Leinaas and J. Myrheim, *Nuovo Cimento* **B37** (1977) 123.

[138] J.E. Mayer and M.G. Mayer, *Statistical Mechanics*, J. Wiley, 1940.

[139] T.L. Hill, *Statistical Mechanics*, Dover, 1978.

[140] T.D. Lee and C.N. Yang, *Phys. Rev.* **87** (1952) 410, and **113** (1959) 1165.

[141] On generalizations of the circle theorem, see T. Asano, *J. Phys. Soc. Japan* **29** (1970) 350; C.M. Newman, *Commun. Math. Phys.* **41** (1975) 1; E.H. Lieb and A.D. Sokal, *Commun. Math. Phys.* **80** (1981) 153. For a negative result in the 1-dimensional Potts model, L. Mittag and M.J. Stephen, *J. Stat. Phys.* **35** (1984) 303. On the position of the zeros H. Nishimori and R.B. Griffiths, *J. Math. Phys.* **24** (1983) 2637, and H. Tasaki, *J. Math. Phys.* **28** (1987) 1164. References can be traced from D. Ruelle, *Bull. Am. Math. Soc.* **19** (1988) 259, and B. Simon, *The Statistical Mechanics of Lattice Gases* Princeton University Press, Princeton, 1993. In a more general context, E.H. Lieb, *Bull. Am. Math. Soc.* **22** (1990) 1, is of great interest. Recent works are J.C.A. Barata and D.H.U. Marchetti, *J. Stat. Phys.* **88** (1997) 231, and M. Biskup *et al*, *Phys. Rev. Lett.* **84** (2000) 4794.

[142] H.L. Frisch and J.L. Lebowitz, *The Equilibrium Theory of Classical Fluids*, Benjamin, 1964.

[143] S. Fishman, B. Georgeot, and R.E. Prange, *J. Phys.* **A29** (1996) 919–937.

[144] T.D. Lee, *Particle Physics and Introduction to Field Theory*, Harwood, New York, 1981.

[145] C.M. Bender, D.C. Brody and B.K. Meister, *J. Math. Phys.* **40** (1999) 3239.

[146] M. Abramowitz and I.A. Stegun, *Handbook of Mathematical Functions*, Dover, New York, 1968.

[147] A.P. Prudnikov, Yu. A Brychkov and O.I. Marichev, *Integrals and Series*, 3 vols., Gordon and Breach, New York, 1986.

[148] A.A. Karatsuba, *Basic Analytic Number Theory*, Springer, Berlin, 1991.

[149] Gottfried Helms: *Continuous iteration of power series-defined functions*, (draft) 2008 for discussion in the "Tetration-Forum" http://go.helms-net.de/math/tetdocs/ContinuousfunctionalIteration.pdf

[150] Tetration Forum: http://math.eretrandre.org/tetrationforum/index.php

[151] Henryk Trappmann: http://andydude.5gigs.net/tetration.co.cc/htdocs/index.html

[152] Daniel Geisler web page: http://www.tetration.org/Dynamics/

[153] R. Aldrovandi, "Tetration: an iterative approach", arXiv:1410.3896 [math-ph], 14 Oct 2014, address: https://arxiv.org/abs/1410.3896v1

Index

www.ingramcontent.com/pod-product-compliance
Lightning Source LLC
Chambersburg PA
CBHW072354300825
31649CB00004BA/289
9789819813889